Preference Learning

Johannes Fürnkranz • Eyke Hüllermeier
Editors

Preference Learning

Springer

Editors
Prof. Dr. Johannes Fürnkranz
Knowledge Engineering Group
Fachbereich Informatik
Technische Universität Darmstadt
Hochschulstr. 10
64289 Darmstadt
Germany
juffi@ke.tu-darmstadt.de

Prof. Dr. Eyke Hüllermeier
Knowledge Engineering & Bioinformatics
Fachbereich Mathematik und Informatik
Philipps-Universität Marburg
Hans-Meerwein-Str.
35032 Marburg
Germany
eyke@mathematik.uni-marburg.de

ISBN 978-3-642-14124-9 e-ISBN 978-3-642-14125-6
DOI 10.1007/978-3-642-14125-6
Springer Heidelberg Dordrecht London New York

Library of Congress Control Number: 2010937568

ACM Computing Classification (1998): I.2.6, H.2.8

Cover design: KuenkelLopka GmbH

Printed on acid-free paper

Springer is part of Springer Science+Business Media (www.springer.com)

Preface

The topic of preferences has attracted considerable attention in Artificial Intelligence (AI) research in previous years. Recent special issues of the *AI Magazine* (December 2008) and the *Artificial Intelligence Journal* (announced for 2010), both devoted to preferences, highlight the increasing importance of this area for AI. Representing and processing knowledge in terms of preferences is appealing as it allows one to specify desires in a declarative way, to combine qualitative and quantitative modes of reasoning, and to deal with inconsistencies and exceptions in a quite flexible manner.

Like in other subfields of AI, including autonomous agents, nonmonotonic reasoning, constraint satisfaction, planning and qualitative decision theory, researchers in machine learning have started to pay increasing attention to the topic of preferences. In fact, as witnessed by a number of dedicated events, notably several workshops on preferences, ranking, and related topics (held, e.g., at NIPS 2004 and 2005, ECML/PKDD 2008 and 2009, SIGIR 2008 and 2009), we currently observe the formation of "preference learning" as a new branch of machine learning and data mining. For the time being, there is still no stipulated demarcation of this emerging subfield, neither in terms of a list of relevant topics nor in terms of an intentional "definition". Roughly, *preference learning* refers to the problem of learning from observations which reveal, either explicitly or implicitly, information about the preferences of an individual (e.g., a user of a computer system) or a class of individuals; the acquisition of this kind of information can be supported by methods for *preference mining*. Generalizing beyond the training data given, the models learnt are typically used for preference prediction, i.e., to predict the preferences of a new individual or the same individual in a new situation. The problem of "learning to rank" is a good example and an important special case; here, the goal is to predict preferences in the form of total orders of a set of alternatives (e.g., a personalized ranking of documents retrieved by a search engine).

This book, which is the first volume specifically dedicated to the topic of preference learning, distinguishes itself through the following features:

- It gives a comprehensive overview of the state-of-the-art in the field of preference learning.

- By including a number of survey chapters, it offers an introduction to the most important subfields of preference learning.
- By proposing a systematic categorization according to learning task and learning technique, along with a unified notation, it helps structuring the field; thereby, it is supposed to have a positive impact on future research.
- Through the selection of contributions, it emphasizes the interdisciplinary character of preference learning and establishes connections to related research fields, such as multicriteria decision-making and operations research.
- Last but not least, it highlights important applications of preference learning in different areas, such as information retrieval and recommender systems, thereby demonstrating its practical relevance.

Some chapters of the book are based on contributions selected from two successful workshops on preference learning that we organized as part of the ECML/PKDD conferences in 2008 and 2009. Besides, however, the material is complemented by a number of chapters that have been solicited explicitly for this book. Overall, we are quite confident that the book provides both a broad coverage and comprehensive survey of the field of preference learning as well as a useful guideline and good introduction to the most important directions in current research.

The origination of this book is largely due to our close collaboration in recent years, which in turn has greatly benefited from a joint research project funded by the German Science Foundation (DFG). This support is gratefully acknowledged. Moreover, we would like to thank Ronan Nugent and the Springer for providing excellent assistance and ready advice during the final stages of preparation.

Darmstadt Marburg
December 2009

Johannes Fürnkranz
Eyke Hüllermeier

Contents

Part VI Preferences in Recommender Systems

Preference Learning: An Introduction

Johannes Fürnkranz and Eyke Hüllermeier

Abstract This introduction gives a brief overview of the field of preference learning and, along the way, tries to establish a unified terminology. Special emphasis will be put on learning to rank, which is by now one of the most extensively studied problem tasks in preference learning and also prominently represented in this book. We propose a categorization of ranking problems into object ranking, instance ranking, and label ranking. Moreover, we introduce these scenarios in a formal way, discuss different ways in which the learning of ranking functions can be approached, and explain how the contributions collected in this book relate to this categorization. Finally, we also highlight some important applications of preference learning methods.

1 Introduction

Reasoning with preferences has been recognized as a particularly promising research direction for artificial intelligence (AI) [15]. A preference can be considered as a relaxed constraint which, if necessary, can be violated to some degree. In fact, an important advantage of a preference-based problem solving paradigm is an increased flexibility, as nicely explained in [6]:

> "Early work in AI focused on the notion of a goal – an explicit target that must be achieved – and this paradigm is still dominant in AI problem solving. But as application domains become more complex and realistic, it is apparent that the dichotomic notion of a goal, while adequate for certain puzzles, is too crude in general. The problem is that in many contemporary application domains ... the user has little knowledge about the set of possible solutions or feasible items, and what she typically seeks is the best that's out there. But

---

J. Fürnkranz (✉)
Technical University Darmstadt, Germany
e-mail: juffi@ke.tu-darmstadt.de

E. Hüllermeier
Philipps-Universität Marburg, Germany
e-mail: eyke@mathematik.uni-marburg.de

J. Fürnkranz and E. Hüllermeier (eds.), *Preference Learning*,
DOI 10.1007/978-3-642-14125-6_1,

> since the user does not know what is the best achievable plan or the best available document or product, she typically cannot characterize it or its properties specifically. As a result, she will end up either asking for an unachievable goal, getting no solution in response, or asking for too little, obtaining a solution that can be substantially improved."

Drawing on past research on knowledge representation and reasoning, AI offers qualitative and symbolic methods for treating preferences that can reasonably complement traditional approaches that have been developed for quite a while in fields such as economic decision theory [37]. Needless to say, however, the acquisition of preferences is not always an easy task. Therefore, not only modeling languages and representation formalisms, but also methods for the automatic learning, discovery, and adaptation of preferences are needed. For example, computerized methods for discovering the preferences of individuals are useful in e-commerce and various other fields, where an increasing trend toward personalization of products and services can be recognized.

It is hence hardly surprising that methods for learning and predicting preferences in an automatic way are among the very recent research topics in disciplines, such as machine learning, knowledge discovery, and recommender systems. Approaches relevant to this area range from approximating the utility function of a single agent on the basis of an as effective as possible question-answer process (often referred to as *preference elicitation*) to collaborative filtering where a customer's preferences are estimated from the preferences of other customers. In fact, problems of preference learning can be formalized within various settings, depending, e.g., on the underlying type of preference model or the type of information provided as an input to the learning system.

Roughly speaking, *preference learning* is about inducing predictive preference models from empirical data. In the literature on choice and decision theory, two main approaches to modeling preferences can be found, namely in terms of *utility functions* and in terms of *preference relations*. From a machine learning point of view, these two approaches give rise to two kinds of learning problems: learning utility functions and learning preference relations. The latter deviates more strongly than the former from conventional problems such as classification and regression, as it involves the prediction of complex structures, such as rankings or partial order relations, rather than single values. Moreover, training input in preference learning will not, as it is usually the case in supervised learning, be offered in the form of complete examples but may comprise more general types of information, such as relative preferences or different kinds of indirect feedback and implicit preference information.

This book tries to give a comprehensive overview of the state-of-the-art in the field of preference learning. Some of its chapters are based on selected contributions to two successful workshops on this topic [29,30], but the material is complemented with chapters that have been solicited explicitly for this book. Most notably, several survey chapters give a detailed account on ongoing research in various subfields of preference learning. Thus, we are confident that the book succeeds in giving a comprehensive survey of work on all aspects of this emerging research area.

In the remainder of this chapter, we shall briefly sketch some important branches of preference learning and, along the way, give pointers to the contributions in this volume. References to these contributions are indicated by capitalized author names, for example FÜRNKRANZ & HÜLLERMEIER.

2 Preference Learning Tasks

Among the problems in the realm of preference learning, the task of "learning to rank" has probably received the most attention in the machine learning literature in recent years. In fact, a number of different ranking problems have been introduced so far, though a commonly accepted terminology has not yet been established. In the following, we propose a unifying and hopefully clarifying terminology for the most important types of ranking problems, which will also serve as a guideline for organizing the chapters of the book. AIOLLI & SPERDUTI give an alternative unifying framework for learning to rank from preferences.

In general, a preference learning task consists of some set of items for which preferences are known, and the task is to learn a function that predicts preferences for a new set of items, or for the same set of items in a different context. Frequently, the predicted preference relation is required to form a total order, in which case we also speak of a ranking problem. In this book, we will frequently use the term "ranking" for categorizing different types of preference learning problems, but we note that the characterization mainly depends on the form of the training data and the required predictions, and not on the fact that a total order is predicted.[1]

In the notation used in the remainder of this chapter (and throughout most of the book), our goal is to stick as much as possible to the terminology commonly used in supervised learning (classification), where a data object typically consists of an *instance* (the input, also called predictive or independent variable in statistics) and an associated *class label* (the output, also called target or dependent variable in statistics). The former is normally denoted by $\boldsymbol{x}$, and the corresponding instance space by $\mathcal{X}$, while the output space is denoted by $\mathcal{Y}$. Instances are often represented in the form of feature vectors, which means that $\boldsymbol{x}$ is a vector

$$\boldsymbol{x} = (x_1, x_2, \ldots, x_m) \in \mathcal{X} = \mathcal{X}_1 \times \mathcal{X}_2 \times \ldots \times \mathcal{X}_m .$$

We distinguish three types of ranking problems, namely label ranking, instance ranking, and object ranking, which are described in more detail in the following.

[1] Besides, one should be aware of conflicts between terminology in different fields. In the field of operations research, for example, the term "ranking" is used for arranging a set of objects in a total order, while "sorting" refers to the assignment of objects to an ordered set of categories, a problem closely related to what is called "ordered classification" in machine learning.

2.1 Label Ranking

In label ranking, we assume to be given an instance space $\mathcal{X}$ and a finite set of labels $\mathcal{Y} = \{y_1, y_2, \ldots, y_k\}$. The goal is to learn a "label ranker" in the form of an $\mathcal{X} \to S_{\mathcal{Y}}$ mapping, where the output space $S_{\mathcal{Y}}$ is given by the set of all total orders (permutations) of the set of labels $\mathcal{Y}$ (the notation is leaned on the common notation S_k for the symmetric group of order k). Thus, label ranking can be seen as a generalization of conventional classification, where a complete ranking

$$y_{\pi_x^{-1}(1)} \succ_x y_{\pi_x^{-1}(2)} \succ_x \ldots \succ_x y_{\pi_x^{-1}(k)}$$

is associated with an instance $\boldsymbol{x}$ instead of only a single class label. Here, π_x is a permutation of $\{1, 2, \ldots, k\}$ such that $\pi_x(i)$ is the position of label y_i in the ranking associated with $\boldsymbol{x}$.

The training data $\mathcal{T}$ of a label ranker typically consist of a set of pairwise preferences of the form $y_i \succ_x y_j$, suggesting that, for instance $\boldsymbol{x}$, y_i is preferred to y_j. In other words, an "observation" consists of an instance $\boldsymbol{x}$ and an ordered pair of labels (y_i, y_j). The label ranking problem is summarized in Fig. 1.

This learning scenario has a large number of practical applications. For example, it is relevant for the prediction of every sort of ordering of a fixed set of elements, such as the preferential order of a fixed set of products (e.g., different types of holiday apartments) based on demographic properties of a person, or the ordering of a set of genes according to their expression level (as measured by microarray analysis) based on features of their phylogenetic profile [1]. Another application scenario is meta-learning, where the task is to rank learning algorithms according to their suitability for a new dataset, based on the characteristics of this dataset [7]. Finally, every preference statement in the well-known CP-nets approach [3], a qualitative graphical representation that reflects conditional dependence and independence of

---

Given:

- a set of training instances $\{\boldsymbol{x}_\ell \mid \ell = 1, 2, \ldots, n\} \subseteq \mathcal{X}$ (each instance typically though not necessarily represented by a feature vector)
- a set of labels $\mathcal{Y} = \{y_i \mid i = 1, 2, \ldots, k\}$
- for each training instance $\boldsymbol{x}_\ell$: a set of *pairwise preferences* of the form $y_i \succ_{x_\ell} y_j$

Find:

- a ranking function that maps any $\boldsymbol{x} \in \mathcal{X}$ to a ranking $\succ_x$ of $\mathcal{Y}$ (permutation $\pi_x \in \mathcal{S}_k$)

Performance measures:

- ranking error (e.g., based on rank correlation measures) comparing predicted ranking with target ranking
- position error comparing predicted ranking with a target label

---

Fig. 1 Label ranking

preferences under a *ceteris paribus* interpretation, formally corresponds to a label ranking.

In addition, it has been observed by several authors [14, 18, 25] that many conventional learning problems, such as classification and multilabel classification, may be formulated in terms of label preferences:

- *Classification*: A single class label y_i is assigned to each example $\boldsymbol{x}_\ell$. This implicitly defines the set of preferences $\{y_i \succ_{\boldsymbol{x}_\ell} y_j \mid 1 \le j \ne i \le k\}$.
- *Multilabel classification*: Each training example $\boldsymbol{x}_\ell$ is associated with a subset $P_\ell \subseteq \mathcal{Y}$ of possible labels. This implicitly defines the set of preferences $\{y_i \succ_{\boldsymbol{x}_\ell} y_j \mid y_i \in L_\ell, y_j \in \mathcal{Y} \setminus P_\ell\}$.

A general framework encompassing these and other learning problems can be found in the chapter by AIOLLI & SPERDUTI.

In each of the former scenarios, a ranking model $f : \mathcal{X} \to \mathcal{S}_k$ is learned from a subset of all possible pairwise preferences. A suitable projection may be applied to the ranking model (which outputs permutations) as a post-processing step, for example, a projection to the top-rank in classification learning where only this label is relevant.

To measure the predictive performance of a label ranker, a loss function on rankings is needed. In principle, any distance or correlation measure on rankings (permutations) can be used for that purpose, for example, the number of pairs of labels that are incorrectly ordered (i.e., the number of label pairs y_i and y_j such that y_i precedes y_j in the predicted ranking although y_j is actually preferred to y_i). Apart from this type of *ranking loss*, which compares a predicted ranking with a given target ranking, it is also possible to compare a predicted ranking with a single class label. For example, if this class label is the target one is looking for, then it makes sense to evaluate a predicted ranking by the position it assigns to the label; in [28], this type of error (measuring the distance of the assigned position from the top-rank) is called the *position error*.

A general survey of label ranking is given by VEMBU & GÄRTNER. Another discussion of label ranking and related problems is given by FÜRNKRANZ & HÜLLERMEIER. This chapter is specifically focused on approaches that are based on the idea of *learning by pairwise comparison*, i.e., of decomposing the original problem into a set of smaller binary classification problems. YU, WAN & LEE show how decision-tree learning algorithms such as CART can be adapted to tackle label ranking learning problems by extending the concept of purity to label ranking data. TSIVTSIVADZE et al. show how an approach for minimizing an approximation of a ranking loss function can be extended with a semi-supervised learning technique that tries to improve predictions by minimizing the disagreement of several ranking functions, which have been learned from different views of the training data.

2.2 Instance Ranking

This setting proceeds from the setting of *ordinal classification*, where an instance $\boldsymbol{x} \in \mathcal{X}$ belongs to one among a finite set of classes $\mathcal{Y} = \{y_1, y_2, \ldots, y_k\}$ and, moreover, the classes have a natural order: $y_1 < y_2 < \ldots < y_k$. Training data consists of a set $\mathcal{T}$ of labeled instances. As an example, consider the assignment of submitted papers to categories *reject*, *weak reject*, *weak accept*, and *accept*.

In contrast to the classification setting, the goal is not to learn a classifier but a ranking function $f(\cdot)$. Given a subset $X \subset \mathcal{X}$ of instances as an input, the function produces a ranking of these instances as an output (typically by assigning a score to each instance and then sorting by scores).

For the case $k = 2$, this problem is well-known as the bipartite ranking problem. The case $k > 2$ has recently been termed *k-partite* [42] or *multipartite ranking* [19]. As an example, consider the task of a reviewer who has to rank the papers according to their quality, possibly though not necessarily with the goal of partitioning this ranking into the above four categories.

Thus, the goal of *instance ranking* – our proposal for a generic term of bipartite and multipartite ranking – is to produce a ranking in which instances from higher classes precede those from lower classes; see Fig. 2 for a formalization of this task. Different types of accuracy measures have been proposed for predictions of this kind. Typically, they count the number of ranking errors, that is, the number of pairs $(\boldsymbol{x}, \boldsymbol{x}') \in X \times X$ such that $\boldsymbol{x}$ is ranked higher than $\boldsymbol{x}'$ even though the former belongs to a lower class than the latter. In the two-class case, this amounts to the well-known AUC, the area under the ROC-curve [5], which is equivalent to the Wilcoxon–Mann–Whitney statistic [38, 47]. Its generalization to multiple (ordered) classes is known as the concordance index or C-index in statistics [22].

These measures and the multipartite ranking scenario are discussed in more detail by WAEGEMAN & DE BAETS. ZHANG et al. discuss different methods for employing rule learning algorithms for learning bipartite rankings. This scenario has been studied for decision-tree learning, but not yet for rule learning, where several additional problems have to be considered, such as how to combine estimates from overlapping rules into a single probability estimate.

Given:

- a set of training instances $\{\boldsymbol{x}_\ell \mid \ell = 1, 2, \ldots, n\} \subseteq \mathcal{X}$ (each instance typically though not necessarily represented by a feature vector)
- a set of labels $\mathcal{Y} = \{y_1, y_2, \ldots, y_k\}$ endowed with an order $y_1 < y_2 < \cdots < y_k$
- for each training instance $\boldsymbol{x}_\ell$ an associated label y_ℓ

Find:

- a ranking function that allows one to order a new set of instances $\{\boldsymbol{x}_j\}_{j=1}^{t}$ according to their (unknown) preference degrees

Performance measures:

- the area under the ROC-curve (AUC) in the dichotomous case ($k = 2$)
- generalizations such as the C-index in the polychotomous case ($k > 2$)

Fig. 2 Instance ranking

2.3 Object Ranking

In the setting of object ranking, there is no supervision in the sense that no output or class label is associated with an object. The goal in object ranking is to learn a ranking function $f(\cdot)$ which, given a subset Z of an underlying referential set $\mathcal{Z}$ of objects as an input, produces a ranking of these objects as an output. Again, this is typically done by assigning a score to each instance and then sorting by scores.

Objects $z \in \mathcal{Z}$ are commonly though not necessarily described in terms of an attribute-value representation. As training information, an object ranker has access to exemplary rankings or pairwise preferences of the form $z \succ z'$ suggesting that z should be ranked higher than z'. This scenario, summarized in Fig. 3, is also known as "learning to order things" [12].

As an example consider the problem of learning to rank query results of a search engine [33, 41]. The training information is provided implicitly by the user who clicks on some of the links in the query result and not on others. This information can be turned into binary preferences by assuming that the selected pages are preferred over nearby pages that are not clicked on [34].

The performance of an object ranker can again be measured in terms of a distance function or correlation measure on rankings. In contrast to the setting of label ranking, however, the number of items to be ordered in the context of object ranking is typically much larger. Therefore, one often prefers measures that put more emphasis on the top of a ranking while paying less attention to the bottom [17]. In Web search, for example, people normally look at the top-10 results while ignoring the rest. Besides, the target is often not a "true" ranking but instead a single object or a subset of relevant objects, for example a set of documents relevant to a query. Evaluation measures especially tailored toward these types of requirements have been proposed in information retrieval. Typical examples include precision and recall as well as normalized discounted cumulative gain (NDCG) [32, 39].

Given:

- a (potentially infinite) reference set of objects $\mathcal{Z}$ (each object typically though not necessarily represented by a feature vector)
- a finite set of pairwise preferences $\boldsymbol{x}_i \succ \boldsymbol{x}_j$, $(\boldsymbol{x}_i, \boldsymbol{x}_j) \in \mathcal{Z} \times \mathcal{Z}$

Find:

- a ranking function $f(\cdot)$ that assumes as input a set of objects and returns a permutation (ranking) of this set

Performance measures:

- ranking error (e.g., based on rank correlation measures) comparing the predicted ranking with the target ranking
- top-K measures comparing the top-positions of the rankings
- retrieval measures such as precision, recall, NDCG

Fig. 3 Object ranking

An extensive survey of object ranking approaches is given by KAMISHIMA, KAZAWA & AKAHO. Subsequently, KAMISHIMA & AKAHO discuss dimensionality reduction methods for object ranking tasks, which retain the preference information as much as possible. They assume a scenario (which they call *supervised ordering*) in which total orders for multiple subsets of objects are given, and the goal is to predict an ordering of the full set of objects. DEMBCZYŃSKI et al. compare different approaches for rule-based learning of object ranking functions, namely one utility-based approach and one approach that directly learns the binary preference predicate (cf. also Sect. 3.3). An application to learning to rank documents in biomedical information retrieval is described by ARENS.

3 Preference Learning Techniques

All three of the basic learning tasks discussed in the previous section can be tackled by very similar basic techniques. In agreement with the distinction between using utility functions and binary relations for modeling preferences, two general approaches to preference learning have been proposed in the literature, the first of which is based on the idea of learning to evaluate individual alternatives by means of a utility function (Sect. 3.1), while the second one seeks to compare (pairs of) competing alternatives, that is, to learn one or more binary preference predicate (Sect. 3.2). Making sufficiently restrictive model assumptions about the structure of a preference relation, one can also try to use the data for identifying this structure (Sect. 3.3). Finally, local estimation techniques à la nearest neighbor can be used, which mostly leads to aggregating preferences in one way or the other (Sect. 3.4).

3.1 Learning Utility Functions

As mentioned previously, an established approach to modeling preferences resorts to the concept of a *utility function*. Such a function assigns an abstract degree of utility to each alternative under consideration. From a machine learning point of view, an obvious problem is to learn utility functions from given training data. Depending on the underlying utility scale, which is typically either numerical or ordinal, the problem becomes one of regression learning or ordered classification. Both problems are well-known in machine learning. However, utility functions often implicate special requirements and constraints that have to be taken into consideration such as, for example, monotonicity in certain attributes (DEMBCZYŃSKI et al.).

Besides, as mentioned earlier, training data are not necessarily given in the form of input/output pairs, i.e., alternatives (instances) together with their utility degrees, but may also consist of qualitative feedback in the form of pairwise comparisons, stating that one alternative is preferred to another one and therefore has a higher utility degree. More generally, certain types of preference information can be formalized in terms of constraints on one or more underlying utility functions. This idea forms the basis of the general framework presented by AIOLLI & SPERDUTI. Sometimes, of course, training data are less generic and more application-specific. In collaborative filtering, for example, it simply consists of an incomplete set of product ratings given by a set of users (see DE GEMMIS et al. in this volume).

In the instance and object preferences scenario, a utility function is a mapping $f : \mathcal{X} \rightarrow \mathbb{R}$ that assigns a utility degree $f(\boldsymbol{x})$ to each instance (object) $\boldsymbol{x}$ and, hence, induces a complete order on $\mathcal{X}$. In the label preferences scenario, a utility function $f_i : \mathcal{X} \rightarrow \mathbb{R}$ is needed for each of the labels y_i $(i = 1, \ldots, k)$; alternatively, the functions can be summarized into a single function $f : \mathcal{X} \times \mathcal{Y} \rightarrow \mathbb{R}$ that maps instance/label tuples $(\boldsymbol{x}, y)$ to real-valued scores (see AIOLLI & SPERDUTI).[2] Here, $f_i(\boldsymbol{x})$ is the utility assigned to alternative y_i by instance $\boldsymbol{x}$. To obtain a ranking for $\boldsymbol{x}$, the alternatives are sorted according to these utility scores, i.e., $\succ_{\boldsymbol{x}}$ is such that $y_i \succ_{\boldsymbol{x}} y_j \Rightarrow f_i(\boldsymbol{x}) \geq f_j(\boldsymbol{x})$.

In the setting of instance ranking , the training data consist of instances for which the sought utility scores are given. Thus, the learning problem can, in principle, be approached by means of classification or (ordinal) regression methods. As an important difference, however, note that the goal is not to maximize classification accuracy but ranking performance. Thus, conventional learning algorithms have to be adapted correspondingly. Approaches of this kind have, e.g., been proposed in [27,33]. In this book, WAEGEMAN & DE BAETS discuss approaches that are based on the optimization of an extension of the binary AUC to a loss function for ordinal data.

In object and label ranking, training data typically originate from a kind of indirect supervision. Instead of the target scores of the utility function, the learner is given the constraints on the function, which are derived from comparative preference information of the form "This object (or label) should have a higher utility score than that object (or label)". Thus, the challenge for the learner is to find a function which is as much as possible in agreement with these constraints.

For object ranking approaches, this idea has first been formalized by Tesauro under the name *comparison training* [44]. He proposed a symmetric neural network architecture that can be trained with representations of two states and a training signal that indicates which of the two states is preferable. The elegance of this approach comes from the property that one can replace the two symmetric components of the network with a single network, which can subsequently provide a real-valued evaluation of single states. Later works on learning utility function from object preference data include [24,33,46].

For the case of label ranking, a method for learning the functions $f_i(\cdot)$ $(i = 1, \ldots, k)$ has been proposed in the framework of *constraint classification* [25,26]. Here, the authors proceed from linear utility functions and find a way to express a constraint of the form $f_i(\boldsymbol{x}) - f_j(\boldsymbol{x}) > 0$ (suggesting that $y_i \succ_{\boldsymbol{x}} y_j$) in the

---

[2] In a sense, this alternative is not just a formally equivalent rewriting. Instead, by considering an instance/label pair as an *object*, it suggests a natural way to unify the problems of object and label ranking.

form of a binary classification example in a newly constructed, higher-dimensional space. In other words, the original label ranking problem is transformed into a single binary classification problem. This problem is solved by fitting a separating hyperplane, and from this hyperplane, the linear utility functions (identified by corresponding weight vectors) can be reconstructed. An alternative approach, so-called *log-linear models for label ranking*, has been proposed in [14]. This approach is essentially equivalent to constraint classification, as it also amounts to learning linear utility functions for all labels. Algorithmically, however, the underlying optimization problem is approached in a different way, namely by means of a boosting-based algorithm that seeks to minimize a (generalized) ranking error in an iterative way. In this book, TSIVTSIVADZE et al. present an approach for learning a utility function for label ranking via minimization of a loss function that is based on a least-squares approximation of the ranking error.

3.2 Learning Preference Relations

The key idea of this approach is to learn a binary preference relation that compares pairs of alternatives (e.g., objects or labels). The training of a model thus becomes simpler, mainly because comparative training information (suggesting that one alternative is better than another one) can be used directly instead of translating it into constraints on a (latent) utility function. On the other hand, the prediction step may become more difficult, since a binary preference relation learned from data is not necessarily consistent in the sense of being transitive and, therefore, does normally not define a ranking in a unique way.

Binary preference relations can be turned into a ranking by finding a ranking that is maximally consistent with the corresponding pairwise preferences. The difficulty of this optimization problem depends on the concrete criterion, though many natural objectives (e.g., minimizing the number of object pairs whose ranks are in conflict with their pairwise preference) lead to NP-hard problems [12]. Fortunately, efficient techniques such as simple voting (known as the Borda count procedure in social choice theory) often deliver good approximations, sometimes even with provable guarantees [13]. Of course, one can also derive other, possibly more complex preference structures from a preference relation, for example weak instead of strict linear orders. In [45], a linear order with ties (indifference between two alternatives) is called a *bucket order* (a total order of "buckets", where each bucket corresponds to an equivalence class), and a method is proposed to find an order of this type, which is maximally consistent with the data.

For object ranking problems, the relational approach has been pursued in [12]. The authors propose to solve object ranking problems by learning a binary preference predicate $\mathcal{Q}(\boldsymbol{x},\boldsymbol{x}')$, which predicts whether $\boldsymbol{x}$ is preferred to $\boldsymbol{x}'$ or vice versa. This predicate is trained on the basis of exemplary preferences of the form $\boldsymbol{x} \succ \boldsymbol{x}'$. A final ordering is found in a second phase by deriving (an approximation of) a ranking that is maximally consistent with these predictions. DEMBCZYŃSKI et al.

discuss this setting for rule learning and propose to combine the predictions using the Net Flow score proposed in [4]. They also compare this setting with an alternative approach that directly learns a utility function based on the preferences and monotonicity constraints.

For label ranking problems, the pairwise approach has been introduced by [18, 31], where it is referred to as *ranking by pairwise comparison*. The key idea is to learn, for each pair of labels (y_i, y_j), a binary predicate $\mathcal{M}_{i,j}(\boldsymbol{x})$ that predicts whether $y_i \succ_{\boldsymbol{x}} y_j$ or $y_j \succ_{\boldsymbol{x}} y_i$ for an input $\boldsymbol{x}$. A label ranking is then derived from these pairwise preferences via weighted voting (generalized Borda counting).

Pairwise learning techniques for instance ranking have been proposed in [19]. More specifically, two approaches were developed and compared in that paper, one which trains binary models $\mathcal{M}_{i,j}$, one for each pair of labels y_i and y_j, and another one that trains models $\mathcal{M}_i$ $(i = 1, \ldots, k-1)$ to separate classes $y_1, \ldots, y_i$ from classes $y_{i+1}, \ldots, y_k$. Given a new query instance $\boldsymbol{x}$, both approaches submit this instance to all models that have been learned and aggregate the corresponding predictions into an overall score. A set of instances is then ranked according to these scores.

An overview of work on learning binary preference relations for label and instance ranking is given by FÜRNKRANZ & HÜLLERMEIER.

3.3 *Model-Based Preference Learning*

Another approach to learning ranking functions is to proceed from specific model assumptions, that is, assumptions about the structure of the preference relations. This approach is less generic than the previous ones, as it strongly depends on the concrete assumptions made.

An example is the assumption that the target ranking of a set of objects described in terms of multiple attributes can be represented as a *lexicographic order*. YAMAN et al. address the learning of lexicographic orders in the context of object ranking. From a machine learning point of view, assumptions of the above type can be seen as an inductive bias restricting the hypothesis space. Provided the bias is correct, this is clearly an advantage, as it may simplify the learning problem. In the case of lexicographic orders, for example, a complete ranking of all objects is uniquely identified by a total order of the attributes plus a total order of each of the attribute domains. For example, suppose objects to be described in terms of (only) four binary attributes. Thus, there are $2^4 = 16$ objects and hence $16! \approx 2 \cdot 10^{13}$ rankings in total. However, only $(2^4) \cdot 4! = 384$ of these rankings can be expressed in terms of a lexicographic order.

Needless to say, the bias induced by the assumption of a lexicographic order is very strong and will be rarely justified in practical applications. In particular, preferences on individual attribute values will normally not be independent of each other. For example, red wine might be preferred as a beverage if the main dish is meat, while white wine might be preferred in the case of fish. As mentioned earlier,

CP-nets [3] offer a language for expressing preferences on the values of single attributes and provide a suitable formalism for modeling dependencies of this type. A compact representation of a complex preference (partial order) relation is achieved by making use of conditional independence relations between such preferences (which are interpreted in terms of a *ceteris paribus* semantics), in much the same way as Bayesian networks reduce complexity of probability models by exploiting conditional independence between random variables. The CP-net itself is a graphical representation of these (in)dependencies, and each node (belonging to a single variable) is associated with a function that assigns a preference relation on the values of that attribute to each combination of values of the parent attributes. In this volume, CHEVALEYRE et al. discuss the learnability of CP-networks, both in a passive and an active learning scenario.

If a ranking function is defined implicitly via an underlying utility (scoring) function, the latter is normally also restricted by certain model assumptions. For example, the approaches outlined in Sect. 3.1 make use of linear functions to represent scores, although mostly for algorithmic reasons. There are other approaches in which the choice of the underlying utility function is more intrinsically motivated and addressed in a more explicit way. For example, GIESEN et al. describe an approach for conjoint analysis, also called multiattribute compositional models, which originated in mathematical psychology and is nowadays widely used in the social sciences and operations research. Proceeding from the description of objects in terms of a set of attributes, they assume that an underlying utility function can be decomposed into a linear sum of individual utility functions, one for each attribute. These utility functions can then be learned efficiently from the data.

Like in the case of lexicographic orders, this model assumption is obviously quite restrictive. TORRA presents a complementary and more general approach. Starting with a discussion of general properties that aggregation operators for attribute-based utility functions should fulfill, he surveys different approaches for learning a complex aggregation operator in the form of a nonadditive integral.

3.4 Local Aggregation of Preferences

Yet another alternative is to resort to the idea of local estimation techniques as prominently represented, for example, by the nearest neighbor estimation principle: Considering the rankings observed in similar situations as representative, a ranking for the current situation is estimated on the basis of these "neighbored" rankings, typically using an averaging-like aggregation operator. This approach is in a sense orthogonal to the previous model-based one, as it is very flexible and typically comes with no specific model assumption (except the regularity assumption underlying the nearest neighbor inference principle).

For label ranking, the nearest neighbor (instance-based learning) approach was first used in [9, 10]. Roughly speaking, the idea is to identify the query's k nearest neighbors in the instance space $\mathcal{X}$, and then to combine the corresponding rankings into a prediction using suitable aggregation techniques. In [11], this approach was developed in a theoretically more sophisticated way, realizing the aggregation step in the form of maximum likelihood estimation based on a statistical model for rank data. Besides, this approach is also able to handle the more general case in which the rankings of the neighbored instances are only partially known.

In the same paper, the authors propose to use this estimation principle for decision tree learning, too, namely for aggregating the label rankings associated with the instances grouped in a leaf node of the tree. Indeed, decision tree learning can also be seen as a kind of local learning method, namely as a piecewise constant approximation of the target function.[3] In this book, YU, WAN & LEE propose a similar method. They grow a decision tree and propose two splitting measures for label ranking data. The rankings that are predicted at the leaves of the trees are derived by aggregating the rankings of all training examples arriving at this leaf.

Aggregation techniques are also used for other types of preference learning problems, including object ranking. For example, assume the rankings of several subsets X_i of a reference set $\mathcal{X}$ to be given. The learning task is to combine these rankings into a complete ranking of all objects in $\mathcal{X}$. A practical application of this setting occurs, e.g., in information retrieval, when different rankings of search results originating from different search engines should be combined into an overall ranking of all retrieved pages [16]. Among other things, the learning problem may involve the determination of suitable weights for the information sources (search engines), reflecting their performance or agreement with the preferences of the user [35]. Another example is the ranking of sports teams or players, where individual tournament results with varying numbers of participants have to be combined into an overall ranking [2], or where different rankings by different judges have to be aggregated into an overall ranking of the participants of a competition [23].

4 Applications of Preference Learning

Preference learning problems in general, and ranking problems in particular, arise quite naturally in many application areas. For example, a search engine should rank Web pages according to a user's preferences. Likewise, cars can be ranked according to a customer's preferences on characteristics that discriminate different models. Another example is the ranking of possible keywords according to their relevance for an article. A few more examples have already been given in this article, and many more could be found.

In particular in the field of information retrieval, ranking applications occur quite naturally. Two particularly interesting problems are learning to rank the results of a query to a search engine, and learning to rank possible recommendations for new products. We will briefly discuss research in these areas below.

[3] More general approximations can be realized by labeling a leaf node with a nonconstant function, for example a linear function in regression learning.

4.1 Learning to Rank Search Results

A widely studied preference learning problem is the ranking of retrieval results of a search engine. Roughly, the problem is the following: given a query $\mathbf{q}$ and a set of documents $\mathcal{D}$, find a ranking of the documents in $\mathcal{D}$ that corresponds to their relevance with respect to $\mathbf{q}$. This ranking is based on an unknown preference relation $\succ_{\mathbf{q}}$, which ought to be learned from user feedback on past rankings of retrieval results for different queries. An elaborate survey of current research in this area can be found in [36].

Current research focuses particularly on suitable ways of characterizing the queries that allow one to transfer the ranking from one query to another [20, 40]. In some sense, a query may be considered as a context for a ranking, and the task is to learn a function that allows one to transfer the ranking from one context to the other. Much of the research is conducted on the LETOR (LEarning TO Rank for information retrieval) collection, a package of datasets containing a wide variety of queries with user feedback in several domains.[4]

Users can provide explicit feedback by labeling the retrieved pages with their degree of relevance. However, users are only willing to do this for a limited number of pages. It would be better if feedback could be collected in a way that is transparent to the user. RADLINSKI & JOACHIMS discuss a variety of techniques that allow one to collect the user feedback implicitly via their clicking behavior. Alternatively, ARENS proposes the use of active learning techniques which help minimizing the burden on the user by a careful automatic selection of suitable training examples for the ranking algorithm. He illustrates his technique in an application which learns a ranking function for the PubMed search engine for the MEDLINE database of biomedical literature.

4.2 Recommender Systems

Nowadays, *recommender systems* [43] are frequently used by online stores to recommend products to their customers. Such systems typically store a data table with products over users, which records the degree of preference of a user for this product. A customer can provide this preference degree explicitly by giving some sort of feedback (e.g., by assigning a rating to a movie) or implicitly (e.g., by buying the DVD of the movie). The elegant idea of collaborative filtering systems [21] is that recommendations can be based on user similarity, and that user similarity can in turn be defined by the similarity of their recommendations. Alternatively, recommender systems can also be based on item similarities, which are defined via the recommendations of the users that recommended the items in question. Yet other approaches try to learn models that capture the preference information contained in the matrix. A very good comparison of these approaches can be found in [8].

[4] http://research.microsoft.com/en-us/um/beijing/projects/letor/

In this book, DE GEMMIS et al. given an extensive survey of recommender systems. Subsequently, KARATZOGLOU & WEIMER describe the use of matrix factorization methods for compressing the information contained in the user/product matrix into a product of two lower-dimensional matrices. The dimensions over which the product is computed may be viewed as hidden concepts, which can be used to categorize the interests of a user. Interestingly, only very few (in the order of 10) such concepts are enough for a sufficiently accurate representation of large numbers of users and products. Finally, BELLOGIN et al. describe an approach that uses decision tree learning for identifying features of recommendation models that influence the quality of the predicted preference ranking.

5 Conclusions

In this introductory chapter, we have tried to give an overview of different approaches to preference learning, categorized by the learning task (label, instance, or object ranking) and the learning technique (learning utility functions, learning binary preference relations, learning preference models having a specific structure, or using local estimation and preference aggregating methods). In principle, all task/technique combinations are conceivable and can be found in the literature. We also highlighted important application areas, in particular in ranking search results and product recommendations.

Throughout the chapter, pointers to the remaining articles in this book were given. They could be characterized along a variety of dimensions, including all of those mentioned above. We have adopted a grouping that more or less corresponds to the sections in this survey. As the categorization is multi-dimensional, most articles fit into several of these categories, and, in some cases, the choice was not entirely clear. This is why we do not have separate parts on learning of a utility function or a binary preference relation, for example. In fact, almost all approaches presented in this book follow either of the two approaches.

Acknowledgements This research has been supported by the German Science Foundation (DFG).

References

1. R. Balasubramaniyan, E. Hüllermeier, N. Weskamp, J. Kämper, Clustering of gene expression data using a local shape-based similarity measure. Bioinformatics **21**(7), 1069–1077 (2005)
2. A. Birlutiu, T. Heskes, Expectation propagation for rating players in sports competitions, in *Proceedings of the 11th European Symposium on Principles of Knowledge Discovery in Databases (PKDD-07)*, ed. by J.N. Kok, J. Koronacki, R. López de Mántaras, S. Matwin, D. Mladenić, A. Skowron (Springer, Warsaw, Poland, 2007), pp. 374–381
3. C. Boutilier, R. Brafman, C. Domshlak, H. Hoos, D. Poole, CP-nets: A tool for representing and reasoning with conditional ceteris paribus preference statements. J. Artif. Intell. Res. **21**, 135–191 (2004)
4. D. Bouyssou, Ranking methods based on valued preference relations: A characterization of the net flow method. Eur. J. Oper. Res. **60**(1), 61–67 (1992)

5. A.P. Bradley, The use of the area under the ROC curve in the evaluation of machine learning algorithms. Pattern Recognit. **30**(7), 1145–1159 (1997)
6. R.I. Brafman, Preferences, planning and control, in *Proceedings of the 11th Conference on Principles of Knowledge Representation and Reasoning (KR-08)*, ed. by G. Brewka, J. Lang (AAAI, Sydney, Australia, 2008), pp. 2–5
7. P.B. Brazdil, C. Soares, J.P. da Costa, Ranking learning algorithms: Using IBL and meta-learning on accuracy and time results. Mach. Learn. **50**(3), 251–277 (2003)
8. J.S. Breese, D. Heckerman, C. Kadie, Empirical analysis of predictive algorithms for collaborative filtering, in *Proceedings of the 14th Conference on Uncertainty in Artificial Intelligence (UAI-98)*, ed. by G.F. Cooper, S. Moral (Morgan Kaufmann, Madison, WI, 1998), pp. 43–52
9. K. Brinker, E. Hüllermeier, Case-based label ranking, in *Proceedings of the 17th European Conference on Machine Learning (ECML-06)*, ed. by J. Fürnkranz, T. Scheffer, M. Spiliopoulou (Springer, Berlin, Germany, 2006), pp. 566–573
10. K. Brinker, E. Hüllermeier, Case-based multilabel ranking, in *Proceedings of the 20th International Joint Conference on Artificial Intelligence (IJCAI-07)*, ed. by M.M. Veloso (Hyderabad, India, 2007), pp. 702–707
11. W. Cheng, J. Hühn, E. Hüllermeier, Decision tree and instance-based learning for label ranking, in *Proceedings of the 26th International Conference on Machine Learning (ICML-09)*, ed. by A. Danyluk, L. Bottou, M.L. Littman (Montreal, Canada, 2009)
12. W.W. Cohen, R.E. Schapire, Y. Singer, Learning to order things. J. Artif. Intell. Res. **10**, 243–270 (1999)
13. D. Coppersmith, L. Fleischer, A. Rudra, Ordering by weighted number of wins gives a good ranking for weighted tournaments, in *Proceedings of the 17th ACM-SIAM Symposium on Discrete Algorithms (SODA-06)* (2006), pp. 776–782
14. O. Dekel, C.D. Manning, Y. Singer, Log-linear models for label ranking, in *Advances in Neural Information Processing Systems (NIPS-03)*, ed. by S. Thrun, L.K. Saul, B. Schölkopf (MIT, Cambridge, MA, 2004), pp. 497–504
15. J. Doyle, Prospects for preferences. Comput. Intell. **20**(2), 111–136 (2004)
16. C. Dwork, R. Kumara, M. Naor, D. Sivakumar, Rank aggregation methods for the Web, in *Proceedings of the 10th International World Wide Web Conference (WWW-01)* (ACM, Hong Kong, China, 2001), pp. 613–622
17. R. Fagin, R. Kumar, D. Sivakumar, Comparing top k lists. SIAM J. Discrete Math. **17**(1), 134–160 (2003)
18. J. Fürnkranz, E. Hüllermeier, Pairwise preference learning and ranking, in *Proceedings of the 14th European Conference on Machine Learning (ECML-03)*, vol. 2837, *Lecture Notes in Artificial Intelligence*, ed. by N. Lavrač, D. Gamberger, H. Blockeel, L. Todorovski (Springer, Cavtat, Croatia, 2003), pp. 145–156
19. J. Fürnkranz, E. Hüllermeier, S. Vanderlooy, Binary decomposition methods for multipartite ranking, in *Proceedings of the European Conference on Machine Learning and Principles and Practice of Knowledge Discovery in Databases (ECML/PKDD-09)*, vol. Part I, ed. by Wray L. Buntine, M. Grobelnik, D. Mladenic, J. Shawe-Taylor (Springer, Bled, Slovenia, 2009), pp. 359–374
20. X. Geng, T.-Y. Liu, T. Qin, A. Arnold, H. Li, H.-Y. Shum, Query dependent ranking using k-nearest neighbor, In *Proceedings of the 31st Annual International ACM SIGIR Conference on Research and Development in Information Retrieval (SIGIR-08)*, ed. by S.-H. Myaeng, D.W. Oard, F. Sebastiani, T.-S. Chua, M.-K. Leong (ACM, Singapore, 2008), pp. 115–122
21. D. Goldberg, D. Nichols, B.M. Oki, D. Terry, Using collaborative filtering to weave and information tapestry. Commun. ACM **35**(12), 61–70 (1992)
22. M. Gönen, G. Heller, Concordance probability and discriminatory power in proportional hazards regression. Biometrika **92**(4), 965–970 (2005)
23. S. Gordon, M. Truchon, Social choice, optimal inference and figure skating. Soc. Choice Welfare **30**(2), 265–284 (2008)
24. P. Haddawy, V. Ha, A. Restificar, B. Geisler, J. Miyamoto, Preference elicitation via theory refinement. J. Mach. Learn. Res. **4**, 317–337 (2003)

25. S. Har-Peled, D. Roth, D. Zimak, Constraint classification: A new approach to multiclass classification, in *Proceedings of the 13th International Conference on Algorithmic Learning Theory (ALT-02)*, ed. by N. Cesa-Bianchi, M. Numao, R. Reischuk (Springer, Lübeck, Germany, 2002), pp. 365–379
26. S. Har-Peled, D. Roth, D. Zimak, Constraint classification for multiclass classification and ranking, in *Advances in Neural Information Processing Systems 15 (NIPS-02)*, ed. by S. Becker, S. Thrun, K. Obermayer (2003), pp. 785–792
27. R. Herbrich, T. Graepel, K. Obermayer, Large margin rank boundaries for ordinal regression, in *Advances in Large Margin Classifiers*, ed. by P.J. Bartlett, B. Schölkopf, D. Schuurmans, A.J. Smola (MIT, 2000), pp. 115–132
28. E. Hüllermeier, J. Fürnkranz, Learning label preferences: Ranking error versus position error, in *Advances in Intelligent Data Analysis: Proceedings of the 6th International Symposium (IDA-05)* (Springer, Madrid, Spain, 2005), pp. 180–191
29. E. Hüllermeier, J. Fürnkranz (eds.), *Proceedings of the ECML/PKDD-08 Workshop on Preference Learning* (Antwerp, Belgium, 2008)
30. E. Hüllermeier, J. Fürnkranz (eds.), *Proceedings of the ECML/PKDD-09 Workshop on Preference Learning* (Bled, Slovenia, 2009)
31. E. Hüllermeier, J. Fürnkranz, W. Cheng, K. Brinker, Label ranking by learning pairwise preferences. Artif. Intell. **172**, 1897–1916 (2008)
32. K. Järvelin, J. Kekäläinen, Cumulated gain-based evaluation of IR techniques. ACM Trans. Inf. Syst. **20**(4), 422–446 (2002)
33. T. Joachims, Optimizing search engines using clickthrough data, in *Proceedings of the 8th ACM SIGKDD International Conference on Knowledge Discovery and Data Mining (KDD-02)* (ACM, 2002), pp. 133–142
34. T. Joachims, L. Granka, B. Pan, H. Hembrooke, G. Gay, Accurately interpreting clickthrough data as implicit feedback, in *Proceedings of the 28th Annual International ACM Conference on Research and Development in Information Retrieval (SIGIR-05)* (2005), pp. 154–161
35. G. Lebanon, J. Lafferty, Cranking: Combining rankings using conditional probability models on permutations, in *Proceedings of the 19th International Conference on Machine Learning (ICML-02)*, ed. by C. Sammut, A. Hoffmann (Sydney, Australia, 2002), pp. 363–370
36. T.-Y. Lu, Learning to rank for information retrieval. Found. Trends Inf. Retrieval **3**(3), 225–331 (2009)
37. R. Duncan Luce, H. Raiffa, *Games and Decisions: Introduction and Critical Survey* (Wiley, New York, NY, 1957)
38. H.B. Mann, D.R. Whitney, On a test of whether one of two random variables is stochastically larger than the other. *Ann. Math. Stat.* **18**(50), 50–60 (1947)
39. C.D. Manning, P. Raghavan, H. Schütze, *Introduction to Information Retrieval* (Cambridge University Press, 2008)
40. W. Ni, Y. Huang, M. Xie, A query dependent approach to learning to rank for information retrieval, in *Proceedings of the 9thh International Conference on Web-Age Information Management (WAIM-08)* (IEEE, Zhangjiajie, China, 2008), pp. 262–269
41. F. Radlinski, T. Joachims, Learning to rank from implicit feedback, in *Proceedings of the ACM Conference on Knowledge Discovery and Data Mining (KDD-05)* (2005), pp. 239–248
42. S. Rajaram, S. Agarwal, Generalization bounds for k-partite ranking, in *Proceedings of the NIPS 2005 Workshop on Learning to Rank*, ed. by S. Agarwal, C. Cortes, R. Herbrich, (Whistler, BC, Canada, 2005), pp. 28–23
43. P. Resnick, H.R. Varian, Special issue on recommender systems. Commun. ACM **40**(3), (1997)
44. G. Tesauro, Connectionist learning of expert preferences by comparison training. in *Advances in Neural Information Processing Systems 1 (NIPS-88)*, ed. by D. Touretzky (Morgan Kaufmann, 1989), pp. 99–106
45. A. Ukkonen, K. Puolamäki, A. Gionis, H. Mannila, A randomized approximation algorithm for computing bucket orders. Inf. Process. Lett. **109**, 356–359 (2009)
46. J. Wang, Artificial neural networks versus natural neural networks: A connectionist paradigm for preference assessment. Decision Support Syst. **11**, 415–429 (1994)
47. F. Wilcoxon, Individual comparisons by ranking methods. Biometrics Bull. **1**, 80–83 (1945)

A Preference Optimization Based Unifying Framework for Supervised Learning Problems

Fabio Aiolli and Alessandro Sperduti

Abstract Supervised learning is characterized by a broad spectrum of learning problems, often involving structured types of prediction, including classification, ranking-based predictions (label and instance ranking), and (ordinal) regression in its various forms. All these different learning problems are typically addressed by specific algorithmic solutions.

In this chapter, we propose a general preference learning model (GPLM), which gives an easy way to translate any supervised learning problem and the associated cost functions into sets of preferences to learn from. A large margin principled approach to solve this problem is also proposed.

Examples of how the proposed framework has been effectively used by us to address non-standard real-world applications are reported showing the flexibility and effectiveness of the approach.

1 Introduction

Supervised learning is probably the most commonly used learning paradigm and a large spectrum of learning algorithms have been devised for different learning tasks in the last decades. The need for such a large spectrum of learning algorithms is, in part, due to the many real-world learning problems, that are characterized by heterogeneous tasks and problem-specific learning algorithms for their solution. These include classification and regression problems (including multilabel and multiclass classification, and multivariate regression), as well as ranking-based (either label or instance ranking) and ordinal regression problems. Typically, the approach followed to deal with a nonstandard problem is to map it into a series of simpler, well-known problems and then to combine the resulting predictions. Often, however, this type

F. Aiolli (✉) and A. Sperduti
Department of Pure and Applied Mathematics - Padova - Italy, Via Trieste 63, 35131 Padova, Italy
e-mail: aiolli@math.unipd.it, sperduti@math.unipd.it

J. Fürnkranz and E. Hüllermeier (eds.), *Preference Learning*,
DOI 10.1007/978-3-642-14125-6_2, © Springer-Verlag Berlin Heidelberg 2010

of methodology lacks a principled theory supporting it and/or requires too much computational resources to be practical for real-world applications.

In this chapter, we give a survey of a quite general framework, which is able to generalize different types of supervised learning settings into a common preference optimization task. In particular, this is done by considering supervision as a set of order preferences over the predictions of the learner. More generally, we show that supervised learning problems can be characterized by considering two main dimensions, the type of prediction and the type of supervision involved in the problem to be solved. Then, based on this characterization, we are able to map any of these learning problems into a simple preference learning task. From a practical point of view, we show how all these supervised tasks can also be addressed in a simple linear setting, where any problem formulation can be transformed into a binary problem defined on an augmented space, thus allowing the exploitation of very simple optimization procedures available for the binary case. We also stress the flexibility of the preference model, which allows a user to optimize the parameters on the basis of a proper evaluation function. In fact, while in general the goal of a problem in terms of its evaluation function is clear, a crucial issue in the design of a learning algorithm is how to get a theoretical guarantee that the defined learning procedure actually minimizes the target cost function. One advantage of the framework reviewed in this chapter is that it defines a very natural and uniform way to devise and code a cost function into a learning algorithm.

Examples of real-world applications are then discussed. In particular, two recent applications are discussed in more detail. The first application concerns the problem to select the best candidate for a job role. This is an instance ranking problem, where, however, only binary supervision from the past history is available. The second application concerns a patent classification task, where patent applications have to be associated with primary categories as well as secondary categories. This is an example of a label ranking task, which cannot be properly addressed by an ordinal regression approach.

In Sect. 2, we review the general preference learning model (GPLM). Specifically, we show how the preference model generalizes the supervised learning setting by considering supervision as a partial order of (soft) constraints over the learner predictions. In addition, we show (Sect. 2.2) how the suggested generalization can be instantiated to well-known supervised learning problems. In the same section, we also discuss how cost functions for learning problems can be cast by using preferences (Sect. 2.3) and a simple linear model for the learner (Sect. 2.4). Quite general optimization procedures for training models within the proposed framework are also presented (Sect. 2.5). In Sect. 3, different application scenarios are described and discussed. In particular, it is discussed how the GPLM applies to a job candidate selection task and to a patent classification task. In Sect. 4, related works are sketched and a discussion about the proposed approach is given. Finally, in Sect. 5, some future extensions to the preference framework are suggested and final conclusions are drawn.

2 GPLM: A General Model for Supervised Learning

2.1 The Learning Domain

Let us consider a very general domain with a space of instances $\mathcal{X}$ and a space of class labels $\mathcal{Y}$. For example, this could be the domain of a recommender system where instances might correspond to customers while labels to products, or the domain of an information retrieval system where instances could correspond to documents while labels to queries.

The basic idea underpinning our general preference learning model is that we want to learn the set of parameters of a real valued relevance (or scoring) function defined on instance label pairs

$$f : \mathcal{X} \times \mathcal{Y} \rightarrow \mathbb{R},$$

which should approximate the actual *target* function. In a recommender system task, for example, this target function would represent the actual rating (a real value) a customer would give to a given product. Similarly, in the information retrieval example, the target function could represent the log-ratio of the probability of relevance of a document given a query.

We can easily note that, once such a scoring function is computed, a predictor will be able to order instances in $\mathcal{X}$ based on their relevance once any label $y \in \mathcal{Y}$ is selected, and similarly, to order class labels in $\mathcal{Y}$ based on their relevance once any instance $x \in \mathcal{X}$ is selected.

2.2 Prediction and Supervision

In supervised learning, supervision is assumed to be provided according to an unknown probability distribution $\mathcal{D}$ over pairs, where the first member is a description of a domain object (instance) and the second member is the corresponding expected prediction (target label). We generalize this setting by considering supervision as (soft) constraints over the learner predictions, that is constraints whose violation entails a cost, or penalty, for the solution. Specifically, we assume a learner makes its predictions on the basis of a set of parameters Θ, characterizing its *hypothesis space*. Each supervision constraint S, that cannot be satisfied makes the learner suffer a cost $c(S|\Theta)$. It is easy to notice that this generalizes the above-mentioned case of supervision as instance-label pairs. In fact, this is obtained back when a unitary cost is given to hypotheses generating incorrect labeling.

Now, we are able to show that, by using the setting presented above, it is possible to cast the main types of supervised learning tasks into a taxonomy on the basis of their expected prediction and supervision feedback. To this end, let us first recall the definition of order relations.

2.2.1 Definition

A *partial order* is a pair $(\mathcal{P}, \succeq)$ in which $\mathcal{P}$ is a set and $\succeq$ is a reflexive, antisymmetric, and transitive binary relation. A *partial ranking* of length r is a partial order in which the set $\mathcal{P}$ can be partitioned in r sets $\mathcal{P}_1, \ldots, \mathcal{P}_r$ such that $z \in \mathcal{P}_i, z' \in \mathcal{P}_j$, $i < j$, implies $z \succeq z'$ and no further information is conveyed about the ordering within subsets $\mathcal{P}_k$. A *full order* on $\mathcal{P}$ is defined as a partial ranking of length $|\mathcal{P}|$. We denote by $PO(\mathcal{P})$, $PR(\mathcal{P})$, and $FO(\mathcal{P})$ the set of partial orders, partial rankings, and full orders over the set $\mathcal{P}$, respectively.

2.2.2 Label Rankings as Qualitative Preferences

A first important family of supervised learning tasks is related to the ordering of the classes on the basis of their relevance for an instance, and thus they are characterized by the fact that predictions should be based on a full order over the labels. This family of problems is referred to as *label rankings*. Supervision is in the form of partial orders over the classes. In our notation, we have supervision $S \in PO(\mathcal{Y})$ and predictions in $FO(\mathcal{Y})$. Different settings can be obtained corresponding to different types of supervision. A few well-known instances are listed in the following:

Category Ranking (CR)

In this setting, the goal is to order categories on the basis of their relevance for an instance. As an example, in a collaborative filtering setting, users could correspond to our instances and the different movies to our classes. Then, one could be interested in the ordering (by relevance) of the set of movies based on user preferences. This is trivially a particular case of label ranking where supervision is given as full orders over $\mathcal{Y}$.

Bipartite Category Ranking (BCR)

In this task, supervision is given as two groups of classes and it is required to predict full orders in which the first group of classes is ranked over the second. As a leading example, in information retrieval, given a document, one might have to rank the available topics with the aim to return the most relevant topics on the top of the list. This is again a specific case of label ranking where supervision is given as partial rankings of length two. This task has been also referred to as category ranking in literature [10]. Here a different terminology is adopted to avoid confusion between these two different tasks.[1]

[1] Note that this task and the two that follow are conceptually different from the task to decide about the membership of an instance. Here, supervision only gives *qualitative* information about the fact that some classes are more relevant than others.

We might also be interested in predictions consisting of the most relevant classes, that is, of a prefix of the full order induced by the relevance function $f(x, y)$. This family of tasks is commonly referred to as *classification* problems. They can, however, be considered as subcases of the BCR ranking task. A few examples of this kind of problems, listed by increasing specificity, is given here:

Q-Label Classification (QC)

In this task, the goal is to select the Q most appropriate classes for a given instance, with Q fixed. The supervision here is a partial ranking of length two where a set of exactly Q labels are preferred over the rest.

Single-Label Classification (SC)

In this well-known classification task, the goal is to select exactly one class (the most relevant) for an instance. This is a trivial subcase of QC with $Q = 1$.

2.2.3 Instance Rankings as Qualitative Preferences

Another interesting family of tasks is *instance rankings*, where the goal is to order instances on the basis of the relevance of a given class. In our notation, predictions are in $FO(\mathcal{X})$ and supervision is given in the form $S \in PO(\mathcal{X})$.

The duality with respect to label rankings is self-evident. In principle, a corresponding problem setting could be defined for each of the label ranking settings. We can easily see that the well-known *(Bipartite) Instance Ranking* (IR) task, corresponds to BCR and is the one to induce an order such that a given set of instances is top-ranked. A natural application of this kind of prediction is in information retrieval, e.g., when listing the results returned by a search engine. Another interesting application is the one presented in Sect. 3 for job role selections. As in BCR, here supervision consists of partial rankings (this time over the set $\mathcal{X}$) of length two. Another task, which can also be considered in this family, is learning preference relations from a given set of ranked instances. For example, in information retrieval, the task to learn preference relations on the basis of basic preferences given as pairs of documents [19].

The two families of tasks above can be considered *qualitative tasks* since they are concerned with order relations between instance-class pairs. On the other side, *quantitative tasks* are the ones that are more concerned with the absolute values of the relevance of instance-class pairs.

2.2.4 Quantitative Predictions

Sometimes there is the necessity to do quantitative predictions about data at hand. For example, in binary classification, one has to decide about the membership of an instance to a class as opposed to rank instances by relevance. These settings are not directly subsumed by the settings presented above. As we will see, this can be overcome by adding a set of thresholds and doing predictions based on these thresholds.

Multivariate *Ordinal Regression* (MOR)

There are many settings where it is natural to rate instances according to an ordinal scale, including collaborative filtering, where there is the need to predict people ratings on unseen items. Borrowing the movie-related application introduced above, suitable rates for movies could be given as "bad", "fair", "good", and "recommended". With no loss in generality, we can consider the target space as the integer set $\mathcal{Z} = \{0, \ldots, R-1\}$ of R available rates. Following an approach similar to the one in [26], rates are made corresponding to intervals of the real line. Specifically, a set of thresholds $T = \{\tau_0 = -\infty, \tau_1, \ldots, \tau_{R-1}, \tau_R = +\infty\}$ can be defined and the prediction based on the rule

$$\hat{z} = \{\mathrm{i} : f(\mathbf{x}, y) \in (\tau_i, \tau_{i+1})\}.$$

In a typical (*instance-pivoted*) version of the MOR problem, given the target rate z_y w.r.t. the label y, a correct prediction will be consistent with the conditions: $f(\mathbf{x}, y) > \tau_i$ when $i \leq z_y$ and $f(\mathbf{x}, y) < \tau_i$ when $i > z_y$. Note that, a different threshold set could also be used for different labels. The well-known *(Univariate) Ordinal Regression*(OR) [20, 31] task is a trivial subcase of MOR when a single class is available. A dual (*label-pivoted*) version of the MOR problem is also possible which can raise when one has to rate classes according to an ordinal scale, and the instance is fixed in this case. An example of this situation is given in Sect. 3.2.

Multilabel Classification (MLC)

In this task, it is required to classify instances with a subset (the cardinality of which is not specified) of the available classes. For us, it is convenient to consider this task as an MOR problem, where only two ranks are available, relevant and irrelevant, and $\mathcal{Z} = \{0, 1\}$. The well-known *Binary Classification* (BC) can be considered a subcase of OR with two ranks $\mathcal{Z} = \{0, 1\}$. Note that this task is considered here conceptually different from SC with two classes. An alternative way to look at the multilabel problem is to add an artificial label, which is always considered less relevant than relevant labels and more relevant than irrelevant labels. In this way, supervision of the same type as for label ranking problems can be given. This approach, named *Calibrated Label Ranking*, has been recently proposed in [17].

Clearly, the taxonomy presented above is not exhaustive but well highlights how many different kinds of structured predictions can be seen as simple constraints over the predictions of a learner. Specifically, they consist of constraints in conjunctive

Table 1 Supervision of problems in Sect. 2.2. Label and instance rankings (LR and IR, respectively) have a preference for each order relation induced by the supervision S. In ordinal regression (MOR), a preference is associated with each threshold and $z \in \mathcal{Z}$ is the rank given by the supervision

Setting	Supervision P-sets
LR	$\{(\mathbf{x}, y_r) \succ (\mathbf{x}, y_s)\}_{(\mathbf{x},y_r)\succeq_S(\mathbf{x},y_s)}$
IR	$\{(\mathbf{x}_i, y) \succ (\mathbf{x}_j, y)\}_{(\mathbf{x}_i,y)\succeq_S(\mathbf{x}_j,y)}$
MOR	$\{(\mathbf{x}, y) \succ \tau_i\}_{i<z} \cup \{\tau_i \succ (\mathbf{x}, y)\}_{i\geq z}$

form where each basic preference is defined over the scoring values and/or a set of threshold values. In particular, we can differentiate between two types of order preferences: *qualitative* preferences in the form

$$(\mathbf{x}_i, y_r) \succ (\mathbf{x}_j, y_s)$$

telling that the value of $f(\mathbf{x}_i, y_r)$ should be higher than the value of $f(\mathbf{x}_j, y_s)$, and *quantitative* preferences in the form

$$(\mathbf{x}, y) \succ \tau \text{ or } \tau \succ (\mathbf{x}, y), \tau \in \mathbb{R}$$

relating the value of $f(\mathbf{x}, y)$ to a given threshold τ. In Table 1, a summary of supervision obtained for the most general settings are presented. Particular instantiations to more specific problems are immediate.

2.3 *Definition of the Preference Problem*

We have seen how supervision of typical supervised learning problems can be decomposed in terms of sets of qualitative and/or quantitative preferences over the scoring function of a learner. Here, we show that preferences also give us a flexible way to express cost functions, which can be directly utilized to optimize a learner. Specifically, we consider preference graphs, i.e., directed graphs where nodes take values on the set $\mathcal{H} \equiv (\mathcal{X} \times \mathcal{Y}) \cup \mathcal{T}$ and edges $(h_1, h_2) \in \mathcal{H} \times \mathcal{H}$ represent preferences $h_1 \succ h_2$. We say that a scoring function is consistent with a preference graph whenever it is consistent with all the preferences in the graph. The evaluation of any scoring function can then be performed by checking for how many graphs the scoring function is not consistent with.

Even more general cost functions can be obtained by associating weights (or costs) with the edges of the graphs. In this case, given a preference graph, the cost incurred by a hypothesis is defined as the maximum cost of its unfulfilled preferences (edges). When not explicitly indicated, we assume the weight associated with an edge to be 1. Summarizing, the total cost suffered by a scoring function f for supervision S, which is given as a set of preference graphs G, is defined as the

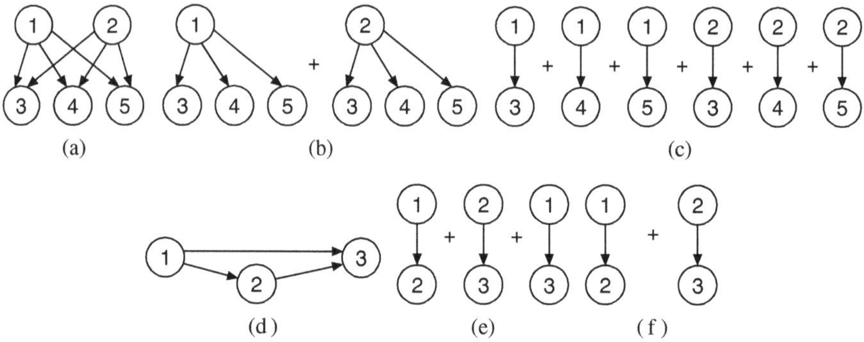

Fig. 1 Examples of label mappings for 2-label classification (**a**–**c**) and ranking (**d**–**f**)

cumulative cost over all the preference graphs. More formally, we have

$$c(G|f) = \sum_{g \in G} c(g|f) \quad \text{and} \quad c(g|f) = \max\{\gamma(\lambda)|\lambda \in E(g) \text{ not fulfilled by } f\},$$

where $\gamma(\lambda)$ represents the weight associated with the preference λ and $E(g)$ is the set of edges in g.

Cost Functions for Label Rankings

Many different cost functions which can be, useful for label ranking problems can be easily reproduced in this way. The reader can see [2, 11] for several examples on how to map this kind of supervision into sets of preference graphs. In Fig. 1, there are graphically presented very simple examples of how supervision for a 2-label classification and a ranking problem can be differently mapped into preference graphs thus obtaining different preference optimization problems. Note that only the labels and not the instances, which are fixed in this case, are indicated into the nodes.

The map in Fig. 1a defines a cost function indicating *if* any of the relevant labels are wrongly classified as irrelevant. The map in Fig. 1b would define a different cost function counting *how many* relevant labels are wrongly classified as irrelevant. Finally, the map in Fig. 1c would give the so-called *ranking loss*, i.e., the number of pairs that are not correctly ordered.

Allowing edges with different weights gives further flexibility to the model. A typical example where costs associated with edges can turn out to be useful is in the classification setting, where misclassifications can have different costs. This can be the case in single-label classification when categories are not represented with the same frequencies in the training and the test set. Another interesting case is when there is some structure between the available classes and a different metric for misclassification costs should be introduced. For example, in hierarchical classification, it makes sense to pay costs proportional to the path length in the tree between the true class and the predicted one. In all these cases, a cost matrix Δ is used to have a better control over the learning algorithm, where the element $\Delta(y_r, y_s)$ represents the cost of classifying a pattern as y_r when it is actually in y_s.

Cost Functions for Instance Rankings

Concerning the instance ranking setting and BCR-like predictions, a common cost function used in many different domains including information retrieval is the so-called AUC (Area under ROC curve) measure. It can be shown that this is proportional to the number of instance pairs incorrectly ordered and thus it can be trivially represented in our model by using a mapping similar to the one given in Fig. 1c.

Cost Functions for Prediction of Ratings

A natural definition of a cost function for ordinal regression problems is $c = |\hat{z}(\mathbf{x}) - z(\mathbf{x})|$, where $\hat{z}(\mathbf{x})$ is the rate given as output by the hypothesis and $z(\mathbf{x})$ the correct rate. In this setting, we assume that the evaluation function is somewhat proportional to the distance between ordered rates. Two different maps can be defined to use GPLM for the solution of an ordinal regression problem which are able to mimic this cost function. The easiest way is to consider the number of thresholds that are not correctly ordered w.r.t. $f(\mathbf{x}, y)$. This can be obtained in our framework by mapping this kind of supervision into $R - 1$ graphs where each graph consists of a single preference of type $(\mathbf{x}, y) \succ \tau_r$, whenever $r \leq z(\mathbf{x})$, and $(\mathbf{x}, y) \succ \tau_r$, otherwise. A second way is by using costs associated with different preferences, i.e., the rth preference is set to $((\mathbf{x}, y) \succ \tau_r)_{z-i+r}$ whenever $r \leq z$, and $(\tau_r \succ (\mathbf{x}, y))_{r-z}$, otherwise. Note that, with this last, we have a greater flexibility on the definition of the cost function. For example, it can be used when the above assumptions about the distance between different rates are not appropriate for the task at hand.

As an example of application of the second model, consider a $R = 4$ univariate ordinal regression problem. Then, we have three thresholds $T = \{\tau_1, \tau_2, \tau_3\}$ and cost mappings defined as in the following:

$$\begin{aligned}
\mathcal{G}(r = 0) &= \{(\tau_1 \succ (\mathbf{x}, y))_1, (\tau_2 \succ (\mathbf{x}, y))_2, (\tau_3 \succ (\mathbf{x}, y))_3\} \\
\mathcal{G}(r = 1) &= \{((\mathbf{x}, y) \succ \tau_1)_1, (\tau_2 \succ (\mathbf{x}, y))_1, (\tau_3 \succ (\mathbf{x}, y))_2\} \\
\mathcal{G}(r = 2) &= \{((\mathbf{x}, y) \succ \tau_1)_2, ((\mathbf{x}, y) \succ \tau_2)_1, (\tau_3 \succ (\mathbf{x}, y))_1\} \\
\mathcal{G}(r = 3) &= \{((\mathbf{x}, y) \succ \tau_1)_3, ((\mathbf{x}, y) \succ \tau_2)_2, ((\mathbf{x}, y) \succ \tau_3)_1\}
\end{aligned}$$

It is easy to verify that this mapping respects the costs as they could be obtained by the natural cost definition given above. For example, considering the instance $\mathbf{x}$ with target rate 1 being rated 3. Then, it means that the scoring function is such that $f(\mathbf{x}, y) \in (\tau_3, +\infty)$, i.e.,

$$-\infty \leq \tau_1 \leq \tau_2 \leq \tau_3 \leq f(\mathbf{x}, y) \leq +\infty,$$

and hence the cost suffered by the hypothesis is correctly computed by

$$c(r = 1) = \max\{0, +1, +2\} = +2.$$

Trivial extensions of these maps which are suitable for the multivariate ordinal regression problem can also be defined but they are omitted here.

Multilabel Classification

The standard evaluation measure for multilabel classification, the so-called Hamming loss, is defined by the number of incorrect decisions of the classifier. For the instance $\mathbf{x}$, missing one of the target classes $Y(\mathbf{x}) \subseteq \mathcal{Y}$ causes an algorithm to incur in a loss smaller than when missing more target classes. This seems quite natural in many real-world situations. In our setting, this cost function directly derives from the one defined for MOR problems when considering only two rates $\{\text{irrelevant} = 0, \text{relevant} = 1\}$. Finally, cost functions for BC problems can be obtained as a trivial subcase when a single class is available.

2.4 A Linear Embedding for Preference Optimization

In this section, we show that, using a linear form of the scoring function, the preference optimization problems defined by our framework become very simple. Consider a simple form of the relevance function, that is

$$f(\mathbf{x}, y) = w \cdot \phi(\mathbf{x}, y),$$

where $\phi(\mathbf{x}, y) \in \mathbb{R}^d$ is a joint representation of instance-class pairs and $w \in \mathbb{R}^d$ is a weight vector [30]. Note that this form generalizes the more standard form $f(\mathbf{x}, y) = w_y \cdot \phi(\mathbf{x})$, where different weight vectors are associated with different labels. In fact, let $|\mathcal{Y}| = m$, we can write:

$$w = (w_1, \ldots, w_m) \text{ and } \phi(\mathbf{x}, y) = (\underbrace{\mathbf{0}, \ldots, \mathbf{0}}_{y-1}, \phi(\mathbf{x}), \underbrace{\mathbf{0}, \ldots, \mathbf{0}}_{m-y}).$$

With this assumption, it is possible to conveniently reformulate an order constraint as a linear constraint. Let $T = \{\tau_1, \ldots, \tau_{R-1}\}$ be the set of available thresholds, then in the qualitative case, given $\lambda \equiv (\mathbf{x}_i, y_r) \succ (\mathbf{x}_j, y_s)$, we obtain

$$f(\mathbf{x}_i, y_r) > f(\mathbf{x}_j, y_s) \Leftrightarrow (w, \tau_1, \ldots, \tau_{R-1}) \cdot \underbrace{(\phi(\mathbf{x}_i, y_r) - \phi(\mathbf{x}_j, y_s), \underbrace{0, \ldots, 0}_{R-1})}_{\psi(\lambda)} > 0$$

while, in the quantitative case when either $\lambda \equiv (\mathbf{x}, y) \succ \tau_r$ or $\lambda \equiv \tau_r \succ (\mathbf{x}, y)$, and using a suitable $\delta \in \{-1, +1\}$ for shortness, we have

$$\delta(f(\mathbf{x}, y) - \tau_r) > 0 \Leftrightarrow (w, \tau_1, \ldots, \tau_{R-1}) \cdot \underbrace{(\delta\phi(\mathbf{x}, y), \underbrace{0, \ldots, 0}_{r-1}, -\delta, \underbrace{0, \ldots, 0}_{R-r-1})}_{\psi(\lambda)} > 0.$$

In general we can see that supervision constraints of all the above-mentioned problems, can be reduced to sets of linear constraints of the form $\mathbf{w} \cdot \psi(\lambda) > 0$, where $\mathbf{w} = (w, \tau_1, \ldots, \tau_{R-1})$ is the vector of weights augmented with the set of available thresholds, and $\psi(\lambda)$ is a suitable representation of the preference under consideration. The quantity

$$\rho_A(\lambda|\mathbf{w}) = \mathbf{w} \cdot \psi(\lambda)$$

will be also referred to as the margin of the hypothesis w.r.t. the preference. Note that this value is greater than zero when the preference is satisfied and less than zero otherwise. We will say that a preference λ is *consistent* with an hypothesis when $\rho_A(\lambda|\mathbf{w}) > 0$. Similarly, for a preference graph g, which represents a conjunction of simple preferences, it is required that $\rho_A(\lambda|\mathbf{w}) > 0$ for all $\lambda \in E(g)$. The margin of an hypothesis w.r.t. the whole preference graph g can be consequently defined as the minimum of the margins of preferences contained in g, i.e.,

$$\rho(g|\mathbf{w}) = \min_{\lambda \in E(g)} \rho_A(\lambda|\mathbf{w}).$$

Summarizing, all the problems defined in the taxonomy in Sect. 2.2 can be seen as an homogeneous linear problem in a opportune augmented space. Specifically, any algorithm for linear classification (e.g., perceptron or linear programming) can be used to solve it, provided the problem has a solution.

2.5 *Learning with Preferences*

In earlier sections, we have discussed the structure behind the supervision, how cost functions can be modeled using preference graphs, and how preferences can be linearly embedded by using a linear form for the scoring function. Now, we see how to give learning algorithms that are able to optimize these kind of preference optimization problems.

The goal in a batch learning algorithm is to optimize the parameters $\mathbf{w}$ so as to minimize the expected cost over $\mathcal{D}$, the actual distribution ruling the supervision feedback. More formally, the following has to be minimized

$$R[\mathbf{w}] = \mathbb{E}_{S \sim \mathcal{D}}[c(S|\mathbf{w})].$$

Table 2 Examples of approximation losses as a function of the margin. $\beta > 0, \theta \in \mathbb{R}$ are intended to be external parameters

Methods	$l(\rho)$
Perceptron	$\max(0, -\rho)$
β-margin	$\max(0, \beta - \rho)$
Mod. Least Square	$[1-\rho]_+^2$
Logistic Regression	$\log_2(1 + e^{-\beta\rho})$
Exponential	$e^{-\beta\rho}$
Sigmoidal	$(1 + e^{\beta(\rho-\theta)})^{-1}$

Although $\mathcal{D}$ is unknown, we can still try to minimize this function by exploiting the same structure of supervision and as much of the information we can gather from the available training set $\mathcal{S}$.

Specifically, the purpose of a GPLM based algorithm will be to find the hypothesis $\mathbf{w}$ that is able to minimize costs $c(S|\mathbf{w})$. As these are not continuous w.r.t. the parameter vector $\mathbf{w}$, they are approximated by introducing a continuous non-increasing loss function $l : \mathbb{R} \to \mathbb{R}^+$ approximating the indicator function. The (approximate) cost will be then defined by

$$\tilde{c}(S|\mathbf{w}) = \sum_{g \in \mathcal{G}(S)} \max_{\lambda \in g} \gamma(\lambda) l(\rho_A(\lambda|\mathbf{w})).$$

Examples of losses one can use are presented in Table 2.

The general problem can be given as in the following:

- Given a set $\mathcal{V}(\mathcal{S}) = \bigcup_{S \in \mathcal{S}} \mathcal{G}(S)$ of preference graphs
- Find a set of parameters $\mathbf{w}$ in such a way to minimize the functional

$$\mathcal{Q}(\mathbf{w}) = \mathcal{R}(\mathbf{w}) + \mu \mathcal{L}(\mathcal{V}(\mathcal{S})|\mathbf{w}), \quad (1)$$

where $\mathcal{L}(\mathcal{V}(\mathcal{S})|\mathbf{w}) = \sum_{S \in \mathcal{S}} \tilde{c}(S|\mathbf{w})$ is related to the empirical cost and $\mathcal{R}(\mathbf{w})$ is a regularization term over the set of parameters. Note that for the solution to be admissible when multiple thresholds are used and there are constraints defined over their values (as in the ordinal regression settings), these constraints should be explicitly enforced.

The use of a regularization term in problems of this type has different motivations, including the theory on regularization networks (see e.g., [12]). Moreover, we can see that by choosing a convex loss function and a convex regularization term (let say the quadratic term $\mathcal{R}(\mathbf{w}) = \frac{1}{2}||\mathbf{w}||^2$) it guarantees the convexity of the functional $\mathcal{Q}(\mathbf{w})$ in (1) and then the uniqueness of the solution. Indeed, current kernel-based approaches defined for basic supervised learning tasks can be seen in this form when using the β-margin with $\beta = 1$. This suggests a new *universal* kernel method, which is able to solve many complex learning tasks [1].

3 GPLM Applications

In the following sections, two recent applications of GPLM are presented: for a job candidate selection task [4] and a patent classification task [3]. These real-world applications are discussed in some detail with the aim to give two examples of how a potential user can approach nonstandard supervised learning problems using a GPLM-based strategy.

3.1 Job Candidate Selection as a Preferential Task

In a candidate selection task for filling a job role, one or more candidates have to be selected from a pool of candidates. Without loss of generality, let assume that the $k \geq 1$ most suited candidates for the job are selected. This decision is taken by looking at each candidate profile. Moreover, we may assume that the number k of candidates to select is already known from the beginning. This last point is very important to model the problem. In fact, a candidate will be selected on the basis of which other candidates are in the pool. In other words, no decisions can be taken for a candidate without knowing who else is competing for the same position(s).

Assume the training set consists of past decisions about promotions to a given role. Then, for any of these decisions, we know which candidates were in a selection pool and how many and which candidates were selected for the job. Thus, it seems natural to interpret any past decision as a set of preferences in which the k selected candidates were preferred to the others. More formally, we define $C_t = \{c_1, \ldots, c_{n_t}\}$ to be the set of candidates for the job role (the pool) at time t, $S_t = \{s_1^{(t)}, \ldots, s_{k_t}^{(t)}\}$ the set of candidates which got the promotion, and $U_t = \{u_1^{(t)}, \ldots, u_{n_t - k_t}^{(t)}\}$ the set of candidates which were not selected. Thus, there is evidence that s_i was preferred to u_j for each $i \in \{1, \ldots, k_t\}$ and $j \in \{1, \ldots, n_t - k_t\}$. Using our notation, we can write $s_i \succ u_j$. Note that a selection having a pool of cardinality n_t and k_t candidates selected for the job will introduce exactly $k_t \times (n_t - k_t)$ preferences. However, since $k_t \ll n_t$, the order of magnitude is still linear in the number of candidates.

Why not a Simple Binary Task?

One could think of a job role selection as a setting where for each candidate an independent decision is taken. In this case, at any time t, we would have exactly n_t independent decisions (e.g., a $+1$ decision, representing that the candidate was selected for the job role, and a -1 decision representing that the candidate was not selected for the job role). This could be modeled as a typical binary task where any of the 2^{n_t} different outcomes are possible. However, a job role selection is competitive in its nature, i.e., the choice of one candidate instead of another is not independent on the other's candidates potentials and only a fixed number of candidates can get the promotion. For this reason, the binary task does not seem to be the best choice. This will be confirmed in the experimental section where we have compared the GPLM model against a binary SVM implementation. Finally, it should be noted that the problem tends to be highly unbalanced when considered as a binary problem. In fact, the number of promoted candidates is a very small percentage of the number of candidates, who compete for the promotion. On the other hand, GPLM makes no additional assumption on the sign of the relevance function for different candidates only on the order it induces. This should make the problem easier and more balanced.

3.1.1 GPLM with SVM

In Sect. 2.4, we have shown how the preferential problem, i.e., the task to find a linear function, which is consistent with a set of preferences, can be cast as a binary problem. Examples in this case become $\psi(\lambda) = \mathbf{s}_i - \mathbf{u}_j$ for each $\lambda \equiv s_i \succ u_j$. Thus, a standard SVM algorithm applied to this new set of examples can be used to find a solution to the preferential problem.

Specifically, let $\Lambda = \{(S_1, U_1), \ldots, (S_T, U_T)\}$ be the sets involved in past promotions given as a training set for a given role, thus the SVM dual problem will be posed as

$$\begin{aligned} &\arg\max_{\alpha} \sum_t \sum_{s_i \in S_t} \sum_{u_j \in U_t} \alpha_{ij}^{(t)} - \frac{1}{2} \left\| \sum_t \sum_{s_i \in S_t} \sum_{u_j \in U_t} \alpha_{ij}^{(t)} \psi(s_i \succ u_j) \right\|^2 \\ &\text{s.t.} \quad 0 \le \alpha_{ij}^{(t)} \le \mu, \end{aligned} \tag{2}$$

and the (primal) SVM solution which solves (1) will be in the form

$$\mathbf{w}_{SVM} = \sum_t \sum_{s_i \in S_t} \sum_{u_j \in U_t} \alpha_{ij}^{(t)} \psi(s_i \succ u_j).$$

Note that the kernel computation in this case consists in computing a kernel between preferences (i.e., dot product between their vectorial representations). Nevertheless, this kernel can be easily reduced to a combination of simpler kernels between candidate profiles in the following way:

$$\begin{aligned} \tilde{k}(c_i^1 \succ c_j^1, c_i^2 \succ c_j^2) &= \langle \mathbf{c}_i^1 - \mathbf{c}_j^1, \mathbf{c}_i^2 - \mathbf{c}_j^2 \rangle = \langle \mathbf{c}_i^1, \mathbf{c}_i^2 \rangle - \langle \mathbf{c}_i^1, \mathbf{c}_j^2 \rangle - \langle \mathbf{c}_j^1, \mathbf{c}_i^2 \rangle + \langle \mathbf{c}_j^1, \mathbf{c}_j^2 \rangle \\ &= k(c_i^1, c_i^2) - k(c_i^1, c_j^2) - k(c_j^1, c_i^2) + k(c_j^1, c_j^2), \end{aligned}$$

where $k(c_i, c_j) = \langle \mathbf{c}_i, \mathbf{c}_j \rangle$ is the kernel function associated with the mapping used for the candidate profiles. We have then reduced a preferential task into a binary task which can be easily solved by a standard SVM by just redefining the kernel function suitably.

Furthermore, using the SVM decision function $f_{\mathrm{SVM}}(\lambda) = \mathrm{sgn}(\langle \mathbf{w}_{\mathrm{SVM}}, \psi(\lambda)\rangle)$ it is possible to determine whether a given order relation is verified between any two candidates. However, to decide which candidates should be selected for a new event t, $k_t \times (n_t - k_t)$ calculations of the above-defined function should be computed to obtain the relative order of candidates.

In the following, we show that the selection can actually be computed in linear time. To this end, we can decompose the weight vector computed by the SVM in the following way:

$$\begin{aligned}\mathbf{w}_{\mathrm{SVM}} &= \sum_t \sum_{c_i \in S_t} \sum_{c_j \in U_t} \alpha_{ij}^{(t)} \psi(\mathbf{c}_i \succ \mathbf{c}_j) = \sum_t \sum_{c_i \in S_t} \sum_{c_j \in U_t} \alpha_{ij}^{(t)} (\mathbf{c}_i - \mathbf{c}_j) \\ &= \sum_t \sum_{c_i \in S_t} \sum_{c_j \in U_t} \alpha_{ij}^{(t)} \mathbf{c}_i - \sum_t \sum_{c_i \in S_t} \sum_{c_j \in U_t} \alpha_{ij}^{(t)} \mathbf{c}_j .\end{aligned}$$

This decomposition allows us to decouple, in the computation of the relevance function for a new candidate, the contribution of candidate profiles given in the training set

$$\begin{aligned}f(c) = \langle \mathbf{w}_{\mathrm{SVM}}, \mathbf{c}\rangle &= \sum_t \sum_{c_i \in S_t} \left(\sum_{c_j \in U_t} \alpha_{ij}^{(t)} \right) \langle \mathbf{c}_i, \mathbf{c}\rangle \\ &\quad - \sum_t \sum_{c_j \in U_t} \left(\sum_{c_i \in S_t} \alpha_{ij}^{(t)} \right) \langle \mathbf{c}_j, \mathbf{c}\rangle \\ &= \sum_t \sum_{c_i \in S_t} \left(\sum_{c_j \in U_t} \alpha_{ij}^{(t)} \right) k(c_i, c) - \sum_t \sum_{c_j \in U_t} \left(\sum_{c_i \in S_t} \alpha_{ij}^{(t)} \right) k(c_j, c) \\ &= \sum_t \sum_{c_i \in S_t} \alpha_i^{(t)} k(c_i, c) - \sum_t \sum_{c_i \in S_t} \alpha_j^{(t)} k(c_j, c).\end{aligned}$$

Hence, the relevance function can be directly computed by post-processing the output of the SVM (the α vector) and then building a new model as follows

$$f(c) = \sum_{c_s} \beta_s k(c_s, c),$$

where $\beta_s = \sum_{t:c_s \in S_t} \alpha_s^{(t)} - \sum_{t:c_s \in U_t} \alpha_s^{(t)}$. The new model defined by the β's can directly be used by an SVM, and it returns the correct relevance for any candidate.

3.1.2 Experimental Setting

Our data were collected from the Human Resources data warehouse of a bank. Specifically, we have considered all the events related to the promotion of an employee to the job role of director of a branch office (target job role). The data used ranges from January 2002 to November 2007. Each event involves from a minimum of 1 promotion up to a maximum of 7 simultaneous promotions. Since for each event a short list of candidates was not available, we were forced to consider as candidates competing for the promotion(s) all the employees which at the time of the event were potentially eligible for promotion to the target job role. Because of that, each event t typically involves k_t "positive" examples, i.e., the employees that were promoted, and $n_t \gg k_t$ "negative" examples, i.e., eligible employees that were not promoted. As already stated, k_t ranges from 1 to 7, while n_t ranges (approximately) from 3,700 to 4,200, for a total of 199 events, 267 positive examples, and 809,982 negative examples.[2] Each candidate is represented, at the time of the event, through a profile involving 102 features. Of these features, 29 involve personal data, such as age, sex, title of study, zone of residence, etc., while the remaining 73 features codify information about the status of service, such as current office, salary, hours of work per day, annual assessment, skills self-assessment, etc. The features, and the way they are numerically coded, were chosen in such a way that it is impossible to recognize the identity of an employee from a profile. Moreover, we were careful in preserving, for each numerical feature, its inherent metric if present, e.g., the ZIP codes where redefined so that the geographic degree of proximity of two areas is preserved in the numerical proximity of the new codes associated with these two areas.

3.1.3 Results

To test whether learning preferences was better than using a binary classifier where binary supervision is used for training and the score of the resulting classifier used to rank the instances belonging to the same event, we have performed a set of experiments on a representative subset of the whole dataset. The binary classifier was an SVM with gaussian kernel and the values to use for the hyperparameters were decided through a validation set. The gaussian kernel was used also for learning preferences. The results showed that it is better to learn preferences as the SVM obtained a total accuracy of 61.88% versus an accuracy of 76.20% obtained for the approach based on learning preferences. The accuracy measures how many ranking relations are correctly predicted. The cost mapping we used for the GPLM is the one described in Sect. 3.1 that is each training selection was mapped into the set of

[2] Note that the same employee can play the role of negative example in several events. Moreover, it might also be a positive example.

preferences obtained between any "selected" profile and any "not selected" profile. The SVMlight [22] implementation has been used for all the experiments.

3.2 *Three-Layered Patent Classification as a Preferential Task*

In many applicative contexts in which textual documents are labeled with thematic categories, a distinction is made between the primary and the secondary categories that are attached to a given document. The primary categories represent the topics that are central to the document, while the secondary categories represent topics that the document somehow touches upon, albeit peripherally. For instance, when a patent application is submitted to the European Patent Office (EPO), a primary category from the International Patent Classification (IPC) scheme[3] is attached to the application, and that category determines the expert examiner who will be in charge of evaluating the application. Secondary categories are instead attached for the only purpose of identifying related prior art, since the appointed expert examiner will need to determine the novelty of the proposed invention against existing patents classified under either the primary or any of the secondary categories. For the purposes of EPO, failing to recognize the true primary category of a document is thus a more serious mistake than failing to recognize a true secondary category.

We now propose GPLM models for the principled solution of the three-layered classification task. Let d denote a document having the set $P(d) = \{c_p\}$ (a singleton) as the set of its primary categories, $S(d) = \{c_{s_1}, \ldots, c_{s_k}\}$ as the (possibly empty) set of its secondary categories, and $N(d) = \{c_{n_1}, \ldots, c_{n_l}\}$ as the set of its noncategories, such that $C = P(d) \cup S(d) \cup N(d)$.

GPLM: Ordinal Regression for Three-Layered Classification

One could be tempted to interpret the three-layered classification problem as a label-pivoted (multivariate) ordinal regression (MOR) problem, i.e., the problem to give a rank from the ordered set {primary, secondary, noncategory} to each category for a given instance. In the following, we first give a GPLM mapping already presented in [2] which can be demonstrated to be equivalent to the ordinal regression method in [7]. Then, we discuss why, in our opinion, this setting does not exactly match the three-layered classification in the patent classification application. Our experiments, which will be summarized in the following, will support this claim.

For ordinal regression, a GPLM model is built by considering two thresholds (see Fig. 2), e.g., τ_p and τ_s. For each training document, the relevance function of a primary category should be above the threshold τ_p, while the relevance function for any other category (either secondary or non-category) should be below the threshold τ_p.

[3] http://www.wipo.int/classifications/en/

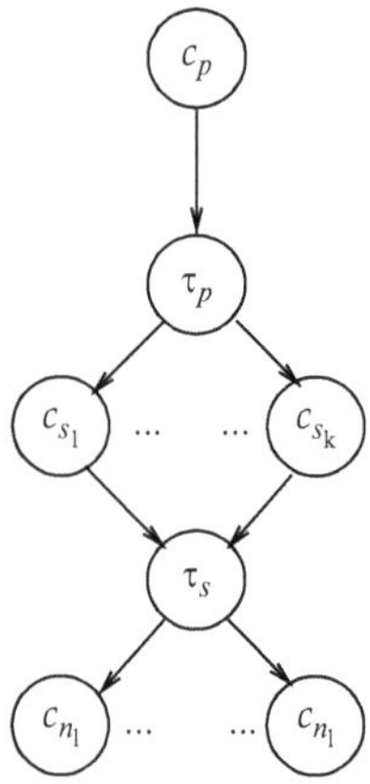

Fig. 2 GPLM mapping for ordinal-regression supervision

On the other hand, the relevance function of any secondary category should be above the threshold τ_s, while any noncategory should be below the threshold τ_s. Summarizing, the preference graph for a given training document will be as in Fig. 2. As a simple example, consider the set of categories $C = \{c_1, c_2, c_3, c_4, c_5\}$ and a training document d such that $P(d) = \{c_1\}$, $S(d) = \{c_2, c_3\}$, and $N(d) = \{c_4, c_5\}$. The set of preferences we generate is

$$\Lambda = \{(c_1 \succ_d \tau_p), (\tau_p \succ_d c_2), (\tau_p \succ_d c_3), (c_2 \succ_d \tau_s), (c_3 \succ_d \tau_s),$$
$$(\tau_s \succ_d c_4), (\tau_s \succ_d c_5)\}$$

Finally, three-layered classification will be performed by selecting the category reaching the highest relevance score as primary category, and among the others, all the categories reaching a relevance score above the threshold τ_s, as secondary categories.

At this point, we can discuss a little more about the OR-based preference model. In particular, in (multivariate) ordinal regression, it is assumed that, for each document, the rate given to a category is independent from the rate given to other categories. This assumption would be reasonable when discriminating between relevant categories (primary and secondaries) and noncategories, since this is not a "competitive" decision, but is far less reasonable when one has to choose exactly one (the most relevant) among relevant categories as the primary category for a document, since in this case we actually have a "competitive" decision. Thus, in this last case, the choice of the primary category is strongly dependent on which are the relevant categories. This difference recalls the difference between single-label classification (which is competitive) and multilabel classification (which is not competitive) in multiclass classification tasks. In other words, requiring the relevance score for the primary category to be higher than a given threshold seems

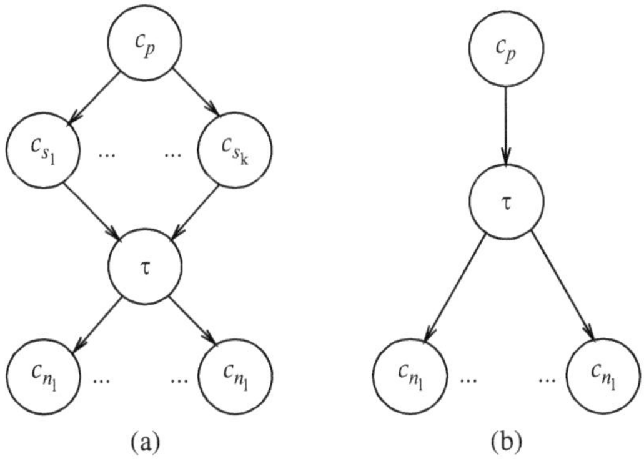

Fig. 3 GPLM mapping for supervision with (**a**) nonempty secondary category set and (**b**) empty secondary category set

an unnecessary constraint which eventually could lead to a deteriorate overall performance.

GPLM: Ad-Hoc Mapping for Three-Layered Classification

A variant of the ordinal regression scheme, which seems more suitable for the task of three-layered classification, can be built as follows. Let us interpret the primary category as the most relevant among the relevant categories. This constraint is introduced by the insertion of a set of qualitative preferences between the primary and all the secondary categories. Moreover, given the multilabel nature of the problem to discern the secondary categories with respect to the remaining categories, a single threshold τ on the relevance scores has to be added between the secondary categories and the noncategories. The categories reaching a relevance score above the threshold (apart from the one recognized as the primary category) will be predicted as secondary categories. See Fig. 3a for a graphical representation of this kind of preference model. Note that whenever $S(d) = \emptyset$, this means that the relevance values for categories in $C \backslash P(d)$ are all below the threshold. To cope with this situation, the qualitative preferences can be collapsed into a direct quantitative preference between the primary category and the threshold. See Fig. 3b for a graphical description of this kind of preference. As a simple example, consider the set of categories $C = \{c_1, c_2, c_3, c_4, c_5\}$ and a training document d such that $P(d) = \{c_1\}$, $S(d) = \{c_2, c_3\}$, and $N(d) = \{c_4, c_5\}$. The set of preferences we generate is

$$\Lambda = \{(c_1 \succ_d c_2), (c_1 \succ_d c_3), (c_2 \succ_d \tau), (c_3 \succ_d \tau), (\tau \succ_d c_4), (\tau \succ_d c_5)\}.$$

Similarly, if d is instead such that $P(d) = \{c_1\}$, $S(d) = \emptyset$, $N(d) = \{c_2, c_3, c_4, c_5\}$, this will generate the set of preferences

$$\Lambda = \{(c_1 \succ_d \tau), (\tau \succ_d c_2), (\tau \succ_d c_3), (\tau \succ_d c_4), (\tau \succ_d c_5)\}$$

3.2.1 Experimental Setting

We have evaluated our method on the WIPO-alpha Intellectual Property Organization (WIPO) in 2003. The dataset consists of 75,250 patents classified according to version 8 of the International Patent Classification scheme (IPC). Each document d has one primary category (known as the *main IPC symbol* of d), and a variable (possibly null) number of secondary categories (the *secondary IPC symbols* of d). To avoid problems due to excessive sparsity, and consistently with previous literature [13], we only consider categories at the subclass level of the IPC scheme; each of the 630 IPC subclasses is thus viewed as containing the union of the documents contained in its subordinate groups.

WIPO-alpha comes partitioned into a training set *Tr* of 46,324 documents and a test set *Te* of 28,926 documents. In our experiments, we used the entire WIPO-alpha set of 75,250 documents. Each document includes a title, a list of inventors, a list of applicant companies or individuals, an abstract, a claims section, and a long description. As in [13], we have only used the title, the abstract, and the first 300 words of the "long description". Pre-processing has been obtained by performing stop word removal, punctuation removal, down-casing, number removal, and Porter stemming. Vectorial representations have been generated for each document by the well-known "ltc" variant of cosine-normalized *tfidf* weighting. We refer the reader to [3] for a complete description of the experimental setting and the dataset.

Two additional baseline methods have been defined. In the first baseline (dubbed "Baseline1"), a binary classifier is built for each $c \in C$ (by using as positive examples of category c_i all the documents that have c_i either as a primary or as a secondary category) and use the real-valued scores returned by each classifier for d: the category for which the largest score has been obtained are selected as the primary category, while the set of secondary categories are identified by optimizing a threshold for each individual category and selecting the categories whose associated classifier has returned a score above its associated threshold. We have indeed implemented this approach (by using standard binary SVMs). A slightly stronger approach (dubbed "Baseline2") consists in performing two different classification tasks, a first one (by means of an SVM-DDAG [28] single-label classifier h_P) aimed at identifying the primary category of d, and a second one (by means of a multilabel classifier h_S consisting of m SVM-based binary classifiers h_S^i, one for each category $c_i \in \{c_1, \ldots, c_m\}$) aimed at identifying, among the remaining categories, the secondary categories of d. The h_P classifier is trained by using, as positive examples of each c_i, only the training documents that have c_i as primary category. Each of the h_S^i is instead trained by using as positive examples only the training documents that have c_i as secondary category, and as negative examples only the training documents that have c_i as noncategory (those that have c_i as primary category are discarded).

3.2.2 Results

The results obtained for the different classifiers are summarized in Table 3. Ad-hoc evaluation measures have been used. In particular, the F_1 measure is computed

Table 3 Micro-averaged F_1^3 values obtained by the classifiers

	F_1^{PS}	F_1^{SN}	F_1^{PN}	F_1^3
Baseline1	0.851	0.180	0.482	0.499
Baseline2	**0.886**	0.200	0.464	0.504
Ordinal regression	0.7847	0.1774	**0.5343**	0.5077
GPLM Adatron	0.8433	**0.2138**	0.5129	**0.5206**

for each pair of layers and then combined to form a single measure F_1^3. The first two rows report the performances of the two baseline classifiers. It can be observed that they have almost identical F_1^3 and are not so good in telling apart secondary categories from noncategories (F_1^{SN}). The third row reports the performance of the ordinal regression classifier, which turns out to have the best separation between primary and noncategories (F_1^{PN}) but a quite low performance on separating primary and secondary categories (F_1^{PS}). These results seem coherent with the analysis we have given in Sect. 3.2 as the separation between primary categories and noncategories is overconstrained by the ordinal regression model. The overall performance (F_1^3) slightly improves over the baseline classifiers. The fourth row reports the performance of the GPLM using an own implementation of the Kernel–Adatron [15] as optimizer. With respect to the baselines and the ordinal regression classifier, there is a clear improvement on F_1^{SN}, while F_1^{PS} decreases. Overall, however, there is a significant improvement in F_1^3.

4 Related Work and Discussion

Some other efforts have been made to generalize label ranking tasks. The first work on this we are aware of is [18] where the authors show how different label ranking problems can be cast into a linear problem which is solvable by a perceptron in an augmented feature space. In [21], a variant is presented in which the ranking is performed based on a voting strategy on classifiers discriminating between label pairs. In [11], the authors propose a setting in which a label ranking problem is map into a set of preference graphs and a convex optimization problem is defined to solve it. Our preference model proposed in [5] generalizes on these two previous approaches, by proposing a more flexible way to model cost functions for the same problems, and giving a kernel-based large margin solution for these kind of tasks. More recently, in [30], a large margin method to solve single-label problems with structured (e.g., hierarchical) output has been proposed. This last approach is not, however, directly applicable to solve general label ranking tasks as it requires the solving of an optimization problem with a different constraint for each possible (label) ranking and also the decoding step can show exponential complexity for general cost functions. In [24] it has been shown that this technique is, however, feasible when applied to certain cost functions that are relevant for information retrieval ranking tasks.

The general task of instance ranking is gaining a large popularity especially in the information retrieval community where the typical need is to rank documents based on their relevance for a query. In this context, this task is commonly referred to as *learning to rank*. The approaches to this general task can be divided into three categories: *point-wise*, *pair-wise*, and *list-wise*. This taxonomy is similar to the one presented in this chapter. In the point-wise approach, see for example [9, 16, 25, 27, 29], the input space are single documents and the output space are real values or ordinal categories. This kind of settings are a subset of the tasks that have been referred to as quantitative tasks in this chapter, namely the class of instance-pivoted Multivariate Ordinal Regression. In the pair-wise approach, see for example [6, 8, 14, 20, 23], the input space are document pairs (basically preferences) and the output space are documents ordered by relevance. This kind of settings are those tasks which have been here referred to as qualitative tasks, and Instance Rankings in particular. Finally, in the list-wise approach, see for example [32], the input space is the whole document set and typically a direct optimization of the evaluation function is required. This last approach is very challenging as these evaluation functions are often noncontinuous and nondifferentiable.

One clear advantage of the approach presented in this chapter, with respect to all the ones sketched in this section, is its ability to treat uniformly the label and the instance ranking settings as well as the regression setting, exploiting a preference-centric point of view. Reducing all of these problems to preference optimization implies that any optimization technique for preference learning can be used to solve them all.

5 Conclusion and Future Extensions

We have discussed a general preference model for supervised learning and applications to complex prediction problems, as job selection and patent classification. The first application is an instance-ranking problem while the second application is a label-ranking problem where categories have to be associated with patents according to a three-layered structure (primary, secondary, non-category).

An interesting aspect of the proposed preference model is that it allows to codify cost functions as preferences and naturally plug them into the same training algorithm. In this view, the role of the cost functions resembles the role of kernels in kernel-machines. Moreover, the proposed method gives a tool for comparing different algorithms and cost functions on a same learning problem.

In the future, it would be interesting to explore extensions to the model, including: (a) Considering models with disjunctive preferences as it would increase the flexibility of the model. (b) Studying new fast (approximate) algorithms when the number of examples/preferences are simply too large to be coped with by standard learning algorithms. (c) Extending the concept of preferences to preferences to a given degree, i.e., when a preference constraint have to be fulfilled with a given margin.

References

1. F. Aiolli, Large margin multiclass learning: models and algorithms. Ph.D. thesis, Department of Computer Science, University of Pisa, 2004. http://www.di.unipi.it/phd/tesi/tesi_2004/PhDthesisAiolli.ps.gz
2. F. Aiolli, A preference model for structured supervised learning tasks, in *Proceedings of the IEEE International Conference on Data Mining (ICDM)* (2005), pp. 557–560
3. F. Aiolli, R. Cardin, F. Sebastiani, A. Sperduti, Preferential text classification: Learning algorithms and evaluation measures. Inf. Retr. **12**(5), 559–580 (2009)
4. F. Aiolli, M. De Filippo, A. Sperduti, Application of the preference learning model to a human resources selection task, in *Proceedings of the IEEE Symposium on Computational Intelligence and Data Mining (CIDM)* (Amsterdam, NL, 2009), pp. 203–210
5. F. Aiolli, A. Sperduti, Learning preferences for multiclass problems, in *Advances in Neural Information Processing Systems* (MIT, Cambridge, MA, 2005) pp. 17–24
6. C.J.C. Burges, T. Shaked, E. Renshaw, A. Lazier, M. Deeds, N. Hamilton, G.N. Hullender, Learning to rank using gradient descent, in *Proceedings of the International Conference on Machine Learning (ICML)* (2005), pp. 89–96
7. W. Chu, S. Sathiya Keerthi, Support vector ordinal regression. Neural Comput. **19**(3), 792–815 (2007)
8. W.W. Cohen, R.E. Schapire, Y. Singer, Learning to order things. J. Artif. Intell. Res. **10** 243–270 (1999)
9. K. Crammer, Y. Singer, Pranking with ranking, in *Advances in Neural Information Processing Systems (NIPS)* (2002), pp. 641–647
10. K. Crammer, Y. Singer, A family of additive online algorithms for category ranking. J. Mach. Learn. Res. **3**, 1025–1058 (2003)
11. O. Dekel, C.D. Manning, Y. Singer, Log-linear models for label ranking, in *Advances in Neural Information Processing Systems* (2003)
12. T. Evgeniou, M. Pontil, T. Poggio, Regularization networks and support vector machines. Adv. Comput. Math. **13**, 1–50 (2000)
13. C.J. Fall, A. Törcsvári, K. Benzineb, G. Karetka, Automated categorization in the International Patent Classification. SIGIR Forum **37**(1), 10–25 (2003)
14. Y. Freund, R.D. Iyer, R.E. Schapire, Y. Singer, An efficient boosting algorithm for combining preferences. J. Mach. Learn. Res. **4**, 933–969 (2003)
15. T.T. Friess, N. Cristianini, C. Campbell, The kernel adatron algorithm: a fast and simple learning procedure for support vector machines, in *Proceedings of International Conference of Machine Learning (ICML)* (1998), pp. 188–196
16. T.T. Friess, N. Cristianini, C. Campbell, Subset ranking using regression, in *Proceedings of the International Conference on Learning Theory (COLT)* (Springer Berlin/Heidelberg, 2006), pp. 605–619
17. J. Fürnkranz, E. Hüllermeier, E. Mencía, K. Brinker, Multilabel classification via calibrated label ranking. Mach. Learn. **73**(2), 133–153 (2008)
18. S. Har-Peled, D. Roth, D. Zimak, Constraint classification for multiclass classification and ranking, in *Advances in Neural Information Processing Systems* (2002), pp. 785–792
19. R. Herbrich, T. Graepel, P. Bollmann-Sdorra, K. Ober-mayer, Learning a preference relation for information retrieval, in *Proceedings of the AAAI Workshop Text Categorization and Machine Learning* (1998)
20. R. Herbrich, T. Graepel, K. Obermayer, Large margin rank boundaries for ordinal regression, in *Advances in Large Margin Classifiers* (MIT, 2000), pp. 115–132
21. E. Hüllermeier, J. Fürnkranz, W. Cheng, K. Brinker, Label ranking by learning pairwise preferences. *Artif. Intell.* **172**(16–17), 1897–1916 (2008)
22. T. Joachims, Making large-scale svm learning practical, in *Advances in Kernel Methods - Support Vector Learning* ed. by B. Schölkopf, C. Burges, A. Smola (MIT, 1999)
23. T. Joachims, Optimizing search engines using clickthrough data, in *Proceedings of the Conference on Knowledge Discovery and Data Mining (KDD)* (2002) pp. 133–142

24. Q. Le, A. Smola, Direct optimization of ranking measures. Technical report, NICTA, Canberra, Australia, 2007
25. P. Li, C. Burges, Q. Wu, Mcrank: Learning to rank using multiple classification and gradient boosting, in *Advances in Neural Information Processing Systems (NIPS)* (MIT, 2008), pp. 897–904
26. P. McCullagh, J.A. Nelder, *Generalized Linear Models* (Chapman & Hall, 1983)
27. R. Nallapati, Discriminative models for information retrieval, in *Proceedings of the Conference on Research and Development in Information Retrieval (SIGIR)* (ACM, 2004), pp. 64–71
28. J.C. Platt, N. Cristianini, J. Shawe-Taylor, Large margin DAGs for multiclass classification, in *Advances in Neural Information Processing Systems (NIPS)* (1999), pp. 547–533
29. A. Shashua, A. Levin, Ranking with large margin principle: Two approaches, in *Advances in Neural Information Processing Systems (NIPS)* (2002), pp. 937–944
30. I. Tsochantaridis, T. Hofmann, T. Joachims, Y. Altun, Support vector machine learning for interdependent and structured output spaces, in *Proceedings of the International Conference on Machine learning (ICML)* (2004), pp. 1453–1484
31. H. Wu, H. Lu, S. Ma, A practical svm-based algorithm for ordinal regression in image retrieval, in *Proceedings of the ACM international conference on Multimedia* (2003), pp. 612–621
32. F. Xia, T. Liu, J. Wang, W. Zhang, H. Li, Listwise approach to learning to rank: theory and algorithm, in *Proceedings of the International Conference on Machine Learning (ICML)* (2008), pp. 1192–1199

Part I
Label Ranking

Label Ranking Algorithms: A Survey

Shankar Vembu and Thomas Gärtner

Abstract Label ranking is a complex prediction task where the goal is to map instances to a total order over a finite set of predefined labels. An interesting aspect of this problem is that it subsumes several supervised learning problems, such as multiclass prediction, multilabel classification, and hierarchical classification. Unsurprisingly, there exists a plethora of label ranking algorithms in the literature due, in part, to this versatile nature of the problem. In this paper, we survey these algorithms.

1 Introduction

Binary classification [20, 63] is a well-studied problem in supervised machine learning. Often, in real-world applications such as object recognition, document classification etc., we are faced with problems where there is a need to predict multiple labels. Label ranking is an example of such a complex prediction problem where the goal is to not only predict labels from among a finite set of predefined labels, but also to rank them according to the nature of the input. A motivating application is document categorization where categories are topics (e.g., sports, entertainment, politics) within a document collection (e.g., news articles). It is very likely that a document may belong to multiple topics, and the goal of the learning algorithm is to order (rank) the relevant topics above the irrelevant ones for the document in question. Label ranking is an interesting problem as it subsumes several supervised learning problems such as multiclass, multilabel, and hierarchical classification [27]. This survey is intended to provide an overview of label ranking algorithms.

S. Vembu (✉) and T. Gärtner
Fraunhofer IAIS, Schloß Birlinghoven, 53754 Sankt Augustin, Germany
e-mail: shankar.vembu@iais.fraunhofer.de, thomas.gaertner@iais.fraunhofer.de

J. Fürnkranz and E. Hüllermeier (eds.), *Preference Learning*,
DOI 10.1007/978-3-642-14125-6_3,

2 Preliminaries

We begin with some definitions from order theory, and describe distance metrics and kernels that will be used in this paper.

A binary relation $\succ$ on a (finite) set Σ is a *partial order* if $\succ$ is asymmetric ($a \succ b \Rightarrow \neg b \succ a$) and transitive ($a \succ b \wedge b \succ c \Rightarrow a \succ c$). The pair $(\Sigma, \succ)$ is then called a *partially ordered set* (or *poset*).

We denote the set $\{(u, v) \in \Sigma \mid u \succ v\}$ by $p(\succ)$ and the set of all partial orders over Σ by $\mathcal{P}_\Sigma$. Note that every partially ordered set $(\Sigma, \succ)$ defines a directed acyclic graph $G_\succ = (\Sigma, p(\succ))$. This graph is also called as *preference graph* in the label ranking literature.

A partially ordered set $(\Sigma, \succ)$ such that $\forall u, v \in \Sigma : u \succ v \vee v \succ u$ is a *totally ordered set* and $\succ$ is called a *total order*, a *linear order*, a *strict ranking* (or simply *ranking*), or a *permutation*. A *partial ranking* is a total order with ties.

A partial order $\succ'$ *extends* a partial order $\succ$ on the same Σ if $u \succ v \Rightarrow u \succ' v$. An extension $\succ'$ of a partial order $\succ$ is a *linear extension* if it is totally ordered (i.e., a total order $\succ'$ is a *linear extension* of a partial order $\succ$ if $\forall u, v \in \Sigma$, $u \succ v \Rightarrow u \succ' v$). A collection of linear orders $\succ_i$ *realizes* a partial order $\succ$ if $\forall u, v \in \Sigma, u \succ v \Leftrightarrow (\forall i : u \succ_i v)$. We denote this set by $\ell(\succ)$. The *dual* of a partial order $\succ$ is the partial order $\bar{\succ}$ with $\forall u, v \in \Sigma : u \bar{\succ} v \Leftrightarrow v \succ u$.

2.1 Distance Metrics

Spearman's rank correlation coefficient (ρ) [70] is a nonparametric measure of correlation between two variables. For a pair of rankings π and π' of length k, it is defined as

$$\rho = 1 - \frac{6D(\pi, \pi')}{k(k^2 - 1)},$$

where $D(\pi, \pi') = \sum_{i=1}^{k} (\pi(i) - \pi'(i))^2$ is the sum of squared rank distances. The sum of absolute differences $\sum_{i=1}^{k} |\pi(i) - \pi'(i)|$ defines the *Spearman's footrule distance metric*.

Kendall tau correlation coefficient (τ) [51] is a nonparametric statistic used to measure the degree of correspondence between two rankings. For a pair of rankings π and π', it is defined as

$$\tau = \frac{n_c - n_d}{\frac{1}{2}k(k-1)},$$

where n_c is the number of concordant pairs, and n_d is the number of discordant pairs in π and π'. The number of discordant pairs defines the *Kendall tau distance metric*.

2.2 *Kernels*

We now define kernels on partial orders and describe their properties.

2.2.1 Position Kernel

Define

$$k_{\#}: \mathcal{P} \times \mathcal{P} \to \mathbb{R} \text{ by } k_p(\succ, \succ') = \sum_{u \in \Sigma} \kappa\left(|\{v \in \Sigma \mid v \succ u\}|, |\{v \in \Sigma \mid v \succ' u\}|\right),$$

where κ is a kernel on natural numbers.

- This function is a kernel and can be computed in time polynomial in Σ.
- It is injective (in the sense that $k(\succ, \cdot) = k(\succ', \cdot) \Leftrightarrow \succ = \succ'$) for linear orders but not for partial orders.

2.2.2 Edge Kernel

Define

$$k_p: \mathcal{P} \times \mathcal{P} \to \mathbb{R} \text{ by } k_p(\succ, \succ') = |p(\succ) \cap p(\succ')|.$$

- This function is a kernel and can be computed in time polynomial in $|p(\succ)|$.
- This kernel is injective (in the sense that $k(\succ, \cdot) = k(\succ', \cdot) \Leftrightarrow \succ = \succ'$).

A downside of this kernel is that $a \succ b$ is as similar to $b \succ a$ as it is to $a \succ c$. However, we can overcome this problem easily. Let $\bar{\succ}$ be the dual partial order of $\succ$. Define

$$k_{\bar{p}}(\succ, \succ') = k_p(\succ, \succ') - k_p(\bar{\succ}, \succ').$$

- This function is a kernel (the feature space has one feature for every pair of elements and the value of feature uv is $+\sqrt{2}$ iff $u \succ v$, $-\sqrt{2}$ iff $v \succ u$, and 0 otherwise).
- It can be computed in time polynomial in $|p(\succ)|$.

2.2.3 Extension Kernel

Define

$$k_\ell: \mathcal{P} \times \mathcal{P} \to \mathbb{R} \text{ by } k(\succ, \succ') = |\ell(\succ) \cap \ell(\succ')|.$$

- This function is a kernel.
- It is injective (in the sense that $k(\succ, \cdot) = k(\succ', \cdot) \Leftrightarrow \succ = \succ'$).
- The kernel cannot be computed in polynomial time as counting linear extensions (or, equivalently, computing $k(\succ, \succ)$) is #P-complete. However, it can possibly

be approximated as (i) the number of linear extensions can be approximated, and (ii) the set of linear extensions can be enumerated almost uniformly.

- We have $k_\ell(\succ, \succ') = 0 \Leftrightarrow \exists u, v \in \Sigma : u \succ v \wedge v \succ' u$. We call such partial orders *contradicting*.
- For noncontradicting partial orders $\succ, \succ'$ define the partial order $\succ \cup \succ'$ such that $\forall u, v \in \Sigma : u(\succ \cup \succ')v \Leftrightarrow u \succ v \vee u \succ' v$.

2.3 Label Ranking: Problem Definition

Let $\mathcal{X} \subseteq \mathbb{R}^m$ be the input (instance) space, $\Sigma = \{1, \ldots, k\} = [\![k]\!]$ be a set of labels, and $\mathcal{Y}$ be the output space of all possible partial orders over Σ. Let $\mathcal{T} = \{x_i, y_i\}_{i=[\![n]\!]} \subseteq \mathcal{X} \times \mathcal{Y}$ be a set of training examples. Let $G_i = (V_i, E_i)$ denote the preference graph corresponding to y_i, for all $i \in [\![n]\!]$. The goal of a label ranking algorithm is to learn a mapping $f : \mathcal{X} \to \mathcal{Y}$, where f is chosen from a hypothesis class $\mathcal{F}$, such that a predefined loss function $\ell : \mathcal{F} \times \mathcal{Y} \times \mathcal{Y} \to \mathbb{R}$ is minimized. In general, the mapping f is required to output a total order, but it is also possible to envisage settings where the desired output is a partial order. Let $k_\mathcal{X} : \mathcal{X} \times \mathcal{X} \to \mathbb{R}$ and $k_\mathcal{Y} : \mathcal{Y} \times \mathcal{Y} \to \mathbb{R}$ denote positive semidefinite kernels on $\mathcal{X}$ and $\mathcal{Y}$, respectively.

2.4 Related Problems

We now describe some problems that are related to label ranking. A comprehensive survey of literature for these problems is beyond the scope of this paper. Nevertheless, we refer to, what we believe, are important (and classical), and possibly also recent contributions.

2.4.1 Multilabel Classification

Multilabel classification [30, 37, 65, 66] is a generalization of multiclass prediction where the goal is to predict a set of labels that are relevant for a given input. It is a special case of multilabel ranking [11] where the preference graph is bipartite with directed edges between relevant and irrelevant labels.

2.4.2 Object Ranking

In this setting, the preference information is given on a subset of the input space and not on the labels. The goal is to learn a ranking function $f : \mathcal{X} \to \mathbb{R}$ such that for any $a, b \in \mathcal{X}$, $f(a) > f(b)$ iff $a \succ b$. Thus, a total order is induced on the input space. This setting has attracted a lot of attention recently, in particular, in information retrieval applications. Various approaches to solve this problem have been proposed in the literature [12, 13, 18, 23, 32, 34, 44, 50, 64]. Learning object ranking functions on structured inputs (graphs) was proposed recently in [1, 2, 78]. A survey on object ranking algorithms also appears as a chapter in this book.

2.4.3 Ordinal Regression

Ordinal regression [16, 17, 43, 44] is a form of multiclass prediction where the labels are defined on an ordinal scale and therefore cannot be treated independently. It is closely related to the object ranking problem where the preference information on (a subset of) the input space is a directed k-partite graph where k is the number of ordinal values (labels).

3 Learning Reductions

Learning reductions are an efficient way to solve complex prediction problems using simple models such as (binary) classifiers as primitives. Such techniques have been applied to solve problems such as ranking [7], regression [54], and structured prediction [26], just to name a few.

Label ranking can be reduced to binary classification using Kesler's construction [60]. This approach was proposed in [40, 41] under the name of constraint classification. The idea is to construct an expanded example set $\mathcal{T}'$ in which every example $(x, y) \in \mathbb{R}^m \times \mathcal{Y}$ with its corresponding preference graph $G = (V, E)$ is embedded in $\mathbb{R}^{km} \times \{-1, 1\}$, with each preference $(p, q) \in E$ contributing a single positive and a single negative example. The Kesler mapping P is defined as follows:

$$P_+(x, y) = \{(x \otimes \mathbf{0}_p, 1) \mid (p, q) \in E\} \subseteq \mathbb{R}^{km} \times \{1\}$$
$$P_-(x, y) = \{(-x \otimes \mathbf{0}_q, -1) \mid (p, q) \in E\} \subseteq \mathbb{R}^{km} \times \{-1\},$$

where $\mathbf{0}_j$ is a k-dimensional vector whose jth component is one and the rest are zeros. Let $P(x, y) = P_+(x, y) \cup P_-(x, y)$. The expanded set of examples is then given by

$$\mathcal{T}' = P(\mathcal{T}) = \bigcup_{(x,y)\in\mathcal{T}} P(x, y) \subseteq \mathbb{R}^{km} \times \{-1, 1\}.$$

A binary classifier (linear separating hyperplane) trained on this expanded set can be viewed as a sorting function over k linear functions, each in $\mathbb{R}^m$. The sorting function is given as $\operatorname{argsort}_{j \in [\![k]\!]} \langle w_j, x \rangle$, where w_j is the j-th chunk of the weight vector $w \in \mathbb{R}^{km}$, i.e., $w_j = (w_{(j-1)m+1}, \ldots, w_{jm})$.

A well-known reduction technique known as pairwise classification [35] can be used to reduce the problem of multiclass prediction to learning binary classifiers. An extension of this technique known as ranking by pairwise comparison (RPC) was proposed in [36,46] to solve the label ranking problem. The central idea is to learn a binary classifier for each pair of labels in Σ resulting in $k(k-1)/2$ models. Every individual model $\mathcal{M}_{pq}$ with $p, q \in \Sigma$ learns a mapping that outputs 1 if $p \succ_x q$ and 0 if $q \succ_x p$ for an example $x \in \mathcal{X}$. Alternatively, one may also learn a model that maps into the unit interval [0, 1] instead of $\{0, 1\}$. The resulting model assigns a valued preference relation R_x to every example $x \in \mathcal{X}$:

$$R_x(p,q) = \begin{cases} \mathcal{M}_{pq}(x) & \text{if } p < q \\ 1 - \mathcal{M}_{pq}(x) & \text{if } p > q \end{cases}$$

The final ranking is obtained by using a ranking procedure that basically tries to combine the results of these individual models to induce a total order on the set of labels. A simple ranking procedure is to assign a score $s_x(p) = \sum_{p \neq q} R_x(p,q)$ to each label p and obtain a final ordering by sorting these scores. This strategy exhibits desirable properties such astransitivity of pairwise preferences. Furthermore, the RPC algorithm minimizes the sum of squared rank distances and an approximation to the Kendall tau distance metric under the condition that the binary models $\mathcal{M}_{pq}$ provide correct probability estimates, i.e., $R_x(p,q) = \mathcal{M}_{pq}(x) = \Pr[p \succ_x q]$.

4 Boosting Methods

A boosting [33] algorithm for label ranking was proposed by Dekel et al. [27]. A label ranking function $f : \mathcal{X} \times \Sigma \to \mathbb{R}$ is learned such that for any given $x \in \mathcal{X}$, a total order is induced on the label set by $p \succ_x q \iff f(x,p) > f(x,q)$. The label ranking function is represented as a linear combination of a set of L base ranking functions, i.e, $f(x,p) = \sum_{l=1}^{L} \lambda_l h_l(x,p)$, where $\{\lambda_l\}_{l \in [\![L]\!]}$ are parameters that are estimated by the boosting algorithm. We denote the label ranking induced by f for x by $f(x)$ (with a slight abuse of notation). A graph decomposition procedure $\mathcal{D}$, which takes a preference graph $G_i = (V_i, E_i)$ for any $x_i \in \mathcal{X}$ as its input and outputs a set of S_i subgraphs $\{G_{i,s}\}_{s \in [\![S_i]\!]}$, has to be specified as an input to the learning algorithm. A simple example of a graph decomposition procedure is to consider every edge $e \in E_i$ as a subgraph. Other examples include decomposing the graph into bipartite directed graph $G_{i,s} = (U_{i,s}, V_{i,s}, E_{i,s})$ such that $|U_{i,s}| = 1$ or $|V_{i,s}| = 1$ (see Fig. 2 in [27] for an illustration). The generalized loss due to $f(x_i)$ w.r.t. G_i is the fraction of subgraphs in $\mathcal{D}(G_i)$ with which $f(x_i)$ disagrees. The generalized loss over all the training instances is defined as

$$\ell_{\text{gen}}(f, \mathcal{T}, \mathcal{D}) = \sum_{i=1}^{n} \frac{1}{S_i} \sum_{s=1}^{S_i} \delta(f(x_i), G_{i,s}),$$

where $\delta(.,.)$ is a loss function defined on the subgraphs such as the 0-1 loss or the ranking loss [66]. While minimizing such a discrete, nonconvex loss function is NP-hard, it is possible to minimize an upper bound given by

$$\delta(f(x_i), G_{i,s}) \leq \log_2(1 + \sum_{e \in E_{i,s}} \exp(f(x_i, \text{term}(e)) - f(x_i, \text{init}(e)))).$$

where $\text{init}(e)$ (resp. $\text{term}(e)$) is the label corresponding to the initial (resp. terminal) vertex of any directed edge $e \in E_{i,s}$. To minimize this upper bound, Dekel et al. proposed to use a boosting-style algorithm for exponential models [19, 56] to estimate the model parameters λ and also proved a bound on the decrease in loss in every iteration of the algorithm.

5 Label Ranking SVM

Elisseeff and Weston [30] proposed a kernel method for multilabel classification. A straightforward generalization of this approach results in a label ranking algorithm. Define a scoring function for label p and input x as $h_p(x) = \langle w_p, x \rangle$, where w_p is a weight vector corresponding to label p. These scoring functions together will define the mapping f by a sorting operation, i.e., $f(x) = \text{argsort}_{j \in [\![k]\!]} \langle w_j, x \rangle$. The ranking loss [66] w.r.t. to a preference graph $G = (V, E)$ is defined as $\ell(f, x, y) = \frac{1}{|E|}|(p,q) \in E \text{ s.t. } h_p(x) \leq h_q(x)|$. The following optimization problem minimizes the ranking loss:

$$\begin{aligned} &\min_{\{w_j\}_{j=[\![k]\!]}} \quad \sum_{j=1}^{k} ||w_j||^2 + \nu \sum_{i=1}^{n} \frac{1}{|E_i|} \sum_{(p,q) \in E_i} \xi_{ipq} \\ &\text{subject to}: \langle w_p - w_q, x_i \rangle \geq 1 - \xi_{ipq}, \forall (p,q) \in E_i, \forall i \in [\![n]\!] \\ &\qquad \xi_{ipq} \geq 0, \forall (p,q) \in E_i, \forall i \in [\![n]\!], \end{aligned}$$

where $\nu > 0$ is the regularization parameter that tradesoff the balance of the loss term against the regularizer.

Shalev-Shwartz and Singer [67] considered the setting where the training labels take the form a feedback vector $\gamma \in \mathbb{R}^k$. The interpretation is that label p is ranked higher than label q iff $\gamma_p > \gamma_q$. The difference $\gamma_p - \gamma_q$ encodes the importance of label p over label q and this information is also used in the optimization problem. The loss function considered in this work is a generalization of the hinge-loss for label ranking. For a pair of labels $(p,q) \in \Sigma$, the loss w.r.t. f is defined as

$$\ell_{p,q}(f(x), \gamma) = [(\gamma_p - \gamma_q) - (h_p(x) - h_q(x))]_+,$$

where $[a]_+ = \max(a, 0)$. At the heart of the algorithm lies a decomposition framework, similar to the one described in the previous section, that decomposes any given feedback vector into complete bipartite subgraphs, and losses are defined and aggregated over these subgraphs. This decomposition framework makes the approach very general, albeit at the cost of solving a complex optimization problem. Interestingly, the quadratic programming formulation for multilabel classification as proposed by Elisseeff and Weston [30] can be recovered as a special case of this approach.

6 Structured Prediction

Discriminative structured prediction algorithms infer a joint scoring function on input–output pairs and, for a given input, predict the output that maximizes this scoring function. Let $\Phi : \mathcal{X} \times \mathcal{Y} \rightarrow \mathbb{R}^m$ (with a slight abuse of notation) denote a joint feature map of input–output pairs. Note that the mapping can be an infinite dimensional feature space, such as the reproducing kernel Hilbert space (RKHS). The scoring function is parameterized by a weight vector w and for a pair $(x, y) \in \mathcal{X} \times \mathcal{Y}$ is defined as $f(x, y; w) = \langle w, \Phi(x, y)\rangle$. The goal of the learning algorithm is to estimate w typically based on the structural risk minimization principle. The final prediction is performed according to the following rule: $\hat{y} = \text{argmax}_{y \in \mathcal{Y}} f(x, y) = \text{argmax}_{y \in \mathcal{Y}} \langle w, \Phi(x, y)\rangle$. This optimization problem is known as the decoding problem in structured prediction literature.

The motivation behind using a structured prediction framework to solve the label ranking problem stems from the added flexibility to use arbitrary loss functions and kernels, in principle, on the output space. In the following sections, we assume that $\mathcal{Y}$ is the space of all total orders of the label set Σ.

6.1 Large-Margin Methods

In the max-margin framework [73–75], one considers the following optimization problem to estimate the parameters w:

$$\begin{aligned} &\min_{w,\xi} \quad ||w||^2 + \nu \sum_{i=1}^{n} \xi_i \\ &\text{subject to}: \langle w, \Phi(x_i, y_i)\rangle - \langle w, \Phi(x_i, y)\rangle \geq \Delta(y_i, y) - \xi_i, \forall y \in \mathcal{Y}\backslash y_i, \forall i \in [\![n]\!] \\ &\qquad \xi_i \geq 0, \forall i \in [\![n]\!], \end{aligned} \quad (1)$$

where $\Delta : \mathcal{Y} \times \mathcal{Y} \rightarrow \mathbb{R}$ is a loss function on total orders. While there are exponential number of constraints in the optimization problem, it is possible to use cutting-plane optimization technique [75] by designing an oracle that returns the most violated constraint in polynomial time. The most violated constraint w.r.t. a training example (x, y) can be computed using the following optimization problem:

$$\hat{y} = \underset{z \in \mathcal{Y}}{\text{argmax}}\, f(x, y) + \Delta(z, y).$$

Note that the optimization problem is similar to the prediction problem, the only difference being the additional loss term.

The oracle and the decoding subroutine can be designed using techniques described by Le and Smola [55]. The scoring function f takes a slightly different form. Let $g(x, p; w_p) = \langle \phi(x), w_p \rangle$ (ϕ is feature map of inputs) denote the scoring function for an individual label $p \in \Sigma$ parameterized by weight vector w_p. Now define the scoring function f for the pair (x, y) as follows:

$$f(x, y; \mathbf{w}) = \sum_{j=1}^{k} g(x, j) c(y)_j = \sum_{j=1}^{k} \langle \phi(x), w_j \rangle c(y)_j,$$

parameterized by the set $\mathbf{w} = \{w_j\}_{j=[\![k]\!]}$ of weight vectors, where c is a decreasing sequence of reals and $c(y)$ denotes the permutation of c according to y, i.e., $c(y)_j = c_{y(j)}$ for all $j \in [\![k]\!]$. The final prediction $\hat{y} = \operatorname{argmax}_{y \in \mathcal{Y}} f(x, y)$ is obtained by sorting the scores $g(x, p)$ of the individual labels. This is possible due to the Polya–Littlewood–Hardy inequality [55]. The decoding problem is thus solved. We now turn our attention to designing a separation oracle. The goal is to find

$$\begin{aligned} \hat{y} &= \operatorname*{argmax}_{z \in \mathcal{Y}} f(x, y) + \Delta(y, z) \\ &= \operatorname*{argmax}_{z \in \mathcal{Y}} \sum_{j=1}^{k} \langle \phi(x), w_j \rangle c(y)_j + \Delta(y, z). \end{aligned} \tag{2}$$

For certain loss functions that are relevant in information retrieval applications, Le and Smola [55] showed that the above optimization problem can be formulated as a linear assignment problem and can be solved using the Hungarian marriage method (Kuhn–Mungres algorithm) in $O(k^3)$ time. For arbitrary loss functions, it may not be feasible to solve (2) efficiently. Note that the term $\Delta(.,.)$ in the separation oracle and also in the constraint set of (1) specifies an input–output dependent margin. Replacing it with a fixed margin γ $(= 1)$ would greatly simplify the design of separation oracle since it reduces to a sorting operation as in the decoding problem. The optimization problem (1) can be kernelized in its dual form. Let $\{\alpha_{iz}\}_{i \in [\![n]\!], z \in \mathcal{Y}}$ denote the set of dual parameters. Let the joint scoring function on input-output pairs be an element of $\mathcal{H}$ with $\mathcal{H} = \mathcal{H}_{\mathcal{X}} \otimes \mathcal{H}_{\mathcal{Y}}$ where $\mathcal{H}_{\mathcal{X}}, \mathcal{H}_{\mathcal{Y}}$ are the RKHS of $k_{\mathcal{X}}, k_{\mathcal{Y}}$ respectively and $\otimes$ denotes the tensor product. Note that the reproducing kernel of $\mathcal{H}$ is then $k[(x, y), (x', y')] = k_{\mathcal{X}}(x, x') k_{\mathcal{Y}}(y, y')$. The dual optimization problem is then given as

$$\begin{aligned} \min_{\alpha} \quad & \sum_{i,j \in [\![n]\!], z, z' \in \mathcal{Y}} \alpha_{iz} \alpha_{jz'} k_{\mathcal{X}}(x_i, z) k_{\mathcal{Y}}(z, z') - \sum_{i,z} \alpha_{iz} \\ \text{subject to :} \quad & \sum_{z \in \mathcal{Y}} \alpha_{iz} \leq \nu, \ \forall i \in [\![n]\!] \\ & \alpha_{iz} \geq 0, \ \forall i \in [\![n]\!], \ \forall z \in \mathcal{Y}. \end{aligned}$$

This allows one to use arbitrary kernel functions on the output space such as those described in Sect. 2.2.

6.2 Least-Squares Formulation

We now consider the setting where a set of training examples $(x_1, Y_1), \ldots, (x_n, Y_n)$ is given from $\mathcal{X} \times 2^{\mathcal{Y}}$. A least-squares approach for structured prediction similar in the spirit of regularized least-squares regression [62] was recently proposed by Gärtner and Vembu [38,39]. The goal is to learn a scoring function $f : \mathcal{X} \times \mathcal{Y} \to \mathbb{R}$ that, for each $x_i \in \mathcal{X}$, orders (ranks) $\mathcal{Y}$ in such a way that it assigns a higher score to all $y \in Y_i$ than to all $z \in \mathcal{Y}\backslash Y_i$. A kernel method based on the structural risk minimization principle can be used to solve this problem. Let $\phi : \mathcal{Y} \to \mathbb{R}^d$ be a finite dimensional embedding of $\mathcal{Y}$ with a corresponding dot-product kernel $k_{\mathcal{Y}}$. Let $k_{\mathcal{X}} : \mathcal{X} \times \mathcal{X} \to \mathbb{R}$ be a kernel on $\mathcal{X}$. Let the joint scoring function on input–output pairs be an element of $\mathcal{H}$ with $\mathcal{H} = \mathcal{H}_{\mathcal{X}} \otimes \mathcal{H}_{\mathcal{Y}}$ as in the previous section. The goal is to solve the following optimization problem:

$$h^* = \operatorname*{argmin}_{f \in \mathcal{H}} \nu \|h\|^2 + \sum_{i \in [\![n]\!]} \mathcal{R}(f, i), \tag{3}$$

where $\mathcal{R} : \mathcal{H} \times [\![n]\!] \to \mathbb{R}$ is the empirical risk on a training example. Since for each x_i the aim is to order all elements of Y_i before all elements of $\mathcal{Y}\backslash Y_i$, the AUC-loss can be used as the empirical error of (3):

$$\mathcal{R}_{\text{auc}}(f, i) = \sum_{y \in Y_i} \sum_{z \in \mathcal{Y}\backslash Y_i} \sigma[f(x_i, z) - f(x_i, y)], \tag{4}$$

where σ is the modified step function: $\sigma(a) = +1$ if $a > 0$, $\sigma(a) = 0$ if $a < 0$, and $\sigma(a) = 1/2$ if $a = 0$. To obtain a convex function one needs to upper bound it by the exponential loss

$$\mathcal{R}_{\exp}(h, i) = \sum_{y \in Y_i} \sum_{z \in \mathcal{Y}\backslash Y_i} \exp[1 + f(x_i, z) - f(x_i, y)] \geq \mathcal{R}_{\text{auc}}(h, i).$$

Despite being convex the exponential loss does not allow compact formulations, but using its second order Taylor expansion at 0, i.e., $texp(a) = 1 + a + \frac{1}{2}a^2 \approx \exp(a)$, does. Ignoring constants that can be accommodated by the regularization parameter, we get

$$\begin{aligned} \mathcal{R}(h, i) = \sum_{y \in Y_i} \sum_{z \in \mathcal{Y}\backslash Y_i} \Big[& f(x_i, z) - f(x_i, y) + \frac{1}{2} f^2(x_i, z) \\ & - f(x_i, z) f(x_i, y) + \frac{1}{2} f^2(x_i, y) \Big]. \end{aligned} \tag{5}$$

Similar loss functions were considered in previous work on structured prediction problems. Altun et al. [6] introduced the ranking loss (4) for structured prediction and minimized an upper bound given by the exponential loss for discriminative sequence labeling. A closely related loss function to the approximation in (5) is the one minimized by least-squares SVM [72] and also its multiclass extension [71]. The approach described above can therefore be seen as an extension of least-squares SVM for structured prediction. The main reason behind deviating from the standard max-margin hinge loss [73–75] is to make the problem tractable, and indeed, as shown by Gärtner and Vembu [38, 39], using the loss function of (5) results in a polynomially sized unconstrained optimization problem.

The crux of this approach lies in the polynomial time computation of the vector $\boldsymbol{\phi} = \sum_{z \in \mathcal{Y}} \phi(z)$ and the matrix $C = \sum_{z \in \mathcal{Y}} \phi(z)\phi^\top(z)$ leading to a tractable optimization problem for training structured prediction models. Let the finite dimensional embedding of the set of permutations $\mathcal{Y}$ of Σ be defined as $\phi : \mathcal{Y} \rightarrow \mathbb{R}^{\Sigma \times \Sigma}$. Then $|\mathcal{Y}| = |\Sigma|!$ and with $\phi_{(uv)}(z) = 1$ if $u \succ_z v$, $\phi_{(uv)}(z) = -1$ if $v \succ_z u$, and $\phi_{(uv)}(z) = 0$ otherwise, we have $\boldsymbol{\phi} = \mathbf{0}$, and

$$C_{(uv)(u'v')} = \begin{cases} -|\Sigma|! & \text{if } u = v' \wedge u' = v \\ +|\Sigma|! & \text{if } u = u' \wedge v = v' \\ \dfrac{+|\Sigma|!}{3} & \text{if } u = u' \text{ xor } v = v' \\ \dfrac{-|\Sigma|!}{3} & \text{if } u = v' \text{ xor } v = u' \\ 0 & \text{otherwise} \end{cases}$$

For prediction, the decoding problem needs to be solved. While exact decoding is hard, there is an efficient $1/2$-factor z-approximation algorithm [38].

7 Online Methods

Online classification and regression algorithms such as perceptron [63] typically learn a linear model $f(x) = \langle w, x \rangle$ parameterized by a weight vector $w \in \mathbb{R}^m$. The algorithms operate in rounds (iterations). In round t, *nature* provides an instance to the learner; the learner makes a prediction using the current weight vector w^t; nature reveals the true label y^t of x^t; learner incurs a loss $\ell(\langle w^t, x^t \rangle, y^t)$ and updates its weight vector accordingly. Central to any online algorithm is the update rule that is designed in such a way so as to minimize the cumulative loss over all the iterations. In label ranking scenarios, online algorithms [24, 25, 69] maintain a set of weight vectors $\{w_j\}_{j=[\![k]\!]}$, one for every label in Σ, and the update rule is applied to each of these vectors.

Online algorithms for label ranking have been analyzed using two different frameworks: passive–aggressive [22] and primal–dual [68]. Passive-aggressive algorithms for label ranking [25] are based on Bregman divergences and result in multiplicative and additive update rules [53]. A Bregman divergence [9] is similar to a distance metric, but does not satisfy the triangle inequality and the symmetry properties. In every iteration t, the algorithm updates its weights in such a way that it stays close to the previous iteration's weight vector w.r.t. the Bregman divergence, and also minimizes the loss on the current input-output (x^t, y^t) pair. Let $W \in \mathbb{R}^{k \times m}$ denote the set of weight vectors in matrix form. The following optimization problem is considered:

$$W^t = \operatorname*{argmin}_W B_F(W || W^{t-1}) + \nu \ell(f(x^t; W), y^t),$$

where B_F is the Bregman divergence defined through a strictly convex function F. The choice of the Bregman divergence and the loss function result in different update rules. Additive and multiplicative update rules can be derived, respectively, by considering the following optimization problems [25]:

$$W^t = \operatorname*{argmin}_W ||W - W^{t-1}||^2 + \nu \ell(f(x^t; W), y^t),$$

and

$$W^t = \operatorname*{argmin}_W D_{\text{KL}}(W || W^{t-1}) + \nu \ell(f(x^t; W^t), y^t),$$

where D_{KL} is the Kullback–Liebler divergence [21]. The loss functions considered by Crammer and Singer [25] is similar to the ones defined by Dekel et al. [27] (refer also Sect. 4), i.e., a preference graph is decomposed into subgraphs using a graph decomposition procedure, and a loss function such as the 0-1 loss or the ranking loss is defined on every subgraph. The loss incurred by a ranker W for a graph decomposition procedure $\mathcal{D}(G)$ is given as

$$\ell(f(x; W), y) = \sum_{g \in \mathcal{D}(G)} |\{(r, s) \in g : \langle w_r, x \rangle \leq \langle w_s, x \rangle\} \neq \emptyset|.$$

The primal–dual framework [68] was used by Shalev–Shwartz and Singer [69] resulting in a unifying algorithmic approach for online label ranking. The loss function considered in this work is a generalization of the hinge-loss for label ranking. The training labels are assumed to be a set of relevant and irrelevant labels (as in multilabel classification). For a given instance $x \in \mathcal{X}$, let $\Sigma_r \subseteq \Sigma$ denote the set of relevant labels. The hinge-loss for label ranking w.r.t. an example (x^t, Σ_r^t) at iteration t is defined as:

$$\ell^\gamma(W^t; (x^t, y^t)) = \max_{r \in \Sigma_r^t, s \notin \Sigma_r^t} [\gamma - (\langle w_r^t, x^t \rangle - \langle w_s^t, x^t \rangle)]_+.$$

The central idea behind the analysis is to cast online learning as an optimization (minimization) problem consisting of two terms: the complexity of the ranking function and the empirical label-ranking loss. The notion of duality in optimization theory [8] is used to obtain lower bounds on the optimization problem, which in turn yields upper bounds on the number of prediction mistakes made by the algorithm. The reader is referred to [69] that presents several update rules for label ranking, and these are also shown to generalize other update rules such as the ones defined in [24].

8 Instance-Based Learning

In instance-based learning, the idea is to predict a label for a given instance based on local information, i.e., labels of neighboring examples. In label ranking, these labels are rankings (partial orders, total orders, partial rankings) and one has to use aggregation algorithms [3, 4, 29, 31, 76] to combine rankings from neighboring examples. Instance-based learning algorithms for label ranking were proposed recently in [10, 11, 15]. Let $\{y_i\}_{b=[\![B]\!]}$ denote a set of B neighboring rankings for any given instance $x \in \mathcal{X}$. The goal is to compute a ranking $\hat{y}$ that is optimal w.r.t. a loss function $\ell : \mathcal{Y} \times \mathcal{Y} \rightarrow \mathbb{R}$ defined on pairs of rankings. More formally, the following optimization problem needs to be solved:

$$\hat{y} = \underset{y \in \mathcal{Y}}{\operatorname{argmin}} \sum_{i=1}^{B} \ell(y, y_i). \tag{6}$$

This is a very general statement of the problem. Various aggregation algorithms, which we survey in the sequel, can be used to solve this optimization problem depending on the nature of the loss function and also on the inputs (of the optimization problem).

8.1 Aggregating Total Orders

The problem of finding an optimal ranking when the inputs in (6) are total orders can be formulated as a feedback arc set problem in digraphs (specifically in tournaments) [4]. A tournament is a directed graph $G = (V, E)$ such that for each pair of vertices $p, q \in V$, either $(p, q) \in E$ or $(q, p) \in E$. The minimum feedback arc set (FAS) is the smallest set $E' \subseteq E$ such that $(V, E - E')$ is acyclic. The rank aggregation problem can be seen as special case of weighted FAS-tournaments; the weight w_{pq} of an edge (p, q) is the fraction of rankings that rank p before q.

Optimizing the Spearman footrule distance metric in (6) is equivalent to finding the minimum cost maximum matching in a bipartite graph with k nodes [29]. A 2-factor approximation algorithm with time complexity $O(Bk + k \log k)$ was proposed by Fagin et al. [31]. Optimizing the Kendall tau distance metric in (6) is NP-hard [47] and therefore one has to use approximation algorithms [4, 76] to output a Kemeny optimal ranking. There exists a deterministic, combinatorial 8/5-approximation algorithm for aggregating total orders [76]. The approximation ratio can be improved to $11/7$ by using a randomized algorithm [4] and to $4/3$ by using a deterministic LP-based algorithm [76]. A polynomial time approximation scheme was proposed by Mathieu and Schudy [52].

8.2 Aggregating Partial Rankings

A typical application of this setting is multilabel ranking. There exists a deterministic, combinatorial $8/5$-approximation algorithm for aggregating partial rankings [76]. The running time of this algorithm is $O(k^3)$. A slightly better approximation guarantee of $3/2$ can be obtained by using a deterministic, LP-based algorithm [76]. These algorithms minimize the Kemeny distance between the desired output and the individual partial rankings. An exact method for aggregating partial rankings using (generalized) sum of squared rank distance metric was proposed by Brinker and Hüllermeier [11].

8.3 Aggregating Partial Orders

Let the labels be partial orders and the desired output is a total order. To the best of our knowledge, there are no approximation algorithms to aggregate partial orders, but it is possible to reduce the problem to that of aggregating total orders as follows: given a partial order, sample a set (of some fixed cardinality) of linear extensions [45] and use existing approximation algorithms for aggregating total orders. If the desired output is a partial order and not a total order, one can consider the following optimization problem:

$$\hat{y} = \underset{y \in \mathcal{Y}}{\operatorname{argmax}} \sum_{i=1}^{n} k_{\mathcal{X}}(x_i, x) k_{\mathcal{Y}}(y_i, y).$$

Under the assumption that $k_{\mathcal{X}} \geq 0$ and $k_{\mathcal{Y}} \geq 0$, and if the edge kernel (defined in Sect. 2.2) on partial orders is used, the above optimization problem can be approximately solved using the maximum acyclic subgraph algorithm [42, 58].

9 Probabilistic Methods

In this section, we describe discriminative probabilistic methods for label ranking where the goal is to estimate the conditional $y|x$ from the training data.

9.1 Instance-Based Method

An instance-based probabilistic method for label ranking was recently proposed by Cheng et al. [14]. It uses the standard Mallows model [57] given by $p(y|\theta, z) = \exp(-\theta D(y, z))/Z(\theta, z)$ to model the conditional distribution, where $Z(\theta, z) = \sum_{y \in \mathcal{Y}} \exp(-\theta D(y, z))$, z is the location parameter, $\theta \geq 0$ is the spread parameter, and $D(.,.)$ is a distance metric on permutations. In instance-based learning, predictions are performed using local neighborhood information as described in the previous section. Let $\{y_i\}_{i=[\![B]\!]}$ denote a set of B neighboring rankings for any given instance $x \in \mathcal{X}$. Under the assumption that the conditional distribution $p(.|x)$ on $\mathcal{Y}$ is locally constant around the query x, and by further assuming independence of the observations (neighbors), the probability to observe $\mathbf{y} = \{y_i\}_{i=[\![B]\!]}$ given the parameters (θ, z) is $p(\mathbf{y}) = \prod_{i=1}^{B} p(y_i|\theta, z)$. The maximum-likelihood estimate (MLE) of z is given by $\hat{z} = \operatorname{argmin}_{z \in \mathcal{Y}} \sum_{i=1}^{B} D(y_i, z)$, which is the (generalized) median of the rankings $\{y_i\}_{i=[\![B]\!]}$. The MLE of the spread parameter θ is derived from an estimate of the expected distance $\mathbf{E}[D(y,z)|\theta, z]$:

$$\frac{1}{B}\sum_{i=1}^{B} D(y_i, \hat{z}) = \frac{k\exp(-\theta)}{1-\exp(-\theta)} - \sum_{j=1}^{k} \frac{j\exp(-j\theta)}{1-\exp(-j\theta)}.$$

The above model can be extended to the setting with incomplete preference information such as partial orders. In this setting, the probability of y_i is given by $p(E(y_i)) = \sum_{y \in E(y_i)} p(y|\theta, z)$, where $E(y_i)$ denotes the set of linear extensions of y_i. Computing the MLE of (θ, z) is nontrivial because the number of linear extensions is possibly exponential in the size of the input. However, Cheng et al. [14] proposed an expectation-maximization algorithm [28] to estimate the parameters where the main idea is to replace the incomplete ranking y_i by its most probable linear extension $y_i^* \in E(y_i)$.

9.2 Structured Prediction Method

Probabilistic discriminative methods for structured prediction were recently analyzed by Vembu et al. [77] with an emphasis on predicting combinatorial structures. In label ranking, the combinatorial output space is the set of permutations. The goal is to estimate the conditional $y|x$ using exponential families via $p(y|x,\theta) = \exp(\langle\phi(x,y),\theta\rangle - \ln Z(\theta|x))$, where $\phi(x,y)$ are the joint sufficient statistics of x and y, and $Z(\theta|x) = \sum_{y \in \mathcal{Y}} \exp(\langle\phi(x,y),\theta\rangle)$ is the partition function that takes care of the normalization. Maximum-a-posteriori estimation is then performed by imposing a normal prior on θ. This leads to optimizing the negative joint likelihood in θ and the conditional $y|x$:

$$\hat{\theta} = \operatorname*{argmin}_{\theta}\left[\nu\|\theta\|^2 + \frac{1}{n}\sum_{i=1}^{n}[\ln Z(\theta|x_i) - \langle\phi(x_i, y_i), \theta\rangle]\right].$$

The difficulty in solving this optimization problem lies in the computation of the partition function. The optimization is typically performed using gradient descent techniques (and advancements thereof). Therefore, it also needed to compute the gradient of the log partition function, which is the first order moment of the sufficient statistics, i.e., $\nabla_\theta \ln Z(\theta|x) = \mathbf{E}_{y \sim p(y|x,\theta)}[\phi(x,y)]$.

Computing the log partition function and its gradient are in general NP-hard. Fortunately, the latter can be approximated using concentration inequalities. Vembu et al. [77] showed how to approximate the partition function using a well-known reduction from counting to sampling [49]. The technique is a Markov chain Monte Carlo-based approximation and is widely used to solve #P-complete problems. The standard approach [48] is to express the quantity of interest, i.e., the partition function $Z(\theta|x)$, as a telescoping product of ratios of parameterized variants of the partition function. Let $0 = \beta_0 < \beta_1 \cdots < \beta_l = 1$ denote a sequence of parameters also called as *cooling schedule* and express $Z(\theta|x)$ as a telescoping product

$$\frac{Z(\theta|x)}{Z(\beta_{l-1}\theta|x)} \times \frac{Z(\beta_{l-1}\theta|x)}{Z(\beta_{l-2}\theta|x)} \times \cdots \frac{Z(\beta_1\theta|x)}{Z(\beta_0\theta|x)} \times Z(\beta_0\theta|x).$$

The next step is to design an appropriate cooling schedule and also to define a random variable f_i, for all $i \in [\![l]\!]$, such that it is an unbiased estimator of each of the above ratios. These ratios can now be estimated by sampling according to the distributions $p(y|x, \beta_i\theta)$ and by computing the sample mean of f_i. The desideratum is an upper bound on the variance of the estimator. Having a low variance implies a small (and polynomial) number of samples to approximate each ratio. The final estimator $Z(\theta|x)$ is then the product of the reciprocals of the individual ratios.

The next step is to design sampling algorithms to sample according to distributions $p(y|x, \beta_i\theta)$. Under the assumption that it is possible to obtain exact samples uniformly at random from the output space $\mathcal{Y}$, Vembu et al. designed a Markov chain using Metropolis process [59] that can be used to sample structures from $p(y|x, \beta_i\theta)$. The Markov chain is very simple: In any state y, select the next state z uniformly at random and move to z with probability $\min[1, p(z|x,\theta)/p(y|x,\theta)]$. Vembu et al. also analyzed the mixing time of this chain using coupling [5] and coupling from the past [61].

The final step is to design an algorithm that can be used to obtain exact samples uniformly at random. For permutations, an exact sample can be obtained uniformly at random by generating a sequence (of length k, the number of labels) of integers where each integer is sampled uniformly from the set $[\![k]\!]$ *without* replacement.

10 Conclusion

The problem of label ranking has attracted a lot of interest in recent years as evidenced by the increasing number of algorithms attempting to solve it. A detailed description of these algorithms is beyond the scope of this paper. However, we hope that by giving an overview of existing literature on label ranking, the reader would be able to delve deeper into the analysis of individual algorithms using the references in this survey paper as a starting point.

References

1. A. Agarwal, S. Chakrabarti, Learning random walks to rank nodes in graphs, in *Proceedings of the Twenty-Fourth International Conference on Machine Learning* (2007)
2. S. Agarwal, Ranking on graph data, in *Proceedings of the Twenty-Third International Conference on Machine Learning* (2006)
3. N. Ailon, Aggregation of partial rankings, p-ratings and top-m lists, in *Proceedings of the Eighteenth Annual ACM-SIAM Symposium on Discrete Algorithms* (2007)
4. N. Ailon, M. Charikar, A. Newman, Aggregating inconsistent information: Ranking and clustering, in *Proceedings of the Thirty-Seventh Annual ACM Symposium on Theory of Computing* (2005)
5. D. Aldous, Random walks on finite groups and rapidly mixing markov chains. Séminaire de probabilités de Strasbourg **17**, 243–297 (1983)
6. Y. Altun, T. Hofmann, M. Johnson, Discriminative learning for label sequences via boosting, in *Advances in Neural Information Processing Systems*, vol. 15 (2002)
7. N. Balcan, N. Bansal, A. Beygelzimer, D. Coppersmith, J. Langford, G. Sorkin, Robust reductions from ranking to classification. Mach. Learn. J. **72**(1–2), 139–153 (2008)
8. S. Boyd, L. Vandenberghe, *Convex Optimization* (Cambridge Univeristy Press, 2004)
9. L.M. Bregman, The relaxation method of finding the common points of convex sets and its application to the solution of problems in convex programming. *USSR Comput. Mathe. Math. Phys.* **7**, 200–217 (1967)
10. K. Brinker, E. Hüllermeier, Case-based label ranking, in *Proceedings of the Seventeenth European Conference on Machine Learning* (2006)
11. K. Brinker, E. Hüllermeier, Case-based multilabel ranking, in *Proceedings of the Twentieth International Joint Conference on Artificial Intelligence* (2007)
12. C.J.C. Burges, R. Ragno, Q.V. Le, Learning to rank with nonsmooth cost functions, in *Advances in Neural Information Processing Systems*, vol. 19 (2006)
13. C.J.C. Burges, T. Shaked, E. Renshaw, A. Lazier, M. Deeds, N. Hamilton, G. Hullender, Learning to rank using gradient descent, in *Proceedings of the Twenty-Second International Conference on Machine Learning* (2005)
14. W. Cheng, J.C. Huhn, E. Hüllermeier, Decision tree and instance-based learning for label ranking, in *Proceedings of the Twenty-Sixth Annual International Conference on Machine Learning* (2009)
15. W. Cheng, E. Hüllermeier, Instance-based label ranking using the Mallows model, in *Proceedings of the Preference Learning Workshop at the European Conference on Machine Learning and Principles and Practice of Knowledge Discovery in Databases* (2008)
16. W. Chu, Z. Ghahramani, Gaussian processes for ordinal regression. J. Mach. Learn. Res. **6**, 1019–1041 (2005)
17. W. Chu, S.S. Keerthi, New approaches to support vector ordinal regression, in *Proceedings of the Twenty-Second International Conference on Machine Learning* (2005)
18. W.W. Cohen, R.E. Schapire, Y. Singer, Learning to order things. J. Artif. Intell. Res. **10**, 243–270 (1999)
19. M. Collins, R.E. Schapire, Y. Singer, Logistic regression, AdaBoost and Bregman distances. Mach. Learn. **48**(1–3), 253–285 (2002)
20. C. Cortes, V. Vapnik, Support-vector networks. Mach. Learn. **20**(3), 273–297 (1995)
21. T.M. Cover, J.A. Thomas, *Elements of information theory* (Wiley-Interscience, New York, NY, USA, 1991)
22. K. Crammer, O. Dekel, J. Keshet, S. Shalev-Shwartz, Y. Singer, Online passive-aggressive algorithms. J. Mach. Learn. Res. **7**, 551–585 (2006)

23. K. Crammer, Y. Singer, Pranking with ranking, in *Advances in Neural Information Processing Systems*, vol. 14 (2001)
24. K. Crammer, Y. Singer, A family of additive online algorithms for category ranking. J. Mach. Learn. Res. **3**, 1025–1058 (2003)
25. K. Crammer, Y. Singer, Loss bounds for online category ranking, in *Proceedings of the Eighteenth Annual Conference on Learning Theory* (2005)
26. H. Daumé III, *Practical Structured Learning Techniques for Natural Language Processing*. PhD thesis, University of Southern California, Los Angeles, CA, August 2006
27. O. Dekel, C.D. Manning, Y. Singer, Log-linear models for label ranking, in *Advances in Neural Information Processing Systems*, vol. 16 (2003)
28. A. Dempster, N. Laird, D. Rubin, Maximum likelihood from incomplete data via the EM algorithm. J.R. Stat. Soc. B Methodological **39**(1), 1–38 (1977)
29. C. Dwork, R. Kumar, M. Naor, D. Sivakumar, Rank aggregation methods for the web, in *Proceedings of the Tenth International World Wide Web Conference* (2001)
30. A. Elisseeff, J. Weston, A kernel method for multi-labelled classification, in *Advances in Neural Information Processing Systems*, vol. 14 (2001)
31. R. Fagin, R. Kumar, M. Mahdian, D. Sivakumar, E. Vee, Comparing and aggregating rankings with ties, in *Proceedings of the Twenty-third ACM SIGACT-SIGMOD-SIGART Symposium on Principles of Database Systems* (2004)
32. Y. Freuend, R. Iyer, R.E. Schapire, Y. Singer, An efficient boosting algorithm for combining preferences. J. Mach. Learn. Res. **4**, 933–969 (2003)
33. Y. Freund, R.E. Schapire, A decision-theoretic generalization of on-line learning and an application to boosting. J. Comput. Syst. Sci. **55**(1), 119–139 (1997)
34. G. Fung, R. Rosales, B. Krishnapuram, Learning rankings via convex hull separation, in *Advances in Neural Information Processing Systems*, vol. 18 (2005)
35. J. Fürnkranz, Round robin classification. J. Mach. Learn. Res. **2**, 721–747 (2002)
36. J. Fürnkranz, E. Hüllermeier, Pairwise preference learning and ranking, in *Proceedings of the Fourteenth European Conference on Machine Learning* (2003)
37. J. Fürnkranz, E. Hüllermeier, E.L. Mencía, K. Brinker, Multilabel classification via calibrated label ranking. Mach. Learn. **73**(2), 133–153 (2008)
38. T. Gärtner, S. Vembu, Learning to predict combinatorial structures, in *Proceedings of the NIPS Workshop on Structured Inputs and Structured outputs* (2008)
39. T. Gärtner, S. Vembu, On structured output training: Hard cases and an efficient alternative. Mach. Learn. J. **76**(2–3), 227–242 (2009)
40. S. Har-Peled, D. Roth, D. Zimak, Constraint classification: A new approach to multiclass classification, in *Proceedings of the Thirteenth International Conference on Algorithmic Learning Theory* (2002)
41. S. Har-Peled, D. Roth, D. Zimak, Constraint classification for multiclass classification and ranking, in *Advances in Neural Information Processing Systems*, vol. 15 (2002)
42. R. Hassin, S. Rubinstein, Approximations for the maximum acyclic subgraph problem. Inf. Process. Lett. **51**(3), 133–140 (1994)
43. R. Herbrich, T. Graepel, K. Obermayer, Support vector learning for ordinal regression,. in *Proceedings of the Ninth International Conference on Artificial Neural Networks* (1999)
44. R. Herbrich, T. Graepel, K. Obermayer, Large margin rank boundaries for ordinal regression, In *Advances in Large Margin Classifiers*, ed. by P.J. Bartlett, B. Schölkopf, D. Schuurmans, A.J. Smola (MIT Press, Cambridge, MA, 2000), pp. 115–132
45. M. Huber, Fast perfect sampling from linear extensions. Discrete Math. **306**(4), 420–428 (2006)
46. E. Hüllermeier, J. Fürnkranz, W. Cheng, K. Brinker, Label ranking by learning pairwise preferences. *Artif. Intell.* **178**, 1897–1916 (2008)
47. J. Bartholdi III, C.A. Tovey, M.A. Trick, Voting schemes for which it can be difficult to tell who won the election. Soc. Choice Welfare **6**(2), 157–165 (1989)
48. M. Jerrum, A. Sinclair, The Markov chain Monte Carlo method: An approach to approximate counting and integration, in *Hochbaum DS(ed) Approximation Algorithms for NP-hard Problems* (PWS Publishing, Boston, Mass., 1996), pp. 482–520

49. M.R. Jerrum, L.G. Valiant, V.V. Vazirani, Random generation of combinatorial structures from a uniform distribution. Theor. Comput. Sci. **32**, 169–188 (1986)
50. T. Joachims, Optimizing search engines using clickthrough data, in *Proceedings of the Eighth ACM SIGKDD International Conference on Knowledge Discovery and Data Mining* (2002)
51. M. Kendall, A new measure of rank correlation. Biometrika **30**, 81–89 (1938)
52. C. Kenyon-Mathieu, W. Schudy, How to rank with few errors: A PTAS for weighted feedback arc set on tournaments, in *Proceedings of the Thirty-Seventh Annual ACM Symposium on Theory of Computing* (2007)
53. J. Kivinen, M.K. Warmuth, Exponentiated gradient versus gradient descent for linear predictors. Inf. Comput. **132**(1), 1–63 (1997)
54. J. Langford, B. Zadrozny, Estimating class membership probabilities using classifier learners, in *Proceedings of the Tenth International Workshop on Artificial Intelligence and Statistics* (2005)
55. Q.V. Le, A. Smola, Direct optimization of ranking measures. Technical report, NICTA, Canberra, Australia, 2007
56. G. Lebanon, J. Lafferty, Boosting and maximum likelihood for exponential models, in *Advances in Neural Information Processing Systems*, vol. 14 (2001)
57. C. Mallows, Non-null ranking models. Biometrika **44**(1/2), 114–130 (1957)
58. R.T. McDonald, K. Crammer, F.C.N. Pereira, Online large-margin training of dependency parsers, in *Proceedings of the Forty-Third Annual Meeting of the Association for Computational Linguistics* (2005)
59. N. Metropolis, A.W. Rosenbluth, M.N. Rosenbluth, A.H. Teller, E. Teller, Equation of state calculation by fast computing machines. J. Chem. Phys. **21**, 1087–1092 (1953)
60. N.J. Nilsson, *Learning Machines: Foundations of Trainable Pattern-Classifying Systems* (McGraw-Hill, New York, NY, USA, 1965)
61. J.G. Propp, D.B. Wilson, Exact sampling with coupled markov chains and applications to statistical mechanics. Random Struct. Algorithms **9**(1–2), 223–252 (1996)
62. R. Rifkin, *Everything Old Is New Again: A Fresh Look at Historical Approaches in Machine Learning*. PhD thesis, Sloan School of Management Science, MIT, 2002
63. F. Rosenblatt, The perceptron: A probabilistic model for information storage and organization in the brain. Psychol. Rev. **65**(6), 386–408 (1958)
64. C. Rudin, Ranking with a p-norm push, in *Proceedings of the 19th Annual Conference on Learning Theory* (2006)
65. R.E. Schapire, Y. Singer, Improved boosting algorithms using confidence-rated predictions, in *Proceedings of the Eleventh Annual Conference on Learning Theory* (1998)
66. R.E. Schapire, Y. Singer, Boostexter: A boosting-based system for text categorization. Mach. Learn. **39**(2/3), 135–168 (2000)
67. S. Shalev-Shwartz, Y. Singer, Efficient learning of label ranking by soft projections onto polyhedra. J. Mach. Learn. Res. **7**, 1567–1599 (2006)
68. S. Shalev-Shwartz, Y. Singer, A primal-dual perspective of online learning algorithms. Mach. Learn. **69**(2–3), 115–142 (2007)
69. S. Shalev-Shwartz, Y. Singer, A unified algorithmic approach for efficient online label ranking, in *Proceedings of the Eleventh International Conference on Artificial Intelligence and Statistics* (2007)
70. C. Spearman, The proof and measurement of association between two things. Am. J. Psychol. **15**, 72–101 (1904)
71. J.A.K. Suykens, Multiclass least squares support vector machines, in *Proceedings of the International Joint Conference on Neural Networks* (1999)
72. J.A.K. Suykens, J. Vandewalle, Least squares support vector machine classifiers. Neural Process. Lett. **9**(3), 293–300 (1999)
73. B. Taskar, V. Chatalbashev, D. Koller, C. Guestrin, Learning structured prediction models: A large margin approach, in *Proceedings of the Twenty-Second International Conference on Machine Learning* (2005)
74. B. Taskar, C. Guestrin, D. Koller, Max-margin Markov networks, in *Advances in Neural Information Processing Systems*, vol. 16 (2003)

75. I. Tsochantaridis, T. Joachims, T. Hofmann, Y. Altun, Large margin methods for structured and interdependent output variables. J. Mach. Learn. Res. **6**, 1453–1484 (2005)
76. A. van Zuylen, D.P. Williamson, Deterministic algorithms for rank aggregation and other ranking and clustering problems, in *In Proceedings of the Fifth International Workshop on Approximation and Online Algorithms* (2007)
77. S. Vembu, T. Gärtner, M. Boley, Probabilistic structured predictors, in *Proceedings of the Twenty-Fifth Conference on Uncertainty in Artificial Intelligence* (2009)
78. S. Vembu, T. Gärtner, S. Wrobel, Semidefinite ranking on graphs, in *Proceedings of the Fifth International Workshop on Mining and Learning with Graphs* (2007)

Preference Learning and Ranking by Pairwise Comparison

Johannes Fürnkranz and Eyke Hüllermeier

Abstract This chapter provides an overview of recent work on preference learning and ranking via pairwise classification. The *learning by pairwise comparison* (LPC) paradigm is the natural machine learning counterpart to the relational approach to preference modeling and decision making. From a machine learning point of view, LPC is especially appealing as it decomposes a possibly complex prediction problem into a certain number of learning problems of the simplest type, namely binary classification. We explain how to approach different preference learning problems, such as label and instance ranking, within the framework of LPC. We primarily focus on methodological aspects, but also address theoretical questions as well as algorithmic and complexity issues.

1 Introduction

Preferences on a set of alternatives can be expressed in two natural ways, namely by evaluating *individual* and by comparing *pairs* of alternatives. As for human decision making, the comparative (relational) approach is intuitively appealing and, moreover, supported by evidence from cognitive psychology. Indeed, instead of directly choosing one alternative from a set of options, or ranking all alternatives directly according to their desirability, it is often much simpler to start by comparing alternatives in a pairwise fashion. The actual problem, whether choosing a single best alternative or ranking all of them, is then solved in a second step on the basis of these pairwise comparisons. Essentially, this is known as Thurstone's *Law of Comparative Judgment* [43]. Modern decision-theoretic methodologies, such as the *Analytic*

J. Fürnkranz (✉)
Technische Universität Darmstadt, Germany
e-mail: juffi@ke.tu-darmstadt.de

E. Hüllermeier
Philipps-Universität Marburg, Germany
e-mail: eyke@mathematik.uni-marburg.de

J. Fürnkranz and E. Hüllermeier (eds.), *Preference Learning*,
DOI 10.1007/978-3-642-14125-6_4,

Hierarchy Process [38], are often based on pairwise comparisons between different options in various stages of complex decision processes [39].

The decomposition of the original problem into a set of presumably simpler subproblems is not only advantageous for human decision making but also useful from a machine learning point of view. In fact, as will be argued in more detail later on, the resulting learning problems can typically be solved in a more accurate and efficient way. The price to pay is a possibly more involved prediction step. Roughly speaking, the pairwise comparisons, being made independently of each other, can be conflicting, so that their aggregation into a solution of the original problem may become nontrivial. On the other hand, such a two-step procedure – learning a binary comparison relation and aggregating the pairwise comparisons into a (complex) prediction afterward – also has another advantage: It is *modular* in the sense that the learning part is decoupled from the prediction part; thus, it becomes possible to solve different types of prediction problems on the basis of the same learning algorithm (ensemble of binary classifiers), simply by changing the aggregation procedure in the second step.

The idea of *learning by pairwise comparison* (LPC), or simply *pairwise learning*, has been explored quite extensively for conventional classification, where it is known under a variety of names, such as *all pairs*, *1-vs-1*, or *round robin* learning. Here, it is used as a special binarization technique, that is, to decompose a polytomous classification problem into a set of pairwise problems, thereby making multiclass problems amenable to binary classification methods. Motivated by its successful use for classification as well as its intuitive appeal from a preference and decision-making perspective, the LPC approach has been extended to different types of preference learning and ranking problems in recent years. The purpose of this chapter is to give an overview of existing work and recent developments in this line of research.

In Sect. 2, we briefly recall the use of pairwise learning in conventional classification. This approach can be generalized in a quite natural way to the setting of label ranking, as will be explained in Sect. 3. The use of LPC for label ranking has in turn motivated its application for a number of generalized classification problems, and these will be discussed in Sect. 4. Section 5 is devoted to the application of pairwise learning for the problem of instance ranking. Section 6 reviews some formal results regarding the optimality of LPC for specific prediction problems, and Sect. 7 addresses the aspect of complexity. Finally, we conclude this chapter with a short summary and an outlook on future work in Sect. 8.

2 LPC for Classification

The use of the pairwise approach to preference learning, the main topic of this chapter, is motivated by its successful application in conventional classification. As a special type of binary decomposition technique, it allows one to tackle multiclass problems with binary classifiers. The key idea is to transform a k-class problem

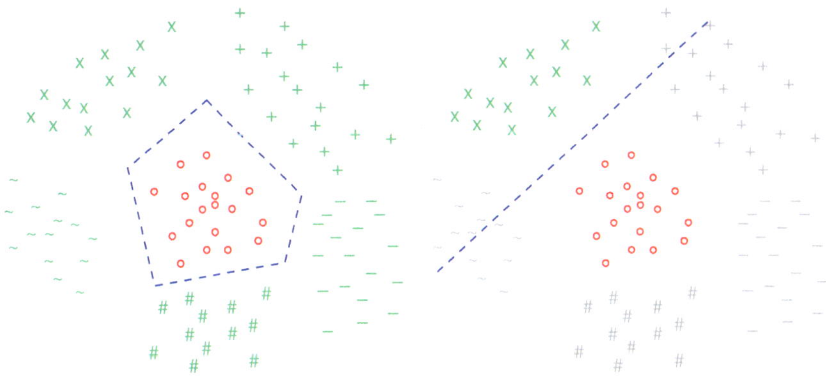

(a) *One-vs-all classification* transforms each k-class problem into k binary problems, one for each class, where each of these problems uses the examples of its class as the positive examples (here o), and all other examples as negatives.

(b) *Pairwise classification* transforms each k-class problem into $k(k-1)/2$ binary problems, one for each pair of classes (here o and x) ignoring the examples of all other classes.

Fig. 1 Decomposition techniques for multiclass classification

involving classes $\mathcal{Y} = \{y_1, y_2, \ldots, y_k\}$ into $k(k-1)/2$ binary problems, one for each pair of classes. More specifically, a separate model (base learner) $\mathcal{M}_{i,j}$ is trained for each pair of labels $(y_i, y_j) \in \mathcal{Y} \times \mathcal{Y}$, $1 \leq i < j \leq k$, using the examples from these two classes as their training set; thus, a total number of $k(k-1)/2$ models is needed. $\mathcal{M}_{i,j}$ is intended to separate the objects with label y_i from those having label y_j. At classification time, a query instance $\boldsymbol{x} \in \mathcal{X}$ is submitted to all models $\mathcal{M}_{i,j}$, and their predictions $\mathcal{M}_{i,j}(\boldsymbol{x})$ are combined into an overall prediction. In the simplest case, each prediction $\mathcal{M}_{i,j}(\boldsymbol{x})$ is interpreted as a vote for either y_i or y_j, and the label with the highest number of votes is proposed as a final prediction.

In comparison to alternative decomposition techniques, such as the one-vs-all approach which learns one model for each label, the pairwise decomposition facilitates effective learning as it leads to maximally simple problems. In particular, the pairwise problems are computationally less complex, since each of them contains fewer training examples (because all examples that do not belong to either of the two classes are ignored). Perhaps even more importantly, these problems typically have simpler decision boundaries. This is illustrated in the example shown in Fig. 1, where each pair of classes can be separated with a linear decision boundary, while more complex functions are required to separate each class from all other classes. Evidence supporting the conjecture that the decision boundaries of the binary problems are indeed simpler can also be found in practical applications: In [26], it was observed that the classes of a digit recognition task were pairwise linearly separable, while the corresponding one-vs-all task was not amenable to single-layer networks. Similarly, in [16] the authors obtained a larger advantage of pairwise classification over one-vs-all for support vector machines with a linear kernel than for support vector machines with a nonlinear kernel.

The basic idea of pairwise classification is fairly well-known from the literature. It has been used in the areas of statistics [3, 8], neural networks [25, 26, 31, 37], support vector machines [15,16,28,40], and others. We refer to [9] for a brief survey of the literature on this topic.

3 LPC for Label Ranking

In the label ranking scenario, the problem is to predict, for any instance $\boldsymbol{x}$ (e.g., a person) from an instance space $\mathcal{X}$, a ranking of a finite set $\mathcal{Y} = \{y_1, y_2, \ldots, y_k\}$ of labels or alternatives (e.g. politicians in an election), i.e., a total order relation $\succ_{\boldsymbol{x}} \subseteq \mathcal{Y} \times \mathcal{Y}$ where $y_i \succ_{\boldsymbol{x}} y_j$ means that instance $\boldsymbol{x}$ prefers the label y_i to the label y_j.

The training information consists of a set of instances for which (partial) knowledge about the associated preference relation is available. More precisely, each training instance $\boldsymbol{x}$ is associated with a subset of all pairwise preferences $y_i \succ_{\boldsymbol{x}} y_j$, $1 \leq i, j \leq k$. The top of Fig. 2 shows a training set consisting of seven examples, each described in terms of three attributes $A1$, $A2$, and $A3$. Each training example is associated with some preferences over the set of possible labels $\mathcal{Y} = \{a, b, c\}$. For example, for the second training instance, we know that $a \succ b$ and $c \succ b$, but we do not know whether a or c is the most preferred option. Thus, even though we assume the existence of an underlying ("true") ranking, we do not expect the training data to provide full information about that ranking. Besides, to increase

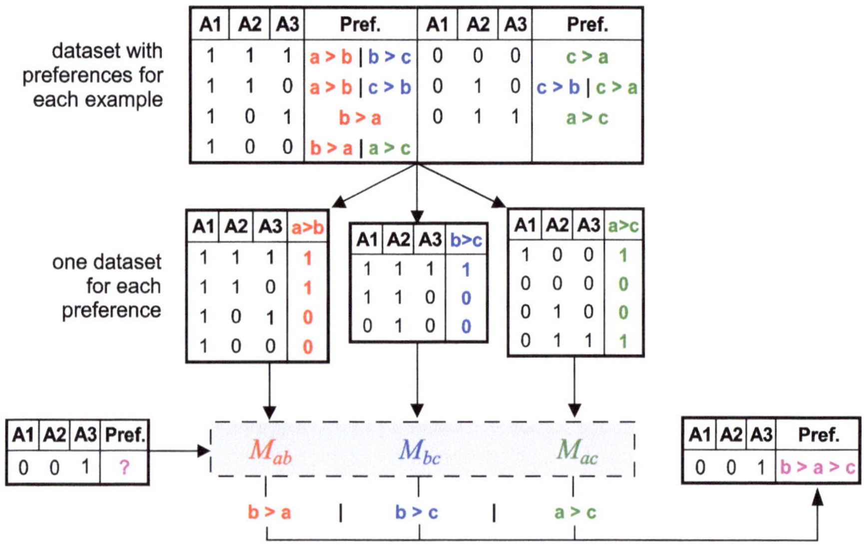

Fig. 2 Schematic illustration of learning by pairwise comparison

the practical usefulness of the approach, we even allow for inconsistencies, such as pairwise preferences which are conflicting (cyclic) due to observation errors.

3.1 Learning Preference Relations

Pairwise classification (cf. Sect. 2) can be extended to the above problem of learning from label preferences in a natural way [11]. To this end, a preference (order) information of the form $y_r \succ_x y_s$ is turned into a training example $(\boldsymbol{x}, z)$ for the learner $\mathcal{M}_{i,j}$, where $i = \min(r, s)$ and $j = \max(r, s)$. Moreover, $z = 1$ if $r < s$ and $z = 0$ otherwise. Thus, $\mathcal{M}_{i,j}$ is intended to learn the mapping that outputs 1 if $y_i \succ_x y_j$ and 0 if $y_j \succ_x y_i$:

$$\boldsymbol{x} \mapsto \begin{cases} 1 & \text{if} \quad y_i \succ_x y_j \\ 0 & \text{if} \quad y_j \succ_x y_i \end{cases} . \tag{1}$$

The model is trained with all examples $\boldsymbol{x}$ for which either $y_i \succ_x y_j$ or $y_j \succ_x y_i$ is known. Examples for which nothing is known about the preference between y_i and y_j are ignored.

The mapping (1) can be realized by any binary classifier. Alternatively, one may also employ base classifiers that map into the unit interval $[0, 1]$ instead of $\{0, 1\}$, and thereby assign a *valued preference relation* $\mathcal{R}_{\boldsymbol{x}}$ to every (query) instance $\boldsymbol{x} \in \mathcal{X}$:

$$\mathcal{R}_{\boldsymbol{x}}(y_i, y_j) = \begin{cases} \mathcal{M}_{i,j}(\boldsymbol{x}) & \text{if} \quad i < j \\ 1 - \mathcal{M}_{ji}(\boldsymbol{x}) & \text{if} \quad i > j \end{cases} \tag{2}$$

for all $y_i \neq y_j \in \mathcal{Y}$. The output of a $[0, 1]$-valued classifier can usually be interpreted as a probability or, more generally, a kind of confidence in the classification: the closer the output of $\mathcal{M}_{i,j}$ to 1, the stronger the preference $y_i \succ_x y_j$ is supported.

3.2 An Example

Figure 2 illustrates the entire process. First, the original training set is transformed into three two-class training sets, one for each possible pair of labels, containing only those training examples for which the relation between these two labels is known. Then three binary models, $\mathcal{M}_{ab}$, $\mathcal{M}_{bc}$, and $\mathcal{M}_{ac}$ are trained. In our example, the result could be simple rules like the following:

$$\begin{aligned} \mathcal{M}_{ab} &: a > b \ \textbf{if} \ A2 = 1. \\ \mathcal{M}_{bc} &: b > c \ \textbf{if} \ A3 = 1. \\ \mathcal{M}_{ac} &: a > c \ \textbf{if} \ A1 = 1 \vee A3 = 1. \end{aligned}$$

Given a new instance with an unknown preference structure (shown in the bottom left of Fig. 2), the predictions of these models are then used to predict a ranking. As we will see in the next section, this is not always as trivial as in this example.

3.3 Combining Predicted Preferences into a Ranking

Given a predicted preference relation $\mathcal{R}_x$ for an instance $\boldsymbol{x}$, the next question is how to derive an associated ranking. This question is nontrivial, since a relation $\mathcal{R}_x$ does not always suggest a unique ranking in an unequivocal way. For example, the learned preference relation is not necessarily transitive. In fact, the problem of inducing a ranking from a (valued) preference relation has received a lot of attention in several research fields, e.g., in fuzzy preference modeling and (multiattribute) decision making [6]. In the context of pairwise classification and preference learning, several studies have empirically compared different ways of combining the predictions of individual classifiers [1, 10, 20, 44].

A simple though effective strategy is a generalization of the aforementioned voting strategy: each alternative y_i is evaluated by the sum of (weighted) votes

$$S(y_i) = \sum_{j \neq i} \mathcal{R}_x(y_i, y_j), \tag{3}$$

and all labels are then ordered according to these evaluations, i.e., such that

$$(y_i \succ_x y_j) \Rightarrow (S(y_i) \geq S(y_j)). \tag{4}$$

Even though this ranking procedure may appear rather ad-hoc at first sight, it does have a theoretical justification (cf. Sect. 6).

4 LPC for Generalized Classification Problems

It has been observed by several authors [5, 11, 14] that, in addition to classification, many learning problems, such as multilabel classification, ordered classification, or ranking may be formulated in terms of label preferences. In this section, we summarize work on using pairwise learning to address such problems.

4.1 Multilabel Classification

Multilabel classification refers to the task of learning a function that maps instances $\boldsymbol{x} \in \mathcal{X}$ to label subsets $\mathcal{P}_x \subset \mathcal{Y}$, where $\mathcal{Y} = \{y_1, y_2, \ldots, y_k\}$ is a finite set of

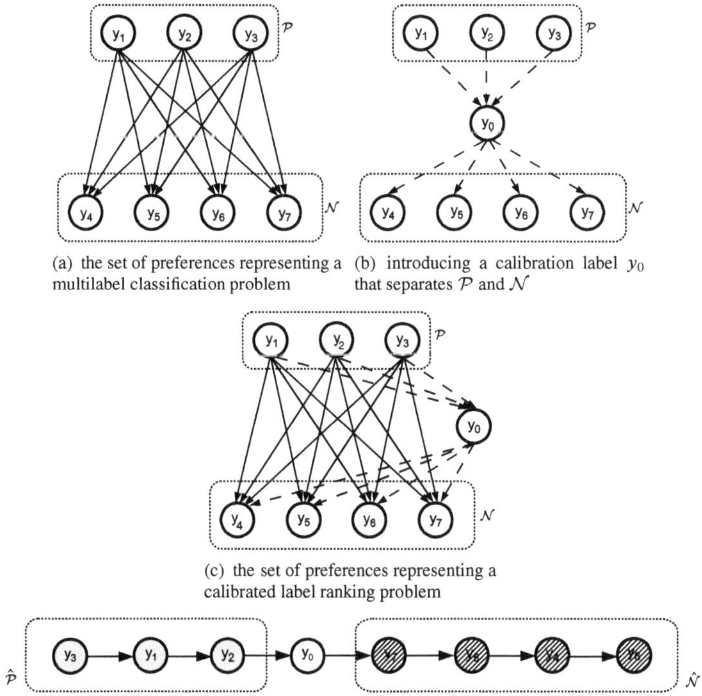

(a) the set of preferences representing a multilabel classification problem

(b) introducing a calibration label y_0 that separates $\mathcal{P}$ and $\mathcal{N}$

(c) the set of preferences representing a calibrated label ranking problem

(d) at prediction time, the calibration label y_0 indicates the split into labels that are predicted relevant ($\hat{\mathcal{P}}$) and labels that are predicted irrelevant ($\hat{\mathcal{N}}$)

Fig. 3 Calibrated Label Ranking

predefined labels, typically with a small-to-moderate number of alternatives. Thus, in contrast to multiclass learning, alternatives are not assumed to be mutually exclusive such that multiple labels may be associated with a single instance. The set of labels $\mathcal{P}_x$ are called *relevant* for the given instance, the set $\mathcal{N}_x = \mathcal{Y} \backslash \mathcal{P}_x$ are the *irrelevant* labels.

In conventional label ranking, a training example typically consists of an instance $\boldsymbol{x} \in \mathcal{X}$, represented in terms of a fixed set of features, and a set of pairwise preferences over labels $\mathcal{R}_x \subset \mathcal{Y} \times \mathcal{Y}$, where $(y, y') \in \mathcal{R}_x$ is interpreted as $y \succ_x y'$. In multilabel classification, the training information consists of a set $\mathcal{P}_x$ of relevant labels and, implicitly, a set $\mathcal{N}_x = \mathcal{Y} \backslash \mathcal{P}_x$ of irrelevant labels. The idea of applying methods for (pairwise) label ranking in the context of multilabel classification is based on the observation that this information can be expressed equivalently in terms of a set of preferences (cf. Fig. 3a):

$$\hat{\mathcal{R}}_x = \{(y, y') \mid y \in \mathcal{P}_x \wedge y' \in \mathcal{N}_x\} \tag{5}$$

In fact, this representation is in a sense even more flexible than the original one; for example, it easily remains applicable in the case where the relevance of some labels is unknown. The preferences (5) can be used to train a pairwise label ranker, which is then able to predict a ranking over all possible labels of a new, unseen example.

Note, however, that a ranking, while determining an order of the labels, does actually not define a partitioning into subsets of relevant and irrelevant labels. A natural way to obtain such a partitioning as additional information is to find an appropriate split-point t in the ranking, suggesting that the first t labels in the ranking are relevant while the remaining ones are irrelevant. A "calibrated" ranking of that kind nicely combines two types of prediction, namely a label ranking and a multilabel classification [4]. From a ranking point of view, it covers additional information about *absolute* preferences; from the point of view of multilabel classification, it offers additional order information, i.e., information about the *relative* preferences within the two sets of relevant and irrelevant labels.

In [45], a few straightforward approaches for determining such split-points are discussed, including methods for determining a fixed threshold for all examples or individual thresholds for each possible label. However, none of these approaches allows one to adjust the thresholds for each individual example.

In [4, 12], we therefore proposed to incorporate the split-point into the learning process. This was achieved by introducing an artificial (neutral) label, which is associated with the split-point and thus calibrates the ranking: For each training instance, this label is preferred to all irrelevant labels, but is less preferable than all relevant labels; see Fig. 3b. A *calibrated label ranker* trained on the enhanced set of preferences (Fig. 3c) is then able to predict a ranking of all labels, *including the artificial one*. The position of this label indicates where the ranking has to be split into a relevant and an irrelevant part. Experiments have shown that this approach outperforms standard methods for multilabel classification, despite a slight tendency to underestimate the number of relevant labels.

4.2 Ordered and Hierarchical Classification

Ordered classification and *hierarchical classification* are problems in which the target label set has an inherent structure. In ordered classification, this structure is a total order, such as *small* < *medium* < *large*. In hierarchical problems, the structure is a partial order in the form of a hierarchy, typically defined by various subconcept/superconcept relations; for example, *Hessen* < *Germany* < *Europe* and *Bayern* < *Germany* < *Europe* (while *Hessen* and *Bayern* are incomparable, i.e., neither *Hessen* < *Bayern* nor *Bayern* < *Hessen*).

The use of conventional loss functions, such as the 0/1 loss, is obviously questionable in the context of ordered and hierarchical classification, as it does not take the relation between class labels into account. If *small* is the true class, for instance, then *medium* is a better prediction than *large*, despite the fact that both are

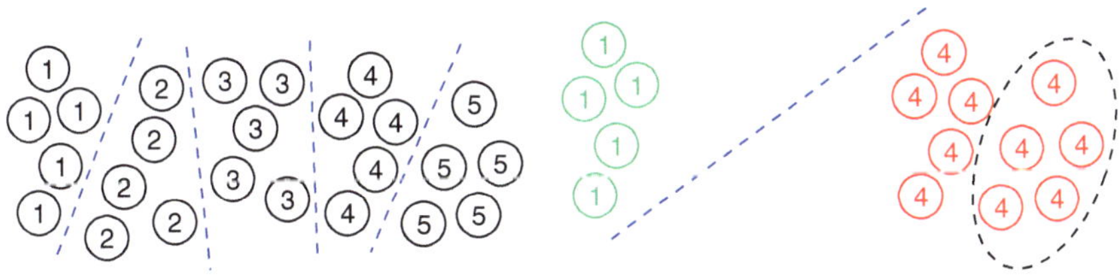

(a) an ordered classification problem where the order of the classes is reflected in the instance space.

(b) the training set for the classifier $\mathcal{M}_{1,4}$ is enriched with the examples of class 5.

Fig. 4 Enriched pairwise classification in ordered domains (figures taken from [33])

incorrect. An obvious idea, then, is to express this type of information in the form of *preferences* on labels.

Within the framework of LPC, this can be done by associating a training instance with all pairwise preferences that can be *inferred* from the label structure. For example, if we know that an instance is of class *small*, we not only know that the label *small* is preferred to all other labels, but we can also infer that *medium* would be a better classification than *large*. In other words, the set of training examples for which it is known that *medium* $\succ$ *large* could be *enriched* by instances from the class *small*. Similarly, if we know that an object belongs to the class *Hessen*, we can infer that the label *Rheinland-Pfalz* would be preferred over *Lower Austria*, because the former is also a German state, while the latter is in a different country.

One idea is to directly add such inferred preferences to the pairwise training data [33]. Figure 4 illustrates this for a problem with five classes, which are labeled from 1 to 5. Here, the classifier $\mathcal{M}_{1,4}$ is enriched with examples from class 5, for which the prediction 4 is clearly preferred to the prediction 1. Note that the examples of class 3 are *not* added to class 4: Since the underlying scale is only ordinal but not numerical, it is not legitimate to assume that 3 is "closer" to 4 than to 1 and, therefore, that predicting 4 for an example from class 3 is preferred to predicting 1.

An analogous technique can be used for hierarchical classification, where the preferences can be defined via common ancestors in the hierarchy. Interestingly, in [41] it was shown that this method may be viewed as a generalization of the so-called Pachinko-machine classifier [27], which contains one binary classifier for each internal node of the label hierarchy.

Even though the above idea of using pairwise preference learning for ordered and hierarchical classification has not yet been explored in full depth, first experimental results suggest that it does not improve the performance of the conventional pairwise classifier. There is a number of possible reasons and explanations for this. Notably, enriching the training data comes with a loss of a key advantage of the pairwise approach, namely the simplicity of the binary problems and the decision boundaries of the binary classifiers. The more inferred preferences are added, the more complex the decision boundary of a single binary classifier will become, eventually approaching the complex decision boundaries of a one-vs-all classifier. Moreover, unlike the example in Fig. 4a, it seems that in many hierarchical and ordered classification problems, the structure on the labels is not so clearly reflected in the topology of the instance space. This means that examples that are "close" in label space are not necessarily neighbored in instance space.

This aspect has been investigated in more detail in [17]. This work also aimed at answering the question to what extent existing techniques and learning algorithms for ordered classification are able to exploit order information, and which properties of these techniques are important in this regard. The main conclusions that could be drawn from this study are as follows: Most learning techniques for ordered classification are indeed able to exploit order information about classes if such information is present, which is often the case though not always. An important factor in this regard is the flexibility of the learner. Roughly speaking, the less flexible a learner is, the more it benefits from an ordinal structure. Interestingly enough, it was found that pairwise classification is fully competitive to other meta-learning techniques specifically designed for ordered classification problems [7]. This result is surprising, since pairwise classification, in its original form, does not explicitly exploit an ordinal structure (and, compared to the other techniques, even uses a smaller amount of training information in terms of the total number of training examples). However, by training only on pairs of classes, it is trivially consistent with each ordinal structure. In a sense, one can argue that it exploits ordinal structure in an implicit way whenever this is possible, but as its binary problems are not explicitly tailored toward the assumption of an ordinal structure, it does not deteriorate when this assumption is invalid.

5 LPC for Instance Ranking

Recall that the term *instance ranking* is used in this book as a generic term of bipartite and multipartite ranking. Multipartite ranking proceeds from the setting of ordinal classification, where an instance $\boldsymbol{x} \in \mathcal{X}$ belongs to one among a finite set of classes $\mathcal{Y} = \{y_1, y_2, \ldots, y_k\}$ and, moreover, the classes have a natural order: $y_1 < y_2 < \ldots < y_k$. Training data consist of a set $\mathcal{T}$ of labeled instances.

In contrast to the classification setting, however, the goal is not to learn a classifier but a ranking function $f(\cdot)$. Given a subset $X \subset \mathcal{X}$ of instances as an input, the function produces a ranking of these instances as an output (typically by assigning a score to each instance and then sorting by scores). A prediction of this type is evaluated in terms of ranking measures, such as the C-index. Ideally, a ranking is produced in which instances from higher classes precede those from lower classes.

The use of LPC for multipartite ranking was recently proposed in [13]. As usual, a separate model $\mathcal{M}_{i,j}$ is induced for each pair of classes $(y_i, y_j) \in \mathcal{Y} \times \mathcal{Y}$, $1 \leq i < j \leq k$, using the subset $\mathcal{T}_{i,j} \subset \mathcal{T}$ of examples from these classes as training data. At classification time, a query $\boldsymbol{x}$ is submitted to all models.

To predict a ranking of a set of query instances $X \subset \mathcal{X}$, each instance $\boldsymbol{x} \in X$ is scored in terms of an aggregation of the predictions

$$\{\mathcal{M}_{i,j}(\boldsymbol{x}) \mid 1 \leq i, j \leq k\}.$$

More specifically, the score is defined as the (weighted) sum of the predictions "in favor of a higher class", that is

$$f(\boldsymbol{x}) = \sum_{1 \leq i < j \leq k} (p_i + p_j) f_{j,i}(\boldsymbol{x}) = \sum_{1 \leq i < j \leq k} (p_i + p_j) \mathcal{M}_{j,i}(\boldsymbol{x}), \qquad (6)$$

where p_i is the probability of class y_i (estimated by the relative frequency in the training data).

In first experimental studies, this approach has been compared to state-of-the-art ranking methods such as SVMRank [24] which, instead of decomposing the original problem into a set of small binary classification problems, transforms it into a single, large classification problem. The results suggest that LPC is competitive in terms of predictive accuracy while being much more efficient from a computational point of view. Roughly speaking, the reason is that solving several small problems is typically more efficient than solving a single large one (cf. Sect. 7). In this particular case, the transformation into a single classification problem is especially critical, as it may come with a considerable increase of the number of original training examples.

6 Theoretical Foundations

Despite their intuitive appeal and practical success, it is of course important to justify pairwise learning methods from a theoretical point of view. In particular, one may wonder whether LPC is provably able to produce predictions that are optimal in the sense of minimizing (the expectation of) a given loss function.

Corresponding results have recently been derived, not only for classification (Sect. 6.1) but also for ranking (Sects. 6.2 and 6.3). Apart from further technical assumptions, which are not detailed here, these results typically rely on the idealized assumption that the prediction of a pairwise model, $\mathcal{M}_{i,j}(\boldsymbol{x})$, can be interpreted as a *probability*, for example the probability that, in the label ranking associated with $\boldsymbol{x}$, label y_i precedes label y_j. Needless to say, assumptions of that kind are not always easy to satisfy in practice, especially because probability estimation is a challenging problem potentially more difficult than classification.

Besides, the pairwise approach suffers from an inherent limitation, which is sometimes called the "noncompetence" problem in the literature. Roughly speaking, this problem is caused by the fact that a pairwise model is only trained on parts of the instance space and, therefore, possibly not competent for inputs coming from other parts. In conventional classification, for example, a pairwise model $\mathcal{M}_{i,j}$ is only trained on examples from classes y_i and y_j and, therefore, arguably noncompetent for classifying instances from other classes. Several proposals for addressing this problem can be found in the literature. In [32], it was proposed to combine the pairwise models $\mathcal{M}_{i,j}$ with a separate set of models $\mathcal{M}_{ij,\bar{\mathcal{Y}}_{i,j}}$, which predict whether an example belongs to the classes y_i or y_j, or to the remaining classes $\bar{\mathcal{Y}}_{i,j} = \mathcal{Y} \backslash \{y_i, y_j\}$. A similar proposal is to learn ternary models $\mathcal{M}_{i,j,\bar{\mathcal{Y}}_{i,j}}$, which directly discriminate between the three options y_i, y_j, or none of the two [2]. Both approaches have to sacrifice one of the key advantages of the pairwise approach, namely the simplicity of the learned binary models.

6.1 Classification

Despite the existence of more sophisticated methods, such as *pairwise coupling* [15, 44], the most popular strategy for aggregating the predictions of pairwise classifiers is "voting". In binary voting, each classifier $\mathcal{M}_{i,j}$ can give a "vote" for either y_i or y_j. In weighted voting, it may split its vote among the two classes, for example according to its probability estimate. Having queried all models, the class with the highest number of votes (sum of weighted votes) is eventually predicted.

Empirically, weighted voting is known to perform very well. A theoretical justification for this performance has been derived in [23]. Under some technical assumptions, it was shown there that weighted voting may be considered as an approximation of a generalized voting strategy, called *adaptive voting*, which in turn was shown to be optimal in the sense of yielding a maximum posterior probability (MAP) prediction of the true class. Besides, weighted voting appears to be even more robust than adaptive voting in the sense of being less sensitive toward violations of the underlying model assumptions.

Moreover, it was shown that the pairwise approach to learning, at least in a slightly generalized form, is able to produce Bayes-optimal decisions in the context of conventional classification [42].

6.2 Label Ranking

In the context of label ranking, it was shown that many standard loss functions on rankings can be minimized in expectation [22]. For example, under some technical assumptions, straightforward weighted voting is a risk minimizer for the sum of squared rank distances as a loss function (thus, it maximizes the expected Spearman rank correlation between the true and the predicted label ranking). Replacing weighted voting by another aggregation strategy, one can also minimize the number of pairwise inversions (i.e., maximize Kendall's tau); the aggregation problem itself, however, is NP-hard.

On the other hand, there are also loss of functions that cannot be minimized by LPC. Roughly speaking, this is due to a loss of information caused by decomposing the original problem into a set of pairwise problems. Examples include the Spearman footrule and Ulam's distance [21].

6.3 Position Error

It may also be reasonable to compare a predicted ranking with a single target label instead of target ranking. For example, given a predicted label ranking, the *position error* is defined by the position on which the true class label is found or, stated differently, the number of labels that are ranked ahead of this target label. In a normalized form, this measure directly generalizes the conventional 0/1-loss for classification, assuming a value of 0 (1) if the target label is put on the first (last) position. The problem of predicting a ranking that minimizes the expected position loss can again be solved in a theoretically optimal way by LPC [21].

Practically, it was shown that the problem can be reduced to an iterated classification problem, which in turn can be solved effectively through a procedure called *empirical conditioning*. This procedure amounts to iteratively predicting a label and re-training the classifier on the remaining labels. While this is practically infeasible in the general case (essentially one needs to train one classifier for each label subset), it can be realized efficiently by means of LPC. This is because, in the pairwise approach, re-training of the classifiers is not necessary; instead, only the aggregation phase needs to be changed [21].

7 Complexity

At first sight, it seems that LPC is very inefficient for high numbers of labels because one has to train a quadratic number of classifiers. However, a closer look reveals that this is often outweighed by a positive effect, namely that the individual problems are much smaller. In fact, an ensemble of pairwise models can often be trained much more efficiently than a single classifier, even when both have the same total number of training examples. In particular, this holds true for expensive learning algorithms whose time complexity is super-linear in the number of training examples. Thus, the key problem of LPC is typically not the training time, but instead the prediction time and storage capacity. In the following, we will recapitulate some important results concerning these problems.

7.1 Training Time

For the pairwise classification scenario, it is known that even though the number of binary classifiers is quadratic in the number of labels, their joint training time is only $O(k \cdot n)$, i.e., linear in the number labels [9]. The reason is that the individual training sets are much smaller because each of the n examples will only occur in $k-1$ different training sets. This distribution of the training effort on a large number of comparably smaller problem increases the advantage in particular for expensive classifiers with a super-linear time complexity. Obviously, these results also hold for ordered and hierarchical classification.

For multilabel classification [12], the crucial factor determining the efficiency of the approach is the average number d of labels per training example (which is often small in comparison to the total number of labels). The training complexity in this case is $O(d \cdot k \cdot n) = O(k \cdot l)$ when l is the total number of labels in the training set.

In the worst case, when all training examples are associated with a complete ranking of all labels, the training complexity is quadratic in the number of labels [22].

7.2 Prediction Time

A more interesting problem is the efficiency at prediction time. In principle, one has to query a quadratic number of classifiers to derive a final ranking of the classes. However, when one is only interested in classification, it is not necessary to query all classifiers to determine the winning class. For example, if one class has received more votes than every other class can possibly achieve in their remaining evaluations, this class can be safely predicted without querying the remaining classifiers. The QWeighted algorithm [34] tries to enforce this situation by always focusing on the class that has lost the least amount of voting mass. Experiments showed that QWeighted has an average runtime of $O(k \cdot \log k)$ instead of the $O(k^2)$ that would be required for computing the same prediction with all evaluations. Because it nevertheless produces the same predictions as regular weighted voting and thus has the same theoretical guarantees (cf. Sect. 6.1), it is preferred to algorithms like the pairwise DAGs [36], which only approximate the correct prediction. The algorithm can also be generalized to multilabel prediction [30] and to general ternary error-correcting output codes [35].

7.3 Memory Requirements

Even if training is quite efficient, and classification can be handled without the need to query all classifiers, we still have to store all $k \cdot (k-1)/2$ binary classifiers because each classifier will be needed for some examples (unless some labels are never predicted). For example, we have recently tackled a large-scale real-world text categorization problem, namely the annotation of the EUR-Lex database of legal documents of the European Union with labels that are taken from the EUROVOC ontology [29]. This multilabel classification task involved around 20,000 documents, each of which was labeled with about 5 out of 4,000 possible labels. The key problem that had to be solved for this task was that the ensemble of almost 8,000,000 binary classifiers (perceptrons) that were trained for tackling this task could no longer be kept in main memory. The problem was solved by resorting to the dual representation of the perceptron, which reduced the total number of weights that had to be stored at the expense of a somewhat higher classification time. This made the pairwise ranking approach feasible for problems of this size. This solution, however, is only applicable to classifiers that can re-formulate their hypothesis as a linear combination of the input examples, such as perceptrons or SVMs.

For concept descriptions with varying sizes, such as rule sets or decision trees, one can expect that the learned theories for the pairwise classifiers are much smaller than the theories for larger problems such as those of a one-vs-all classifier. However, it is unclear whether these savings on individual theories can compensate for the higher number of classifiers that have to be stored. Moreover, a general solution not restricted to any specific classifier is still an open research problem. Presumably, one will have to resort to fixed approximation techniques, which allow that some classifiers are not trained at all.

8 Conclusions and Outlook

This chapter has reviewed recent work on preference learning and ranking via pairwise classification. The *learning by pairwise comparison* (LPC) paradigm has a natural motivation in the context of preference learning and goes hand-in-hand with the relational approach to preference modeling. Roughly speaking, a pairwise classification corresponds to a pairwise comparison between two alternatives, which in turn can be seen as a basic building block of more complex decision-making procedures. Eventually, complex prediction problems can thus be reduced to the solution of a set of "simple" binary classification problems, which makes the LPC approach especially appealing from a machine learning point of view.

It was shown that LPC can be used in a quite general way to solve different types of preference learning problems. It disposes of sound theoretical foundations and has shown very strong performance in a number of empirical studies. Despite the high number of binary classifiers needed (quadratic in the number of class labels), the training time of the approach seems to be competitive with alternative approaches. Storing and querying such a large number of classifiers, however, is a more significant problem, and an active research area.

Besides, there are several other open questions and promising lines of research. From a practical point of view, for example, it would be interesting to develop a unified framework of LPC for label, instance, and object ranking. The task in label ranking, discussed in Sect. 3, is to learn a model that orders a fixed set of labels, where the ordering depends on the context as specified by an instance. This instance, which constitutes the input of the model, is characterized by properties (e.g., in the form of an attribute-value representation) that can be exploited by the learner, whereas the labels are mere identifiers. In object ranking, the task is to learn a model that accepts a set of objects as input and outputs a preferential order of these objects. As opposed to label ranking, the objects to be ordered are now characterized by properties, but the sought ranking is not context-dependent. Instance ranking may be viewed as a special case of both cases. In many applications, it would be desirable to combine the above two settings, i.e., to have a unified framework in which both the instances (e.g., users of a recommender system) and the labels/objects (e.g., the items to be purchased) can be described in terms of properties. While this can be achieved in a quite straightforward way for other approaches (see e.g., Aiolli and Sperduti, this volume), it is much less obvious for LPC, mainly because the set of "labels" may become very large and change from prediction to prediction.

Another line of research concerns the prediction of preference relations more general than rankings, in particular relations which are not necessarily total (i.e., partial orders) or not strict (i.e., allow for the indifference between alternatives). As an interesting point of departure for a generalization of this type, we mention recent work on the learning of *valued preference structures* [18, 19]. Instead of producing, in the first step, a single binary relation $\mathcal{R}_x$ from which the final prediction (e.g., a ranking) is then derived, the idea is to predict a complete preference structure consisting of three such relations: a strict preference relation, an indifference relation, and an incomparability relation. A structure of that kind conveys much more information which can then be used, among other things, for predicting generalized preferences such as weak or partial orders.

Acknowledgements This research has been supported by the German Science Foundation (DFG). We would like to thank our collaborators Klaus Brinker, Weiwei Cheng, Jens Hühn, Eneldo Loza Mencía, Sang-Hyeun Park, and Stijn Vanderlooy.

References

1. E.L. Allwein, R.E. Schapire, Y. Singer, Reducing multiclass to binary: A unifying approach for margin classifiers. J. Mach. Learn. Res. **1**, 113–141 (2000)
2. C. Angulo, F.J. Ruiz, L. González, J.A. Ortega, Multi-classification by using tri-class SVM. Neural Process. Lett. **23**(1), 89–101 (2006)
3. R.A. Bradley, M.E. Terry, The rank analysis of incomplete block designs – I. The method of paired comparisons. Biometrika **39**, 324–345 (1952)
4. K. Brinker, J. Fürnkranz, E. Hüllermeier, A unified model for multilabel classification and ranking, in *Proceedings of the 17th European Conference on Artificial Intelligence (ECAI-06)*, ed. by G. Brewka, S. Coradeschi, A. Perini, P. Traverso (2006), pp. 489–493
5. O. Dekel, C.D. Manning, Y. Singer, Log-linear models for label ranking, in *Advances in Neural Information Processing Systems (NIPS-03)*, ed. by S. Thrun, L.K. Saul, B. Schölkopf (MIT, Cambridge, MA, 2004), pp. 497–504
6. J. Fodor, M. Roubens, *Fuzzy Preference Modelling and Multicriteria Decision Support* (Kluwer, 1994)
7. E. Frank, M. Hall, A simple approach to ordinal classification, in *Proceedings of the 12th European Conference on Machine Learning (ECML-01)*, ed. by L. De Raedt, P. Flach (Springer, Freiburg, Germany, 2001), pp. 145–156
8. J.H. Friedman, Another approach to polychotomous classification. Technical report, Department of Statistics, Stanford University, Stanford, CA, 1996
9. J. Fürnkranz, Round robin classification. J. Mach. Learn. Res. **2**, 721–747 (2002)
10. J. Fürnkranz, Round robin ensembles. Intell. Data Anal. **7**(5), 385–404 (2003)
11. J. Fürnkranz, E. Hüllermeier, Pairwise preference learning and ranking, in *Proceedings of the 14th European Conference on Machine Learning (ECML-03)*, vol. 2837, *Lecture Notes in*

Artificial Intelligence, ed. by N. Lavrač, D. Gamberger, H. Blockeel, L. Todorovski (Springer, Cavtat, Croatia, 2003), pp. 145–156
12. J. Fürnkranz, E. Hüllermeier, E. Loza Mencía, K. Brinker, Multilabel classification via calibrated label ranking. Mach. Learn. **73**(2), 133–153 (2008)
13. J. Fürnkranz, E. Hüllermeier, S. Vanderlooy, Binary decomposition methods for multipartite ranking, in *Proceedings of the European Conference on Machine Learning and Principles and Practice of Knowledge Discovery in Databases (ECML/PKDD-09)*, vol. Part I, ed. by W.L. Buntine, M. Grobelnik, D. Mladenic, J. Shawe-Taylor (Springer, Bled, Slovenia, 2009), pp. 359–374
14. S. Har-Peled, D. Roth, D. Zimak, Constraint classification: A new approach to multiclass classification, in *Proceedings of the 13th International Conference on Algorithmic Learning Theory (ALT-02)*, ed. by N. Cesa-Bianchi, M. Numao, R. Reischuk (Springer, Lübeck, Germany, 2002), pp. 365–379
15. T. Hastie, R. Tibshirani, Classification by pairwise coupling, in *Advances in Neural Information Processing Systems 10 (NIPS-97)*, ed. by M.I. Jordan, M.J. Kearns, S.A. Solla (MIT, 1998), pp. 507–513
16. C.-W. Hsu, C.-J. Lin, A comparison of methods for multi-class support vector machines. IEEE Trans. Neural Netw. **13**(2), 415–425 (2002)
17. J. Hühn, E. Hüllermeier, Is an ordinal class structure useful in classifier learning? Intl. J. Data Min. Model. Manage. **1**(1), 45–67 (2008)
18. J. Hühn, E. Hüllermeier, FR3: A fuzzy rule learner for inducing reliable classifiers. IEEE Trans. Fuzzy Syst. **17**(1), 138–149 (2009)
19. E. Hüllermeier, K. Brinker, Learning valued preference structures for solving classification problems. Fuzzy Sets Syst. **159**(18), 2337–2352 (2008)
20. E. Hüllermeier, J. Fürnkranz, Comparison of ranking procedures in pairwise preference learning, in *Proceedings of the 10th International Conference on Information Processing and Management of Uncertainty in Knowledge-Based Systems (IPMU-04)*, Perugia, Italy, 2004
21. E. Hüllermeier, J. Fürnkranz, On predictive accuracy and risk minimization in pairwise label ranking. J. Comput. Syst. Sci. **76**(1), 49–62 (2010)
22. E. Hüllermeier, J. Fürnkranz, W. Cheng, K. Brinker, Label ranking by learning pairwise preferences. Artif. Intell. **172**, 1897–1916 (2008)
23. E. Hüllermeier, S. Vanderlooy, Combining predictions in pairwise classification: An optimal adaptive voting strategy and its relation to weighted voting. *Pattern Recognit.* **43**(1), 128–142 (2010)
24. T. Joachims, Optimizing search engines using clickthrough data, in *Proceedings of the 8th ACM SIGKDD International Conference on Knowledge Discovery and Data Mining (KDD-02)*, (ACM, 2002), pp. 133–142
25. S. Knerr, L. Personnaz, G. Dreyfus, Single-layer learning revisited: A stepwise procedure for building and training a neural network, in *Neurocomputing: Algorithms, Architectures and Applications*, vol. F68, *NATO ASI Series*, ed. by F. Fogelman Soulié, J. Hérault (Springer, 1990), pp. 41–50
26. S. Knerr, L. Personnaz, G. Dreyfus, Handwritten digit recognition by neural networks with single-layer training. IEEE Trans. Neural Netw. **3**(6), 962–968 (1992)
27. D. Koller, M. Sahami, Hierarchically classifying documents using very few words, in *Proceedings of the 14th International Conference on Machine Learning (ICML-97)* (Nashville, 1997), pp. 170–178
28. U.H.-G. Kreßel, Pairwise classification and support vector machines, in *Advances in Kernel Methods: Support Vector Learning*, Chap. 15, ed. by B. Schölkopf, C.J.C. Burges, A.J. Smola (MIT, Cambridge, MA, 1999), pp. 255–268
29. E. Loza Mencía, J. Fürnkranz, Efficient pairwise multilabel classification for large-scale problems in the legal domain, in *Proceedings of the European Conference on Machine Learning and Principles and Practice of Knowledge Disocvery in Databases (ECML-PKDD-2008), Part II*, ed. by W. Daelemans, B. Goethals, K. Morik (Springer, Antwerp, Belgium, 2008), pp. 50–65

30. E. Loza Mencía, S.-H. Park, J. Fürnkranz, Efficient voting prediction for pairwise multilabel classification, in *Proceedings of the 11th European Symposium on Artificial Neural Networks (ESANN-09)* (d-side publicationsBruges, Belgium, 2009), pp. 117–122
31. B.-L. Lu, M. Ito, Task decomposition and module combination based on class relations: A modular neural network for pattern classification. IEEE Trans. Neural Netw. **10**(5), 1244–1256 (1999)
32. M. Moreira, E. Mayoraz, Improved pairwise coupling classification with correcting classifiers, in *Proceedings of the 10th European Conference on Machine Learning (ECML-98)*, ed. by C. Nédellec, C. Rouveirol (Springer, Chemnitz, Germany, 1998), pp. 160–171
33. G.H. Nam, Ordered pairwise classification. Master's thesis, TU Darmstadt, Knowledge Engineering Group, 2007
34. S.-H. Park, J. Fürnkranz, Efficient pairwise classifciation, in *Proceedings of 18th European Conference on Machine Learning (ECML-07)*, ed. by J.N. Kok, J. Koronacki, R. Lopez de Mantaras, S. Matwin, D. Mladenič, A. Skowron (Springer, Warsaw, Poland, 2007), pp. 658–665
35. S.-H. Park, J. Fürnkranz, Efficient decoding of ternary error-correcting output codes for multiclass classification, in *Proceedings of the European Conference on Machine Learning and Principles and Practice of Knowledge Discovery in Databases (ECML/PKDD-09)*, ed. by W.L. Buntine, M. Grobelnik, D. Mladenic, J. Shawe-Taylor, vol. Part I (Springer, Bled, Slovenia, 2009), pp. 189–204
36. J.C. Platt, N. Cristianini, J. Shawe-Taylor, Large margin DAGs for multiclass classification, in *Advances in Neural Information Processing Systems 12 (NIPS-99)*, ed. by S.A. Solla, T.K. Leen, K.-R. Müller (MIT, 2000), pp. 547–553
37. D. Price, S. Knerr, L. Personnaz, G. Dreyfus, Pairwise neural network classifiers with probabilistic outputs, in *Advances in Neural Information Processing Systems 7 (NIPS-94)*, ed. by G. Tesauro, D. Touretzky, T. Leen (MIT, 1995), pp. 1109–1116
38. T.L. Saaty, *Decision Making for Leaders: The Analytic Hierarchy Process for Decisions in a Complex World* (RWS Publications, Pittsburgh, Pennsylvania, 1999)
39. T.L. Saaty, Relative measurement and its generalization in decision making: Why pairwise comparisons are central in mathematics for the measurement of intangible factors – the analytic hierarchy/network process. Revista de la Real Academia de Ciencias Exactas Físicas y Naturales A Matemáticas (RACSAM) **102**(2), 251–318 (2008)
40. M.S. Schmidt, H. Gish, Speaker identification via support vector classifiers, in *Proceedings of the 21st IEEE International Conference Conference on Acoustics, Speech, and Signal Processing (ICASSP-96)* (Atlanta, GA, 1996), pp. 105–108
41. J.F. Sima, Präferenz-Lernen für Hierarchische Multilabel Klassifikation. Master's thesis, TU Darmstadt, Knowledge Engineering Group, 2007
42. J.-N. Sulzmann, J. Fürnkranz, E. Hüllermeier, On pairwise naive bayes classifiers, in *Proceedings of 18th European Conference on Machine Learning (ECML-07)*, ed. by J.N. Kok, J. Koronacki, R. Lopez de Mantaras, S. Matwin, D. Mladenič, A. Skowron (Springer, Warsaw, Poland, 2007), pp. 371–381
43. L.L. Thurstone, A law of comparative judgement. Psychol. Rev. **34**, 278–286 (1927)
44. T.-F. Wu, C.-J. Lin, R.C. Weng, Probability estimates for multi-class classification by pairwise coupling. J. Mach. Learn. Res. **5**, 975–1005 (2004)
45. Y. Yang, An evaluation of statistical approaches to text categorization. *Inf. Retr.* **1**, 69–90 (1999)

Decision Tree Modeling for Ranking Data

Philip L.H. Yu, Wai Ming Wan, and Paul H. Lee

Abstract Ranking/preference data arises from many applications in marketing, psychology, and politics. We establish a new decision tree model for the analysis of ranking data by adopting the concept of classification and regression tree. The existing splitting criteria are modified in a way that allows them to precisely measure the impurity of a set of ranking data. Two types of impurity measures for ranking data are introduced, namely g-wise and top-k measures. Theoretical results show that the new measures exhibit properties of impurity functions. In model assessment, the area under the ROC curve (AUC) is applied to evaluate the tree performance. Experiments are carried out to investigate the predictive performance of the tree model for complete and partially ranked data and promising results are obtained. Finally, a real-world application of the proposed methodology to analyze a set of political rankings data is presented.

1 Introduction

Ranking data are frequently collected when individuals are asked to rank a set of items based on certain pre-defined criterion. It is a simple and efficient way to understand judges' perception and preferences on the ranked alternatives. In many preference studies, ranking responses and additional information about the investigated raters are observed, e.g., socioeconomic characteristics. It is often of great interest to determine how these covariates affect the perceived rankings.

Our aim in this paper is to develop new decision tree model to analyze ranking data for discovering the factors that affect the judgement process by which people make choice. It will serve as a complement to existing parametric ranking models (see review in [6, 20] for more details), and algorithms in label ranking and preference learning (see [9, 13] for more details). Decision tree models are

P.L.H. Yu (✉), W.M. Wan, and P.H. Lee
Department of Statistics and Actuarial Science, University of Hong Kong, Pokfulam Road, Hong Kong
e-mail: plhyu@hkucc.hku.hk,wwmminky@yahoo.com.hk,honglee@graduate.hku.hk

J. Fürnkranz and E. Hüllermeier (eds.), *Preference Learning*,
DOI 10.1007/978-3-642-14125-6_5,

nonparametric statistical methodology designed for classification and prediction problems. It produces a set of decision rules for predicting the class of a categorical response variable on the basis of the *input attributes/features/predictors/covariates*. This classification technique is widely used in statistics, machine learning, pattern recognition, and data mining because of its ease of interpretability comparing with other statistical models, and it can handle input attributes in both categorical and interval measurement. Comparing to parametric ranking models, the merit of decision tree lies in its ease of interpretability of nonlinear and interaction effects. Additionally, learning a decision tree can be seen as a process of variable selection for the data. Questions on adding explanatory variables and interaction terms between variables are handled automatically.

A variety of algorithms have been proposed to construct a decision tree for a single discrete/continuous response in a top-down recursive divide-and-conquer manner, such as ID3 [21], C4.5 [22], CHAID [18], and QUEST [19]. More decision tree algorithms are available in the literature, many of them are a variation of the algorithmic framework mentioned above (see [24] for details). Among all the tree building methodologies, the most popular one is the CART procedure [3]. Construction of CART comprises two stages: *growing* and *pruning*. Detailed review of CART will be provided later in Sect. 2.

Nominal data, ordinal data as well as continuous data can be handled by the decision tree model. It was extended to cope with multivariate data recently through building the tree with a two-stage splitting criteria [25] and through a so-called output kernel trees that are based on a kernelization of the output space of regression trees [10]. Karlaftis [17] used the recursive partitioning models to predict individual mode choice decisions by considering both univariate and multivariate splits. It has been found that trees performed surprisingly well and were comparable to discrete choice logit models. Therefore, we propose extension of this machine learning method for ranking data and it has been presented in the Preference Learning Workshop in ECML PKDD 2008.

In principle, existing tree models for discrete choice data can be applied to preference data by two approaches. The first approach is to build tree based on the top choice of the given ranking data. Another approach is to treat each ranking of m items as a discrete choice. So each possible ranking outcome contributes to one target level, resulting a total of $m!$ levels. For instance, given three alternatives (y_1, y_2 and y_3), a top-choice tree with 3 target levels or a tree with 6 target levels ($y_1 \succ y_2 \succ y_3$, $y_1 \succ y_3 \succ y_2$, $y_2 \succ y_1 \succ y_3$, $y_2 \succ y_3 \succ y_1$, $y_3 \succ y_1 \succ y_2$ and $y_3 \succ y_2 \succ y_1$) can be constructed.

However, in observational studies, discrete choice tree can provide only limited insights about the underlying behavioral processes that give rise to the data. For the second approach, it will be too heavy-handed in practical, because even moderate values of m would lead to overwhelmingly large number of ranking outcomes ($4! = 24$ and $5! = 120$). Moreover, these nominal trees are restricted only for consistent ranking responses, which all individuals rank the same number of given items. They also are not suitable to handle data with tie ranks because more rank combinations would be involved and this would tremendously increase the number of target levels.

Another drawback of this method is ignorance of the ordinal structure in rankings, which is often useful in explaining individuals' preference judgement. Therefore, it is impractical to build tree for ranking data using conventional algorithms.

In view of all the limitations and inappropriateness of existing decision tree models for rankings, we are interested to develop a tree model specifically for preference data. In this article, binary tree is considered. Following the landmark CART procedure, we extend the splitting criteria Gini and entropy to accommodate complete and partial ranking data by utilizing the rank-order structure of preference data in the tree growing process. Other distance-based splitting criterion is considered in [4], which uses a weighted average of the within-leaf variances as a goodness-of-split measure.

Another issue that will be addressed in this paper is the performance assessment of the built decision tree model. The most frequently used performance measure is misclassification rate, which equals to the number of misclassified samples divided by the total number of samples. However, we will not consider it to be the performance measure of our tree model because a sample can be classified either correctly or incorrectly, overlooking the fact that a ranking can be partially agreed with the predicted ranking. That means some items in the rank permutation, but not all, are in the correct ordered position.

We consider goodness-of-fit measures for parametric ranking models for our tree model, such as log-likelihood or other likelihood-based statistics (e.g., BIC). But these approaches may not be suitable because maximizing entropy or deviance is equivalent to maximize the log-likelihood [23]. This will lead to bias toward decision tree that is built on entropy. Therefore, an assessment method independent of the splitting criteria will be more favorable.

The *Receiver Operating Characteristic (ROC) curve* provides a visualization of the performance of scoring classifier by plotting sensitivity versus 1-specificity at all possible values of the classification threshold. It starts at the bottom-left corner and rises to the top-right corner. Moving along the ROC curve represents trading off false positives for false negatives. In the worst case, random models will run up the diagonal, and the performance of classifier improves as the ROC curve gets near the top-left corner of the plot. Unfortunately, in a comparison of two classifiers, one classifier may not always outperform another at all thresholds. Ambiguous conclusion would be drawn when the two curves intersect. More inadequacies of the ROC curve were discussed in [7].

The *area under the ROC curve (AUC)* provides a single measure of overall performance of a classifier based on the ROC curve. It is simple and attractive because it is not susceptible to the threshold choice and it is regardless of the costs of the different kinds of misclassification and class priors. The calculation of AUC can be referred to [2] and [11]. The value of AUC always fall within [0.5, 1.0] – it equals 0.5 when the instances are predicted at random and equals 1.0 for perfect accuracy. Statistically, the AUC of a classifier can be seen as the probability that the classifier will rank a randomly chosen positive instance higher than a randomly chosen negative instance. This is equivalent to Mann–Whitney–Wilcoxon test statistic [12].

Traditional ROC curves analysis mainly focus on data with binary target, recently it is extended to multiple class data [11]. In this paper, we adopt the approach of multiclass AUC and generalize the performance measure to pairwise ranking data. The choice of extension to pairwise data over top-2 data is because pairwise data concentrates on two items, while keeping away other irrelevant alternatives.

The remainder of this paper is organized as follows. Section 2 reviews the CART methodology. In Sect. 3, the framework of growing and pruning a decision tree for ranking data is presented. Two new impurity measures, top-k and g-wise measure, are introduced and they are shown to possess the properties of impurity function. Methods for assessing the performance of the tree-structured classifier are discussed in Sect. 4. To illustrate the feasibility of the proposed algorithm, some examples and evaluation studies, and an application on real data are presented in Sect. 5. Finally, some concluding remarks and further research are given in Sect. 6.

2 Review of Classification and Regression Tree (CART)

Suppose we have a learning sample of size n with measurements $(C_i, \boldsymbol{\mathcal{A}}_i)$, $i = 1, \ldots, n$, where C is our target variable and $\boldsymbol{\mathcal{A}}$ is the vector of Q attributes $\mathcal{A}^q, q = 1, \ldots, Q$. $\boldsymbol{\mathcal{A}}$ and C can be interval, ordinal or categorical variables. The goal is to predict C based on $\boldsymbol{\mathcal{A}}$ via tree-structured classification.

CART is a binary decision tree that is constructed by recursively partitioning the n learning sample into different subsets, beginning with the root node that contains the whole learning sample. Each subset is represented by a node in the tree. In a binary tree structure, all internal nodes have two child nodes whereas the nodes with no descendants are called *terminal/leaf* nodes. At each partition process, a splitting rule $s(t)$, comprised of a splitting variable $\mathcal{A}^q$ and a split point, is used to split a group of $N(t)$ cases in node t to left node $N_L(t)$ and right node $N_R(t)$. Decision tree identifies the best split by exhaustive search. The number of possible splits of a categorical attribute $\mathcal{A}^q$ of I categories is $2^{I-1} - 1$. For an interval $\mathcal{A}^q$ with F distinct values or an ordinal attribute with F ordered categories, $F - 1$ possible splits will be produced on $\mathcal{A}^q$.

2.1 Growing Stage of Decision Tree

The key step of tree growing is to choose a split among all possible splits at each node so that the resulting child nodes are the "purest". To measure the purity of a node t, [3] proposed a measure called impurity function $i(t)$. Let $p(j|t), j \in 1, \ldots, J$ be the conditional probability of having class j in the learning sample in node t, $\sum_{j=1}^{J} p(j|t) = 1$. Impurity function should satisfy the following three properties: (a) It is minimum when the node is pure ($p(j|t) = 1$ for one $j \in \{1, \ldots, J\}$); (b) it is maximum when the node is the most impure ($p(1|t) = \ldots = p(J|t) = \frac{1}{J}$); (c) renaming of items does not change the node impurity.

It can be shown that if the impurity function is concave, properties (a) and (b) will be satisfied. Property (c) is required because labeling of classes is arbitrary. CART includes various impurity criteria for classification trees, namely the Gini criterion $1 - \sum_{j=1}^{J} p(j \mid t)^2$ and Two-ing criterion. Another frequently used impurity-based criterion applied is entropy $-\sum_{j=1}^{J} p(j \mid t) \log_2 p(j \mid t)$. Modification of existing measures of node homogeneity is essential for building decision tree model for ranking data and they will be discussed in Sect. 3.1.

Based on the impurity measure for a node, a splitting criterion $\Delta i(s,t)$ can be defined as the reduction in impurity resulting from the split s of node t.

$$\Delta i(s,t) = i(t) - p_L i(t_L) - p_R i(t_R), \tag{1}$$

where $p_L = N_L(t)/N(t)$ and $p_R = N_R(t)/N(t)$ are the proportion of data cases in t to the left child node t_L and to the right child node t_R, respectively. The best split is chosen to maximize a splitting criterion. The concavity property of $i(t)$ assures that further splitting does not increase the impurity, so we can continue growing a tree until every node is pure, and some may contain only one observation. This would lead to a very large tree that would overfit the data. To eliminate nodes that are overspecialized, pruning is required so that the best pruned subtree can be obtained.

2.2 *Pruning Stage of Decision Tree*

The minimal cost-complexity pruning method is developed by Breiman et al. in 1984 [3]. Before proceeding to the algorithmic framework, some notations are first defined. Let $\tilde{T}$ be the set of terminal nodes of tree T, and the number of terminal nodes, denoted by $|\tilde{T}|$, is defined as the complexity of T. Define $R(t)$ to be the misclassification cost of node t. An obvious candidate of $R(t)$ is the misclassification rate; there are also other choices for the cost function. In a class probability tree, [3] considered pruning with the mean square error, which corresponds to take $R(t)$ as the Gini diversity index. For entropy tree, it is natural to take $R(t)$ as deviance. Chou [5] developed a class of divergences in the form of expected loss function and it was shown that Gini, entropy and misclassification rate can be written in the proposed form. In this paper, we specify the cost functions $R(t)$ such that they coincide with impurity functions for ranking data. More details will be given in Sect. 3.3.

For any tree T, the cost-complexity function $R_\alpha(T)$ is formulated as a linear combination of the cost of T and its complexity: $R(T) + \alpha|\tilde{T}|$. The complexity parameter α measures how much additional accuracy a split must add to the entire tree to warrant the addition of one more terminal node. Now consider $T_{t'}$ as the subtree with root t'. As long as $R_\alpha(T_{t'}) < R_\alpha(t')$, the branch $T_{t'}$ contributes less complexity cost to tree T than node t'. This occurs for small α. When α increases to a certain value, the equality of the two cost-complexities is achieved. At this point, the subtree $T_{t'}$ can be removed since it no longer helps improving the classification. The strength of the link from node t, $g(t)$, is therefore defined as $\frac{R(t)-R(T_t)}{|\tilde{T}_t|-1}$.

The V-fold cross-validation cost-complexity pruning algorithm works as follows. The full learning dataset L is divided randomly into V equal-size subsets $L_1, L_2, \ldots, L_V$ and the vth learning sample is denoted to be $L^v = L - L_v$. Using the full learning dataset L, an overly large tree T^0 is built. $g(t)$ are calculated for all internal nodes in T^0 and the node with the minimum value $g(t^1)$ is located. A pruned tree T^1 is created by turning the weakest-linked internal node t^1 into a leaf node. This process is repeated until T^0 is pruned up to the root T^m. Denote by α_i the value of $g(t)$ at the ith stage. A sequence of nested trees $T^0 \supseteq T^1 \supseteq T^2 \supseteq \ldots \supseteq T^\theta$ is generated, such that each pruned tree T^i is optimal for $\alpha \in [\alpha_i, \alpha_{i+1})$. Here, the word "nested" means that each subsequent tree in the sequence is obtained from its predecessor by cutting one or more subtrees, and thus the accuracy of the sequence of progressively smaller pruned trees decreases monotonically.

Next, for $v = 1, \ldots, V$, the vth auxiliary maximal tree T_v^0 is constructed based on L^v and the nested sequence of pruned subtrees of T_v^0 is generated $(T_v^0 \supseteq T_v^1 \supseteq T_v^2 \supseteq \ldots \supseteq T_v^\theta)$. The cross-validation estimate of the misclassification rate $R^{CV}(T^i)$ is then evaluated as $\frac{1}{V}\sum_{v=1}^{V} R(T_v(\sqrt{\alpha_i\,\alpha_{i+1}}))$, where $T_v(\alpha)$ is equal to the pruned subtree T_v^i in the ith stage such that $\alpha_i \leq \alpha \leq \alpha_{i+1}$. Note that the misclassification cost of the pruned subtree $T_v(\sqrt{\alpha_i\,\alpha_{i+1}})$ is estimated by the independent subset L_v. The simplest subtree T^* is selected as the final tree model from $\{T^0, T^1, \ldots, T^\theta\}$ by the following rule $R^{CV}(T^*) \leq \min_i R^{CV}(T^i) + SE(R^{CV}(T^i))$. The 1-SE rule is adopted because the position of the minimum $R^{CV}(T^*)$ is uncertain [3]. As a consequence, we get a more conservative estimate for the cross-validated $R^{CV}(T^*)$.

2.3 Class Assignment of Terminal Nodes of Decision Tree

Each terminal node of the final selected tree T^* carries with it a class label $j^* \in \{1, \ldots, J\}$, which represents the predicted class for target C, of the samples which fall within this node. The class label is usually determined by the plurality rule, so that the misclassification rate of the tree is minimized. Decision tree classifies an instance by passing it down the tree from the root node till it ends up in a class j^* leaf node and obviously the instance will be assigned to class j^*.

3 Decision Tree Model for Ranking Data

In this section, we describe our methodology for constructing decision tree using a learning dataset of rankings. Following the idea of the CART method, our algorithm involves two stages – growing and pruning – to generate the final best subtree. Mathematically, in a completely ranked data of k items, a ranking can be described by a permutation function $\boldsymbol{\pi} = (\pi(1), \ldots, \pi(m))$ from $\{1, \ldots, m\}$ onto $\{1, \ldots, m\}$

such that $\pi(j)$, $j = 1, \ldots, m$ is the rank assigned to item j. The convention that smaller ranks correspond to the more preferred items is adopted.

Let $\boldsymbol{\mathcal{A}}$ be a vector of Q attributes $\mathcal{A}^q, q = 1, \ldots, Q$ observed in the data of a preference study and $\boldsymbol{\pi}$ be the observed ranking responses. We are interested to examine how the covariates affects the n individuals' choice behavior on the basis of the learning sample $(\boldsymbol{\pi}_i, \boldsymbol{\mathcal{A}}_i), i = 1, \ldots, n$, via tree-based method. Input attributes $\boldsymbol{\mathcal{A}}$ can be measured in continuous, ordinal, or nominal scale.

3.1 Impurity Measures for Ranking Data

In tree construction, our approach searches for the best-splitting rule based on an impurity function. It is not easy to compute the impurity of ranking data based on the permutation function, therefore we introduce two new measures, namely top-k and g-wise measures. Before proceeding, we first define some notations and terminology of choice probabilities and measures for ranking data.

Definition 1. For top-k measured data ($k \leq m$) in node t, $p_\tau(y_1, \ldots, y_k | t)$ and $N^\tau_{y_1,\ldots,y_k}(t)$ indicates, respectively, the proportion and number of judges who rank item y_1 first, item y_2 the second, and so on, and y_k in the kth place. Rankings of the remaining $m - k$ items are not considered.

Definition 2. For g-wise measured data ($g \leq m$) in node t, let $p_w(y_1, \ldots, y_g | t)$ and $N^w_{y_1,\ldots,y_g}(t)$ to be the proportion and number of judges which item y_1 ranks higher than y_2, which in turn higher than y_3, and so on. Items other than $y_1, y_2, \ldots, y_g$ are not taken into consideration.

For example, in node t, $p_\tau(1, 2| t)$ denotes the proportion of data which item 1 ranks first, and item 2 ranks second, whereas $p_w(1, 2| t)$ is the proportion of data which item 1 is preferable to item 2, regardless on the ranks of the two items.

Nevertheless, there are advantages and disadvantages for both methods. The advantage of top-k measure is that existing tree methods for nominal response can be directly applied, owing to the fact that the sum of all proportions of top-k measured data equals one. Therefore, impurity measures such as Gini and entropy can still be employed. However, g-wise measured data do not satisfy this property, therefore we need to modify the impurity measures such that they can estimate the node heterogeneity for ranking data based on g-wise comparison. The advantage of g-wise measure is that it takes account of the ordinal nature of ranking. Top-k measures treat every combination of top-k ranking equally, thereby treating preference data as nominal data. For g-wise measure, it models the rankings by making g-wise comparison for all items.

Definition 3. Given m items, denote by $\mathcal{R}^{m,d}$ a set of rankings with members from all m-choose-d permutations in $\{1, 2, \ldots, m\}$. $\mathcal{R}^{m,d}$ contains P^m_d ($C^m_d \times d!$) rankings coming from C^m_d possible ranked d-item subsets and each d-item subset gives $d!$ possible rank permutation. Furthermore, denote a subset of rankings

$\mathcal{R}^{m,d}_{\{y_1,y_2,\dots y_d\}}$ to represent the d ! rank permutations for item subset $\{y_1, y_2, \dots, y_d\}$ and $\boldsymbol{\Omega}^m_d$ to indicate all C^m_d d-item subsets based on m items.

For example, we have an arbitrary item set $\{1,2,3,4\}$, then $\mathcal{R}^{4,2}$ includes $\{(1,2)$, $(2,1)$, $(1,3)$, $(3,1)$, $(1,4)$, $(4,1)$, $(2,3)$, $(3,2)$, $(2,4)$, $(4,2)$, $(3,4)$, $(4,3)\}$ and all members of $\mathcal{R}^{4,2}_{\{1,2\}}$ are $\{(1,2),(2,1)\}$, whereas $\boldsymbol{\Omega}^4_2$ represents all pairwise combinations $\{(1,2),(1,3),(1,4),(2,3),(2,4),(3,4)\}$.

3.2 Growing Stage of Decision Tree for Rankings

In Sect. 2, impurity functions for unordered categorical responses are described. Following this reasoning, we provide extension of impurity functions Gini and entropy to deal with ranking data. As mentioned before, top-k ranking data can be viewed as a kind of nominal data, the corresponding impurity functions thus have similar properties with those for nominal target. Properties of g-wise impurity functions are different: (a) it is minimum when there is only one ranking observed for each of C^m_g ranked item subsets; (b) it attains maximum when all $g!$ rank permutations are equally distributed in each of C^m_g ranked item subsets; (c) renaming of items do not change the value of impurity.

Theorem 1 proves that an impurity measure for nominal data can be extended to handle g-wise measured data if it satisfies certain conditions. A definition is given before theorem 1:

Definition 4. If an impurity function $i(t) = \phi(p(1|t), p(2|t), \dots, p(J|t))$, satisfying $\sum_{j=1}^{J} p(j|t) = 1$ can be written as $\sum_{j=1}^{J} f(p(j|t))$, then it can be generalized to g-wise impurity measure, denoted as $i^{(g)}_w(t) = \phi^{(g)}_w(p_w(\pi|t), \forall\pi \in \mathcal{R}^{m,g})$, with value equals to $\sum_{\pi\in\mathcal{R}^{m,g}} f(p_w(\pi|t))$.

Theorem 1. *The g-wise impurity function $i^{(g)}_w(t)$ satisfies the following conditions:*

1.1 Concave $(\frac{\partial\phi^{(g)}_w(p_w(\pi|t))}{\partial p_w(\pi_a|t)\partial p_w(\pi_b|t)}) \le 0, \forall\pi_a, \pi_b \in \mathcal{R}^{m,g}$.
1.2 Minimum when one of $p_w(\pi|t), \pi \in \mathcal{R}^{m,g}_{\{y_1,\dots,y_g\}}$ *equals 1* $\forall$ $\{y_1, \dots, y_g\} \in \boldsymbol{\Omega}^m_g$.
1.3 Maximum when all $p_w(\pi|t) = 1/g!$.
1.4 Symmetric with respect to $p_w(\pi|t)$.

The proof is given in Appendix.

Using g-wise and top-k measures defined above, we can write down, for example, the Gini index of a node t. Given a ranking dataset of m items,

$$\text{top-}k \text{ Gini:} \quad i_{\tau}^{(k)}(t) = 1 - \sum_{\pi \in \mathcal{R}^{m,k}} [p_{\tau}(\pi \mid t)]^2 \tag{2}$$

$$g\text{-wise Gini:} \quad i_{w}^{(g)}(t) = \frac{1}{C_g^m} \sum_{B_g \in \Omega_g^m} \left(1 - \sum_{\pi \in \mathcal{R}_{B_g}^{m,g}} [p_w(\pi \mid t)]^2 \right) \tag{3}$$

The normalizing term $1/C_g^m$ is to bound $i_w^{(g)}(t)$ in the range of 0 and 1. In g-wise impurity measure, $\mathcal{R}_{B_g}^{m,g}$ denotes the set of permutations for each of all C_g^m ranked item subset in $\mathcal{R}^{m,g}$. Top-k and g-wise splitting criteria can thus be constructed based on $i_\tau^{(k)}(t)$ and $i_w^{(g)}(t)$ correspondingly to measure the reduction of heterogeneity between two subnodes. The split that best separates the parent node into two subgroups having the highest consensus in ranking should be chosen. The node will continue splitting until the node size is less than the user-specified minimum node size value. In our case studies, the minimum node size is set to $1/10$ of the training sample size.

3.3 Pruning Stage of Decision Tree for Rankings

We consider pruning the model in a bottom-up manner, using the minimal cost-complexity algorithm introduced in Sect. 2.2 with tenfold cross-validation to obtain the final tree that minimizes the misclassification cost. With reference to [5], our cost function $R(t) = P(t) \cdot E_\pi \left[\ell(\pi, \hat{p}(\pi \mid t)) \mid t \right]$ is expressed as an expected loss function based on the impurity function of the partition, where $P(t)$ is the proportion of judges classified into node t in testing data. The loss functions arising from top-k Gini and entropy are

$$\text{top-}k \text{ Gini:} \quad \ell(\pi, \hat{p}_\tau(\pi \mid t)) = 1 + \sum_{\pi \in \mathcal{R}^{m,k}} [\hat{p}_\tau(\pi \mid t)]^2 - 2\hat{p}_\tau(\pi \mid t) \tag{4}$$

$$\text{top-}k \text{ entropy:} \quad \ell(\pi, \hat{p}_\tau(\pi \mid t)) = -\log_2 \hat{p}_\tau(\pi \mid t) \tag{5}$$

It should be aware that $\hat{p}_\tau(\pi \mid t)$ and $\hat{p}_w(\pi \mid t)$ are evaluated by the learning data, whereas $N_\pi^\tau(t)$ and $N_\pi^w(t)$ are obtained from the testing data. The cost function of g-wise impurity measure can be extended analogously, by taking the expectation over all possible C_g^m item subsets.

3.4 Assignment of Terminal Nodes of Decision Tree for Rankings

We consider various approaches to make the assignment. For every leaf node, (a) mean rank of each item is calculated and the predicted ranking is obtained by ordering the mean ranks; (b) top-choice frequency of each item is calculated and is ordered to give the predicted ranking; (c) the most frequently observed ranking represents the predicted ranking; (d) look at the paired comparison probabilities of each item pair or the top-5 most frequently observed ranking responses. The first three approaches reveal the predicted ranking of the items. However, in some situations, the predicted rankings are not of primary concern, when the tree plays a role in facilitating investigation of attributes, which influence individuals' difference in item evaluation. For this kind of exploration purpose, method (d) will give us a more general idea of how the preference orders distributed within a terminal node.

4 Performance Assessment

To compare the performance of the tree models generated by different splitting criteria, we apply the area under the ROC curve in a testing dataset of size N_{ts}. Suppose we have grown a decision tree T with z terminal nodes, the AUC of an item pair (i, j) is calculated as follows:

1. Calculate the pairwise probability $\hat{p}_w(i, j \mid t), t = 1, \ldots, z$ for every leaf node.
2. Assign $\hat{p}_w(i, j \mid t)$ to judges who fall in terminal node t.
3. Rank the judges in the testing dataset in increasing order according to $\hat{p}_w(i, j \mid t)$ and assign rank π_v for the vth individual who prefer item i over item j, $v = 1, \ldots, N_{ij}^w$. Note that equal rank is assumed when tied.
4. Calculate the number of judges who rank item i higher than item j (c_t), and the number of judges who rank item j higher than item i (d_t) for $t = 1, \ldots, z$.
5. Compute the sum of the ranks (S) for individuals with preference order $i \succ j$, where $S = \sum_{t=1}^{z} c_t \pi_t$.
6. Evaluate the AUC of item pair (i, j), A_{ij} by $A_{ij} = \frac{S - c_0(c_0-1)/2}{c_0 d_0}$, where $c_0 = \sum_{t=1}^{z} c_t$ is the total number of judges who rank item i higher than item j, and $d_0 = \sum_{t=1}^{z} d_t$ is the total number of judges who rank item j higher than item i.

The overall performance measure for tree T is defined as the average of AUC over all item pairs $\text{AUC}(T) = \frac{2}{m(m-1)} \sum_{i<j} A_{ij}$ for $i, j = 1, \ldots, m$, where m is the number of items to be ranked. Tree model with larger AUC reflects better predictive ability. Standard error of the AUC for a two-class problem is given in [11]. However, for multiclass measure of AUC, they recommended using bootstrap method to estimate the standard error because the derivation is difficult.

5 Case Studies

In this section, we illustrate the tree methodology for ranking data described in Sects. 3 and 4. The first example involves a toy dataset and a simulation ranking data generated for the second case study. The third section details the performance evaluation of the tree algorithm on complete and partially ranked dataset. The last part is a real data analysis of political values priority among Europeans.

5.1 Example

A toy example is given to illustrate the performance of different impurity measures. Ranking data is a high dimension data, especially when the number of items is large. Top-k and g-wise measures reduce the dimension of ranking data, and it may lead to information loss. For example, for a ranking dataset of m items, pairwise measure reduces the dataset from $m!--1$ parameters into C_2^m parameters and top-k measure reduces it into $P_k^m --1$ parameters. Generally speaking, information loss due to pairwise measure is larger because number of parameters is less. However, it may not always be the case.

Suppose we have 32 observations in a ranked dataset of three items. The Gini index of top-3 and pairwise measured data in the parent node t are $i_\tau^{(3)}(t) = 0.8320$ and $i_w^{(2)}(t) = 0.4987$, respectively. Now consider 2 candidate splits by variable A and B that partition the data into the left and right node as below. The Gini reductions of the two splits based on different measures are computed. No difference is observed in the two splits by viewing the data using top-3 measure as both splits give the same Gini reduction of 0.0977.

However, if the impurity reduction is evaluated by pairwise measure, the difference between them will stand out ($\Delta i(A,t) = 0.0977$ and $\Delta i(B,t) = 0.0326$). It is trivial when the preference is presented in paired rankings based on the two splits. Clearly, a split based on variable A is preferred. The pairwise measure selects the correct splitting variable, whereas top-3 measure cannot distinguish between the two splits.

Ranking (π)	$N_\pi^\tau(t)$ in Node t	$N_\pi^\tau(t_L)$ in Left Node		$N_\pi^\tau(t_R)$ in Right Node	
		Split A	Split B	Split A	Split B
1≻2≻3	5	5	5	0	0
1≻3≻2	5	5	0	0	5
2≻1≻3	5	0	5	5	0
2≻3≻1	5	0	0	5	5
3≻1≻2	6	3	3	3	3
3≻2≻1	6	3	3	3	3

Paired Ranking (π)	$N_\pi^w(t_L)$ in Left Node		$N_\pi^w(t_R)$ in Right Node	
	Split A	Split B	Split A	Split B
1≻2	13	8	3	8
1≻3	10	10	5	5
2≻3	5	10	10	5

5.2 Simulation Study

In this study, pairwise and top-3 measures are compared using a simulation ranking dataset of 3 items. There are two independent variables, namely A and B, which both have two levels 0 and 1. A total of 100 simulation trials has been carried out. In each trial, 40 samples are simulated. Each sample has equal chance of having A to be 0 or 1. If $A = 0$, then the sample must have B to be 0. If $A = 1$, then B will have half chance to be 0 and half chance to be 1. This results in three possible independent variables combinations $(0, 0)$, $(1, 0)$ and $(1, 1)$ with probability 0.5, 0.25 and 0.25.

The ranking responses are generated by ordering the random utility U_i of item i. It is assumed that U_i depends on the two independent variables via $\lambda_{i0} + \lambda_{i1}A + \lambda_{i2}AB + \epsilon_i$ for $i = 1, 2, 3$. Here ϵ_i is a random noise and follows iid $N(0, 1)$. The simulation study is carried out with $(\lambda_{10}, \lambda_{11}, \lambda_{12}, \lambda_{20}, \lambda_{21}, \lambda_{22}, \lambda_{30}, \lambda_{31}, \lambda_{32}) = (0, 2, 0, 1, 0, -2, 2, -3, 2)$. In this setting, the corresponding modal rankings of (A, B) = (0, 0), (1, 0), and (1, 1) are $(3 \succ 2 \succ 1)$, $(1 \succ 2 \succ 3)$, and $(1 \succ 3 \succ 2)$, respectively.

It is trivial that the root node should be split according to variable A. For every simulation trial, Gini reduction of the two candidate splits based on pairwise and top-3 measures are calculated to determine which split is preferred. It is found that pairwise measure gives a perfect selection of variable A but top-3 measure mistakenly chooses variable B to split for 19 times. This indicates that pairwise measure performs better in this simulation study. The main reason is that pairwise measure describes a dataset in three parameters $p_w(1, 2|\, t)$, $p_w(1, 3|\, t)$, and $p_w(2, 3|\, t)$, but top-3 measures use five parameters $p_\tau(1, 2, 3|\, t)$, $p_\tau(1, 3, 2|\, t)$, $p_\tau(2, 1, 3|\, t)$, $p_\tau(2, 3, 1|\, t)$, and $p_\tau(3, 1, 2|\, t)$, and thus the standard error of the impurity is larger for top-3 measured data.

5.3 Experiments and Results

In this section, we test the performance of our proposed tree model on both complete and incomplete preference/ranking data. Due to the lack of suitable real-world datasets, artificial data is generated by adopting a setting described in [14] from a Naïve Bayes classifier. Four synthetic ranking datasets are simulated on the basis of the following multiclass datasets – iris, car, dermatology and thyroid from the UCI

Table 1 Summary of four synthetic ranking datasets

Dataset	No. of instances	No. of classes	No. of attributes	Attribute characteristics
Iris	150	3	4	real
Car	1,728	4	6	categorical
Dermatology	358	6	34	categorical, integer
Thyroid	3,772	3	21	categorical, real

Respiratory of machine learning databases [1]. Table 1 provides the description of the datasets, their characteristics and attributes in this experimental evaluation.

In a nutshell, the data generation works as follows: A Naive Bayes classifier is first trained on each multiclass dataset. Next, for each instance, the set of ranked alternatives is ordered with respect to the predicted class probabilities. The most preferred alternative is the one with the highest predicted class probability and so on. In the case of ties, alternatives with lower indices are ranked higher. For the sake of comparison, partially ranked datasets are further produced based on the complete ranked datasets. We independently sample some instances and at the same time determine the number of missing ranks for those sampled instances from a uniform distribution. In this study, tree performance was evaluated with six scenarios of different proportions of missing data, including the noise-free scenario (0% of missing data) and 10–50% missing data with an increment of 10%.

For each scenario, the dataset is randomly partitioned into learning data (70%) for building the decision tree and testing data (30%) for evaluating the algorithm. We train the tree classifier using top-3 Gini, top-3 Entropy, pairwise Gini, and pairwise Entropy as the impurity measures. The accuracy is derived in terms of the averaged

Table 2 Summary of experimental results on four synthetic Datasets

	Proportion of missing data					
	0%	10%	20%	30%	40%	50%
Iris						
G	0.980(0.006)	0.963(0.007)	0.966(0.010)	0.961(0.007)	0.952(0.009)	0.945(0.008)
E	0.993(0.004)	0.987(0.004)	0.979(0.004)	0.972(0.006)	0.966(0.004)	0.955(0.006)
PG	0.975(0.012)	0.962(0.008)	0.961(0.009)	0.956(0.009)	0.953(0.007)	0.948(0.005)
PE	0.992(0.005)	0.987(0.004)	0.979(0.004)	0.971(0.006)	0.965(0.004)	0.959(0.006)
Car						
G	0.873(0.013)	0.853(0.009)	0.841(0.005)	0.826(0.006)	0.810(0.012)	0.801(0.008)
E	0.885(0.007)	0.866(0.007)	0.855(0.005)	0.839(0.005)	0.828(0.003)	0.813(0.005)
PG	0.877(0.009)	0.853(0.005)	0.841(0.007)	0.826(0.005)	0.814(0.007)	0.800(0.006)
PE	0.886(0.006)	0.854(0.009)	0.841(0.006)	0.828(0.004)	0.818(0.007)	0.804(0.006)
Dermatology						
G	0.834(0.014)	0.828(0.018)	0.808(0.019)	0.782(0.017)	0.779(0.029)	0.743(0.023)
E	0.857(0.011)	0.841(0.013)	0.819(0.015)	0.801(0.012)	0.792(0.010)	0.789(0.010)
PG	0.853(0.011)	0.836(0.011)	0.826(0.019)	0.816(0.014)	0.802(0.013)	0.787(0.018)
PE	0.873(0.008)	0.847(0.015)	0.839(0.007)	0.822(0.014)	0.814(0.012)	0.798(0.014)
Thyroid						
G	0.902(0.005)	0.875(0.004)	0.858(0.006)	0.835(0.006)	0.821(0.006)	0.805(0.009)
E	0.910(0.005)	0.879(0.008)	0.862(0.007)	0.841(0.010)	0.831(0.011)	0.806(0.009)
PG	0.911(0.005)	0.881(0.004)	0.857(0.008)	0.836(0.004)	0.815(0.008)	0.793(0.015)
PE	0.909(0.004)	0.879(0.010)	0.864(0.007)	0.842(0.010)	0.831(0.011)	0.809(0.009)

Remark: For each of the four synthetic datasets, the mean AUC of four impurity measures (*G*: Top-3 Gini, *E*: Top-3 Entropy, *PG*: Pairwise Gini, *PE*: Pairwise Entropy) are reported and their corresponding standard deviations are shown in the brackets.

AUC over ten repetitions. Table 2 summarizes the detailed AUC numbers and the corresponding standard errors of the 24 trials of each synthetic dataset.

As shown below, our experimental results are encouraging. High means of AUC are achieved for the four complete artificial ranked datasets using various impurity measures: Iris (0.980–0.992); Car (0.873–0.886); Dermatology (0.834–0.873); and Thyroid (0.902–0.911). The box plots in Fig. 1 depict the performance of the four impurity measures toward missing rankings for each of the four datasets. It can be observed that all methods produced comparable classifier performances. As expected, higher proportion of missing data tends to cause greater deteriorations in performance of the decision tree. However, the drop in accuracy is not serious. The average decrease in AUC per 10% increment of missing data for Iris, Car, Dermatology, and Thyroid are 0.007, 0.015, 0.015, and 0.021, respectively.

5.4 European Value Priority Data

The partially ranked dataset was obtained from the International Social Service Programme (ISSP) in 1993 [16]. It mainly focused on value orientations, attitudes, beliefs, and knowledge concerning nature and environmental issues, and included the so-called Inglehart Index, a collection of four indicators of materialism/ post-materialism as well. Respondents were asked to pick the most important and the second most important goals for their Government from the following four alternatives: (a) Maintain order in nation [ORDER]; (b) Give people more to say in Government decisions [SAY]; (c) Fight rising prices [PRICES] (d) Protect freedom of speech [SPEECH]. The survey gave a ranked dataset of 5,737 observations with top choice and top-2 rankings. In addition, the data provide some judge-specific characteristics and they are applied in tree partitioning. The candidate splitting variables are summarized in Table 3.

Respondents can be classified into value priority groups on the basis of their top 2 choices among the four goals. "Materialist" corresponds to individual who gives priority to ORDER and PRICES regardless of the ordering, whereas those who choose SAY and SPEECH will be termed "post-materialist". The last category comprises judges giving all the other combinations of rankings, and they will be classified as holding "mixed" value orientations.

Inglehart's thesis of generational based values has been influential in political science since the early 1970s. He has argued that value priorities were shifting profoundly in economically developed Western countries, from concern over sustenance and safety needs toward quality of life and freedom of self-expression, thus from a materialist orientation to a post-materialist orientation [15]. In this analysis, we study the Inglehart hypothesis in five European countries by our decision tree approach, which helps identifying the attributes that affecting Europeans' value priority.

The data is divided randomly into 2 sets, 70% to the learning set for growing the initial tree and finding the best pruned subtree for each of the four splitting criteria;

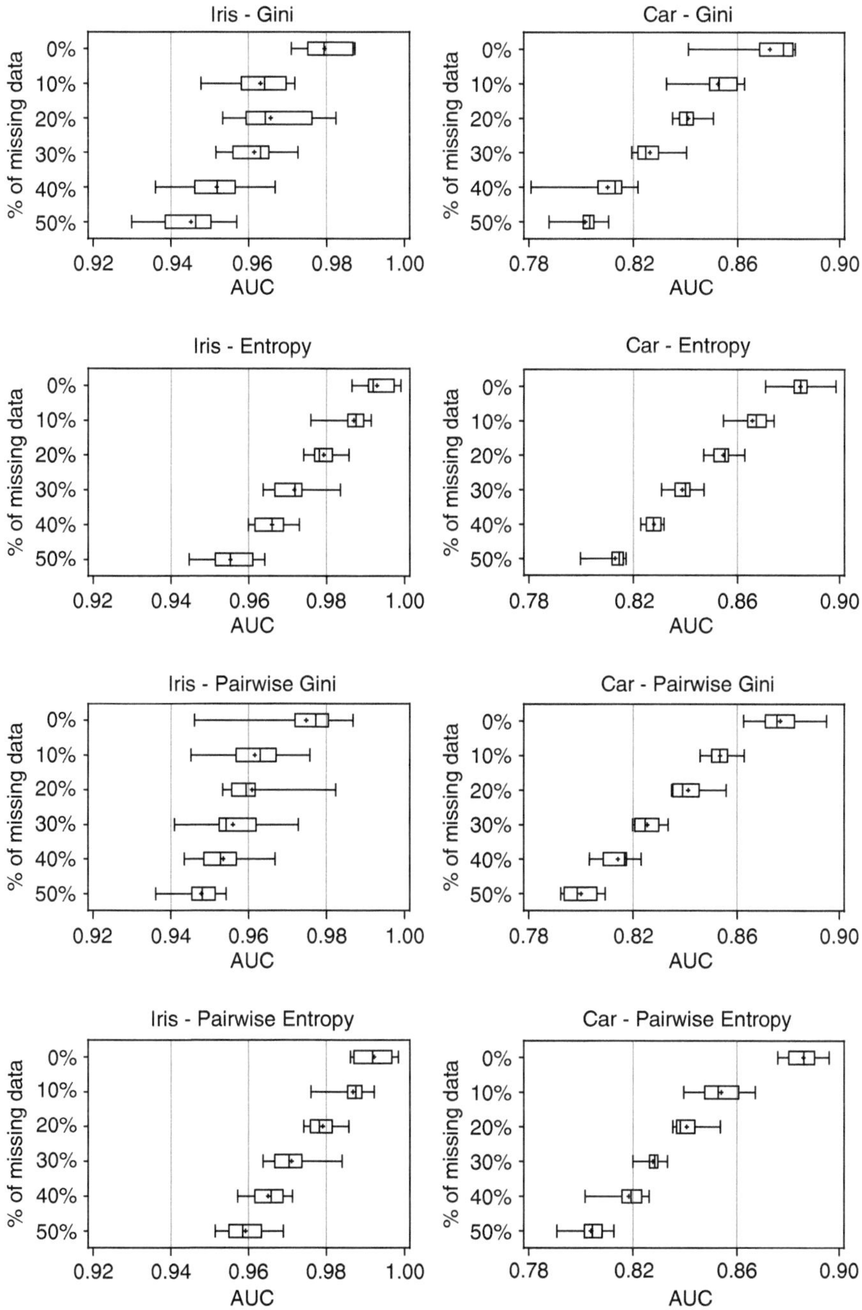

Fig. 1 (Continued)

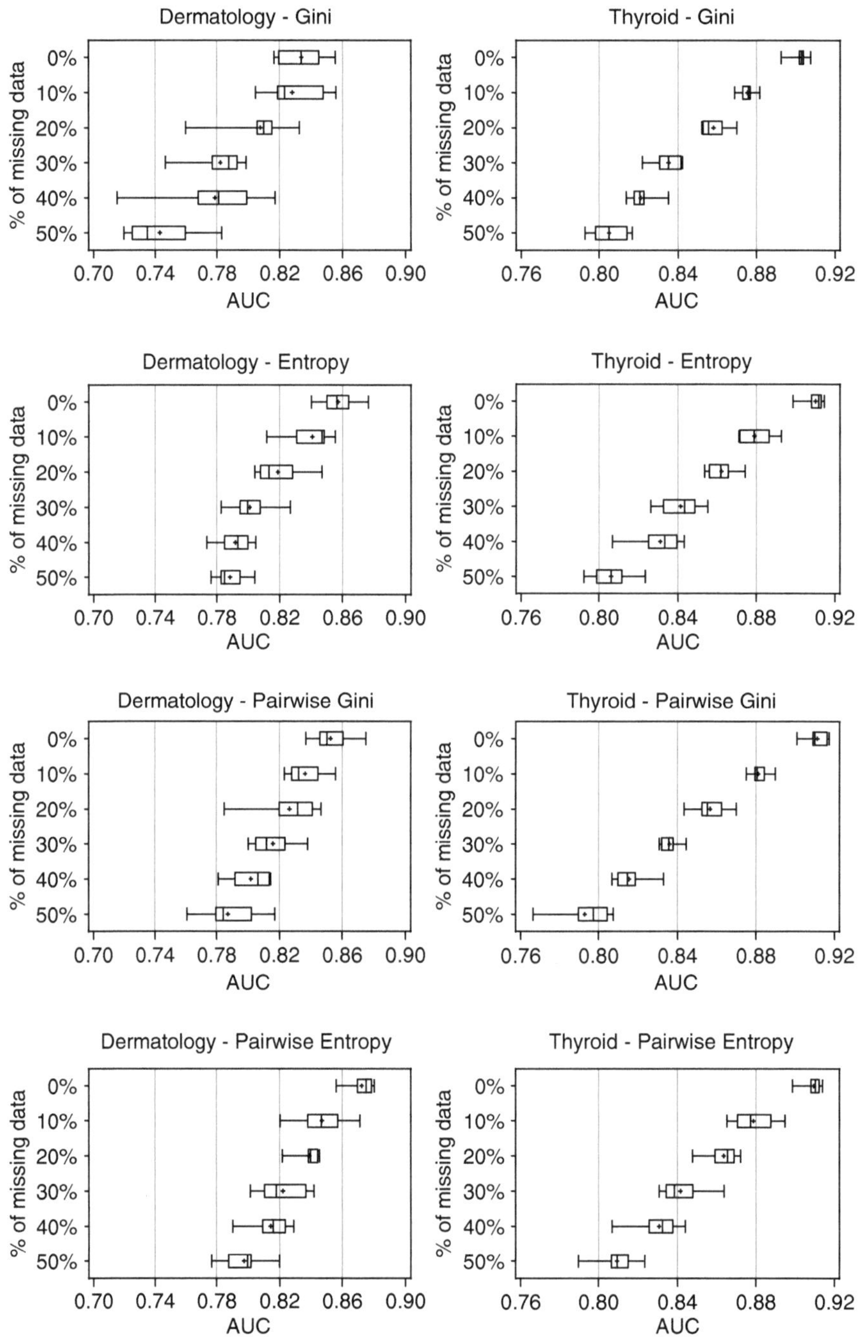

Fig. 1 Comparison of experimental results of four synthetic datasets on various impurity measures. The cross plotted inside the box of the box plot represents the mean AUC of each scenario

Table 3 Description of European ranking data of political values

Attribute	Description/Code	Type	No. of possible values
Country	West Germany = 1, East Germany = 2, Great Britain = 3, Italy = 4, Poland = 5	Nominal	5
Gender	Male = 1, Female = 2	Binary	2
Education	0–10 years = 1, 11–13 years = 2, 14 or more years = 3	Ordinal	2
Age	Value ranges from 15 to 91	Interval	76
Religion	Catholic and Greek Catholic = 1, Protestant = 2, Others = 3, None = 4	Nominal	4

Table 4 Summary of the best pruned subtrees of 4 splitting criteria

Method	Avg. AUC	S.E	AUC	No. of Leaves	Depth
Top-2 entropy	0.61947	0.0056	0.62951	12	5
Pairwise Gini	0.61896	0.0058	0.62902	12	5
Pairwise entropy	0.61857	0.0056	0.62709	11	5
Top-2 Gini	0.61425	0.0063	0.61931	9	4

and 30% to the testing set for performance assessment and selection of the splitting criterion to build the final tree. As decision tree is an unstable classifier, small changes in the learning set can cause major changes in the fitted tree structure; we therefore repeat this procedure 50 times and compare the four splitting criteria with their averaged AUC. Lastly, the final tree model is created using the entire dataset for interpretation. Notice that the testing set is not involved in the tree building process and pruned subtree selection, therefore it serves as an out-of-sample dataset for model comparison. The four splitting criteria for rankings include top-2 and pairwise measure of Gini and entropy. Here, we apply pairwise and top-2 measures as the data only contain individuals' preference orders of the most and the second most desirable goals. Table 4 shows the averaged AUC and their standard error of the best pruned subtrees for each splitting criterion based on 50 repetitions. The tree structure and performance of the final models are also presented in the same table. Figure 2 displays the six ROC curves of each item pairs that arise from the top-2 entropy tree. The tree did a better job of predicting the item pair "SAY vs PRICES", but poor for "SAY vs SPEECH". We do not illustrate the ROC curves of other trees as the performance of the four trees is comparable and it is hard to distinguish them in the graph.

The four tree models are found to have similar node partitions. The root node is split according to whether the judges came from Poland or not (country $= 5$ vs $\neq 5$). At the second level, the splits are based on age. For Polish, the respondents are divided with the rule "age<59?", while the remaining judges are split according to age<53 or not. Further partitions involved education level, country, and age. The factors religion and gender seem not to be influential. It is observed that in the learning phase, top-2 Gini tends to give a smaller tree, while top-2 entropy gives a more complicated tree on average. Based on the assessment criterion, the top-2

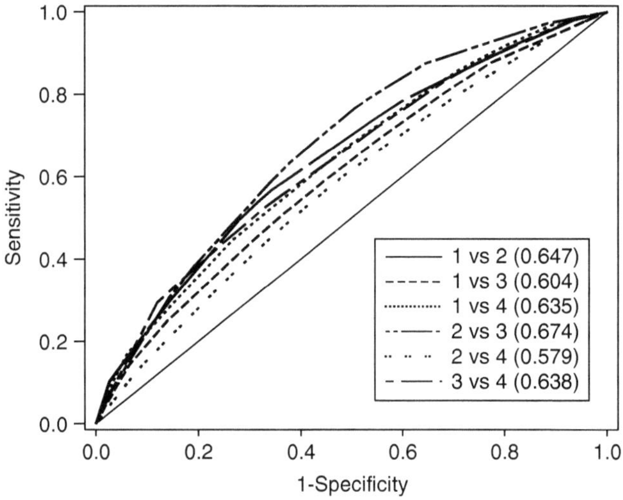

Fig. 2 ROC curves of top-2 entropy tree. The four value items are coded as follows 1 = [ORDER], 2 = [SAY], 3 = [PRICES], and 4 = [SPEECH]. The 45° diagonal line connecting (0,0) and (1,1) is the ROC curve corresponding to random chance. Given next to the legends are the areas under the corresponding dashed ROC curves

entropy tree is chosen as the best model and it is applied for further analysis. As shown in Fig. 3, this tree has 5 levels of depth and 12 leaves. For sake of brevity, we do not show the other three tree structures. A summary of the terminal nodes of the final tree is reported in two tables. Table 5 lists the mean rank of the four political goals and the three most frequent top-2 ranking, whereas Table 6 shows the individuals' value priority and the proportion of six pairs of political goals in each leaf node.

We now turn to examine the attributes and interaction effects based on the final tree model. In Poland, individuals were more likely to favor materialistic items ORDER and PRICES (in leaves 5, 8, and 9). In East Germany, judges appeared to support ORDER and SAY more, particularly those older generations gave higher priority to ORDER (in leaf 12). Respondents of West Germany showed stronger emphasis on SAY. Those better educated West Germans were more post-materialist than the lower educated ones as they preferred SAY and SPEECH, rather than the other two materialist items (in leaf 15). Mixed value orientations were anchored in British because all the related leaf nodes give us a preference prediction of ORDER $\succ$ SAY or SAY $\succ$ ORDER.

The result can be summed up in two observations: (a) Despite some cross-national differences, our findings do not deviate much from Inglehart's theory, which claimed that societies embrace post-materialist values as they move toward more economic security and affluence. The older European generations experienced economic and social insecurity in their pre-adult years during World War II, they thus gave stronger concern on the materialist values compared to the younger

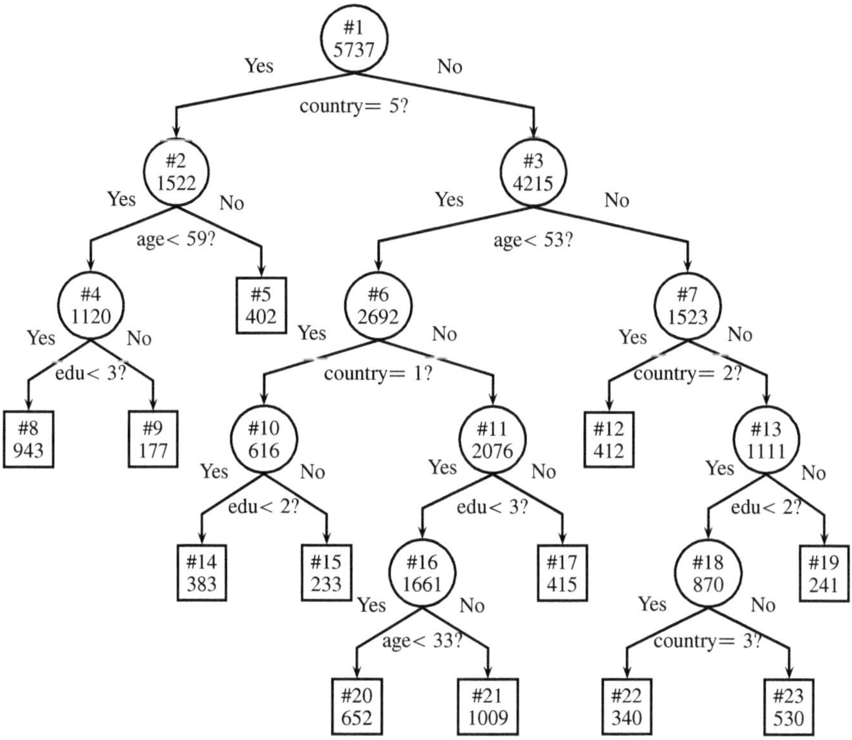

Fig. 3 Tree structure diagram based on top-2 entropy. In each node, the node ID and the number of judges are shown. The splitting rule is given under the node. The abbreviation "edu" stands for the variable education

Table 5 Importance of 4 political values in terminal nodes of top-2 entropy tree

Node(t)	Node	Mean rank				Frequent top-2 ranking		
	Size	Order	Say	Prices	Speech	1st	2nd	3rd
5	402	1.69	3.08	1.99	3.24	1,3 (40.3%)	3,1 (24.9%)	1,2 (7.5%)
8	943	2.10	2.70	1.95	3.26	3,1 (25.7%)	1,3 (22.5%)	3,2 (13.0%)
9	177	2.16	2.49	2.40	2.95	1,3 (17.3%)	3,1 (13.7%)	1,2 (12.2%)
12	412	1.65	2.42	2.63	3.30	1,3 (27.5%)	1,2 (22.2%)	2,1 (18.2%)
14	383	2.33	2.11	2.64	2.92	2,1 (17.0%)	2,3 (12.3%)	2,4 (12.0%)
15	233	2.73	1.86	2.97	2.44	2,4 (25.9%)	4,2 (15.7%)	2,3 (11.7%)
17	415	2.31	2.05	2.80	2.83	2,4 (15.7%)	2,1 (14.9%)	1,2 (14.2%)
19	241	1.88	2.40	2.83	2.89	1,3 (22.1%)	2,1 (14.9%)	1,2 (13.8%)
20	652	2.28	1.95	2.67	3.09	2,1 (20.1%)	2,3 (19.1%)	1,2 (13.4%)
21	1009	2.09	2.20	2.62	3.10	1,3 (18.0%)	1,2 (16.1%)	2,1 (15.8%)
22	340	2.22	2.30	2.40	3.07	1,3 (18.1%)	2,3 (15.9%)	1,2 (12.3%)
23	530	1.83	2.62	2.48	3.07	1,3 (28.2%)	1,2 (16.2%)	3,1 (9.6%)

Remark: In the last three columns, the code 1–4 represents each of the political goals: 1 = [ORDER], 2 = [SAY], 3 = [PRICES], and 4 = [SPEECH]; "i, j" implies goal $i \succ$ goal j and the percentage beside indicates the proportion of instances having the corresponding top-2 ranking in node t.

Table 6 Value priority and pairwise probabilities in leaf nodes of top-2 entropy tree

Node(t)	Node Size	Value	Pairwise probabilities					
			$p_w(1,2\|t)$	$p_w(1,3\|t)$	$p_w(1,4\|t)$	$p_w(2,3\|t)$	$p_w(2,4\|t)$	$p_w(3,4\|t)$
5	402	M	83.1%	60.8%	87.3%	21.0%	53.7%	83.1%
8	943	M	65.3%	45.9%	79.3%	31.0%	64.8%	82.1%
9	177	M	58.8%	55.9%	69.2%	47.7%	61.9%	63.8%
12	412	B	67.5%	77.7%	90.0%	53.9%	72.0%	68.3%
14	383	B	27.4%	32.1%	41.0%	37.6%	40.1%	36.0%
15	233	P	44.1%	57.8%	64.8%	63.6%	69.5%	57.3%
17	415	B	44.5%	63.1%	61.0%	68.3%	71.1%	51.0%
19	241	B	60.8%	76.3%	74.5%	58.3%	62.2%	51.9%
20	652	B	41.6%	59.7%	70.6%	68.9%	77.3%	61.5%
21	1009	B	52.6%	64.5%	74.1%	60.0%	72.8%	62.6%
22	340	B	51.3%	55.6%	71.2%	52.8%	68.2%	67.9%
23	530	M	69.0%	69.1%	78.7%	45.7%	61.8%	66.8%

Remark: The third column "Value" shows the value priority group of judges in each leaf node, where B=Mixed values; M=Materialist and P=Post-materialist. For columns 4 to 9, the four political goals are labeled as: 1=[ORDER], 2=[SAY], 3=[PRICES], and 4=[SPEECH].

cohorts. Younger post-war generations developed post-materialist values as they grew up during periods of relative prosperity. (b) There is a clear tendency in each country for the higher educated to be the more postmaterialist groups. Duch and Taylor [8] stated that the post-materialist items tap certain fundamental democratic values, such as liberty and rights consciousness. The better educated would have had more opportunity to learn to appreciate such principles, and thus they will prefer post-materialist items more.

For comparison, we tried to learn a decision tree in another setting for this dataset, by transforming the top-2 ranking problem into six binary classification problems of pairwise preferences ($1 \succ 2$ vs $2 \succ 1, \ldots, 3 \succ 4$ vs $4 \succ 3$). However, due to large proportion of ties in some pairwise preferences (61.6% for $\{2, 4\}$ and 71.5% for $\{3, 4\}$), not all information can be utilized to build this alternative tree model and so the model comparison is not relevant.

6 Conclusion and Future Work

We have investigated the use of decision tree model for analyzing ranking data, which makes explanation of individuals' rank-order preference differences easier compared to existing parametric ranking models and algorithms in label ranking and preference learning, especially when nonlinearity and high-order interactions are involved in the studied covariates. It is noteworthy that our tree methodology includes the multinomial tree as a special case and it can accommodate inconsistent rankings, as well as tie rankings. We have proposed two impurity measures, namely g-wise and top-k measures, to evaluate the goodness of split for ranking data. Examples and simulations showed that the established impurity functions effectively measure the node heterogeneity. It is interesting to find that pairwise impurity measure in some instances is more preferred than top-k measure. The main reason is that pairwise measure describes a dataset with less parameters, and thus the corresponding standard error of the impurity is smaller.

The tree algorithm is easy to implement and flexible that we can specify the number of ranks used in splitting for the top-k measures and the value of g in the g-wise measure according to the ranking data being analyzed. To assess the predictive performance of the final tree, the AUC is used for the purpose. Experimentation has shown that our tree classifier is capable of dealing with both complete and incomplete preference data without much loss in accuracy. We further illustrate the proposed methodology through a real application of European values priority data. The AUCs of the four competitive trees are compared. Nevertheless, we are not trying to draw any conclusion about which splitting criterion is more superior, as it is definitely related to the types of the observed rankings.

In future work, we plan to generalize the method to other data, such as paired comparison data and rating data. In addition, other impurity measures or splitting criteria will be investigated to further improve the predictive accuracy of the tree classifier. Another more straightforward extension of our technique will be tree-structured ranking model. The two-stage modeling approach helps us to capture those effects of attributes that are not detected by the tree as primary node splitters. This compensates the weakness of tree classifier in representing strong linear covariates structure.

Acknowledgements The research of Philip L.H. Yu was supported by a grant from the Research Grants Council of the Hong Kong Special Administrative Region, China (Project No. HKU 7473/05H).

References

1. A. Asuncion, D.J. Newman, UCI Machine Learning Repository [http://www.ics.uci.edu/~mlearn/MLRepository.html]. Irvine, CA: University of California, School of Information and Computer Science (2007)
2. A.P. Bradley, The use of the area under the ROC curve in the evaluation of machine learning algorithms. Pattern Recognit. **30**, 1145-Ű1159 (1997)
3. L. Breiman, J.H. Friedman, R.A. Olshen, C.J. Stone. *Classification and Regression Trees* (Belmont, California: Wadsworth, 1984)
4. W. Cheng, J. Hühn, E. Hüllermeier, Decision tree and instance-based learning for label ranking, in *Proceedings of the 26th International Conference on Machine Learning (ICML 2009)* (Montreal, Canada, 2009)
5. P.A. Chou, Optimal partitioning for classification and regression trees. IEEE Trans. Pattern Anal. Mach. Intell. **13**, 340–354 (1991)
6. D.E. Critchlow, M.A. Fligner, J.S. Verducci, Probability models on rankings. J. Math. Psychol. **35**, 294–318 (1991)
7. C. Drummond, R.C. Holte, What ROC curves can't do (and cost curves can), in *Proceedings of the 1st Workshop on ROC Analysis in AI* (Valencia, Spain, 2004), pp. 19-Ű26

8. R.M. Duch, M.A. Taylor, Postmaterialism and the economic condition. Am. J. Pol. Sci. **37**, 747–778 (1993)
9. J. Fürnkranz, E. Hüllermeier, Pairwise preference learning and ranking, in *Proceedings of the 14th European Conference on Machine Learning (ECML-03)* (Springer, Cavtat, Croatia, 2003), pp. 145–156
10. P. Geurts, L. Wehenkel, A. Florence, Kernelizing the output of tree-based methods, in *Proceedings of the 23rd International Conference on Machine Learning (ICML-06)*, (Pittsburgh, Pennsylvania, 2006), pp. 345–352
11. D.J. Hand, R.J. Till, A simple generalisation of the area under the ROC curve for multiple class classification problems. Mach. Learn. **45**(2), 171–186 (2001)
12. J.A. Hanley, B.J. McNeil, The meaning and use of the area under a receiver operating characteristic (ROC) curve. Radiology **143**, 29-Ű36 (1982)
13. E. Hüllermeier, J. Fürnkranz, On Minimizing the Position Error in Label Ranking, in *Proceedings of the 17th European Conference on Machine Learning (ECML-07)* (Springer, Warsawa, Poland, 2007), pp. 583–590
14. E. Hüllermeier, J. Fürnkranz, W. Cheng, K. Brinker, Label ranking by learning pairwise preferences. Artif. Intell. **172**(16–17), 1897–1916 (2008)
15. R. Inglehart, *The Silent Revolution: Changing Values and Political Styles among Western Publics* (Princeton Univerity Press, Princeton, 1977)
16. R. Jowell, L. Brook, L. Dowds, *International Social Attributes: the 10th BSA Report* (Dartmouth Publishing, Aldershot, 1993)
17. M.G. Karlaftis, Predicting mode choice through multivariate recursive partitioning. J. Trans. Eng. **130**(22), 245–250 (2004)
18. G.V. Kass, An exploratory technique for investigation large quantities of categorical data. Appl. Stat. **29**, 119–127 (1980)
19. W.-Y. Loh, Y.-S. Shih, Split selection methods for classification trees. Statistica Sinica **7**, 815–840 (1997)
20. J.I. Marden, Analyzing and Modeling Rank Data (Chapman & Hall, 1995)
21. J.R. Quinlan, Induction of decision trees. Mach. Learn. **1**, 81–106 (1986)
22. J.R. Quinlan. *C4.5: Programs for Machine Learning* (Morgan Kaufmann, 1993)
23. B.D. Ripley, *Pattern Recognition and Neural Networks* (Cambridge University Press, Cambridge, 1996)
24. L. Rokach, O. Maimon, Decision trees, in *The Data Mining and Knowledge Discovery Handbook*, ed. by O. Maimon, L. Rokach (Springer, Berlin, 2005), pp. 165–192
25. R. Siciliano, F. Mola, Multivariate data analysis and modeling through classification and regression trees. Comput. Stat. Data Anal. **32**, 285–301 (2000)

Appendix

In this appendix, the proof of Theorems 1, 2, and 3 will be provided. Recall that the size of the ranked item set is m. We hereafter omit t, which stands for node t to simplify the notation of proportion and impurity.

Proof of Theorem 1

Proof. Denote $p_w(\pi), \forall \pi \in \mathcal{R}^{m,g}$ as $\boldsymbol{p}_\gamma$, and thus we have $i_w^{(g)} = \phi_w^{(g)}(\boldsymbol{p}_\gamma)$. Also, define B_g to be a g-item subset from $\{1, \ldots, m\}$

1.1. Let $\pi_a, \pi_b \in \mathcal{R}^{m,g}$. If $\pi_a \neq \pi_b$, it is trivial that $\left(\frac{\partial \phi_w^{(g)}(\boldsymbol{p}_\gamma)}{\partial p_w(\pi_a)\partial p_w(\pi_b)}\right) = 0$.

If $\pi_a = \pi_b$, since $\phi_w^{(g)}(\boldsymbol{p}_\gamma)$ can be written as

$$\sum_{B_g \in \boldsymbol{\Omega}_g^m} \sum_{\pi \in \mathcal{R}_{B_g}^{m,g}} f(p_w(\pi)) \tag{6}$$

and $\sum_{\pi \in \mathcal{R}_{B_g}^{m,g}} p_w(\pi) = 1$, hence $f(p_w(\pi))$ is concave, and the permutation π_a (WLOG assume $\pi_a = \{\pi_{a_1}, \ldots, \pi_{a_g}\}$) will only appear once in the function $\phi_w^{(g)}(\boldsymbol{p}_\gamma)$, therefore $\left(\frac{\partial \phi_w^{(g)}(\boldsymbol{p}_\gamma)}{\partial p_w(\pi_a)\partial p_w(\pi_b)}\right) = \left(\frac{\partial f(p_w(\pi_a))}{\partial p_w(\pi_a)\partial p_w(\pi_a)}\right) \leq 0$ (sum of concave functions is concave).

1.2. Since $\phi_w^{(g)}(\boldsymbol{p}_\gamma)$ can be written as (6), and note that $\sum_{\pi \in \mathcal{R}_{B_g}^{m,g}} p_w(\pi) = 1$, $\forall B_g \in \boldsymbol{\Omega}_g^m$, therefore minimizing $\phi_w^{(g)}(\boldsymbol{p}_\gamma)$ will be equivalent to minimize $\sum_{\pi \in \mathcal{R}_{B_g}^{m,g}} f(p_w(\pi)), \forall B_g \in \boldsymbol{\Omega}_g^m \in \{1, \ldots, m\}$. The condition will be for each of $B_g \in \boldsymbol{\Omega}_g^m$, one of the ranking probabilities $p_w(\pi), \pi \in \mathcal{R}_{B_g}^{m,g}$ equals 1.

Note that g-wise measured data are derived from full ranking, and some combinations of g-wise data are intransitive. For example, it is impossible to have $p_w(1,2) = 1$, $p_w(2,3) = 1$, and $p_w(3,1) = 1$. Another contradictory example is $p_w(1,2,3) = 1$ and $p_w(3,2,4) = 1$. Therefore, by eliminating those intransitive g-wise data combinations, the minimizing condition of g-wise impurity functions can be reduced to: one of the probability $P(y_1 \succ y_2 \succ \ldots \succ y_m)=1$ and all other full ranking probabilities equal to zero. The proof is given in Theorem 2.

1.3. Since $\phi_w^{(g)}(\boldsymbol{p}_\gamma)$ can be written as (6), and note that $\sum_{\pi \in \mathcal{R}_{B_g}^{m,g}} p_w(\pi) = 1$, $\forall B_g \in \boldsymbol{\Omega}_g^m$, therefore maximizing $\phi_w^{(g)}(\boldsymbol{p}_\gamma)$ will be equivalent to maximize $\sum_{\pi \in \mathcal{R}_{B_g}^{m,g}} f(p_w(\pi)), \forall B_g \in \boldsymbol{\Omega}_g^m \in \{1, \ldots, m\}$. The condition will be for each of $B_g \in \boldsymbol{\Omega}_g^m$, all ranking probabilities $p_w(\pi), \pi \in \mathcal{R}_{B_g}^{m,g}$ equal to $1/g!$.

The g-wise impurity function is maximized at all $p_w(\pi)$ equal. However, unlike the above minimum case, it cannot be generalized to the case when all full ranking probabilities are uniformly distributed. For example, a full ranking of 3 items with $p_\tau(1,2,3) = p_\tau(3,2,1) = 0.5$ implied pairwise measure of $p_w(1,2) = p_w(2,1) = p_w(1,3) = p_w(3,1) = p_w(2,3) = p_w(3,2) = 0.5$. However, under special condition (see Theorem 3), $p_w(\pi) \forall \pi \in \mathcal{R}^{m,g}$ equal implies evenly distributed full ranking probabilities.

1.4 When item y_i and y_j are swapped, all values of $p_w(\pi)$ with $y_i, y_j \in \pi$ and $p_w(\pi)$ with $y_i, y_j \notin r$ remain the same. All values of $p_w(\pi)$ with $y_i \in \pi, y_j \notin \pi$ and $p_w(\pi)$ with $y_j \in \pi, y_i \notin \pi$ exchange. Afterall, there is no effect in $i_w^{(g)}$.

Theorem 2. *A set of pairwise measured data with m items to be ranked, with all* $\{p_w(i,j) : i \neq j;\, i, j \in [1,m]\}$ *equals to either 0 or 1 and are transitive, represent a unique full ranking with probability* $P(y_1 \succ y_2 \succ \ldots \succ y_m) = 1$, *for all distinct* $y_1, y_2, \ldots, y_m \in [1,m]$.

Proof. There are two cases induced by the set of paired comparison probabilities $\{p_w(i,j) : i \neq j; i, j \in [1,m]\}$, which have value of either 0 or 1: (i) if there exists distinct $y_1, y_2, \ldots, y_b \in [1, m]$ such that $p_w(y_1, y_2) = p_w(y_2, y_3) = \ldots = p_w(y_b, y_1) = 1$, then intransitive ranking occurs; (ii) no cyclic occurs in the rank structure. The set of $\{p_w(i, j)\}$ with cyclic structure cannot be induced by a full ranking, because a full ranking satisfies $y_1 \succ y_2$, $y_2 \succ y_3$, ..., and $y_{b-1} \succ y_b$ should imply $y_1 \succ y_b$.

Note that $p_w(i, j) = 1$ and $p_w(j, k) = 1$ imply $p_w(i, k) = 1$, then $\{p_w(i, j)\}$ are weakly stochastically transitive. By Theorem 2 of [6], the full ranking induced by $\{p_w(i, j)\}$ is strongly unimodal. Assume $P(y_1 \succ y_2 \succ \ldots \succ y_m), y_1, y_2, \ldots, y_m \in [1, m]$ is implied by $\{p_w(i, j) : i \neq j; i, j \in [1, m]\}$, and $P(y_1 \succ y_2 \succ \ldots \succ y_b \succ \ldots \succ y_c \succ \ldots \succ y_m) > 0$, where $b < c$. For any different full ranking which the bth best item is c and the cth best item is b, the probability should be zero, because $P(y_1 \succ y_2 \succ \ldots \succ y_b \succ \ldots \succ y_c \succ \ldots \succ y_m) > 0 \Rightarrow p_w(y_b, y_c) = 1$ and hence $p_w(y_c, y_b) = 0$. Therefore, only $y_1 \succ y_2 \succ \ldots \succ y_b \succ \ldots \succ y_c \succ \ldots \succ y_m$ is possible. Hence, the transitive pairwise probabilities $\{p_w(i, j)\}$ with equal to either 0 or 1 will lead to one and only one possible full ranking with $P(y_1 \succ y_2 \succ \ldots \succ y_m) = 1$, from some distinct $y_1, y_2, \ldots, y_m \in [1, m]$.

Theorem 3. *A set of pairwise measured data with m items to be ranked, with all $\{p_w(i, j) : i \neq j; i, j \in [1, m]\}$ equal to 0.5 and are transitive, represent a full ranking with probability $P(y_1 \succ y_2 \succ \ldots \succ y_m) = 1/m!$ for all distinct $y_1, y_2, \ldots, y_m \in [1, m]$, if the ranking is derived from an independent random utility (IRU) model, and the cdf of random utility $U_i, i \in [1, m]$ can be written as $G_i(x) = G(x - \mu_i)$.*

Proof. Let $g_i, i \in [1, m]$ be the pdf of U_i. Under IRU model, $p_w(y_1, \ldots, y_m) = P(U_{y_1} > \ldots > U_{y_m})$. Consider three items i, j, and k. $p_w(j, i) = P(U_i < U_j) = P(U_i - U_j < 0) = \int_{-\infty}^{\infty} P(U_i - x < 0) g_j(x) dx = \int_{-\infty}^{\infty} G_i(x) dG_j(x) = \int_{-\infty}^{\infty} G(x - \mu_i) dG(x - \mu_j)$. Similarly, $p_w(j, k) = \int_{-\infty}^{\infty} G(x - \mu_k) dG(x - \mu_j)$. Since $p_w(j, i) = p_w(j, k) \forall i, k \neq j, \mu_i = \mu_k \forall i, k \in [1, m]$, and this implies $U_1, \ldots, U_m$ are iid. Therefore, $P(U_{y_1} > \ldots > U_{y_m})$ equals for all distinct $y_1, y_2, \ldots, y_m \in [1, m]$.

Co-Regularized Least-Squares for Label Ranking

Evgeni Tsivtsivadze, Tapio Pahikkala, Jorma Boberg, Tapio Salakoski, and Tom Heskes

Abstract Situations when only a limited amount of labeled data and a large amount of unlabeled data are available to the learning algorithm are typical for many real-world problems. To make use of unlabeled data in preference learning problems, we propose a semisupervised algorithm that is based on the multiview approach. Our algorithm, which we call Sparse Co-RankRLS, minimizes a least-squares approximation of the ranking error and is formulated within the co-regularization framework. It operates by constructing a ranker for each view and by choosing such ranking prediction functions that minimize the disagreement among all of the rankers on the unlabeled data. Our experiments, conducted on real-world dataset, show that the inclusion of unlabeled data can improve the prediction performance significantly. Moreover, our semisupervised preference learning algorithm has a linear complexity in the number of unlabeled data items, making it applicable to large datasets.

1 Introduction

Semisupervised learning algorithms have gained more and more attention in recent years as unlabeled data is typically much easier to obtain than labeled one. *Multi-view* learning algorithms, such as co-training [1], split the attributes into independent sets and an algorithm is learnt based on these different "views". The goal of the learning process consists in finding for every view a prediction function (for the

E. Tsivtsivadze (✉) and T. Heskes
Institute for Computing and Information Sciences, Radboud University Nijmegen
Toernooiveld 1, 6525 ED Nijmegen, The Netherlands
e-mail: evgeni@science.ru.nl, t.heskes@science.ru.nl

T. Pahikkala, J. Boberg, and T. Salakoski
Turku Centre for Computer Science (TUCS),
Department of Information Technology, University of Turku,
Joukahaisenkatu 3-5 B, 20520 Turku, Finland
e-mail: firstname.lastname@it.utu.fi

J. Fürnkranz and E. Hüllermeier (eds.), *Preference Learning*,
DOI 10.1007/978-3-642-14125-6_6,

learning task) performing well on the labeled data of the designated view such that all prediction functions agree on the unlabeled data. Closely related to this approach is the *co-regularization* framework described in [20], where the same idea of agreement maximization between the predictors is central. Briefly stated, algorithms based upon this approach search for hypotheses from different Reproducing Kernel Hilbert Spaces [19], namely views, such that the training error of each hypothesis on the labeled data is small and, at the same time, the hypotheses give similar predictions for the unlabeled data. Within this framework, the disagreement is taken into account through a co-regularization term. Empirical results show that the co-regularization approach works well for classification [20], regression [2], and clustering [3] tasks. Moreover, theoretical investigations demonstrate that the co-regularization approach reduces the Rademacher complexity by an amount that depends on the "distance" between the views [18, 21].

We consider the problem of learning a function capable of arranging data points according to a given preference relation [8]. Training of existing kernel-based ranking algorithms, such as RankSVM [10], may be infeasible when the size of the training set is large. This is especially the case when nonlinear kernel functions are used. Recently, a sparse preference learning algorithm, called *Sparse RankRLS*, that can take advantage of a large amount of data in the training process, has been proposed [23]. In this paper, we formulate a co-regularized version of RankRLS, called *Sparse Co-RankRLS*, and aim to improve the performance of RankRLS by making it applicable to situations when only a small amount of labeled data, but a large amount of unlabeled data, is available.

We evaluate our algorithm on a *parse ranking task* [5, 24] that is a common problem in natural language processing. In this task, the aim is to rank a set of parses associated with a single sentence, based on some goodness criteria giving a score to the parse. In our experiments, we consider the case when both scored and a large amount of unscored data is available to the learning algorithm. We demonstrate that Sparse Co-RankRLS is computationally efficient when trained on large datasets and the obtained results are significantly better than the ones obtained with the standard RankRLS algorithm. We consider the parse ranking task as label ranking. However, in the parse ranking task the labels (i.e. the parses of a sentence) are instance-specific. That is, for each sentence, we have a different set of labels, while in the conventional label ranking setting labels are not instance specific.

2 Problem Setting

Let $\mathcal{X}$ be a set of instances and $\mathcal{Y}$ be a set of labels. The learning scenario we consider is *label ranking* [6, 8], i.e., we want to predict for any instance $\boldsymbol{x} \in \mathcal{X}$ (e.g., a person) a preference relation $\mathcal{P}_{\boldsymbol{x}} \subseteq \mathcal{Y} \times \mathcal{Y}$ among the set of labels $\mathcal{Y}$, where each label $y \in \mathcal{Y}$ can be thought of as an alternative. An element $(y, y') \in \mathcal{P}_{\boldsymbol{x}}$ means that the instance $\boldsymbol{x}$ prefers the label y compared to y', also written as $y \succ_{\boldsymbol{x}} y'$. [1] We assume that the (true) preference relation $\mathcal{P}_{\boldsymbol{x}}$ is transitive and asymmetric for each instance $\boldsymbol{x} \in \mathcal{X}$. As training information, we are given a finite set $\{(\boldsymbol{q}_i, s_i)\}_{i=1}^n$ of n data points, where each data point $(\boldsymbol{q}_i, s_i) = ((\boldsymbol{x}_i, y_i), s_i) \in (\mathcal{X} \times \mathcal{Y}) \times \mathbb{R}$ consists of an instance-label tuple $\boldsymbol{q}_i = (\boldsymbol{x}_i, y_i) \in \mathcal{X} \times \mathcal{Y}$ and its score $s_i \in \mathbb{R}$. We say that the pair of data points $((\boldsymbol{x}, y), s)$ and $((\boldsymbol{x}', y'), s')$ is *relevant*, iff $\boldsymbol{x} = \boldsymbol{x}'$. Considering relevant pair $((\boldsymbol{x}, y), s)$ and $((\boldsymbol{x}, y'), s')$, we say that instance $\boldsymbol{x}$ *prefers* label y to y', iff $s > s'$. If $s = s'$, the labels are called *tied*. Accordingly, we write $y \succ_{\boldsymbol{x}} y'$ if $s > s'$ and $y \sim_{\boldsymbol{x}} y'$ if $s = s'$.

A *label ranking function* is a function $f : \mathcal{X} \times \mathcal{Y} \to \mathbb{R}$ mapping each instance-label tuple $(\boldsymbol{x}, y)$ to a real value representing the (predicted) relevance of the label y with respect to the instance $\boldsymbol{x}$. This induces for any instance $\boldsymbol{x} \in \mathcal{X}$ a transitive preference relation $\mathcal{P}_{f,\boldsymbol{x}} \subseteq \mathcal{Y} \times \mathcal{Y}$ with $(y, y') \in \mathcal{P}_{f,\boldsymbol{x}} \Leftrightarrow f(\boldsymbol{x}, y) > f(\boldsymbol{x}, y')$. Ties can be broken arbitrarily. Informally, the goal of our ranking task is to find a label ranking function $f : \mathcal{X} \times \mathcal{Y} \to \mathbb{R}$ such that the ranking $\mathcal{P}_{f,\boldsymbol{x}} \subseteq \mathcal{Y} \times \mathcal{Y}$ induced by the function for any instance $\boldsymbol{x} \in \mathcal{X}$ is a good "prediction" for the true (unknown) preference relation $\mathcal{P}_{\boldsymbol{x}} \subseteq \mathcal{Y} \times \mathcal{Y}$.

In order to incorporate the relevance information, we define an undirected preference graph which is determined by its adjacency matrix W such that for each i and j $(1 \leq i, j \leq n, i \lesseqgtr j)$ $[W]_{i,j} = 1$, if $(\boldsymbol{q}_i, \boldsymbol{q}_j)$ is relevant, and $[W]_{i,j} = 0$ if $(\boldsymbol{q}_i, \boldsymbol{q}_j)$ is not relevant. To avoid loops, we set $[W]_{i,i} = 0$ for $i = 1, \ldots, n$, although an instance-label tuple is relevant to itself. Furthermore, let $Q = (\boldsymbol{q}_1, \ldots, \boldsymbol{q}_n)^t \in (\mathcal{X} \times \mathcal{Y})^n$ be the vector of instance-label training tuples and $S = (s_1, \ldots, s_n)^t \in \mathbb{R}^n$ the corresponding vector of scores. Given these definitions, our training set is the triple $\mathcal{T} = (Q, S, W)$.

Let us define $\mathbb{R}^{\mathcal{Q}} = \{f : \mathcal{Q} \to \mathbb{R}\}$ with $\mathcal{Q} = \mathcal{X} \times \mathcal{Y}$ and let $\mathcal{H} \subseteq \mathbb{R}^{\mathcal{Q}}$ be the hypothesis space of possible ranking functions. To measure how well a hypothesis $f \in \mathcal{H}$ is able to predict the preference relations $\mathcal{P}_{\boldsymbol{x}}$ for all instances $\boldsymbol{x} \in \mathcal{X}$, we consider the following cost function that captures the amount of incorrectly predicted pairs of relevant training data points:

$$d(f, \mathcal{T}) = \frac{1}{2} \sum_{i,j=1}^{n} [W]_{i,j} \left| \operatorname{sign}\big(s_i - s_j\big) - \operatorname{sign}\big(f(\boldsymbol{q}_i) - f(\boldsymbol{q}_j)\big) \right|, \tag{1}$$

where sign$(\cdot)$ denotes the signum function

$$\operatorname{sign}(r) = \begin{cases} 1, & \text{if } r > 0 \\ -1, & \text{if } r \leq 0 \end{cases}.$$

It is well-known that the use of cost functions such as (1) leads to intractable optimization problems. Therefore, we consider the following least-squares approximation,

[1] As described in [8], one can distinguish between *weak preference* ($\succeq$) and *strict preference* ($\succ$), where $y \succ_{\boldsymbol{x}} y' \Leftrightarrow (y \succeq_{\boldsymbol{x}} y') \wedge (y' \not\succeq_{\boldsymbol{x}} y)$; furthermore, $y \sim_{\boldsymbol{x}} y' \Leftrightarrow (y \succeq_{\boldsymbol{x}} y') \wedge (y' \succeq_{\boldsymbol{x}} y)$.

which in fact regresses the differences $s_i - s_j$ with $f(\boldsymbol{q}_i) - f(\boldsymbol{q}_j)$ of relevant training data points $\boldsymbol{q}_i$ and $\boldsymbol{q}_j$:

$$c(f, \mathcal{T}) = \frac{1}{2} \sum_{i,j=1}^{n} [W]_{i,j} \left((s_i - s_j) - (f(\boldsymbol{q}_i) - f(\boldsymbol{q}_j)) \right)^2. \tag{2}$$

Note that the above cost function c also takes the extent of discrepancy between the predicted preference $(f(\boldsymbol{q}_i) - f(\boldsymbol{q}_j))$ and the training preference $(s_i - s_j)$ of pairs of relevant training data points into account.

3 Regularized Least-Squares Ranking

The co-regularized ranking algorithm presented in this paper stems from the results developed in [12] and [23]. For completeness, we briefly review these results in this section.

We aim to construct an algorithm that selects a hypothesis f from $\mathcal{H}$ which minimizes (2) and which is, at the same time, not too "complex", i.e., which does not overfit at training phase and is therefore able to generalize to unseen data. We consider the framework of regularized kernel methods [19], in which $\mathcal{H}$ is a so-called *Reproducing Kernel Hilbert Space* (RKHS) defined by a positive definite kernel function.

3.1 Regularization Framework

Let $k : \mathcal{Q} \times \mathcal{Q} \to \mathbb{R}$ be a positive definite kernel defined on the set $\mathcal{Q}$. Then we define $\mathcal{H}$ as

$$\mathcal{H} = \left\{ f \in \mathbb{R}^{\mathcal{Q}} \mid f(\cdot) = \sum_{j=1}^{\infty} \beta_j k(\cdot, \boldsymbol{q}_j), \beta_j \in \mathbb{R}, \boldsymbol{q}_j \in \mathcal{Q}, \|f\|_{\mathcal{H}} < \infty \right\}, \tag{3}$$

where $\|\cdot\|_{\mathcal{H}}$ denotes the norm in $\mathcal{H}$. Using the RKHS $\mathcal{H}$ as our hypothesis space, we consider the optimization problem

$$\mathcal{A}(\mathcal{T}) = \operatorname*{argmin}_{f \in \mathcal{H}} J(f), \tag{4}$$

where $J(f) = c(f, \mathcal{T}) + \lambda \|f\|_{\mathcal{H}}^2$ and $\lambda \in \mathbb{R}^+$ is a regularization parameter controlling the tradeoff between the cost on the training set and the complexity of the hypothesis. By the generalized representer theorem [19], the minimizer of (4) has the form

$$f^*(\cdot) = \sum_{i=1}^{n} a_i k(\cdot, \boldsymbol{q}_i) \tag{5}$$

with appropriate coefficients $a_i \in \mathbb{R}$. Hence, we can focus on functions $f \in \mathcal{H}$ having the above form. Defining the kernel matrix $K \in \mathbb{R}^{n \times n}$ with entries of the form $[K]_{i,j} = k(\boldsymbol{q}_i, \boldsymbol{q}_j)$ and $f(Q) = (f(\boldsymbol{q}_1), \ldots, f(\boldsymbol{q}_n))^t \in \mathbb{R}^n$, we can write $f(Q) = KA$ and $\|f\|_{\mathcal{H}}^2 = A^t KA$, where $A = (a_1, \ldots, a_n)^t \in \mathbb{R}^n$ is a corresponding coefficient vector.[2]

3.2 *RankRLS*

Let $\mathcal{L} = D - W$ be the Laplacian matrix [4], where D denotes the diagonal matrix with elements of the form $[D]_{i,i} = \sum_{j=1}^{n} [W]_{i,j}$. Using a slightly different notation, it is shown in [12] that the cost function (2) can be rewritten as

$$c(f, \mathcal{T}) = (S - KA)^t \mathcal{L}(S - KA). \tag{6}$$

Considering this representation of the cost function c, we get the following optimization problem called *RankRLS* in [12]:

$$\mathcal{A}(\mathcal{T}) = \underset{A \in \mathbb{R}^n}{\operatorname{argmin}}\, J(A), \tag{7}$$

where $J(A) = (S - KA)^t \mathcal{L}(S - KA) + \lambda A^t KA$. Using the fact that $\mathcal{L}$ is positive semidefinite [15] and assuming that K is positive definite, it is easy to see that the Hessian matrix $H(J) = 2K^t \mathcal{L} K + 2\lambda K$ of J is positive definite. Thus, J is strictly convex and the global minimum of J can be obtained by setting the first derivative $\frac{\mathrm{d}}{\mathrm{d}A} J(A) = -2K^t \mathcal{L}(S - KA) + 2\lambda KA$ to zero and by solving the resulting system of equations with respect to A. The optimal solution for (7) is

$$A = (K\mathcal{L}K + \lambda K)^{-1} K\mathcal{L}S = (\mathcal{L}K + \lambda I)^{-1} \mathcal{L}S, \tag{8}$$

where I denotes the identity matrix. The computational complexity of the matrix inversion in (8) is $\mathcal{O}(n^3)$.

Fact 1 *[12] For fixed $\lambda \in \mathbb{R}^+$, the solution of the RankRLS optimization problem (7) can be found in $\mathcal{O}(n^3)$ time.*

3.3 *Sparse RankRLS*

Similarly to [14] and [22], an approximation algorithm aiming at reducing the cubic running time of the RankRLS approach is developed in [23]: The cost function c is evaluated over *all* points, but only a subset of the coefficients $a_1, \ldots, a_n$ is allowed

---

[2] Unless stated otherwise, we assume that a kernel matrix K is positive definite, i.e., $B^{'} KB > 0$ for all $B \in \mathbb{R}^n$, $B \neq 0$. This can be ensured, for example, by performing a small diagonal shift.

to be nonzero, thus an approximation of the optimization problem is considered. Let $R = \{i_1, \ldots, i_r\} \subseteq \{1, \ldots, n\}$ be a subset of indices. Then, we only allow the coefficients $a_{i_1}, \ldots, a_{i_r}$ to be nonzero in (5), i.e., we search for minimizers $\widehat{f} \in \mathcal{H}$ having the form

$$\widehat{f}(\cdot) = \sum_{j=1}^{r} a_{i_j} k(\cdot, \boldsymbol{q}_{i_j}). \tag{9}$$

By defining $\bar{K} \in \mathbb{R}^{n \times r}$ to be the submatrix of $K \in \mathbb{R}^{n \times n}$ that only contains the columns indexed by R and by defining $\widehat{K} \in \mathbb{R}^{r \times r}$ to be the submatrix of $\bar{K}$ only containing the rows indexed by R, we can express $\widehat{f}(Q) = (\widehat{f}(\boldsymbol{q}_1), \ldots, \widehat{f}(\boldsymbol{q}_n))^t \in \mathbb{R}^n$ as $\widehat{f}(Q) = \bar{K}\widehat{A}$ and $\|\widehat{f}\|^2_{\mathcal{H}} = \widehat{A}^t \widehat{K} \widehat{A}$, where $\widehat{A} = (a_{i_1}, \ldots, a_{i_r})^t \in \mathbb{R}^r$. Given these notations, the approximation presented in [23], called *Sparse RankRLS*, can be formulated as

$$\mathcal{A}(\mathcal{T}) = \underset{\widehat{A} \in \mathbb{R}^r}{\operatorname{argmin}}\, \widehat{J}(\widehat{A}), \tag{10}$$

where $\widehat{J}(\widehat{A}) = (S - \bar{K}\widehat{A})^t \mathcal{L} (S - \bar{K}\widehat{A}) + \lambda \widehat{A}^t \widehat{K} \widehat{A}$. Setting the derivative of $\widehat{J}$ to zero and solving the resulting system of equations with respect to $\widehat{A}$ leads to

$$\widehat{A} = (\bar{K}^t \mathcal{L} \bar{K} + \lambda \widehat{K})^{-1} \bar{K}^t \mathcal{L} S. \tag{11}$$

The overall training complexity of the Sparse RankRLS algorithm is $\mathcal{O}(nr^2)$, see [23] for more details.

Fact 2 *([23]) For fixed $\lambda \in \mathbb{R}^+$, the solution of the Sparse RankRLS optimization problem (10) can be found in $\mathcal{O}(nr^2)$ time.*

Hence, selecting r to be much smaller than n results in a significant acceleration of the training procedure. Clearly, the selection of the index set R may have an influence on results obtained by the above approximation approach. Various methods for subset selection have been proposed (see e.g.[17, 25]), however, for simplicity and computational efficiency, we consider random selection of data points contained in R.

3.4 Constructing Kernels with Subsets of Regressors

Considering the Sparse RankRLS algorithm, the label predictions for the training data points can be obtained by $\bar{K}\widehat{A}$. Using the Woodbury matrix identity [9] and (11) and by defining $\widetilde{K} = \frac{1}{\lambda} \bar{K} \widehat{K}^{-1} \bar{K}^t$, we can reformulate this expression as follows:

$$\begin{aligned}\bar{K}\widehat{A} &= \bar{K}(\bar{K}^t \mathcal{L} \bar{K} + \lambda \widehat{K})^{-1} \bar{K}^t \mathcal{L} S \\ &= \bar{K}\left(\frac{1}{\lambda}\widehat{K}^{-1} - \frac{1}{\lambda}\widehat{K}^{-1}\bar{K}^t\left(\frac{1}{\lambda}\mathcal{L}\bar{K}\widehat{K}^{-1}\bar{K}^t + I\right)^{-1}\frac{1}{\lambda}\mathcal{L}\bar{K}\widehat{K}^{-1}\right)\bar{K}^t \mathcal{L} S\end{aligned}$$

$$= (\widetilde{K} - \widetilde{K}(\mathcal{L}\widetilde{K} + I)^{-1}\mathcal{L}\widetilde{K})\mathcal{L}S$$
$$= (\widetilde{K}(I - (\mathcal{L}\widetilde{K} + I)^{-1}\mathcal{L}\widetilde{K})\mathcal{L}S$$
$$= (\widetilde{K}((\mathcal{L}\widetilde{K} + I)^{-1}(\mathcal{L}\widetilde{K} + I) - (\mathcal{L}\widetilde{K} + I)^{-1}\mathcal{L}\widetilde{K})\mathcal{L}S$$
$$= \widetilde{K}(\mathcal{L}\widetilde{K} + I)^{-1}(\mathcal{L}\widetilde{K} + I - \mathcal{L}\widetilde{K})\mathcal{L}S$$
$$= \widetilde{K}(\mathcal{L}\widetilde{K} + I)^{-1}\mathcal{L}S.$$

Note that because $\mathcal{L}$ is positive semidefinite and $\widetilde{K}$ is positive definite, their product $\mathcal{L}\widetilde{K}$ contains only nonnegative eigenvalues [11]. Hence, $\mathcal{L}\widetilde{K} + I$ is invertible. Further, the last term can be rewritten as $\widetilde{K}(\mathcal{L}\widetilde{K} + I)^{-1}\mathcal{L}S = \check{K}(\mathcal{L}\check{K} + \lambda I)^{-1}\mathcal{L}S$, where $\check{K} = \bar{K}\widehat{K}^{-1}\bar{K}^t \in \mathbb{R}^{n \times n}$. These derivations show that the Sparse RankRLS algorithm operating with a kernel function k is essentially equivalent to the standard RankRLS algorithm operating with a modified kernel $\check{k}$. In the following section, we use this fact for constructing different Hilbert spaces by taking different sets of basis vectors.

4 Co-Regularized Least Squares Ranking

The co-regularization approach is based on the idea of constructing M prediction functions from M different RKHSs such that the error of each function on the labeled data is small and, at the same time, the functions give similar predictions for the unlabeled data. These RKHSs can stem from different data point descriptions (i.e., different features), from different kernel functions, and/or from different subsets of the data. Note that the case of different data point descriptions can be obtained by applying the kernel functions only to appropriate subsets of features. Further, as depicted in Sect. 3.4, taking different subsets of the data leads to different RKHSs. Hence, we will consider M RKHSs $\mathcal{H}_1, \ldots, \mathcal{H}_M$ along with their corresponding kernel functions $k_\nu : \mathcal{Q} \times \mathcal{Q} \to \mathbb{R}, 1 \leq \nu \leq M$, to incorporate the co-regularization approach.

4.1 Co-Regularized RankRLS

Considering our ranking task, we have a training set $\mathcal{T} = (Q, S, W)$ originating from a set $\{(\boldsymbol{q}_i, s_i)\}_{i=1}^n$ of data points *with* scoring information, where $Q = (\boldsymbol{q}_1, \ldots, \boldsymbol{q}_n)^t \in \mathcal{Q}^n$, $S = (s_1, \ldots, s_n)^t \in \mathbb{R}^n$, and W is the matrix incorporating the relevance information. Moreover, we have a training set $\widetilde{\mathcal{T}} = (\widetilde{Q}, \widetilde{W})$ from a set $\{\boldsymbol{q}_{n+i}\}_{i=1}^l$ of data points *without* scoring information, $\widetilde{Q} = (\boldsymbol{q}_{n+1}, \ldots, \boldsymbol{q}_{n+l})^t \in \mathcal{Q}^l$, and an appropriate adjacency matrix $\widetilde{W}$. To avoid misunderstandings with the definition of the label ranking task, we will use the terms "scored" instead of "labeled" and "unscored" instead of "unlabeled".

In the ranking task, we search for a vector $\mathbf{f} = (f_1, \ldots, f_M) \in \mathcal{H}_1 \times \ldots \times \mathcal{H}_M$ of prediction functions which minimizes

$$J(\mathbf{f}) = \sum_{v=1}^{M} c(f_v, \mathcal{T}) + \lambda \sum_{v=1}^{M} \| f_v \|_{\mathcal{H}_v}^2 + \nu \sum_{v,u=1}^{M} \tilde{c}(f_v, f_u, \widetilde{\mathcal{T}}), \tag{12}$$

where $\lambda, \nu \in \mathbb{R}^+$ are regularization parameters and $\tilde{c}$ is the loss function measuring the disagreement between the prediction functions of the views on the unscored data:

$$\tilde{c}(f_v, f_u, \widetilde{\mathcal{T}}) = \frac{1}{2} \sum_{i,j=1}^{l} [\widetilde{W}]_{i,j} \Big(\big(f_v(\boldsymbol{q}_{n+i}) - f_v(\boldsymbol{q}_{n+j})\big) - \big(f_u(\boldsymbol{q}_{n+i}) - f_u(\boldsymbol{q}_{n+j})\big) \Big)^2 .$$

Applying the representer theorem [19] in this context shows that the minimizers $f_v^* \in \mathcal{H}_v$ of (12) for $v = 1, \ldots, M$ have the form

$$f_v^*(\cdot) = \sum_{i=1}^{n} a_i^{(v)} k_v(\cdot, \boldsymbol{q}_i) + \sum_{i=1}^{l} a_{n+i}^{(v)} k_v(\cdot, \boldsymbol{q}_{n+i}) \tag{13}$$

with adequate coefficients $a_1^{(v)}, \ldots, a_{n+l}^{(v)} \in \mathbb{R}$. Using matrix notations, we can reformulate (12) as

$$\begin{aligned} J(\mathbf{A}) = & \sum_{v=1}^{M} (S - L_v A_v)^t \mathcal{L}_L (S - L_v A_v) + \lambda \sum_{v=1}^{M} A_v{}^t K_v A_v \\ & + \nu \sum_{v,u=1}^{M} (U_v A_v - U_u A_u)^t \mathcal{L}_U (U_v A_v - U_u A_u), \end{aligned} \tag{14}$$

where $A_v = (a_1^{(v)}, \ldots, a_{n+l}^{(v)})^t \in \mathbb{R}^{n+l}$, and $\mathbf{A} = (A_1^t, \ldots, A_M^t)^t \in \mathbb{R}^{M(n+l)}$. The matrix $L_v \in \mathbb{R}^{n \times (n+l)}$ has entries of the form $[L_v]_{i,j} = k_v(\boldsymbol{q}_i, \boldsymbol{q}_j)$ and the matrix $U_v \in \mathbb{R}^{l \times (n+l)}$ has entries of the form $[U_v]_{i,j} = k_v(\boldsymbol{q}_{n+i}, \boldsymbol{q}_j)$. Stacking both matrices up gives the matrix K_v:

$$K_v = \begin{pmatrix} L_v \\ U_v \end{pmatrix} \in \mathbb{R}^{(n+l) \times (n+l)}.$$

Further, $\mathcal{L}_L \in \mathbb{R}^{n \times n}$ and $\mathcal{L}_U \in \mathbb{R}^{l \times l}$ denote the Laplacian matrices corresponding to W and $\widetilde{W}$, respectively. Hence, we have the following optimization problem:

$$\mathcal{A}(\mathcal{T}, \widetilde{\mathcal{T}}) = \underset{\mathbf{A} \in \mathbb{R}^{M(n+l)}}{\operatorname{argmin}} \; J(\mathbf{A}). \tag{15}$$

4.2 Sparse Co-Regularized RankRLS

Similar to the non-co-regularized case, the above optimization problem could be difficult to solve due to the computations involving the complete kernel matrices. Hence, as in Sect. 3, we aim at solving an approximation of the above optimization problem by only allowing a subset of the coefficients in (13) to be nonzero for each view. This corresponds to taking submatrices of the original matrices, i.e., for each view v we define $\bar{L}_v \in \mathbb{R}^{n\times r}$ to be the submatrix of L_v that only contains the columns corresponding to r selected basis vectors $\boldsymbol{q}_{c_v(1)}, \ldots, \boldsymbol{q}_{c_v(r)}$. Here, the number $c_v(i) \in \{1, \ldots, n+l\}$ denotes the index (column) of the ith selected vector of view v. Accordingly, we define $\bar{U}_v \in \mathbb{R}^{l\times r}$ to be the submatrix of U_v that only contains the columns corresponding to $\boldsymbol{q}_{c_v(1)}, \ldots, \boldsymbol{q}_{c_v(r)}$. Finally, we define $\widehat{K}_v \in \mathbb{R}^{r\times r}$ to be the kernel matrix with elements $\left[\widehat{K}_v\right]_{i,j} = k_v(\boldsymbol{q}_{c_v(i)}, \boldsymbol{q}_{c_v(j)})$. Hence, we obtain the following optimization problem, which we call *Sparse Co-RankRLS*:

$$\mathcal{A}(\mathcal{T}, \widetilde{\mathcal{T}}) = \underset{\widehat{\mathbf{A}}\in\mathbb{R}^{Mr}}{\operatorname{argmin}}\, \widehat{J}(\widehat{\mathbf{A}}), \tag{16}$$

where

$$\widehat{J}(\widehat{\mathbf{A}}) = \sum_{v=1}^{M}\left(S - \bar{L}_v\widehat{A}_v\right)^t \mathcal{L}_L\left(S - \bar{L}_v\widehat{A}_v\right) + \lambda\sum_{v=1}^{M}\widehat{A}_v{}^t\widehat{K}_v\widehat{A}_v \tag{17}$$
$$+\nu\sum_{v,u=1}^{M}\left(\bar{U}_v\widehat{A}_v - \bar{U}_u\widehat{A}_u\right)^t\mathcal{L}_U\left(\bar{U}_v\widehat{A}_v - \bar{U}_u\widehat{A}_u\right),$$

$\widehat{A}_v = (a^{(v)}_{c_v(1)}, \ldots, a^{(v)}_{c_v(r)})^t \in \mathbb{R}^r$, and $\widehat{\mathbf{A}} = (\widehat{A}_1^t, \ldots, \widehat{A}_M^t)^t \in \mathbb{R}^{Mr}$. For ease of notation, we consider the same number of basis vectors for each view. Given this matrix formulation of our optimization problem, we can follow the framework described in [2] to find a closed form for the solution: Taking the partial derivative of $\widehat{J}(\widehat{\mathbf{A}})$ with respect to $\widehat{A}_v$ we get

$$\frac{\mathrm{d}}{\mathrm{d}\widehat{A}_v}\widehat{J}(\widehat{\mathbf{A}}) = -2\bar{L}_v^t\mathcal{L}_L(S - \bar{L}_v\widehat{A}_v) + 2\lambda\widehat{K}_v\widehat{A}_v$$
$$-4\nu\sum_{u=1,u\neq v}^{M}\bar{U}_v^t\mathcal{L}_U(\bar{U}_u\widehat{A}_u - \bar{U}_v\widehat{A}_v).$$

By defining $G_v^\nu = 2\nu(M-1)\bar{U}_v^t\mathcal{L}_U\bar{U}_v$, $G_v^\lambda = \lambda\widehat{K}_v$ and $G_v = \bar{L}_v^t\mathcal{L}_L\bar{L}_v$, we can rewrite the above term as

$$\frac{\mathrm{d}}{\mathrm{d}\widehat{A}_v}\widehat{J}(\widehat{\mathbf{A}}) = 2(G_v + G_v^{\nu} + G_v^{\lambda})\widehat{A}_v - 2\bar{L}_v^t\mathcal{L}_L S$$
$$-4\nu \sum_{u=1,u\neq v}^{M} \bar{U}_v^t\mathcal{L}_U\bar{U}_u\widehat{A}_u.$$

At the optimum we have $\frac{\mathrm{d}}{\mathrm{d}\widehat{A}_v}\widehat{J}(\widehat{\mathbf{A}}) = 0$ for all views, thus we get the exact solution by solving

$$\begin{pmatrix} \bar{G}_1 & -2\nu\bar{U}_1^t\mathcal{L}_U\bar{U}_2 & \dots \\ -2\nu\bar{U}_2^t\mathcal{L}_U\bar{U}_1 & \bar{G}_2 & \dots \\ \vdots & \vdots & \ddots \end{pmatrix}\begin{pmatrix} \widehat{A}_1 \\ \widehat{A}_2 \\ \vdots \end{pmatrix} = \begin{pmatrix} \bar{L}_1^t\mathcal{L}_L S \\ \bar{L}_2^t\mathcal{L}_L S \\ \vdots \end{pmatrix}$$

with respect to $\widehat{A}_1, \dots, \widehat{A}_M$, where $\bar{G}_v = G_v + G_v^{\nu} + G_v^{\lambda}$. The left-hand side matrix is positive definite and therefore invertible (see Appendix). By defining

$$B = \begin{pmatrix} G_1 & 0 & \dots \\ 0 & G_2 & \dots \\ \vdots & \vdots & \ddots \end{pmatrix} \quad D = \begin{pmatrix} G_1^{\lambda} & 0 & \dots \\ 0 & G_2^{\lambda} & \dots \\ \vdots & \vdots & \ddots \end{pmatrix} E = \begin{pmatrix} \bar{L}_1^t\mathcal{L}_L S \\ \bar{L}_2^t\mathcal{L}_L S \\ \vdots \end{pmatrix}$$

$$C = \begin{pmatrix} G_1^{\nu} & -2\nu\bar{U}_1^t\mathcal{L}_U\bar{U}_2 & \dots \\ -2\nu\bar{U}_2^t\mathcal{L}_U\bar{U}_1 & G_2^{\nu} & \dots \\ \vdots & \vdots & \ddots \end{pmatrix}$$

we can formulate the solution of the system as follows:

$$\widehat{\mathbf{A}} = (B + C + D)^{-1}E. \tag{18}$$

The computational complexity of constructing the vector E is $\mathcal{O}(Mnr)$. Further, the matrices B, C, and D can be constructed in $\mathcal{O}(Mr^2n)$, $\mathcal{O}(M^2r^2l)$, and $\mathcal{O}(Mr^2)$, respectively. The resulting matrix $(B + C + D) \in \mathbb{R}^{Mr \times Mr}$ can be inverted in $\mathcal{O}(M^3r^3)$. Hence, our algorithm scales linearly in the number of unscored data items. Note that the multiplications involving the Laplacian matrices $\mathcal{L}_L$ and $\mathcal{L}_U$ can be accelerated using the approach described in [23]. Assuming $l \geq n$, we have shown the following theorem:

Theorem 1. *For fixed parameters $\lambda, \nu \in \mathbb{R}^+$ and assuming $l \geq n$, the solution of the Sparse Co-RankRLS optimization problem (16) can be found in $\mathcal{O}(M^3r^3 + M^2r^2l)$ time.*

5 Efficient Regularization Parameter Selection

When performing experiments, the recurrent matrix inversion in (18) for each combination of the regularization parameters λ and ν could be time-consuming. Therefore, we propose a procedure which accelerates this parameter selection process. Writing D as $D = \lambda \acute{D}$ with an appropriate (positive definite) matrix $\acute{D}$ and rewriting $\acute{D}$ as $\acute{D} = PP^t$ using the Cholesky decomposition [9], we obtain

$$\begin{aligned}(B + C + D)^{-1} &= (B + C + \lambda \acute{D})^{-1} \\ &= \left(PP^{-1}(B + C)(P^t)^{-1} P^t + \lambda PP^t\right)^{-1} \\ &= (P^t)^{-1}\left(P^{-1}(B + C)(P^t)^{-1} + \lambda I\right)^{-1} P^{-1}.\end{aligned}$$

Further, the matrix $P^{-1}(B + C)(P^t)^{-1}$ can be eigen decomposed to $V \Lambda V^t$, where Λ is a diagonal matrix containing the eigenvalues and V is the matrix composed of the eigenvectors [9]. Hence, we get

$$\begin{aligned}(B + C + D)^{-1} &= (P^t)^{-1}\left(V \Lambda V^t + \lambda I\right)^{-1} P^{-1} \\ &= (P^t)^{-1} V\left(\Lambda + \lambda I\right)^{-1} V^t P^{-1}\end{aligned}$$

and the solution in (18) can be rewritten as

$$\widehat{\mathbf{A}} = (P^t)^{-1} V\left(\Lambda + \lambda I\right)^{-1} V^t P^{-1} E.$$

Thus, by fixing the parameter ν, we can efficiently search for the second regularization parameter λ. The decompositions and the inversion of P can be calculated in $\mathcal{O}(M^3 r^3)$ time, and hence, the overall training complexity is not increased. The computational cost of calculating $(\Lambda + \lambda I)^{-1}$ is $\mathcal{O}(Mr)$, since it is a diagonal matrix. If the matrices $V^t P^{-1} E \in \mathbb{R}^{Mr \times 1}$ and $(P^t)^{-1} V \in \mathbb{R}^{Mr \times Mr}$ are stored in memory, the subsequent training with different values of λ can be performed in $\mathcal{O}(M^2 r^2)$ time.

6 Experiments

We evaluate the performance of the Sparse Co-RankRLS algorithm[3] on the task of ranking given parses for an unseen sentence. For this purpose, we use the BioInfer corpus [16] which consists of 1,100 manually annotated sentences. A detailed description of the parse ranking problem and the data used in the experiments is given in [24]. Each sentence is associated with a set of candidate parses. The manual

[3] Python implementation of the algorithm and the dataset are available on request.

annotation of the sentence, present in the corpus, provides the correct parse. Further, each candidate parse is associated with a goodness score that indicates how close to the correct parse it is. The correct ranking of the parses associated with the same sentence is determined by this score. While the scoring induces a total order over the whole set of parses, the preferences between parses associated with different sentences are not considered in the parse ranking task.

Using the definitions presented in Sect. 2, we consider each sentence as an instance and the parses generated for the sentence as the labels associated with it. The score of an input indicates how well the parse included in the input matches the correct parse of the sentence. We have previously demonstrated that the RankRLS algorithm performs comparably to some state-of-the-art ranking methods [12]. In this section, we will compare the performance of the Sparse Co-RankRLS algorithm with that of the RankRLS algorithm.

6.1 Experimental Setup

From the 1,100 sentences of the BioInfer corpus, we randomly select 500 and 600 sentences for the training and final validation phase, respectively. To simulate a semisupervised setting, we consider that only half of sentence-parse pairs in the training set are scored, while the remaining sentence-parse pairs do not have the scoring information associated with them. For the evaluation of the Sparse Co-RankRLS method, we set the number M of views to 2. Further, we randomly select 20 sentence-parse pairs from the training data set as basis vectors for the first view and repeat this procedure for the second view. According to Sect. 4, we select different basis vectors for each view.

Both algorithms have the regularization parameter λ that controls the tradeoff between the minimization of the training error and the complexity of the learnt function(s). In addition, the Sparse Co-RankRLS algorithm has the regularization parameter ν that controls the agreement between the predictions of the different views. As a similarity measure for parses, we use the best performing graph kernel with the appropriate parameter considered in [13]. The values of the regularization parameters for RankRLS as well as for Sparse Co-RankRLS are estimated during a fivefold cross-validation procedure, with the splits being performed on the sentence level ensuring that all parses associated with the same sentence are present in the same fold. In the semisupervised setting each fold consists of one tenth of scored and unscored data present in the training set. For the cross-validation phases, we randomly select parses for each sentence to be associated with it, out of which 30 parses are used for training the model and 30 for testing. Finally, we use 30 parses per sentence for the final validation procedure.

Table 1 Comparison of the parse ranking performances of the standard RankRLS and the Sparse Co-RankRLS algorithms using a normalized version of the disagreement error (1) as performance evaluation measure. The results of the Sparse Co-RankRLS algorithm are obtained by averaging the predictions of the two views

Standard RankRLS	Sparse Co-RankRLS
0.348	0.326

6.2 Results

The normalized version of the disagreement error (1) is used to measure the performance of the ranking algorithms. The error is calculated for each sentence separately and the performance is averaged over all sentences.

The algorithms are trained on the whole parameter estimation data set with the best found parameter values and tested with the 600 sentences reserved for the final validation. The results of the validation are presented in Table 1. They show that the Sparse Co-RankRLS algorithm notably outperforms the RankRLS method. We note that random selection of the basis vectors for both methods has an influence on the performance of the learning algorithm. To avoid variations in the final results obtained with particular set of basis vectors, we perform complete experiment 5 times, selecting different sets for basis vectors in all of the experiments and report averaged results.

Furthermore, to test the statistical significance of the performance difference between the Sparse Co-RankRLS and RankRLS algorithms, we conduct the Wilcoxon signed-ranks test [7]. The sentences reserved for the final validation are considered as independent trials. We observe that the performance differences are statistically significant ($p < 0.05$).

6.3 Learning Curve

To evaluate performance of the Sparse Co-RankRLS algorithm with respect to the number of unscored sentence-parse pairs used for training, we divide 4,000 sentence-parse pairs into 4 parts containing 1,000, 2,000, 3,000, and 4,000 unscored data, respectively. The algorithm is trained using complete scored training set with best found parameters that were estimated with fivefold cross-validation procedure. The separate test set is used for final validation of the algorithm. Each of the unscored data sets is re-sampled 5 times and obtained results are averaged. The outcomes of the experiments are presented in Fig. 1.

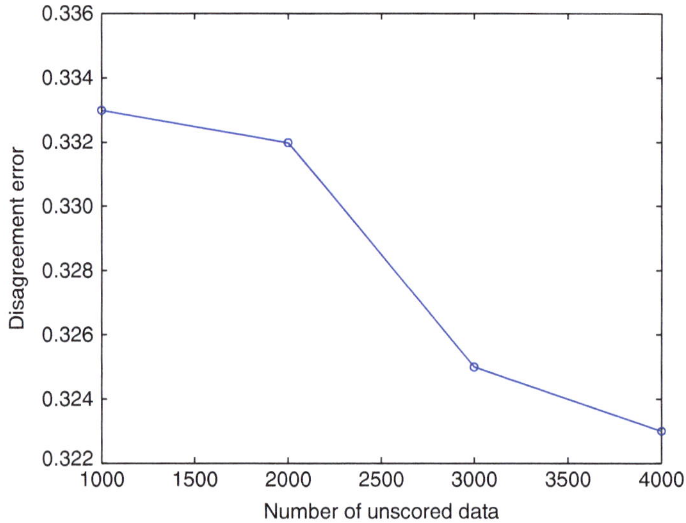

Fig. 1 The plot shows the relation between the disagreement error and the amount of unscored data

7 Conclusions

We propose Sparse Co-RankRLS, a semisupervised regularized least-squares algorithm, for learning preference relations. The computational complexity of the algorithm is $\mathcal{O}(M^3r^3 + M^2r^2l)$, where l is the number of unscored training examples, M is the number of views, and r is the number of basis vectors. We formulate the algorithm within the co-regularization framework, which aims at improving the prediction performance by minimizing the disagreement of all prediction hypotheses on the unscored data. In our experiments, we consider a parse ranking task and show that the Sparse Co-RankRLS algorithm significantly outperforms the standard RankRLS algorithm on this task.

Due to the fact that our semisupervised preference learning algorithm has a linear complexity in the number of unscored examples, it is primarily applicable in cases when only a small amount of scored but a large amount of unscored data is available for training. In the future, we aim to evaluate our Sparse Co-RankRLS algorithm on various tasks where scored data is scarce.

Acknowledgements We acknowledge support from the Netherlands Organization for Scientific Research (NWO), in particular a Vici grant (639.023.604). We also thank CSC, the Finnish IT center for science for providing us with computing resources.

References

1. A. Blum, T. Mitchell, Combining labeled and unlabeled data with co-training, in *Proceedings of the 11th Annual Conference on Computational Learning Theory* (ACM, New York, NY, USA, 1998), pp. 92–100
2. U. Brefeld, T. Gärtner, T. Scheffer, S. Wrobel, Efficient co-regularised least squares regression, in *Proceedings of the 23rd International Conference on Machine learning* (ACM, New York, NY, USA, 2006), pp. 137–144
3. U. Brefeld, T. Scheffer, Co-em support vector learning, in *Proceedings of the 21st International Conference on Machine learning* (ACM, New York, NY, USA, 2004), p. 16
4. R.A. Brualdi, H.J. Ryser, *Combinatorial Matrix Theory* (Cambridge University Press, 1991)
5. M. Collins, N. Duffy, New ranking algorithms for parsing and tagging: kernels over discrete structures, and the voted perceptron, in *ACL '02: Proceedings of the 40th Annual Meeting on Association for Computational Linguistics* (Association for Computational Linguistics, Morristown, NJ, USA, 2001), pp. 263–270
6. O. Dekel, C.D. Manning, Y. Singer, Log-linear models for label ranking, in *Advances in Neural Information Processing Systems*, vol. 16, ed. by S. Thrun, L. Saul, B. Schölkopf (MIT, Cambridge, MA, 2004), pp. 497–504
7. J. Demšar, Statistical comparisons of classifiers over multiple data sets. J. Mach. Learn. Res. **7**, 1–30 (2006)
8. J. Fürnkranz, E. Hüllermeier, Preference learning. Künstliche Intelligenz **19**(1), 60–61 (2005)
9. G.H. Golub, C.F. Van Loan, *Matrix Computations* (Johns Hopkins University Press, 1996)
10. R. Herbrich, T. Graepel, K. Obermayer, Support vector learning for ordinal regression, in *Proceedings of the Ninth International Conference on Articial Neural Networks* (Institute of Electrical Engineers, London, 1999), pp. 97–102
11. R. Horn, C.R. Johnson, *Matrix Analysis* (Cambridge University Press, Cambridge, 1985)
12. T. Pahikkala, E. Tsivtsivadze, A. Airola, J. Järvinen, J. Boberg, An efficient algorithm for learning to rank from preference graphs. Mach. Learn. **75**(1), 129–165 (2009)
13. T. Pahikkala, E. Tsivtsivadze, J. Boberg, T. Salakoski, Graph kernels versus graph representations: a case study in parse ranking, in *Proceedings of the ECML/PKDD'06 workshop on Mining and Learning with Graphs*, ed. by T. Gärtner, G.C. Garriga, T. Meinl, Berlin, Germany (pp. 181–188) (2006)
14. T. Poggio, F. Girosi, Networks for approximation and learning. Proc. IEEE **78**(9), 1481–1497 (1990)
15. A. Pothen, H.D. Simon, K.-P. Liou, Partitioning sparse matrices with eigenvectors of graphs. SIAM J. Matrix Anal. Appl. **11**(3), 430–452 (1990)
16. S. Pyysalo, F. Ginter, J. Heimonen, J. Björne, J. Boberg, J. Järvinen, T. Salakoski, BioInfer: A corpus for information extraction in the biomedical domain. BMC Bioinformatics, **8**, 50 (2007)
17. J. Quinonero-Candela, CE. Rasmussen, A unifying view of sparse approximate Gaussian process regression. J. Mach. Learn. Res. **6**, 1939–1959 (2005)
18. D. Rosenberg, P.L. Bartlett, The rademacher complexity of co-regularized kernel classes, in *Proceedings of the Eleventh International Conference on Artificial Intelligence and Statistics*, ed. by M. Meila, X. Shen (2007), pp. 396–403
19. B. Schölkopf, R. Herbrich, A.J. Smola, A generalized representer theorem, in *Proceedings of the 14th Annual Conference on Computational Learning Theory*, ed. by David P. Helmbold, B. Williamson (Springer, London, 2001), pp. 416–426
20. V. Sindhwani, P. Niyogi, M. Belkin, A co-regularization approach to semi-supervised learning with multiple views, in *Proceedings of ICML Workshop on Learning with Multiple Views* (2005)
21. V. Sindhwani, D. Rosenberg, An rkhs for multi-view learning and manifold co-regularization, in *Proceedings of the 25th Annual International Conference on Machine Learning (ICML 2008)*, ed. by A. McCallum, S. Roweis (Omnipress, Helsinki, Finland, 2008), pp. 976–983
22. A.J. Smola, B. Schölkopf, Sparse greedy matrix approximation for machine learning, in *Proceedings of the 17th International Conference on Machine Learning*, ed. by Pat Langley (Morgan Kaufmann Publishers, San Francisco, Ca, USA, 2000), pp. 911–918

23. E. Tsivtsivadze, T. Pahikkala, A. Airola, J. Boberg, T. Salakoski, A sparse regularized least-squares preference learning algorithm, in *10th Scandinavian Conference on Artificial Intelligence (SCAI 2008)*, vol. 173, ed. by A. Holst, P. Kreuger, P. Funk (IOS, 2008), pp. 76–83
24. E. Tsivtsivadze, T. Pahikkala, S. Pyysalo, J. Boberg, A. Mylläri, T. Salakoski, Regularized least-squares for parse ranking, in *Advances in Intelligent Data Analysis VI*, ed. by A. Fazel Famili, J.N. Kok, J.M. Peña, A. Siebes, A.J. Feelders (Springer, 2005), pp. 464–474
25. P. Vincent, Y. Bengio, Kernel matching pursuit. *Mach. Learn.* **48**(1–3), 165–187 (2002)

Appendix

We will show that the matrix

$$\begin{pmatrix} \bar{G}_1 & -2\nu \bar{U}_1^t \mathcal{L}_U \bar{U}_2 & \dots \\ -2\nu \bar{U}_2^t \mathcal{L}_U \bar{U}_1 & \bar{G}_2 & \dots \\ \vdots & \vdots & \ddots \end{pmatrix}$$

is positive definite. To prove that, we decompose the above matrix into a sum of matrices

$$X_1 = \begin{pmatrix} \bar{G}_1 - 2\nu(M-1)\bar{U}_1^t \mathcal{L}_U \bar{U}_1 & 0 & \dots \\ 0 & \bar{G}_2 - 2\nu(M-1)\bar{U}_2^t \mathcal{L}_U \bar{U}_2 & \dots \\ \vdots & \vdots & \ddots \end{pmatrix}$$

and

$$X_2 = \begin{pmatrix} \nu(M-1)\bar{U}_1^t \mathcal{L}_U \bar{U}_1 & -\nu \bar{U}_1^t \mathcal{L}_U \bar{U}_2 & \dots \\ -\nu \bar{U}_2^t \mathcal{L}_U \bar{U}_1 & \nu(M-1)\bar{U}_2^t \mathcal{L}_U \bar{U}_2 & \dots \\ \vdots & \vdots & \ddots \end{pmatrix}.$$

The matrix X_1 is positive definite as each block matrix is positive definite (we require the matrix $\widehat{K}_\nu$ to be positive definite). Further, the matrix X_2 is positive semidefinite as we can write it as a sum of positive semidefinite matrices of the form

$$\begin{pmatrix} 0 \cdots & 0 & \cdots & 0 & \cdots 0 \\ \vdots & \vdots & & \vdots & \vdots \\ 0 \cdots & \nu \bar{U}_i^t \mathcal{L}_U \bar{U}_i & \cdots & -\nu \bar{U}_i^t \mathcal{L}_U \bar{U}_j & \cdots 0 \\ \vdots & \vdots & \ddots & \vdots & \vdots \\ 0 \cdots & -\nu \bar{U}_j^t \mathcal{L}_U \bar{U}_i & \cdots & \nu \bar{U}_j^t \mathcal{L}_U \bar{U}_j & \cdots 0 \\ \vdots & \vdots & & \vdots & \vdots \\ 0 \cdots & 0 & \cdots & 0 & \cdots 0 \end{pmatrix} = X_{(i,j)}^t X_{(i,j)},$$

where $X_{(i,j)} = \big(0, \ldots, 0, \sqrt{\nu} P \bar{U}_i, 0, \ldots, 0, -\sqrt{\nu} P \bar{U}_j, 0, \ldots, 0\big)$. Here, the positive semidefinite matrix $\mathcal{L}_U$ is decomposed as $\mathcal{L}_U = P^t P$ using the Cholesky decomposition [9].

Part II
Instance Ranking

A Survey on ROC-Based Ordinal Regression

Willem Waegeman and Bernard De Baets

Abstract Ordinal regression can be seen as a special case of preference learning, in which the class labels corresponding with data instances can take values from an ordered finite set. In such a setting, the classes usually have a linguistic interpretation attached by humans to subdivide the data into a number of preference bins. In this chapter, we give a general survey on ordinal regression from a machine learning point of view. In particular, we elaborate on some important connections with ROC analysis that have been introduced recently by the present authors. First, the important role of an underlying ranking function in ordinal regression models is discussed, as well as its impact on the performance evaluation of such models. Subsequently, we describe a new ROC-based performance measure that directly evaluates the underlying ranking function, and we place it in the more general context of ROC analysis as the volume under an r-dimensional ROC surface (VUS) for in general r classes. Furthermore, we also discuss the scalability of this measure and show that it can be computed very efficiently for large samples. Finally, we present a kernel-based learning algorithm that optimizes VUS as a specific case of structured support vector machines.

1 Introduction

In many situations, humans compare items or objects to choose the appropriate item for a specific goal. Think for example of buying clothes, listening to music, the dish one orders in a restaurant, etc. Continually, we evaluate objects on criteria such as appropriateness, beauty, correctness, etc. In research areas like decision making, preference modeling, fuzzy modeling, statistics and machine learning, scientists have proposed various ways to characterize this human behavior with mathematical

W. Waegeman (✉) and B. De Baets
Department of Applied Mathematics, Biometrics and Process Control, Ghent University, Coupure links 653, B-9000 Ghent, Belgium
e-mail: Willem.Waegeman,Bernard.DeBaets@UGent.be

J. Fürnkranz and E. Hüllermeier (eds.), *Preference Learning*,
DOI 10.1007/978-3-642-14125-6_7,

models. In a preference modeling scope, mainly two learning settings can be distinguished: ranking models and pairwise preference models. The connection between both types of models is characterized by the transitivity property, since transitive pairwise preference models can be expressed in terms of a ranking [46, 49]. Here, we will concentrate on a particular type of ranking problems, namely, the ordinal regression problem, where classes typically correspond to quotations or linguistic terms – varying from "very bad" to "brilliant" for example – that express a difference in correctness, quality, beauty, or any other characteristic of the analyzed objects.

We will start in Sect. 2 with a general introduction to ordinal regression as a special case of the more general ranking framework. Subsequently, in following sections, we particularly focus on the important role of ROC analysis in ordinal regression. In binary classification, ROC curves provide a way to select possibly optimal models for discriminating two kinds of objects without the need of specifying the cost or class distribution. The area under the ROC curve (AUC) has a graph–theoretic interpretation and corresponds to the Wilcoxon–Mann–Whitney test statistic. It is nowadays commonly used as performance measure for evaluating binary classifiers. As a consequence, it is straightforward to think of introducing similar techniques for ordinal regression problems as well, by investigating the underlying ranking function.

In the machine learning field, it is known that existing measures for evaluating ordinal regression models suffer from a number of important shortcomings. Because of that, we develop in Sect. 3 alternative measures by extending existing ROC measures for classification to ordinal regression problems, and we investigate whether these measures can compete with the existing ones. Similar to multiclass classification problems, we generalize the AUC and its underlying probabilistic interpretation to ordinal regression problems such that it now corresponds to the volume under an r-dimensional surface (VUS) for r ordered classes. By counting the number of correctly ranked r-tuples consisting of one data object of each class, VUS rather evaluates the ranking returned by an ordinal regression model instead of measuring the error rate. This is a way of thinking which is especially advantageous in case of skew class or cost distributions. We give theoretical and experimental evidence of the advantages and different behavior of VUS compared to error rate, mean absolute error, and other ranking-based performance measures for ordinal regression. The results demonstrate that the models produced by ordinal regression algorithms minimizing the error rate or a pairwise error function not necessarily impose a good ranking on the data.

The computation of VUS as well as the U-statistics estimators of its variance and covariance for two models is believed to be complex. That is why we analyze its scalability in Sect. 4, where new algorithms to compute VUS and its (co)variance estimator are presented. In particular, the volume under the ROC surface can be found very efficiently with a simple dynamic program dominated by a single sorting operation on the data set. Simulation experiments confirm that the presented algorithms scale well with respect to the size of the data set and the number of classes. For example, the volume under the ROC surface could be rapidly computed on very large data sets of more than 500,000 instances, while a naive implementation spent much more time on data sets of size less than 1,000.

Finally, in Sect. 5 we discuss how VUS can be embedded as a loss function in a learning algorithm, unlike the conventional approach of minimizing the pairwise error. This leads to a new type of kernel method based on structured support vector machines. This method tries to optimize the fraction of correctly ranked r-tuples. A large number of constraints appear in the resulting quadratic program, but the optimal solution can be computed in $\mathcal{O}(n^3)$ time for samples of size n with a cutting plane algorithm and graph-based techniques. Our approach can offer benefits for applications in various domains. On various synthetic and benchmark data sets, it outperforms the pairwise approach for balanced as well as unbalanced problems. In addition, scaling experiments confirm the theoretically derived time complexity.

2 Ordinal Regression as a Special Case of Ranking

We start by introducing some notations. Let us assume that examples, training data as well as test data, are identically and independently drawn according to an unknown distribution $\mathcal{D}$ over $\mathcal{X} \times \mathcal{Y}$ with $\mathcal{X}$ the object space and $\mathcal{Y}$ the set of labels. As mentioned above, $\mathcal{Y} = \{\mathcal{C}_1, \ldots, \mathcal{C}_r\}$ will be an ordered set of class labels containing r elements. The conditional distribution of an object given that it belongs to class $\mathcal{C}_k$ will be denoted $\mathcal{D}_k$. Furthermore, a data set of size n, i.i.d. according to $\mathcal{D}$, will be denoted $D = \{(\mathbf{x}_1, y_1), \ldots, (\mathbf{x}_n, y_n)\}$ and it will contain n_k elements of class $\mathcal{C}_k$.

During the last decade, a lot of interesting papers on ordinal regression and the related setting of pairwise preference learning have appeared in the machine learning community [3,5–12,15,16,23,24,26,31,32,35,41,44,45,51,52,55,56,58,64,66]. We do not intend to explain all these approaches in detail, but we will rather focus on the common aspects.

2.1 Formal Definition of an Ordinal Regression Model

Formally speaking, an ordinal regression model $h : \mathcal{X} \to \mathcal{Y}$ maps a data object to one of the classes of $\mathcal{Y}$. The vast majority of existing ordinal regression models can be represented in the following general form

$$h(\mathbf{x}) = \begin{cases} \mathcal{C}_1, & \text{if } f(\mathbf{x}) < b_1, \\ \mathcal{C}_k, & \text{if } b_{k-1} < f(\mathbf{x}) \leq b_k, k = 2, \ldots, r-1, \\ \mathcal{C}_r, & \text{if } f(\mathbf{x}) > b_{r-1}, \end{cases} \tag{1}$$

with $b_1, \ldots, b_{r-1}$ free parameters and $f : \mathcal{X} \to \mathbb{R}$ any function that assigns a real value to a data object. It is possible to impose an ordering on a sequence of data objects with f and therefore it is commonly referred to as a *ranking* function. In this way, consecutive quality levels are modeled by consecutive intervals on the real line. This is a direct extension of binary classification, where only two intervals are considered and a single threshold b defines in the linear case a hyperplane perpendicular to the discriminant function. Roughly speaking, we will consider in general $r-1$ hyperplanes for linear ranking functions in an r-class ordinal regression setting.

Ordinal regression models with an underlying ranking function like (1) have less parameters compared to their multiclass classification counterparts: with r classes, r ranking functions are inferred in a one-versus-all ensemble [53] and even $r(r-1)/2$ functions are inferred in a one-versus-one ensemble [25, 33]. Owing to this simplification, one can often interpret ordinal regression models more easily and less data is required to get stable estimates of the parameters and the predictions. In Fig. 1, we give a graphical illustration of the reduced complexity of an ordinal regression model compared to a multiclass classification model.

Yet, can the reduction to a single ranking on the other hand result in a too simple model? The answer is usually no, since in any realistic application of ordinal regression the assumption that the humanly assigned ordinal labels originate from a latent (thus unobserved) continuous variable is made. In some way, the ranking function of the ordinal regression model represents this latent variable. Moreover, by inferring a single function, ordinal regression models also reduce the possibility of making "big" errors, i.e., misclassifying objects into a much lower or higher class than the real one. In exceptional cases, when such an underlying latent variable is missing, the data might be too complex to be modeled with a single ranking function. Then, a multiclass classification scheme would be preferred [38, 56]. We will not consider that kind of problems in this work. On the other hand, one can also wonder whether a multiclass classification model can be simplified to an ordinal regression model in some occasions. The answer to that question can be found in [59, submitted].

The vast majority of existing methods for ordinal regression comply with the assumption of an underlying latent variable and can therefore be represented as (1). This holds for example for traditional statistical methods [2], kernel methods [10, 55], Bayesian approaches [7]), ordinal decision trees [23, 41], neural models [16], etc. Since this work will not concentrate on clarifying the main differences between these approaches, we will not give a detailed explanation of all of them. Hereunder, the proportional odds model and the basic kernel-based ordinal regression model are presented.

2.2 The Proportional Odds Model

In statistics, the proportional odds model [47] is without doubt the best known and most applied technique to represent ordinal responses. This kind of model involves modeling cumulative logits. Given a data set $D \subset \mathcal{X} \times \mathcal{Y}$, the cumulative probability of observing an outcome greater than or equal to $\mathcal{C}_k$ is defined as follows:

$$P_{ik} = \mathbf{Pr}\{y_i \geq \mathcal{C}_k \mid \mathbf{x}_i\},$$

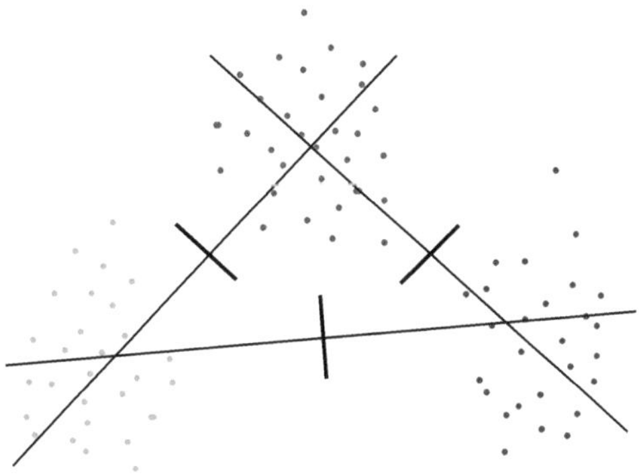

(a) A linear one-versus-one classifier.

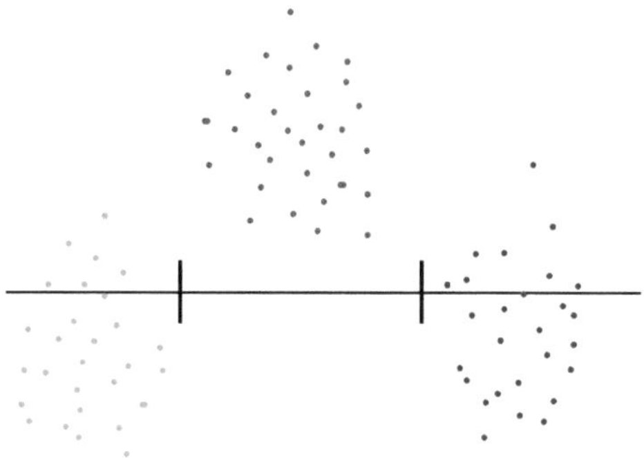

(b) A linear ordinal regression model.

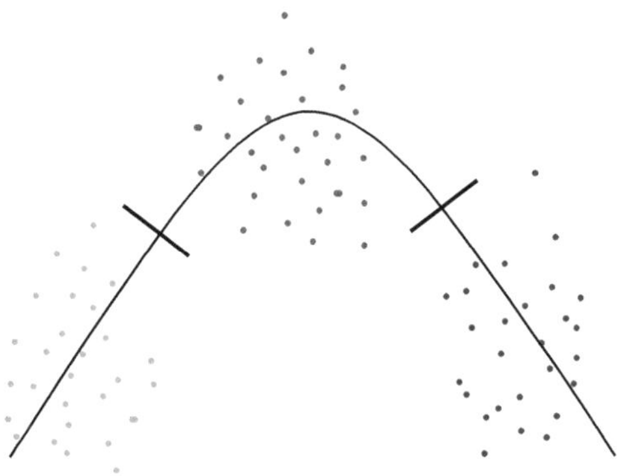

(c) A non-linear ordinal regression model.

Fig. 1 Graphical illustration of the difference between one-versus-one multiclass classification and ordinal regression models. A three-class hypothetical data set is considered

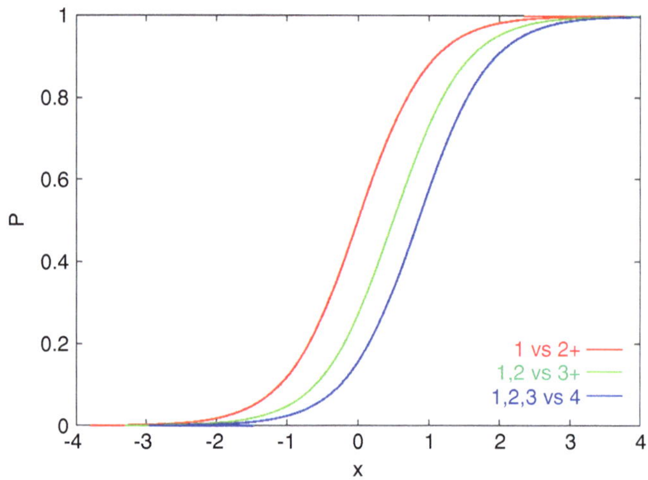

Fig. 2 An example of a proportional odds model with a one-dimensional input and four classes

for $i = 1, \ldots, n$ and $k = 1, \ldots, r$. Similar to the logistic regression model for binary responses, the proportional odds model fits a linear model to the log-odds of these cumulative probabilities, i.e.,

$$\log\left(\frac{P_{ik}}{1 - P_{ik}}\right) = \mathbf{w} \cdot \mathbf{x}_i + b_k,$$

for $i = 1, \ldots, n$ and $k = 1, \ldots, r$. Note that this construction can be translated into (1). Since a single vector $\mathbf{w}$ of parameters is used, the model has the same effect for each class. As a consequence, all response curves for individual classes have the same shape. They share exactly the same rate of increase or decrease but are horizontally displaced by the thresholds b_k. An example is given in Fig. 2. One can see that the proportional odds model is a direct generalization of the logistic regression model by considering a threshold for each class [2]. If a different slope for each class would be considered too, a one-versus-one ensemble with logistic regression models as binary classifiers would be obtained.

Remark that the proportional odds model can easily be generalized to nonlinear ranking functions by putting

$$\log\left(\frac{P_{ik}}{1 - P_{ik}}\right) = f(\mathbf{x}_i) + b_k, \tag{2}$$

but much of its interpretability gets lost.

2.3 Support Vector Ordinal Regression

The support vector ordinal regression algorithm has been introduced in [55] as a direct generalization of SVMs to more than two ordered classes. Later, this algorithm was enhanced in [9, 10] by repairing some shortcomings of the initial algorithm. They actually presented two slightly different support vector approaches to ordinal regression. We will only expound one of them, namely the version with implicit constraints on the thresholds. Similar to other kernel methods, we consider a model of the form

$$f(\mathbf{x}) = \mathbf{w} \cdot \phi(\mathbf{x}),$$

corresponding to a linear function in a given feature space of basis expansions [17, 54]. This model acts as a ranking function that forms the basis of (1). In this way, the thresholds can be interpreted as hyperplanes in kernel space that lie perpendicular to the direction $\mathbf{w}$. According to Tikhonov regularization, the vector $\mathbf{w}$ and $r-1$ thresholds b_k are inferred such that the weighted sum of the error on training data and the regularizer $\frac{1}{2}||\mathbf{w}||^2$ are minimized. Errors occur when an object of class $\mathcal{C}_k$ does not lie in the part of the space defined by the thresholds b_{k-1} and b_k. When more than one hyperplane is located in between the part of the space defined by the real and the predicted class, then all these errors will be taken into account separately. This is known as an all-threshold loss function [52]. Similar to the SVM, individual errors are denoted with slack variables ξ. Let us introduce $\mathbf{x}_i^k$ to denote the ith object of class $\mathcal{C}_k$ and ξ_{ki}^j to denote the slack variable associated with this object on threshold b_j, then we arrive at the following (primal) optimization problem:

$$\min_{\mathbf{w},b_k} \quad \frac{1}{2}||\mathbf{w}||^2 + C\sum_{j=1}^{r-1}\left(\sum_{k=1}^{j}\sum_{i=1}^{n_k}\xi_{ki}^j + \sum_{k=1}^{j}\sum_{i=1}^{n_k}\xi_{ki}^{*j}\right)$$

$$\text{subject to} \begin{cases} \mathbf{w}\cdot\phi(\mathbf{x}_i^k) - b_j \leq -1 + \xi_{ki}^j, \\ \xi_{ki}^j \geq 0, \\ \text{for } k = 1,\ldots,j \text{ and } i = 1,\ldots,n_k; \\ \mathbf{w}\cdot\phi(\mathbf{x}_i^k) - b_j \geq +1 - \xi_{ki}^{*j}, \\ \xi_{ki}^{*j} \geq 0, \\ \text{for } k = j+1,\ldots,r \text{ and } i = 1,\ldots,n_k, \end{cases} \tag{3}$$

where j runs over $1,\ldots,r-1$. A visualization of the all-threshold loss function and the resulting active constraints is given in Fig. 3.

Like for other kernel methods, the transformation to the basis ϕ must not be computed explicitly due to the kernel trick. Optimization problem (3) can be rewritten in terms of dot products by introducing Lagrange multipliers. The dual problem is optimized by a variant of the sequential minimal optimization algorithm for support vector machines [50]. More details about the derivation of the dual problem and the implementation can be found in [10].

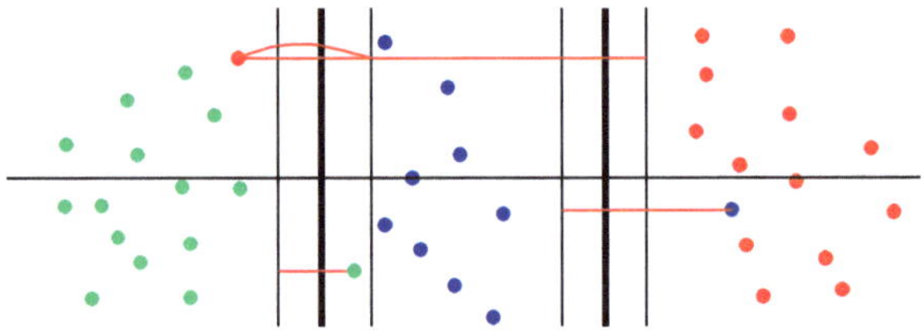

Fig. 3 An example to illustrate the all-threshold loss function of the support vector ordinal regression method. The red lines denote active slack variables ξ. As shown in the figure, the misclassified red data object gives rise to two active slack variables, since two hyperplanes are situated in between its predicted class and its real class. On the other hand, only one slack variable becomes active for the misclassified green and blue data objects

3 ROC Analysis in Ordinal Regression

Different measures are used in the literature to evaluate the performance of an ordinal regression model. Although it is established that accuracy has serious shortcomings, various authors nowadays still report it as a measure of the goodness of fit of an ordinal regression model. Accuracy gives a biased view of the performance, because it does not take into account the magnitude of an error. For example, misclassifying an object of class $\mathcal{C}_4$ into class $\mathcal{C}_1$ is a more serious error than misclassifying it into class $\mathcal{C}_3$. Other frequently used measures that take the magnitude between the real and the predicted class into account are the mean squared error and the *mean absolute error* (MAE), i.e.,

$$\text{MAE}(h, D) = \frac{1}{n} \sum_{i=1}^{n} |h(x_i) - y_i|. \tag{4}$$

However, here one assumes a metric on the output space $\mathcal{Y}$, but no metric can be defined on an ordinal scale. So, the use of mean squared error and mean absolute error in ordinal regression settings should be avoided too.

In the statistical literature, often the C-index or concordance index is reported as a measure of the goodness of fit. This measure is an estimator of the concordance probability of an ordinal regression model by counting the number of (lower-class; higher-class) object couples that are correctly ranked by the model [27]. Compared to the mean squared error, it has the important advantage that no metric on $\mathcal{Y}$ is required. Given a model $h : \mathcal{X} \to \mathcal{Y}$ defined by (1), the C-index ignores the thresholds b_k and assesses the predictive power of f. In machine learning, it is often replaced by the equivalent pairwise error [35]. The C-index is formally given by:

$$\widehat{U}_{\text{pairs}}(f, D) = \frac{1}{\sum_{\mathcal{C}_k < \mathcal{C}_l} n_k n_l} \sum_{y_i < y_j} I_{f(\mathbf{x}_i) < f(\mathbf{x}_j)},$$

with I the indicator function returning 1 when its argument is true and zero otherwise. Thus, $\widehat{U}_{\text{pairs}}(f, D)$ counts the number of couples of objects in the data set D correctly ranked by the franking function f. It yields an unbiased estimate of the probability that the order of two objects (X_1, Y_1) and (X_2, Y_2), independently drawn from $\mathcal{D}$, is consistent with the order of their class labels, i.e.,

$$U_{\text{pairs}}(f) = \mathbf{Pr}_{(X_1,Y_1),(X_2,Y_2)\sim\mathcal{D}}\{f(X_1) < f(X_2) \mid Y_1 <_{\mathcal{Y}} Y_2\}.$$

Remark that $<_{\mathcal{Y}}$ denotes the linear order relation on the classes. One can easily verify that $\widehat{U}_{\text{pairs}}(f, D)$ matches the *area under the ROC curve* (AUC) when $\mathcal{Y}$ contains only two labels. For more than two classes, $\widehat{U}_{\text{pairs}}(f, D)$ is related to the multiclass AUC of [28] for evaluating one-versus-one multiclass classification models. The one-versus-one multiclass AUC simply takes the average over the AUCs measured on binary scoring classifiers for all pairs of classes. Contrary to multiclass classification, let us now employ the same ranking function for each pair of classes, then we get the following performance measure

$$\begin{aligned}\widehat{U}_{\text{ovo}}(f, D) &= \frac{2}{r(r-1)} \sum_{k<l} \widehat{A}_{kl}(f, D),\\ \widehat{A}_{kl}(f, D) &= \frac{1}{n_k n_l} \sum_{y_i=\mathcal{C}_k} \sum_{y_j=\mathcal{C}_l} I_{f(\mathbf{x}_i)<f(\mathbf{x}_j)},\end{aligned} \tag{5}$$

with $\widehat{A}_{kl}(f, D)$ the AUC obtained when only objects of classes $\mathcal{C}_k$ and $\mathcal{C}_l$ are considered. In nonparametric statistics, $\widehat{U}_{\text{ovo}}(f, D)$ is known as the *Jonckheere–Terpstra* statistic, a more powerful alternative to the *Kruskal–Wallis* statistic for simultaneously testing whether more than two ordered populations significantly differ [37]. $\widehat{U}_{\text{ovo}}(f, D)$ yields un unbiased estimate of

$$U_{\text{ovo}}(f) = \frac{2}{r(r-1)} \sum_{k<l} \mathbf{Pr}_{X_1\sim\mathcal{D}_k, X_2\sim\mathcal{D}_l}\{f(X_1) < f(X_2)\}.$$

Like $\widehat{U}_{\text{ovo}}(f, D)$, $\widehat{U}_{\text{pairs}}(f, D)$ can also be expressed in terms of binary AUCs:

$$\widehat{U}_{\text{pairs}}(f, D) = \frac{1}{\sum_{\mathcal{C}_k<\mathcal{C}_l} n_k n_l} \sum_{k<l} n_k n_l \widehat{A}_{kl}(f, D).$$

Evaluating the ranking by pairwise comparisons might in many cases not be the best option. This is for example the case in information retrieval applications, where typically one is primarily interested in the top of the ranking. Measures such as normal discounted cumulative gain and mean average precision better reflect the desired performance in information retrieval applications. To this end, several ranking methods that concentrate on learning the top of the ranking have been proposed recently, see e.g., [5, 11].

The idea of focusing on correctly predicting the top of the ranking might be advisable for other application domains as well, but correctly predicting one end of the ranking does not have to be important in general. One could be interested in correctly predicting the top and the tail simultaneously, or, in other situations, all r classes could be equally important. The first situation, for example, arises when the ordinal levels express a linguistic degree of uncertainty about the class of the object, which is typical for label assignments by humans. When the goal does not consist of predicting the top end of the ranking correctly, the traditional pairwise approach might manifest important shortcomings as well. The main reason is that the pairwise approach to evaluation, as well as the multiclass AUC, subdivides the performance evaluation into evaluations of pairs of classes individually. However, the observed trend might be substantially different for each pair of classes. Let us illustrate this claim with a rather extreme example. Table 1 gives an overview of a hypothetical ranking of a data set containing 18 instances from three ordered classes. Computing the AUC for each pair of classes yields:

$$\widehat{A}_{1,2}(f,D) = 0.556, \quad \widehat{A}_{2,3}(f,D) = 0.694, \quad \widehat{A}_{1,3}(f,D) = 0.419.$$

Intuitively, one would expect that

$$\widehat{A}_{1,3}(f,D) \geq \max\{\widehat{A}_{1,2}(f,D), \widehat{A}_{2,3}(f,D)\}.$$

In preference modeling, this property is known as strong stochastic transitivity. However, the pairwise AUC as defined by (5) is in the example even not weakly stochastically transitive, i.e.,

$$\widehat{A}_{1,3}(f,D) \geq \frac{1}{2},$$

does not hold. The pairwise AUC can be interpreted as a specific type of probabilistic relation that was introduced recently for comparing collections of dice and independent random variables in a pairwise manner. Such probabilistic relations exhibit a specific type of cycle-transitivity, namely dice-transitivity [20, 21], which is much weaker than strong stochastic transitivity. We refer to [18, 19] for an introduction to the general framework of cycle-transitivity and the various types of transitivity it covers. Here, we only want to illustrate that both $\widehat{U}_{\text{pairs}}(f,D)$ and $\widehat{U}_{\text{ovo}}(f,D)$ can yield quite inconsistent performance evaluations with regard to the

Table 1 A hypothetical example of a data set containing 18 objects with three possible labels to illustrate the inconsistency that can occur when the ranking function is evaluated with pairwise comparisons. The ordering of the objects in the table represents the obtained ranking, increasing from left to right (example taken from [20])

i	1	2	3	4	5	6	7	8	9	10	11	12	13	14	15	16	17	18
y_i	C_3	C_1	C_1	C_1	C_2	C_2	C_2	C_2	C_2	C_3	C_3	C_3	C_3	C_3	C_1	C_1	C_2	C_1

performance on each pair of classes individually, because they are constructed from binary AUCs. This effect was, for example, observed for unbalanced data sets in [64]. When the data set is very unbalanced, models minimizing the pairwise error might overfit on the most occurring classes and these classes are often not the classes in which one is the most interested.

We aim to assess the ranking in a global way instead of considering pairwise comparisons. Therefore, let us consider the following probability:

$$U(f) = \mathbf{Pr}_{X_k \sim \mathcal{D}_k}\Big\{f(X_1) < \ldots < f(X_r)\Big\},$$

which is an evident generalization of the concept of *expected ranking accuracy* introduced in [1] for bipartite ranking. Contrary to $U_{\text{pairs}}(f)$, $U(f)$ can be naturally expressed in terms of the conditional distributions $\mathcal{D}_k$. As for the AUC, a nonparametric unbiased estimator of $U(f)$ is given by

$$\widehat{U}(f, D) = \frac{1}{\prod_{k=1}^{r} n_k} \sum_{y_{j_1} < \ldots < y_{j_r}} I_{f(\mathbf{x}_{j_1}) < \ldots < f(\mathbf{x}_{j_r})}.$$

We refer to [43] for an in-depth explanation why the U-statistics estimators presented in this article are unbiased estimators. Instead of counting the number of correctly ranked object couples, we now examine all sequences of r objects, one of each class. Furthermore, $\widehat{U}(f, D)$ has a geometrical interpretation. Several authors [22, 48, 63] remarked that $\widehat{U}(f, D)$ corresponds to the volume under the ROC surface (VUS) for r ordered classes.[1] In this case, the ROC space is spanned by the true positive rates of each class (i.e., the fraction of correctly classified instances of that class).

Theorem 1 ([63]). *Given a ranking function $f : \mathcal{X} \to \mathbb{R}$ that imposes a ranking over a data set $D \subset \mathcal{X} \times \mathcal{Y}$, then $\widehat{U}(f, D)$ corresponds to the volume under the r-dimensional ROC surface (VUS) spanned by the true positive rates of each class.*

For three ordered classes, the ROC surface can be visualized. We have constructed this ROC surface for a synthetic data set. We sampled 3×100 instances from 3 bivariate Gaussian clusters representing consecutive classes. The mean of the clusters was set to (10,10), (20,10) and (20,20) respectively, σ_1 and σ_2 were set to 5 for the first two clusters and σ_3 was set to 7 for the last cluster. ρ was fixed to 0. This data set is visualized in Fig. 4a. We used the support vector ordinal regression algorithm of [9] to estimate the ranking function f, without looking at the thresholds. The obtained ROC surface is shown in Fig. 4b.

[1] Unlike Nakas and Yannoutsos [48] who mainly place ROC analysis in a medical decision-making perspective, we rather focus on its use in a machine learning context. The approach in [22] is limited to the three-class case.

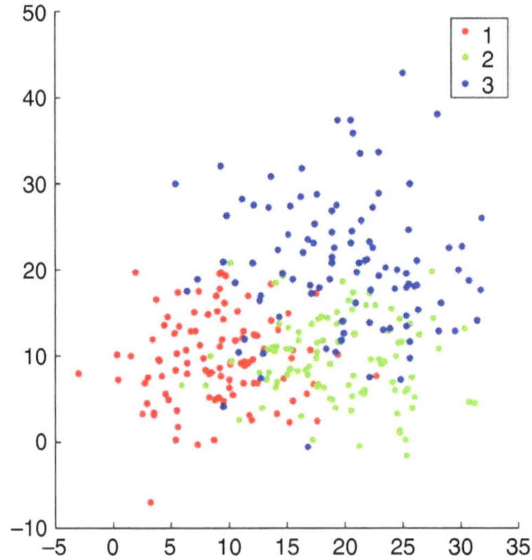

(a) A synthetic data set with three bivariate Gaussian clusters representing three ordered classes with respective means (10,10), (20,10) and (20,20). The standard deviation was set to (5,5) for the first two clusters and to (7,7) for the last cluster, while ρ was fixed to 0.

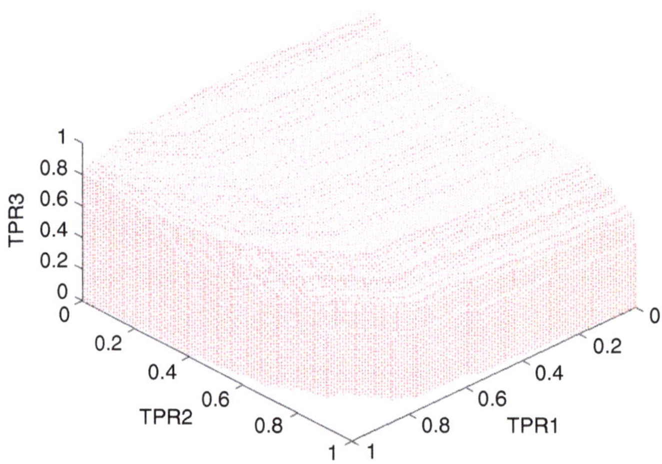

(b) The ROC surface obtained for the synthetic data set given in (a) and the ranking returned by a support vector ordinal regression algorithm.

Fig. 4 The ROC surface visualized for a three-class ordinal regression data set

In [63], several experiments were reported to illustrate that different models will be obtained when different performance measures are optimized in an ordinal regression setting. In one of these experiments, the multiobjective particle swarm optimization algorithm MOPSO was used to find the tradeoff between different performance measures in the region of the hypothesis space where the better solutions

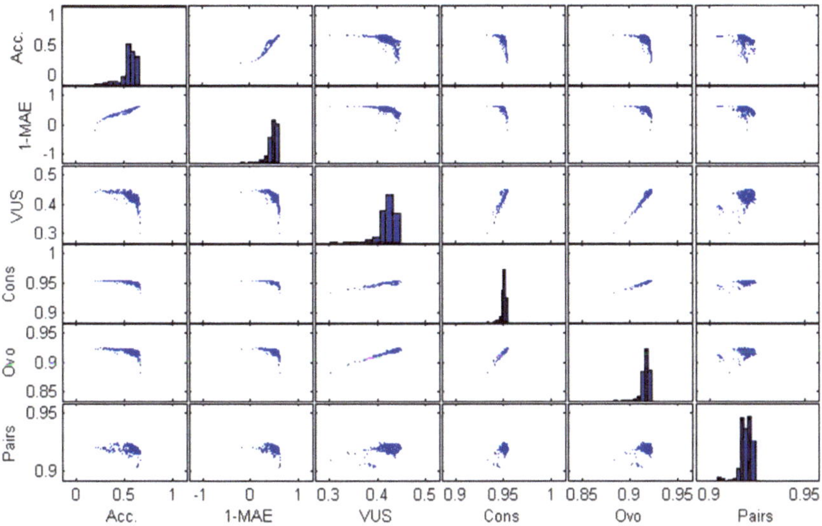

Fig. 5 The set of nondominated solutions aggregated from 20 runs of the MOPSO-algorithm. The six-dimensional Pareto front is plotted as a matrix of two-dimensional scatter plots showing the tradeoff for each pair of objectives

are located. Six different objectives were considered, namely accuracy (or equivalently "1 - mean zero-one error"), "1 - mean absolute error", $\widehat{U}(f, D)$, $\widehat{U}_{\text{ovo}}(f, D)$, $\widehat{U}_{\text{pairs}}(f, D)$ and a sixth objective $\widehat{U}_{\text{cons}}(f, D)$ that is similar to $\widehat{U}_{\text{ovo}}(f, D)$, but not further discussed due to lack of space. The algorithm was executed 20 times for 100 iterations with a population of 500 particles and different seeds for the random generator. In all runs, the nondominated solutions found during the search were stored in a repository and afterward the global nondominated set of these 20 repositories was computed. This set is visualized by a matrix of two-dimensional Pareto fronts in Fig. 5.

One can easily see that none of the six measures manifests a monotone relationship with another. Accuracy and mean absolute error, on the one hand, and the ranking-based measures, on the other hand exhibit a relatively large tradeoff, as almost all solutions lie on the two-dimensional front. The ranking-based performance measures give also rise to tradeoffs, but here the monotonic association is more prominent. The multiclass approaches $\widehat{U}_{\text{cons}}(f, D)$ and $\widehat{U}_{\text{ovo}}(f, D)$ turn out to approximate the behavior of $\widehat{U}(f, D)$ better than simply counting all correctly ordered pairs. Apparently, for $\widehat{U}_{\text{pairs}}(f, D)$ the optimal models are biased toward correctly ranking the biggest classes (due to the skew class distribution, the data set contains only 640 object pairs of classes $\mathcal{C}_4$ and $\mathcal{C}_5$ compared to more than 15,000 object pairs of the biggest classes $\mathcal{C}_1$ and $\mathcal{C}_2$). Methods minimizing the error rate or the number of incorrect instance pairs hence will both overfit on the biggest classes.

4 Scalability Aspects

In the previous section, the concept of an ROC curve was generalized to an r-dimensional surface for r ordered classes so that the volume under this ROC surface (VUS) measures the overall power of an ordinal regression model to classify objects of the various classes. However, the computation of this criterion as well as the U-statistics estimators of its variance and covariance for two models is believed to be complex. New algorithms to compute VUS and its (co)variance estimator are presented in this section. In particular, the volume under the ROC surface can be found very efficiently with a simple dynamic program dominated by a single sorting operation on the data set.

4.1 Variance and Covariance Estimators of VUS

As mentioned, the generalization of ROC analysis presented in the previous chapter has been reported in [22, 48] from a medical decision making perspective, in which the observed numbers are treated as random variables instead of outcomes of an ordinal regression model. Based on U-statistics, these authors derive expressions for the variance of the volume under the ROC surface and the covariance of the volumes obtained for two different ranking functions f_1 and f_2, as well as nonparametric estimators of these quantities. We briefly recapitulate their observations. To this end, let us define Υ as the set of all splits of $\mathcal{Y}$ into two (possibly empty) disjoint parts, i.e.,

$$\Upsilon := \{(\mathcal{Z}_1, \mathcal{Z}_2) \mid \mathcal{Z}_1 \cup \mathcal{Z}_2 = \mathcal{Y} \wedge \mathcal{Z}_1 \cap \mathcal{Z}_2 = \emptyset\}.$$

Let $f : \mathcal{X} \to \mathbb{R}$ be a fixed ranking function and let $D \subset \mathcal{X} \times \mathcal{Y}$ be any data set of size n identically and independently distributed according to $\mathcal{D}$ with n_k instances sampled according to $\mathcal{D}_k$ as defined in Sect. 2, then Nakas and Yiannoutsos define the variance of the volume under the ROC surface as:

$$\sigma_U^2(f) = \frac{1}{\prod_{k=1}^{r} n_k} \sum_{(\mathcal{Z}_1,\mathcal{Z}_2)\in\Upsilon} \Big(\prod_{C_l \in \mathcal{Z}_2} (n_l - 1) \Big) (q_v(f, \mathcal{Z}_1, \mathcal{Z}_2) - U(f)^2).$$

In this expression, $q_v(f, \mathcal{Z}_1, \mathcal{Z}_2)$ is still undefined. For each $(\mathcal{Z}_1, \mathcal{Z}_2) \in \Upsilon$, we introduce random variables X_i and X_i' as follows:

$$\begin{aligned} X_i &\sim \mathcal{D}_i, \ \ \forall i : C_i \in \mathcal{Y}, \\ X_i' &= X_i, \ \ \forall i : C_i \in \mathcal{Z}_1, \\ X_i' &\sim \mathcal{D}_i, \ \ \forall i : C_i \in \mathcal{Z}_2. \end{aligned}$$

Then $q_v(f, \mathcal{Z}_1, \mathcal{Z}_2)$ is defined as:

$$q_v(f, \mathcal{Z}_1, \mathcal{Z}_2) = \mathbf{Pr}\Big\{(f(X_1) < \ldots < f(X_r)) \wedge (f(X_1') < \ldots < f(X_r'))\Big\}.$$

The measure $q_v(f, \mathcal{Z}_1, \mathcal{Z}_2)$ represents the probability that two r-tuples are correctly ranked by the ranking function f, in which each r-tuple is randomly sampled according to $\mathcal{D}_1 \times \ldots \times \mathcal{D}_r$ with the restriction that the objects sampled from the classes $\mathcal{C}_k \in \mathcal{Z}_1$ are identical. Thus, we consider a single random variable for classes $\mathcal{C}_k \in \mathcal{Z}_1$ and two random variables for classes $\mathcal{C}_k \in \mathcal{Z}_2$. This expression for the variance is a natural extension of the one in the binary case [14, 29, 30].

The variance of the volume under the ROC surface still depends on the unknown distribution $\mathcal{D}$ of the data but can be estimated from a fixed data set $D \subset \mathcal{X} \times \mathcal{Y}$, by estimating $U(f)$ and $q_v(f, \mathcal{Z}_1, \mathcal{Z}_2)$ from D, i.e.,

$$\widehat{\sigma}_U^2(f, D) = \frac{1}{\prod_{k=1}^r n_k} \sum_{(\mathcal{Z}_1, \mathcal{Z}_2) \in \Upsilon} \Big(\prod_{\mathcal{C}_l \in \mathcal{Z}_2} (n_l - 1) \Big) (\widehat{q_v}(f, D, \mathcal{Z}_1, \mathcal{Z}_2) - \widehat{U}(f, D)^2).$$

We first introduce

$$\underline{n} := \prod_{k=1}^{r} n_k \prod_{\mathcal{C}_l \in \mathcal{Z}_2} n_l,$$

and a shorter notation for the indicator function of a given ranking function and a k-tuple, i.e.,

$$I_f(j_1, \ldots, j_k) := I_{\{f(\mathbf{x}_{j_1}) < \ldots < f(\mathbf{x}_{j_k})\}}.$$

Using these notations, the following estimator for $q_v(f, \mathcal{Z}_1, \mathcal{Z}_2)$ is proposed:

$$\widehat{q_v}(f, D, \mathcal{Z}_1, \mathcal{Z}_2) = \frac{1}{\underline{n}} \sum_{y_{j_1} < \ldots < y_{j_r}} \sum_{\substack{y_{j'_1} < \ldots < y_{j'_r} \\ \forall \mathcal{C}_k \in \mathcal{Z}_1 : j_k = j'_k}} I_f(j_1, \ldots, j_r) \cdot I_f(j'_1, \ldots, j'_r).$$

This estimator counts the number of couples of r-tuples that are correctly ranked by f, with the restriction that the data objects sampled from $\mathcal{D}_k$ must be identical objects when $\mathcal{C}_k \in \mathcal{Z}_1$.

A similar strategy can be followed to derive an expression for the covariance of the volumes obtained for two ranking functions f_1 and f_2, i.e.,

$$\mathrm{Cov}_U(f_1, f_2) = \frac{1}{\prod_{k=1}^r n_k} \sum_{(\mathcal{Z}_1, \mathcal{Z}_2) \in \Upsilon} \Big(\prod_{\mathcal{C}_l \in \mathcal{Z}_2} (n_l - 1) \Big) \quad (6)$$
$$(q_c(f, \mathcal{Z}_1, \mathcal{Z}_2) - U(f_1) \cdot U(f_2))$$

and

$$q_c(f_1, f_2, \mathcal{Z}_1, \mathcal{Z}_2) = \mathbf{Pr}\Big\{(f_1(X_1) < \ldots < f_1(X_r)) \wedge (f_2(X'_1) < \ldots < f_2(X'_r))\Big\},$$

with $X_1, \ldots, X_r, X'_1, \ldots, X'_r$ random variables as formerly defined. We derive a nonparametric unbiased estimator in the same way as for the variance, i.e.,

$$\widehat{q_c}(f_1, f_2, D, \mathcal{Z}_1, \mathcal{Z}_2) = \frac{1}{n} \sum_{y_{j_1} < \ldots < y_{j_r}} \sum_{\substack{y_{j'_1} < \ldots < y_{j'_r} \\ \forall C_k \in \mathcal{Z}_1 : j_k = j'_k}} I_{f_1}(j_1, \ldots, j_r) \cdot I_{f_2}(j'_1, \ldots, j'_r). \tag{7}$$

Then an unbiased estimator of the covariance $\widehat{\text{Cov}_U}(f_1, f_2, D)$ is obtained by inserting this estimator in (6).

Let us now analyze the computational complexity of the estimators. For the estimation of $U(f)$, all r-tuples are examined and hence, this number rapidly grows when the sample size or the number of classes increases. For a five-class ordinal regression data set with 20 instances per class, 3,200,000 such r-tuples can be formed! In general, the number of different r-tuples to be analyzed on a data set of n elements and r classes is of the order $\mathcal{O}(n^r)$.

For the covariance (and consequently also for the variance as $\sigma^2_U(f) = \text{Cov}_U(f, f)$) all splits of $\mathcal{Y}$ are considered. The procedure for estimating $q_c(f_1, f_2, \mathcal{Z}_1, \mathcal{Z}_2)$ and $q_v(f, \mathcal{Z}_1, \mathcal{Z}_2)$ thus must be repeated 2^r times, but fortunately this is not a big issue since in realistic ordinal regression applications r will only occasionally exceed 10. Again, the inspection of r-tuples gives rise to the biggest bottleneck. Now couples of r-tuples with at most $2r$ different objects are investigated. As a result, the estimation of $q_c(f_1, f_2, \mathcal{Z}_1, \mathcal{Z}_2)$ and $q_v(f, \mathcal{Z}_1, \mathcal{Z}_2)$ by exhaustively examining all couples of r-tuples is here for some splits of $\mathcal{Y}$ of the order $\mathcal{O}(n^{2r})$.

Naive exhaustive implementations of the estimators for $U(f)$, its variance and covariance, in which the indicator function is verified on all r-tuples (or couples of r-tuples), will become computationally heavy for large data sets. Fortunately, much faster algorithms can be constructed by using some combinatorial tricks. To that end, the estimators are reformulated in terms of graphs. We start with recapitulating some basic material on graph theory.

4.2 Algorithm for Computing VUS

In order to present a better algorithm for VUS, we first introduce some basic graph concepts. Formally, let us define a *graph* $G = (V, E)$ as a data structure consisting of a set of *vertices* or nodes V and a set of *edges* E, where an edge $e \in E$ corresponds to a relation between two elements $v, v' \in V$, denoted as $e = (v, v')$. We will only consider *directed* graphs, which means that E contains couples (ordered pairs) of instances. In a *weighted* graph $G = (V, E, w)$, there is an additional function $w : E \to \mathbb{R}$ defined on the edges. Next, in a *bipartite* graph the set of vertices V can be split into two disjoint subsets V_1 and V_2 so that no two nodes from the same subset are connected. In a *multipartite* graph, this property extends to splits of V

into more than two subsets $V_1, \ldots, V_k$. Finally, in a *layered* graph the set of vertices can be split into k ordered subsets (or layers) $V_1, \ldots, V_k$ so that any edge connects nodes from successive layers. We will call a layered graph complete when any two nodes from successive layers are connected.

Definition 1. Let $f : \mathcal{X} \to \mathbb{R}$ be a function that imposes a ranking on an i.i.d. data set D and let $\mathcal{Y}$ contain r categories, then we define the *Δ-graph* $G_\Delta(f, D) = (V, E)$ of f and D as a directed weighted layered graph with $r + 2$ layers $V = (V_0, \ldots, V_{r+1})$ in which a node of layer V_k is associated with any element in the data set of category $\mathcal{C}_k$, for $k = 1, \ldots, r$. The first and last layers contain only a start node (source) and an end node (sink) without a reference to any element of the data set ($V_0 = \{v_s\}, V_{r+1} = \{v_e\}$). All nodes of successive layers are connected and the weight of the edge $e = (v_1, v_2)$ connecting the nodes $v_1 = (\mathbf{x}_1, y_1) \in V_k$ and $v_2 = (\mathbf{x}_2, y_2) \in V_{k+1}$ is defined as:

$$w(e) = \begin{cases} +\infty, & \text{if } k \in \{0, r\} \\ f(\mathbf{x}_2) - f(\mathbf{x}_1), & \text{if } k \in \{1, \ldots, r-1\}. \end{cases}$$

Remark that a layered graph simply means a graph for which the set V of nodes can be subdivided into a number of subsets (in this case $V_1, \ldots, V_r$) such that all edges in the graph only connect nodes of consecutive subsets, thus nodes of layer V_l can only be connected with nodes of layers V_{l-1} and V_{l+1}.

Definition 2. Let $f : \mathcal{X} \to \mathbb{R}$ be a function that imposes a ranking on an i.i.d. data set D, let $G_\Delta(f, D) = (V, E)$ be the corresponding Δ-graph, then we define a path ϖ between v_s and v_e as a sequence of $r + 1$ edges $\varpi = (e_1, \ldots, e_{r+1})$ connecting v_s and v_e, i.e.,

$$\begin{aligned} e_1 &= (v_s, v_1), \\ e_2 &= (v_1, v_2), \\ &\vdots \\ e_r &= (v_{r-1}, v_r), \\ e_{r+1} &= (v_r, v_e). \end{aligned}$$

The set of all paths connecting v_s and v_e will be denoted Ξ.

Lemma 1. *Let $f : \mathcal{X} \to \mathbb{R}$ be a function that imposes a ranking on an i.i.d. data set D, let $G_\Delta(f, D) = (V, E)$ be the corresponding Δ-graph with Ξ the set of paths connecting source and sink, then*

$$\widehat{U}(f, D) = \frac{|\{\varpi \in \Xi \mid \forall e \in \varpi : w(e) > 0\}|}{\prod_{k=1}^{r} n_k}.$$

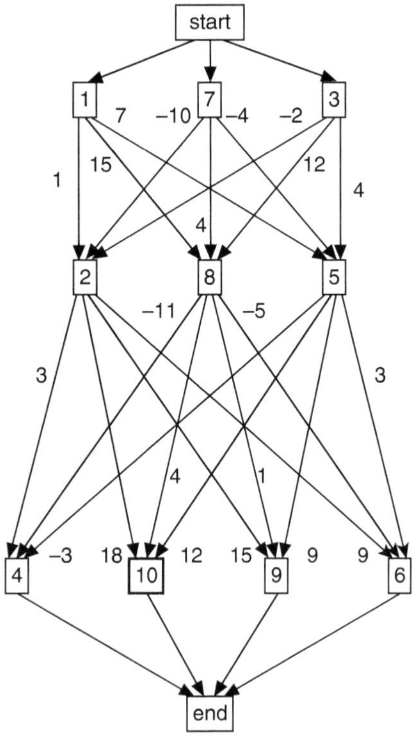

Fig. 6 The Δ-graph visualized for a hypothetical ranking on 10 data objects (see text for further details)

In order to clarify these graph concepts a small example is given. Let us consider a data set of 10 instances for which the following hypothetical ranking is obtained:

i	1	2	3	4	5	6	7	8	9	10
y_i	$\mathcal{C}_1$	$\mathcal{C}_2$	$\mathcal{C}_1$	$\mathcal{C}_3$	$\mathcal{C}_2$	$\mathcal{C}_3$	$\mathcal{C}_1$	$\mathcal{C}_2$	$\mathcal{C}_3$	$\mathcal{C}_3$
$f(\mathbf{x}_i)$	1	2	4	5	8	11	12	16	17	20

The Δ-graph characterizing this data set is visualized in Fig. 6. One can verify by enumeration that for this small data set 16 of the 27 paths from the start to the end node consist of solely positive edges, resulting in a volume under the ROC surface of 16/27. The Δ-graph forms one of the main building blocks in the next section to construct a learning algorithm, but $\widehat{U}(f, D)$ itself can be found very efficiently without constructing the graph.

Definition 3. Let $\mathcal{Z}_k = \{\mathcal{C}_1, \ldots, \mathcal{C}_k\}$ contain the first k elements of $\mathcal{Y}$ and let $D \subset \mathcal{X} \times \mathcal{Y}$ be a data set sorted according to a ranking function $f : \mathcal{X} \to \mathbb{R}$, i.e.,

$$\forall i, j \in \{1, \ldots, n\} : i < j \Rightarrow f(\mathbf{x}_i) < f(\mathbf{x}_j),$$

then, for any $p \leq n$, we define the function Ψ by:

$$\Psi(f, D, \mathcal{Z}_k, p) = \sum_{\substack{y_{j_1} < \ldots < y_{j_k} \\ \forall C_l \in \mathcal{Z}_k : j_l \leq p}} I_f(j_1, \ldots, j_r).$$

$\Psi(f, D, \mathcal{Z}_k, p)$ can be interpreted as the number of correctly ordered k-tuples occurring in the first p elements of the data set D, where a k-tuple, as defined in Sect. 4.1, consists of a sequence of one randomly drawn element of each of the k lowest classes that is part of the first p elements of the data set D.

Proposition 1. *Given a ranking function $f : \mathcal{X} \rightarrow \mathbb{R}$ that imposes a ranking on a data set $D \subset \mathcal{X} \times \mathcal{Y}$, $\widehat{U}(f, D)$ can be computed in $\mathcal{O}(n\log(n))$ time.*

We construct a fast algorithm by writing Ψ as a recurrence equation. Let the data set D be sorted according to f. From the definition of $\Psi(f, D, \mathcal{Z}, p)$ it follows that

$$\widehat{U}(f, D) = \frac{1}{\prod_{k=1}^{r} n_k} \Psi(f, D, \mathcal{Y}, n).$$

We can write $\Psi(f, D, \mathcal{Z}_k, p)$ as a recurrent function:

$$\Psi(f, D, \mathcal{Z}_k, p) = \begin{cases} \Psi(f, D, \mathcal{Z}_k, p-1), & \text{if} y_p \neq C_k, \\ \Psi(f, D, \mathcal{Z}_k, p-1) + \Psi(f, D, \mathcal{Z}_{k-1}, p-1), & \text{if} y_p = C_k, \end{cases} \tag{8}$$

for $k = 2, \ldots, r$ and for $p = 1, \ldots, n$. For $k = 1$ and $p = 1, \ldots, n$, we have:

$$\Psi(f, D, \mathcal{Z}_1, p) = \begin{cases} \Psi(f, D, \mathcal{Z}_1, p-1), & \text{if } y_p \neq C_1, \\ \Psi(f, D, \mathcal{Z}_1, p-1) + 1, & \text{if } y_p = C_1. \end{cases}$$

In addition, the starting values ($p = 0$) are:

$$\Psi(f, D, \mathcal{Z}_k, 0) = 0,$$

for $k \in \{1, \ldots, r\}$. Algorithm 1 computes $\widehat{U}(f, D)$ in $\mathcal{O}(n\log(n))$ time. Instead of implementing $\Psi(f, D, \mathcal{Y}, n)$ recursively, this simple dynamic program computes the desired quantity in one iteration over the sorted data set. The algorithm is dominated by the sorting of D according to f, which takes $\mathcal{O}(n\log(n))$ time.

We demonstrate Algorithm 1 on the ranking function and hypothetical data set of Table 2. In this example, the cardinality of $\mathcal{Y}$ is $r = 5$. The labels y_i of a fictive data set are shown and it is also assumed that the data set is sorted according to a fictive ranking function f_1. Our algorithm computes the number of correctly ordered r-tuples in one iteration over the data. For example, the r-tuple $(1, 2, 4, 7, 12)$ is correctly ranked by f_1 and the r-tuple $(1, 5, 4, 7, 12)$ is incorrectly ranked because the object of class C_3 precedes the object of class C_2. An overview of the subsequent

Algorithm 1 Computation of $\widehat{U}(f, D)$.

Input: data set $D = \{(\mathbf{x}_1, y_1), \ldots, (\mathbf{x}_n, y_n)\}$, ranking function f
Sort D according to f
Initialize $\Psi(f, D, \mathcal{Z}_k) = 0$ for $k = 1, \ldots, r$
for $j = 1$ to n **do**
    y_j corresponds to class $\mathcal{C}_k$
    if $k = 1$ **then**
        $\Psi(f, D, \mathcal{Z}_1) \leftarrow \Psi(f, D, \mathcal{Z}_1) + 1$
    end if
    if $k > 1$ **then**
        $\Psi(f, D, \mathcal{Z}_k) \leftarrow \Psi(f, D, \mathcal{Z}_k) + \Psi(f, D, \mathcal{Z}_{k-1})$
    end if
end for
$\widehat{U}(f, D) = \frac{1}{\prod_{k=1}^{r} n_k} \Psi(f, D, \mathcal{Z}_r)$
Output: $\widehat{U}(f, D)$

Table 2 Algorithm 1 for the computation of $\widehat{U}(f, D)$ demonstrated on a hypothetical data set D and fictive ranking function f. In this example, it is assumed that the data is ranked from left to right. This results in 78 of the 432 different r-tuples that are correctly ranked, giving a volume under the ROC surface of 0.181

	Steps in the algorithm																
i	1	2	3	4	5	6	7	8	9	10	11	12	13	14	15	16	17
y_i	$\mathcal{C}_1$	$\mathcal{C}_2$	$\mathcal{C}_1$	$\mathcal{C}_3$	$\mathcal{C}_2$	$\mathcal{C}_3$	$\mathcal{C}_4$	$\mathcal{C}_1$	$\mathcal{C}_2$	$\mathcal{C}_3$	$\mathcal{C}_4$	$\mathcal{C}_5$	$\mathcal{C}_4$	$\mathcal{C}_5$	$\mathcal{C}_3$	$\mathcal{C}_4$	$\mathcal{C}_5$
$\Psi(f_1, D, \mathcal{Z}_1)$	1	1	2	2	2	2	2	3	3	3	3	3	3	3	3	3	3
$\Psi(f_1, D, \mathcal{Z}_2)$	0	1	1	1	3	3	3	3	6	6	6	6	6	6	6	6	6
$\Psi(f_1, D, \mathcal{Z}_3)$	0	0	0	1	1	4	4	4	4	10	10	10	10	10	16	16	16
$\Psi(f_1, D, \mathcal{Z}_4)$	0	0	0	0	0	0	4	4	4	4	14	14	24	24	24	40	40
$\Psi(f_1, D, \mathcal{Z}_5)$	0	0	0	0	0	0	0	0	0	0	0	14	14	38	38	38	78

steps in the algorithm after sorting the data set is given and one can see that for this example 78 of the $3 \times 3 \times 4 \times 4 \times 3 = 432$ r-tuples are correctly ranked resulting in a volume of 0.181.

So, the construction of the Δ-graph can be avoided for the computation of VUS. In [62], we considered similar graph–theoretic reformulations for the variance and covariance estimators. Unlike the computation of VUS, these reformulations really helped to construct scalable algorithms for the variance and covariance estimators. In a nutshell, the estimators were reformulated in terms of recurrent functions over graphs and we showed that with dynamic programming techniques for evaluating the recurrent functions, much faster algorithms could be constructed compared to exhaustive algorithms that evaluate all r-tuples. In particular, $\mathcal{O}(2^r n^2)$ and $\mathcal{O}(2^r n^4)$ algorithms were designed for estimating the variance and the covariance respectively.

Simulations confirmed our theoretical observations and showed that the algorithms for computing $\widehat{U}(f, D)$ and $\widehat{\sigma}^2_U(f, D)$ are applicable to very large data sets. Our algorithm for computing the covariance estimator resulted in a smaller time gain compared to an exhaustive algorithm, but still scaled well with increasing r.

5 Optimizing VUS with Structured SVMs

As a final step, we develop in this section a learning algorithm that optimizes $\widehat{U}(f, D)$ or, equivalently, that learns a layered graph with a maximal number of paths connecting source and sink. It means that we will convert $\widehat{U}(f, D)$ into a suitable loss function to embed it into the framework of kernel methods. We add the large margin principle to our framework by constructing a loss function from $\widehat{U}(f, D)$. Therefore, we will define a single slack variable for each r-tuple in the data set. Given a data set D we will use the shorthand notation for $\mathbf{j} = (j_1, \ldots, j_r)$ to denote an r-tuple. We introduce $\mathcal{S}$ as the set of all r-tuples in D, i.e.,

$$\mathcal{S} := \{\mathbf{j} = (j_1, \ldots, j_r) \mid y_{j_1} < \ldots < y_{j_r}\},$$

and the symbol $s = |\mathcal{S}| = \prod_{k=1}^{r} n_k$ will represent the cardinality of this set. Let us consider such an r-tuple $\mathbf{j}$ of data objects, then we define the corresponding slack variable $\xi_{\mathbf{j}}$ as

$$\xi_{\mathbf{j}} = \max\left\{0, \max_{k \in \{1,\ldots,r-1\}} \{1 - (f(\mathbf{x}_{j_{k+1}}) - f(\mathbf{x}_{j_k}))\}\right\}.$$

This means that the slack variable $\xi_{\mathbf{j}}$ of an r-tuple $\mathbf{j}$ will differ from zero only when the difference of the function values of any couple of consecutive elements of the r-tuple is less than one. In particular, $\xi_{\mathbf{j}}$ will take 1 minus the lowest value of these $r-1$ differences in function values for any couple of consecutive objects in the r-tuple $\mathbf{j}$, or, in terms of the Δ-graph, $\xi_{\mathbf{j}}$ equals 1 minus the weight of the edge with the lowest weight on the path from source to sink defined by the r-tuple $\mathbf{j}$.

Let us consider learning in reproducing kernel Hilbert spaces $\mathcal{F}$ induced by a kernel function $K : \mathcal{X}^2 \to \mathbb{R}$ and let us minimize a regularized risk functional, then the learning algorithm can be summarized as finding the function $\widehat{f} \in \mathcal{F}$ such that:

$$\widehat{f} = \arg\min_{f \in \mathcal{F}} \left\{\Delta(f, D) + \lambda J(f)\right\}$$

with Δ a sample-based loss which is in our case defined as

$$\Delta(f, D) = \sum_{\mathbf{j} \in \mathcal{S}} \xi_{\mathbf{j}},$$

$J : \mathcal{F} \to \mathbb{R}$ a penalty functional defining the complexity of a particular ranking function and λ a regularization parameter. For the time being, let us suppose that the feature map ϕ for which $K(\mathbf{x}_1, \mathbf{x}_2) = \phi(\mathbf{x}_1) \cdot \phi(\mathbf{x}_2)$ is known, then we are learning ranking functions of the form $f(\mathbf{x}) = \mathbf{w} \cdot \phi(\mathbf{x})$ and the corresponding penalty functional is defined by

$$J(f) = \frac{1}{2}||\mathbf{w}||^2.$$

Similar to the C-formulation of the support vector machine [4] or the kernel-based ordinal regression method of [35], we arrive at the following primal optimization problem:

$$\min_{\mathbf{w},\xi_{\mathbf{j}}} \quad \frac{1}{2}||\mathbf{w}||^2 + C\sum_{\mathbf{j}\in\mathcal{S}} \xi_{\mathbf{j}}$$

$$\text{subject to} \begin{cases} \min_{k\in\{1,\dots,r-1\}} \left\{\mathbf{w}\cdot\delta(\mathbf{j},k)\right\} \geq 1-\xi_{\mathbf{j}} \\ \xi_{\mathbf{j}} \geq 0 \\ \mathbf{j}\in\mathcal{S} \end{cases} \tag{9}$$

in which we used the shorthand notation $\delta(\mathbf{j},k) := \phi(\mathbf{x}_{j_{k+1}}) - \phi(\mathbf{x}_{j_k})$. The constraints of this optimization problem can be interpreted as follows: for each r-tuple (i.e., a path in the graph) we require that all consecutive pairs in this r-tuple are correctly ordered and have a separation of 1 (the margin). We penalize that pair in the r-tuple for which this condition is the most violated (with the minimum operator). Since a constraint pops up for each r-tuple, we have in total n^r constraints.

The quadratic program can be solved by a QP-solver by deriving the dual quadratic program and in the dual formulation the ranking function $f(\mathbf{x}) = \mathbf{w}\cdot\phi(\mathbf{x})$ can be written only in terms of kernels and Lagrange multipliers. Unfortunately, the large number of constraints (exponential in r) gives rise to an identical number of Lagrange multipliers in the dual formulation and, hence, the running time required by a QP-solver will be excessively high, especially when the number of classes increases. However, we will show that the same solution can be found within a reasonable amount of time by using an approximation algorithm that only considers a subset of the constraints.

Recently, in [39, 57] one proposed a general support vector algorithm for structured outputs and multivariate performance measures, which basically consists of the optimization of a quadratic program with a combinatorial number of constraints. Instead of optimizing this huge quadratic program, they start optimizing the unconstrained objective function and iteratively add new constraints to the optimization problem. We will follow a similar approach. Let us therefore consider the following optimization problem:

$$\min_{\mathbf{w},\xi} \quad \frac{1}{2}||\mathbf{w}||^2 + C\xi$$

$$\text{subject to} \begin{cases} \mathbf{w}\cdot\left(\frac{1}{s}\sum_{\mathbf{j}\in\mathcal{S}} c_{\mathbf{j}}\delta(\mathbf{j},t_{\mathbf{j}})\right) \geq \frac{1}{s}\sum_{\mathbf{j}\in\mathcal{S}} c_{\mathbf{j}} - \xi \\ \xi \geq 0 \\ \mathbf{j}\in\mathcal{S} \\ \mathbf{c}\in\{0,1\}^s \\ \mathbf{t}\in\{1,\dots,r-1\}^s \end{cases} \tag{10}$$

Algorithm 2 Optimization of $\widehat{U}(f, D)$.

Input: data set $D = \{(\mathbf{x}_1, y_1), \ldots, (\mathbf{x}_n, y_n)\}, C, \mathbf{K}, \epsilon$
$\mathcal{W} \leftarrow \emptyset$
repeat
  $\widehat{\mathbf{w}} \leftarrow$ optimize $\mathbf{w}$ over working set $\mathcal{W}$
  $\widehat{\xi} \leftarrow$ corresponding ξ for $\widehat{\mathbf{w}}$
  $(\mathbf{c}, \mathbf{t}) \leftarrow \arg\max_{\substack{\mathbf{c}\in\{0,1\}^s \\ \mathbf{t}\in\{1,\ldots,r-1\}^s}} \left\{\frac{1}{s}\sum_{\mathbf{j}\in\mathcal{S}} c_{\mathbf{j}} - \mathbf{w}\cdot\frac{1}{s}\sum_{\mathbf{j}\in\mathcal{S}} c_{\mathbf{j}}\delta(\mathbf{j}, t_{\mathbf{j}})\right\}$
  $\xi_{\max} \leftarrow \max_{\substack{\mathbf{c}\in\{0,1\}^s \\ \mathbf{t}\in\{1,\ldots,r-1\}^s}} \left\{\frac{1}{s}\sum_{\mathbf{j}\in\mathcal{S}} c_{\mathbf{j}} - \mathbf{w}\cdot\frac{1}{s}\sum_{\mathbf{j}\in\mathcal{S}} c_{\mathbf{j}}\delta(\mathbf{j}, t_{\mathbf{j}})\right\}$
  $\mathcal{W} \leftarrow \mathcal{W} \cup \{(\mathbf{c}, \mathbf{t})\}$
until $\xi_{\max} \leq \widehat{\xi} + \epsilon$
Output: $\widehat{\mathbf{w}}$

with $\mathbf{w}$, C and $\mathcal{S}$ as defined before. Remarkably, it contains only one slack variable. The following theorem reveals the connection between both optimization problems.

Theorem 2 ([61]). *The solutions of optimization problems (9) and (10) result in identical models* $\mathbf{w}$ *and their errors are related:* $\xi = \frac{1}{s}\sum_{\mathbf{j}\in\mathcal{S}} \xi_{\mathbf{j}}$.

Optimization problem (10) has even more constraints than the previous one. Nevertheless, it can be optimized much more efficiently with a cutting plane algorithm [40]. Algorithm 2 gives an overview of this iterative procedure. Of crucial importance is that not all constraints are active in the quadratic program. Normally, for only one $(\mathbf{c}, \mathbf{t})$ the inequality constraint will become an equality constraint. To this end, a working set $\mathcal{W}$ of active constraints is introduced, so $\mathcal{W} \subseteq \{0, 1\}^s \times \{1, \ldots, r-1\}^s$.

Given a data set $D = \{(\mathbf{x}_1, y_1), \ldots, (\mathbf{x}_n, y_n)\}$ with corresponding Gram matrix $\mathbf{K}$, regularization parameter C and stopping criterion ϵ, the algorithm starts with $\mathcal{W} = \emptyset$. In each iteration, first the optimal solution $\widehat{\mathbf{w}}$ and corresponding slack $\widehat{\xi}$ are computed. Then, the constraint with maximum violation is added to the working set (this is the constraint for which the inequality becomes an equality with regard to the current solution $\widehat{\mathbf{w}}$). In the next iteration, the objective function is re-optimized over the adjusted working set of constraints. Consequently, the constraints not belonging to $\mathcal{W}$ are not necessarily observed. The algorithm repeatedly adds new constraints to the working set until no constraint is violated more than the stopping criterion ϵ. It has been proven before that such a cutting plane algorithm stops after adding a constant number of constraints, independent of the data set size. The total running time therefore strongly depends on the time required to find the maximum violating constraint. As discussed in [60, 61], this constraint can be found very efficiently by using a graph–theoretic reformulation and dynamic programming techniques, leading to the following theorem.

Theorem 3 ([61]). *For any* $C > 0$, $\epsilon > 0$, *positive definite kernel* $K : \mathcal{X}^2 \to \mathbb{R}$ *and training sample* $D = \{(\mathbf{x}_1, y_1), \ldots, (\mathbf{x}_n, y_n)\}$, *Algorithm 2 has a* $\mathcal{O}(n^3)$ *time complexity.*

Another interesting property of our approach is that the whole procedure can be expressed in terms of kernels such that the transformation function to a high-dimensional feature space must not be modeled explicitly. In our software, both Algorithm 2 and the algorithm to find the maximum violating constraint are implemented in terms of the dual representation. The dual in each iteration is solved with the quadratic solver of the MOSEK-package.[2] This solver implements an interior point algorithm and was for example formerly used by [34, 42] in the context of machine learning. The prediction rule for a new object $\mathbf{x} \in \mathcal{X}$ can be written in terms of kernels as follows:

$$f(\mathbf{x}) = \sum_{(\mathbf{c}^+,\mathbf{c}^-)\in\mathcal{W}} \alpha_{(\mathbf{c}^+,\mathbf{c}^-)} \frac{1}{s} \sum_{i=1}^{n} \left((c_i^+ - c_i^-) K(\mathbf{x}, \mathbf{x}_i)\right),$$

with $\alpha_{(\mathbf{c}^+,\mathbf{c}^-)}$ Lagrange multipliers and $(\mathbf{c}^+, \mathbf{c}^-)$ a compressed version of the active constraints in the working set $\mathcal{W}$. This prediction rule will also serve as input for computing the maximum violating constraint during training. In this way, all parts of the optimization procedure are carried out without explicitly processing $\mathbf{w}$ and ϕ.

Sparsity in the obtained solution can be ensured in two ways:

1. For constraints that are added to $\mathcal{W}$, many elements of the n-dimensional vectors $\mathbf{c}^+$ and $\mathbf{c}^-$ will manifest zeroes. Zeroes are encountered when no ingoing and outgoing edges of the corresponding node in the graph have minimal weight for any path from source to sink. This happens frequently.
2. Once a new constraint $(\mathbf{c}^+, \mathbf{c}^-)$ is added to $\mathcal{W}$ and the dual is re-optimized, the corresponding Lagrange multiplier $\alpha_{(\mathbf{c}^+,\mathbf{c}^-)}$ will typically decrease in further iterations. Because the upper bound C on

$$\sum_{(\mathbf{c}^+,\mathbf{c}^-)\in\mathcal{W}} \alpha_{(\mathbf{c}^+,\mathbf{c}^-)}$$

will normally be reached during any iteration in the cutting plane algorithm, at least one Lagrange multiplier has to decrease when a new constraint is added to $\mathcal{W}$. Many Lagrange multipliers tend to become zero again in practice, few iterations after adding them to the quadratic program.

6 Conclusion

In this chapter, we gave an introduction to ordinal regression and its frequently used performance measures. We argued that evaluating the ranking returned by an ordinal regression model is often more appropriate than looking at accuracy or mean absolute error, especially with skew class or cost distributions. To that end, we extended

[2] The MOSEK-package can be freely downloaded for noncommercial use from www.mosek.com.

the concept of expected ranking accuracy for ordinal labeled data and showed that a nonparametric unbiased estimator $\widehat{U}(f, D)$ of this quantity corresponds to the volume under the ROC surface (VUS) spanned by the true positive rates of each class. We conclude from Sect. 3 that existing methods for ordinal regression, which typically minimize a loss based on error rate or the number of incorrectly ranked object pairs, might not construct appropriate models when the class or cost distributions are skew. ROC analysis offers in this case a valuable alternative allowing to pick a classifier from the surface for a specific setting of cost. The volume under the ROC surface gives a good overall indication of the quality of the model for different costs without favoring the majority classes.

In Sect. 4, we further analyzed the scalability of VUS and we presented a fast exact algorithm to compute this estimator efficiently. In this way, it is possible to perform ROC analysis on large data sets and to estimate $U(f)$ on large ordinal regression models. The proposed algorithms can be directly plugged into the statistical tests developed by [48]. Moreover, new ranking methods can be constructed for using or optimizing the volume under the ROC surface. Such algorithms have been proposed for the binary case and in particular optimizing ranking-based measures can lead to noticeably different models compared to minimizing the error rate [13,36,65].

In Sect. 5, we further exploited these ideas to develop a new algorithm for ordinal regression learning. Inspired by the graph-theoretic reformulations for VUS, we called this setting the layered ranking problem because each correctly ranked r-tuple, consisting of one object of each class, induces a path in the corresponding layered ranking graph. Notwithstanding the increasing complexity of such an approach, we believe that looking at correctly ranked r-tuples instead of correctly ranked pairs has important benefits and leads to better models. Several experiments in [61] clearly support this claim. On synthetic data and several benchmark data sets, a statistically significant improvement in performance was measured, definitely in terms of the VUS, but also in terms of the C-index. For some unbalanced problems, a small tradeoff appeared between optimizing VUS on the one hand and the C-index on the other hand. We argued that the pairwise approach might focus too strongly on some of the categories instead of assessing the ranking model globally. Nevertheless, we acknowledge that the complexity of ranking r-tuples remains a bottleneck for large samples.

Acknowledgements Willem Waegeman has been supported by a grant of the "Institute for the Promotion of Innovation through Science and Technology in Flanders (IWT-Vlaanderen)". He is currently supported by the Research Foundation Flanders.

References

1. S. Agarwal, T. Graepel, R. Herbrich, S. Har-Peled, D. Roth, Generalization bounds for the area under the ROC curve. J. Mach. Learn. Res. **6**, 393–425 (2005)
2. A. Agresti, *Categorical Data Analysis, 2nd version.* (Wiley, 2002)

3. K. Ataman, N. Street, Y. Zhang, Learning to rank by maximizing AUC with linear programming, in *Proceedings of the IEEE International Joint Conference on Neural Networks* (Vancouver, BC, Canada, 2006), pp. 123–129
4. C.J.C. Burges, A tutorial on support vector machines for pattern recognition. Data Min. Knowl. Discov. **2**(2), 121–167 (1998)
5. Z. Cao, T. Qin, T. Liu, M. Tsai, H. Li, Learning to rank: from pairwise approach to listwise approach, in *Proceedings of the International Conference on Machine Learning* (Corvallis, OR, USA, 2007), pp. 129–136
6. K. Cao-Van, *Supervised Ranking, from semantics to algorithms*. PhD thesis, Ghent University, Belgium, 2003
7. W. Chu, Z. Ghahramani, Gaussian processes for ordinal regression. J. Mach. Learn. Res. **6**, 1019–1041 (2005)
8. W. Chu, Z. Ghahramani, Preference learning with Gaussian processes, in *Proceedings of the International Conference on Machine Learning* (Bonn, Germany, 2005), pp. 137–144
9. W. Chu, S. Keerthi, New approaches to support vector ordinal regression, in *Proceedings of the International Conference on Machine Learning* (Bonn, Germany, 2005), pp. 321–328
10. W. Chu, S. Keerthi, Support vector ordinal regression. *Neural Comput.* **19**(3), 792–815 (2007)
11. S. Clémençon, N. Vayatis, Ranking the best instances. J. Mach. Learn. Res. **8**, 2671–2699 (2007)
12. W. Cohen, R. Schapire, Y. Singer, Learning to order things, in *Advances in Neural Information Processing Systems*, vol. 10 (MIT, Vancouver, Canada, 1998), pp. 451–457
13. C. Cortes, M. Mohri, AUC optimization versus error rate minimization, in *Advances in Neural Information Processing Systems*, vol. 16 (MIT, Vancouver, Canada, 2003), pp. 313–320
14. C. Cortes, M. Mohri. Confidence intervals for the area under the ROC curve, in *Advances in Neural Information Processing Systems*, vol. 17 (MIT, Vancouver, Canada, 2004), pp. 305–312
15. C. Cortes, M. Mohri, A. Rastogi, Magnitude-preserving ranking algorithms, in *Proceedings of the International Conference on Machine Learning* (Corvallis, OR, USA, 2007), pp. 169–176
16. K. Crammer, Y. Singer, Pranking with ranking, in *Proceedings of the Conference on Neural Information Processing Systems* (Vancouver, Canada, 2001), pp. 641–647
17. N. Cristianini, J. Shawe-Taylor, *An Introduction to Support Vector Machines* (Cambridge University Press, 2000)
18. B. De Baets, H. De Meyer, Transitivity frameworks for reciprocal relations: cycle-transitivity versus *FG*-transitivity. Fuzzy Sets Syst. **152**, 249–270 (2005)
19. B. De Baets, H. De Meyer, B. De Schuymer, S. Jenei, Cyclic evaluation of transitivity of reciprocal relations. Soc. Choice Welfare **26**, 217–238 (2006)
20. B. De Schuymer, H. De Meyer, B. De Baets, Cycle-transitive comparison of independent random variables. J. Multivar. Anal. **96**, 352–373 (2005)
21. B. De Schuymer, H. De Meyer, B. De Baets, S. Jenei, On the cycle-transitivity of the dice model. Theory Decis. **54**, 164–185 (2003)
22. S. Dreiseitl, L. Ohno-Machado, M. Binder, Comparing three-class diagnostic tests by three-way ROC analysis. Med. Decis. Mak. **20**, 323–331 (2000)
23. E. Frank, M. Hall, A simple approach to ordinal classification, in *Proceedings of the European Conference on Machine Learning*, (Freibourg, Germany, 2001), pp. 146–156
24. Y. Freund, R. Yier, R. Schapire, Y. Singer, An efficient boosting algorithm for combining preferences. J. Mach. Learn. Res. **4**, 933–969 (2003)
25. J. Fürnkranz, Round robin classification. J. Mach. Learn. Res. **2**, 723–747 (2002)
26. J. Fürnkranz, E. Hüllermeier, Pairwise preference learning and ranking. Lect. Notes Comput. Sci. **2837**, 145–156 (2003)
27. M. Gönen, G. Heller, Concordance probability and discriminatory power in proportional hazards regression. Biometrika **92**(4), 965–970 (2005)
28. D. Hand, R. Till, A simple generalization of the area under the ROC curve for multiple class problems. Mach. Learn. **45**, 171–186 (2001)
29. J. Hanley, B. McNeil, The meaning and use of the area under a receiver operating characteristics curve. Radiology **143**, 29–36 (1982)

30. J. Hanley, B. McNeil, A method of comparing receiver operating characteristics curves derived from the same class. Radiology **148**, 839–843 (1983)
31. S. Har-Peled, D. Roth, D. Zimak, Constraint classification: A new approach to multi-class classification and ranking, in *Advances in Neural Information Processing Systems*, vol. 15 (MIT, Vancouver, Canada, 2002), pp. 785–792
32. E. Harrington, Online ranking/collaborative filtering using the perceptron algorithm, in *Proceedings of the 20th International Conference on Machine Learning* (Washington, USA, 2003), pp. 250–257
33. T. Hastie, R. Tibshirani, Classification by pairwise coupling. Ann. Stat. **26**(2), 451–471 (1998)
34. M. Heller, B. Schnörr, Learning sparse representations by non-negative matrix factorization and sequential cone programming. J. Mach. Learn. Res. **7**, 1385–1407 (2006)
35. R. Herbrich, T. Graepel, K. Obermayer, Large margin rank boundaries for ordinal regression, in *Advances in Large Margin Classifiers*, ed. by A. Smola, P. Bartlett, B. Schölkopf, D. Schuurmans (MIT, 2000), pp. 115–132
36. A. Herschtal, B. Raskutti, Optimizing area under the ROC curve using gradient descent, in *Proceedings of the International Conference on Machine Learning* (Bonn, Germany, 2005), pp. 49–57
37. J. Higgins, *Introduction to Modern Nonparametric Statistics* (Duxbury, 2004)
38. E. Hüllermeier, J. Hühn, Is an ordinal class structure useful in classifier learning? Int. J. Data Min. Model. Manag. **1**(1), 45–67 (2009)
39. T. Joachims, A support vector method for multivariate performance measures, in *Proceedings of the International Conference on Machine Learning* (Bonn, Germany, 2005), pp. 377–384
40. J. Kelley, The cutting plane method for convex programs. J. Soc. Ind. Appl. Math. **9**, 703–712 (1960)
41. S. Kramer, G. Widmer, B. Pfahringer, M. Degroeve, Prediction of ordinal classes using regression trees. Fundam. Informaticae **24**, 1–15 (2000)
42. G. Lanckriet, N. Cristianini, P. Bartlett, L. El Gaoui, M. Jordan, Learning the kernel matrix with semidefinite programming. J. Mach. Learn. Res. **5**, 27–72 (2004)
43. L. Lehmann, *Nonparametrics: Statistical Methods based on Ranks* (Holden Day, 1975)
44. S. Lievens, B. De Baets, K. Cao-Van, A probabilistic framework for the design of instance-based supervised ranking algorithms in an ordinal setting. Ann. Oper. Res. **163**, 115–142 (2008)
45. H. Lin, L. Li, Large-margin thresholded ensembles for ordinal regression: Theory and practice. Lect. Notes Comput. Sci. **4264**, 319–333 (2006)
46. R. Luce, P. Suppes, *Handbook of Mathematical Psychology*, chapter Preference, Utility and Subjective Probability (Wiley, 1965), pp. 249–410
47. P. McCullagh, Regression models for ordinal data. J. R. Stat. Soc. B **42**(2), 109–142 (1980)
48. C. Nakas, C. Yiannoutsos, Ordered multiple-class ROC analysis with continuous measurements. Stat. Med. **22**, 3437–3449 (2004)
49. M. Öztürk, A. Tsoukiàs, Ph. Vincke, Preference modelling, in *Multiple Criteria Decision Analysis. State of the Art Surveys*, ed. by J. Figueira, S. Greco, M. Ehrgott (Springer, 2005), pp. 27–71
50. J. Platt, N. Cristianini, J. Shawe-Taylor, Large margin DAGs for multiclass classification. Adv. Neural Process. Syst. **12**, 547–553 (2000)
51. R. Potharst, J.C. Bioch, Decision trees for ordinal classification. *Intell. Data Process.* **4**(2) (2000)
52. J.D. Rennie, N. Srebro, Loss functions for preference levels: Regression with discrete, ordered labels in *Proceedings of the IJCAI Multidisciplinary Workshop on Advances in Preference Handling*, (Edinburgh, Scotland, 2005), pp. 180–186
53. R. Rifkin, A. Klautau, In defense of one-versus-all classification. J. Mach. Learn. Res. **5**, 101–143 (2004)
54. B. Schölkopf, A. Smola. *Learning with Kernels, Support Vector Machines, Regularisation, Optimization and Beyond* (MIT, 2002)
55. A. Shashua, A. Levin, Ranking with large margin principle: Two approaches, in *Advances in Neural Information Processing Systems*, vol. 16 (MIT, Vancouver, Canada, 2003), pp. 937–944

56. V. Torra, J. Domingo-Ferrer, J.M. Mateo-Sanz, M. Ng, Regression for ordinal variables without underlying continuous variables. Inf. Sci. **176**, 465–476 (2006)
57. Y. Tsochantaridis, T. Joachims, T. Hofmann, Y. Altun, Large margin methods for structured and independent output variables. J. Mach. Learn. Res. **6**, 1453–1484 (2005)
58. G. Tutz, K. Hechenbichler, Aggregating classifiers with ordinal response structure. J. Stat. Comput. Simul. **75**(5), (2004)
59. W. Waegeman, B. De Baets, On the ERA ranking representability of multi-class classifiers. Artif. Intell. (2009) submitted
60. W. Waegeman, B. De Baets, L. Boullart, Learning a layered graph with a maximal number of paths connecting source and sink, in *Proceedings of the ICML Workshop on Constrained Optimization and Structured Output Spaces* (Corvallis, OR, USA, 2007)
61. W. Waegeman, B. De Baets, L. Boullart, Learning layered ranking functions with structured support vector machines. Neural Netw. **21**(10), 1511–1523 (2008)
62. W. Waegeman, B. De Baets, L. Boullart, On the scalability of ordered multi-class ROC analysis. Comput. Stat. Data Anal. **52**, 3371–3388 (2008)
63. W. Waegeman, B. De Baets, L. Boullart, ROC analysis in ordinal regression learning. Pattern Recognit. Lett. **29**, 1–9 (2008)
64. J. Xu, Y. Cao, H. Li, Y. Huang, Cost-sensitive learning of SVM for ranking, in *Proceedings of the 17th European Conference on Machine Learning* (Berlin, Germany, 2006), pp. 833–840
65. L. Yan, R. Dodier, M. Mozer, R. Wolniewiecz, Optimizing classifier performance via an approximation to the Wilcoxon-Mann-Whitney statistic, in *Proceedings of the International Conference on Machine Learning* (Washington, DC, USA, 2003), pp. 848–855
66. S. Yu, K. Yu, V. Tresp, H. Kriegel, Collaborative ordinal regression, in *Proceedings of the International Conference on Machine Learning* (Pittsburgh, PA, 2006), pp. 1089–1096

Ranking Cases with Classification Rules

Jianping Zhang, Jerzy W. Bala, Ali Hadjarian, and Brent Han

Abstract Many real-world machine learning applications require a ranking of cases, in addition to their classification. While classification rules are not a good representation for ranking, the human comprehensibility aspect of rules makes them an attractive option for many ranking problems where such model transparency is desired. There have been numerous studies on ranking with decision trees, but not many on ranking with decision rules. Although rules are similar to decision trees in many respects, there are important differences between them when used for ranking. In this chapter, we propose a framework for ranking with rules. The framework extends and substantially improves on the reported methods for ranking with decision trees. It introduces three types of rule-based ranking methods: post analysis of rules, hybrid methods, and multiple rule set analysis. We also study the impact of rule learning bias on the ranking performance. While traditional measures used for ranking performance evaluation tend to focus on the entire rank ordered list, the aim of many ranking applications is to optimize the performance on only a small portion of the top ranked cases. Accordingly, we propose a simple method for measuring the performance of a classification or ranking algorithm that focuses on these top ranked cases. Empirical studies have been conducted to evaluate some of the proposed methods.

1 Introduction

Many real-world machine learning applications require a ranking of cases, in addition to their classification. Such ranking is often based on some measure of reliability or likelihood or a numeric assessment of the quality of each classification (e.g., probability value of a class membership). In other words, the decision-making

J. Zhang (✉), J.W. Bala, A. Hadjarian, and B. Han
The MITRE Corporation, 7515 Colshire Drive McLean, Virginia 22102-7508, USA
e-mail: jzhang@mitre.org, jerzy.w.bala@gmail.com, ahadjarian@mitre.org, bhan@mitre.org

J. Fürnkranz and E. Hüllermeier (eds.), *Preference Learning*,

DOI 10.1007/978-3-642-14125-6_8,

process extends the class membership prediction to include an estimate of the reliability for this prediction. For example, in credit application processing, the goal is to rank applicants in terms of their likelihoods of profitability or loan defaults. This is significantly different than simply classifying them into qualified versus nonqualified groups. Other decision-making applications where case ranking could be of importance include bankruptcy prediction, medical diagnosis, customer targeting for marketing campaigns, and customer churn prediction. Ranking of cases is particularly important for those decision-making applications where it is preferable to abstain from decision making altogether in the absence of sufficient support. Examples include medical and military applications.

While classification rules are not a good representation for ranking, the human comprehensibility aspect of rules makes them an attractive option for many ranking applications where such model transparency is desired or even an essential requirement. There have been numerous studies on ranking with decision trees [1, 8–10], but not many on ranking with decision rules. Although rules are similar to decision trees in many respects, there are important differences between them when used for ranking. Separate and conquer (covering) techniques of rule learning algorithms may generate rules that overlap, whereas divide and conquer techniques of decision trees do not result in such overlapping of decisions. Rules may not cover some areas of a feature space, but the leaf nodes of a decision tree cover the entire area of the feature space (see Fig. 1). In addition, a rule learning algorithm for a two-class problem may only learn rules for one class, but a decision tree always includes leaf nodes for both classes. Rule learning algorithms tend to generate fewer rules than leaf nodes of a decision tree. Such differences bring both research challenges and opportunities for developing methods for ranking cases with rules.

In this chapter, we propose a framework for ranking with rules. The framework extends and substantially improves on the reported methods for ranking with decision trees. It introduces three types of rule-based ranking methods: post-analysis of rules, hybrid methods, and multiple rule set analysis (i.e., rule ensembles and redundant rules). Methods for combining scores from overlapping rules are also proposed and studied.

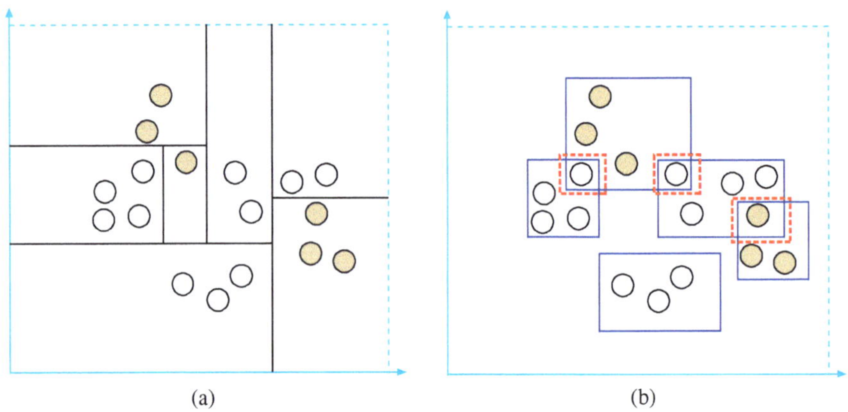

Fig. 1 Decision boundaries of a decision tree (**a**) vs. those of a rule-based classifier (**b**)

We also study the impact of rule learning bias on the ranking performance. While traditional measures used for ranking performance evaluation tend to focus on the entire rank ordered list, the aim of many ranking applications is to optimize the performance on only a small portion of the top ranked cases. Accordingly, we propose a simple method for measuring the performance of a classification or ranking algorithm that focuses on these top ranked cases. More specifically, instead of measuring the entire area under the ROC curve, the proposed method computes only the left most part of it (i.e., the part covering only the top $n\%$ of the ranked cases). Empirical studies have been conducted to evaluate some of the more popular post-analysis methods, methods for combining scores of overlapping rules, and the impact of rule learning bias.

The remainder of the chapter is organized as follows. Related work is discussed in Sect. 2. Section 3 describes the proposed framework. Section 4 describes the empirical results of two separate studies: (1) a real-world application in investigations of companies suspected of financial fraud, and (2) performance evaluation using six of the UCI Machine Learning Repository data sets. Finally, Sect. 5 concludes the chapter with future research.

2 Related Work

2.1 Expert Systems and Fuzzy Logic

Ranking examples using rules is not a novel undertaking. There have been many attempts in the past, including those by the researchers in the fields of expert systems, fuzzy logic, and cognitive science [2]. MYCIN [4] is a well-known rule-based expert system, in which each rule is assigned a certainty factor (CF) by domain experts. CF of different rules may be combined or propagated to produce the CF of a decision inferred by MYCIN. In PROSPECTOR [12], an expert system to assist geologists working in mineral exploration, rules are assigned probability by human experts and are propagated and combined using Bayesian inference.

Development of fuzzy rules has been studied widely in fuzzy logic, e.g., [13]. Here, a general method is developed to generate fuzzy rules from numeric data, using linguistic variables. The degree of class membership of an example depends on the degree of match of the example to a fuzzy rule.

2.2 Machine Learning

There are also several related rule-based machine learning studies, e.g., [13]. These studies generally focus on methods for generating partial matching, whereby the scores for individual examples are computed based on how well they match the rules. In these approaches, examples that satisfy all conditions of a rule share the same score.

Extensive studies have been dedicated to the incorporation of ranking capabilities into the decision tree learning paradigm.

Related work generally falls into the following four groups of methods: learning probability estimation trees [10], geometric methods [1], hybrid trees including: the Perceptron Tree [11] and NBTree [8], and ensembles of trees [3]. Other methods have also been reported in [7, 9].

3 Framework for Ranking with Classification Rules

A rule-based classifier, typically defined as a disjunctive normal form of conditional rules, is a mapping function from a set of m arguments or attributes (which can be either nominal or numeric) to a single nominal value, known as the class. Let us represent by D the set of d classification decisions, generally labeled by numbers $0, 1, 2, \ldots d - 1$, and by E the set of unlabeled examples. A classifier is a function $f : E \rightarrow D$. In a simplified scenario, a classifier with ranking capability computes a number or score for every example $e \in E$ and for every class $i \in D$.

A score may be interpreted differently depending on the application. In statistics, a score ranges from 0 to 1 and indicates the probability of an example belonging to a given class. In expert systems, a score may be interpreted as a certainty factor. In fuzzy logic, a score represents the degree of membership in a class. Similarly, in cognitive science, a score could be defined as the typicality of the example being a member of the class. In other applications, a score is just a measure for ranking cases.

The framework for ranking cases with classification rules presented in this chapter consists of a grouping of methods that can be used independently or jointly. Figure 2 depicts the taxonomy of these methods.

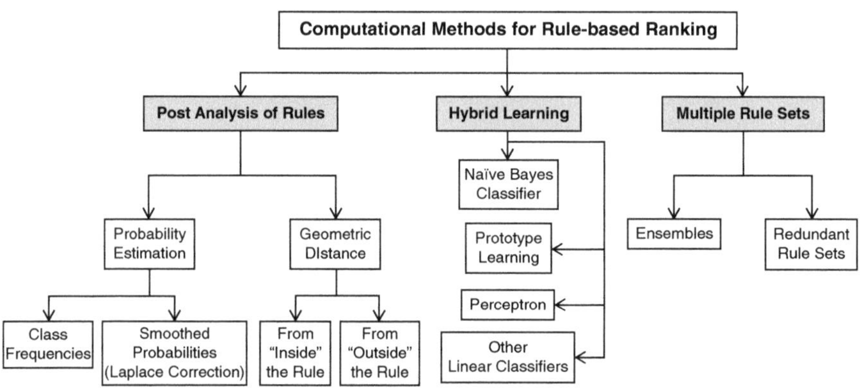

Fig. 2 Computational methods for rule-based ranking

The framework extends and substantially improves on the reported methods for ranking with decision trees. It introduces the following three groups of rule-based ranking computation methods:

- Post-analysis of rules
- Hybrid methods
- Multiple rule set analysis (i.e., rule ensembles and redundant rules)

In addition to the three groups of rule-based ranking methods, we introduce methods for combining the scores of overlapping rules. We also study the impact of the inductive bias on the ranking performance.

3.1 *Post Analysis of Rules*

In this group of methods, ranking scores are computed using the rules generated by some rule induction algorithm plus some additional information such as the number of positive and negative examples covered or the ranges of attribute values. There are two subgroups of methods under this group: probability estimation and geometric methods.

3.1.1 Probability Estimation

With probability estimation, the score of the example covered by a rule is computed as the ratio of the number of positive examples covered and the total number of examples covered by a rule. This approach has been explored in Probability Estimation Trees.

In the simple probability estimation method, scores assigned by two different rules are identical as long as the above-mentioned ratio is the same, no matter how many positive examples are actually covered by each rule. For example, all rules covering no negative examples result in the same score, namely one. A simple method for overcoming this problem is the Laplace correction, in which the score is computed using $\frac{k+1}{n+C}$, where k and n are the numbers of positive examples and the total number of examples covered, respectively, and C is the number of classes. Here, the scores are the same for all the examples covered by a given rule. This becomes a serious problem when the number of rules generated by a rule induction algorithm is small, as many of the examples will end up with identical scores.

The probability estimation method, even with the Laplace correction, ignores the absolute number of examples covered by the rule. A rule covering 9 positive examples and 10 negative examples receives the same score as a rule covering 90 positive examples and 100 negative examples. In real applications, however, even when having similar precisions, rules covering more examples are generally preferred. An alternative option would be to use the F-measure for scoring:

$$F - \text{measure}(r) = \frac{\beta^2 + 1}{\frac{\beta^2}{\text{recall}(r)} + \frac{1}{\text{precision}(r)}}, \tag{1}$$

where β is a parameter for assigning relative weights to recall and precision. When β is set to 1, recall and precision are weighted equally. F-measure favors recall with $\beta > 1$ and favors precision with $\beta < 1$. Namely, an F-measure with a large β value favors more general and less precise rules, while one with a small β value favors more specific and more precise rules. When $\beta = 0$, F-measure score is the same as probability estimation.

3.1.2 Geometric Methods

As indicated above, the probability estimation method assigns the same score to all examples covered by the same rule. When the number of rules generated is small, this causes a problem for applications that need a fine-grained ranking. For example, in one of our data mining applications, only 0.1% of the top ranked cases are selected for further investigations. If all rules cover more than 0.1% of the examples, we have no way to accurately select 0.1% of the top ranked cases. Geometric methods can help generate such fine-grained rankings.

Geometric methods have been used in decision trees [1] and assume that classifications/rankings of the examples near the rule boundary are less certain. In ranking with rules, there are two types of geometric methods, one for examples covered by a rule and the other for examples that are not covered by any rules. The latter is also called partial matching. For an example covered by a rule, we can measure the distance between the example and the rule boundary or the center of the rule. The closer the example is to the boundary (or the farther from the center), the smaller its score gets. The distance may also be weighted by the estimated probability of the rule. The geometric method for covered examples works for numeric attributes.

Partial matching also computes the distance of an uncovered example to the boundary of a rule, but from outside of the rule. The closer to the boundary, the larger the score is. Again, the distance could be weighted by the estimated probability of the rule. Partial matching is different from strict matching, where an example has to satisfy all the conditions of the matched rule. Partial matching computes a degree of match between an example and a rule. The degree of match can vary in the range of 0 (matches no condition) to 1.0 (matches all conditions). Partial matching works for both numeric and nominal attributes.

3.2 *Hybrid Methods*

Hybrid methods integrate rule induction with other learning techniques that have ranking capabilities. These latter techniques include the perceptron algorithm, naïve Bayes, instance-based learning, and prototype-based learning. For example, a separate perceptron may be learned for examples covered by each rule. Figure 3 shows a rule with a linear classifier in a two-dimensional space. The rectangle represents the rule and the line inside the rectangle represents the linear classifier. The white circles represent the positive examples, while the gray ones represent the negative

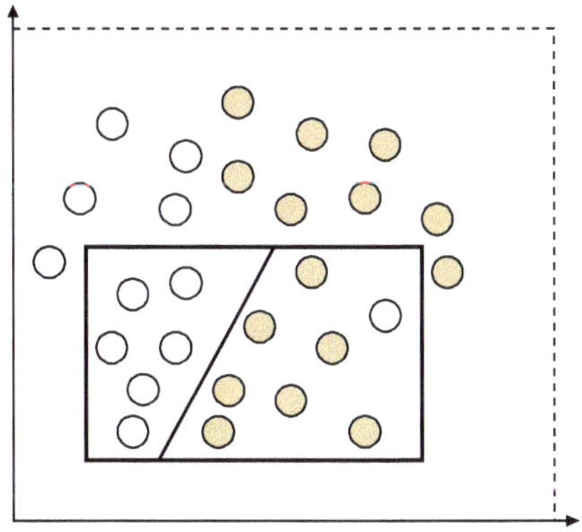

Fig. 3 A rule and the corresponding linear classifier

examples. The linear classifier is used to assign a score to each of the examples covered by the rule.

Rules and linear or naïve Bayes models may be learned together to optimize the performance of the hybrid method. Alternatively, rules may be learned first and other models then generated for each rule. As discussed in Sect. 2, there are many studies on building hybrid decision trees, e.g., Perceptron Tree and NBTree.

3.3 Rule Ensembles and Redundant Rules

Previous studies have shown that ensemble techniques can significantly improve the ranking performance of decision trees [3, 10]. In ensemble learning, multiple classifiers are learned, using approaches such as bagging and boosting. Typically, in such ensembles, a majority vote technique is used to determine the final classification of an example and the average score can be used as the final score for each example. Although little work has been done in building ensembles of rules, we believe such an ensemble can yield improved ranking performance.

Previous studies have also shown how redundant rules could improve classification performance. Since redundant rules allow for finer grained rankings, it is worthwhile to design a rule induction algorithm for generating redundant rules for ranking.

3.4 Combining Scores from Overlapping Rules

A rule induction algorithm usually generates a set of overlapping rules. In this section, we discuss three simple methods: *Max*, *Average*, and *Probabilistic Sum (P-Sum)* for combining the scores of an example covered by more than one rule. The

Max approach simply takes the largest score of all the rules that cover the example. Given an example e and a set of l rules $RS = \{R_1, \ldots, R_l\}$, the combined score of e using *Max* is computed as follows:

$$\text{score}(e, RS) = \max_{i=1}^{l} \text{score}(e, R_i) \tag{2}$$

where $\text{score}(e, R_i)$ is the score of e assigned by Rule R_i. Similarly, the combined score of e using *Average* is computed as follows:

$$\text{score}(e, RS) = \frac{1}{l} \cdot \sum_{i=1}^{l} \text{score}(e, R_i) \tag{3}$$

For rules that do not cover e, $\text{score}(e, R_i) = 0$ unless partial matching is applied. For the *P-Sum* method, the formula can be defined recursively as follows:

$$\begin{aligned}
\text{score}(e, \{R_1\}) &= \text{score}(e, R_1) \\
\text{score}(e, \{R_1, R_2\} &= \text{score}(e, R_1) + \text{score}(e, R_2) \\
&\quad - \text{score}(e, R_1) \times \text{score}(e, R_2) \\
\text{score}(e, \{R_1, \ldots, R_n\} &= \text{score}(e, \{R_1, \ldots, R_{n-1}\}) + \text{score}(e, R_n) \\
&\quad - \text{score}(e, \{R_1, \ldots, R_{n-1}\}) \times \text{score}(e, R_n)
\end{aligned}$$

Both *Average* and *P-Sum* generate a finer grained ranking than *Max*. These three methods may also be used to combine the scores in a rule ensemble.

3.5 Impact of the Rule Induction Algorithm

Most rule induction algorithms have been designed for maximizing the classification accuracy. Recently, some learning algorithms have been proposed that optimize the AUC (Area Under the Curve) of an ROC (Receiver Operating Characteristics) curve [6]. The design of a rule induction algorithm for optimizing the AUC could be an interesting future research objective.

Rule induction algorithms typically include parameter settings that allow the users to trade generality for accuracy. When such parameters are set to favor accuracy, more rules may be generated. On the one hand, more rules result in a finer grained ranking, but on the other hand, the rules themselves tend to be over-specific.

4 Empirical Studies

4.1 Study 1: Detection of Public Companies Suspected of Financial Fraud

This section focuses on a real-world data mining application to prioritize human investigations of companies suspected of engaging in fraudulent behavior based on their financial filings.

4.1.1 The Problem

Incidents of financial statement fraud have increased substantially over the past two decades and affected individuals, especially investors and creditors, have lost billions of dollars as a result. Financial statement fraud involves the misstatement of a company's financial information with the intent to mislead users of such information, such as investors and creditors. Typical falsifications of financial statements include manipulation of assets, sales, profits, liabilities, expenses, or losses. Such frauds are often difficult to detect, because they are typically committed by a company's top management team, who understands the limitations of an audit, with an opportunity and motive to distort the financial statements.

Prevention of financial statement fraud by internal or external auditors is probably the best strategy; however, sometimes the auditors themselves are involved in the fraud activity or make mistakes. Thus, to minimize the potential economic impact, there is a need for early detection of financial statement fraud by regulatory organizations. There have been several past studies to address such a need.

This section focuses on a data mining application to prioritize a list of target companies suspected of engaging in fraudulent behavior based on their financial statements. Such prioritization of potentially large volumes of targets is particularly crucial for applications such as this one where the decisions ultimately rest with human examiners with limited resources. In other words, only companies deemed as highly suspicious, normally a small percentage of the total population, will eventually go through the vigorous task of in-depth examination.

The application involves building predictive models for identifying companies that may have failed to comply with accounting principles or violated securities laws with respect to the accuracy of public disclosure information. The raw data used in this application contains SEC registered public organizations' financial filings (publicly available SEC EDGAR filings), including those of companies with materially misstated financial statements. Each training example corresponds to a company filing and is a vector of about 100 attribute values that are primarily numeric. Positive examples are all filings for companies that had issued material restatements, while negative examples are filings of companies that had not issued such restatements.

As mentioned previously, the learned predictive models are intended to help the human investigators prioritize which companies to look at more closely and to optimize investigative resources and increase efficiency by focusing examiners on highly suspicious companies that warrant the most attention. The investigators use the predictive models as a tool to generate a rank ordered list of target companies to be examined. These predictive models will neither replace human inspectors nor fully automate the suspicious behavior detection processes.

4.1.2 A Ranking Performance Metric

Since only a very small percentage, typically 1–2% depending on the resources available, of top ranked companies may be selected for detailed inspections, it is important to maximize the classification precision (or true positive rate) on these top ranked companies. The classifier's performance on lower ranked companies, as such, becomes rather irrelevant. This problem is similar to that of web search, in which a search engine might return thousands of web pages for a given query, but only a small number of which, typically the top ranked pages, is ever viewed by the user. Thus, it is much more important for an effective search engine to optimize the relevance on top ranked pages than the lower ones.

While existing research offers an array of machine learning algorithms that can accommodate such ranking of classification decisions, these algorithms, and measures such as the area under the ROC curve (AUC) used to evaluate their performance, generally tend to focus on the entire rank ordered list. Based on the need for our application, we propose a simple method for measuring the performance of classification/ranking algorithms that instead of measuring the entire area under the ROC curve, it computes only the left most portion of it (i.e., the part covering only the top $n\%$ of the ranked cases). We have named the aforementioned area as LAUC (Left-most portion of the Area Under the Curve). Figure 4 shows the ROC curves obtained from two different learning algorithms. With LAUC as a measure (i.e., to the left of the cutoff point), Algorithm 2 achieves a better performance than

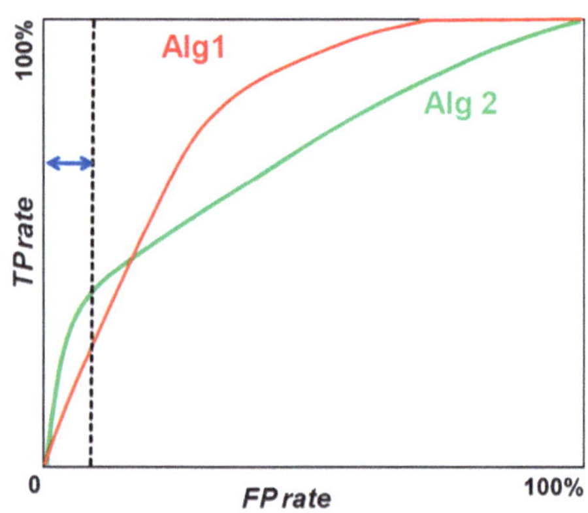

Fig. 4 Example ROC curves. The LAUC cutoff point is depicted by the dashed line

Algorithm 1, whereas Algorithm 1 is preferred over Algorithm 2 with the AUC. The have been some recent studies by machine learning researchers focusing on the notion of LAUC [9].

4.1.3 Data and Setting

For these experiments, we create input records for each company based on cases that are publicly available on the SEC EDGAR filing site. Each input record is a vector of attribute values that are primarily numeric. These are a combination of original data that is contained in sources, such as SEC required financial filings as well as calculated and derived attributes. These input records are split into two datasets based on the year, one for training and one for testing. The training set includes the data from 2003 to 2004, while the test set contains the data from 2005. Examples in both sets are described by 130 attributes, 95 of which are numeric and the other 35 are symbolic. The training set includes 4,932 positive and 38,792 negative examples. The test set is composed of 836 positive and 18,780 negative examples. In our experiments, we randomly selected 50% of the examples from the training set to learn the rules and the process was repeated five times. Rules were then tested on the test set and the results were reported using the average of the five runs. The classification performance has been evaluated with both the AUC and the LAUC as the measure. Here, LAUC is the area under the ROC curve at the left of the 1% false positive rate cutoff point, normalized by the total area to the left of that point. LAUC is 0.005 for random selection.

The classification task is quite challenging due to factors such as noisy data, which is generally the result of the mechanism by which the positive and negative examples have been labeled. Here, negative examples are not truly negative, rather their class memberships are not known. As such, it is not quite possible to learn rules with high recall and precision. For example, when minimum precision and minimum recall were set to 0.6 and 0.03, respectively, no rules were generated. Only a small number of positive examples could be covered by rules with high precision. Since there is no well-established baseline performance data for the comparison of our results, we compared the performance of our model with the random choice model by assuming that target class examples are drawn at random from all the potential examples.

4.1.4 Results

In the first experiment, with minimum recall and minimum precision parameters set to 0.01 and 0.6, respectively, we varied β from 0.001 to 0.1. Accordingly, all the rules learned have a recall larger than 1% and a precision larger than 60%. With $\beta = 0.001$, a set of highly specific and precise rules was generated. Each increase in β resulted in more general and less precise rules. Probability estimation with Laplace correction, using three different score combining methods: *Max*, *Average*, and *P-Sum*, was applied to the learned rules to obtain the scores of all the test examples. Figure 5 shows the AUCs for different β values and different score combining methods. The general trend of the performances of all three score combining methods seems to be upward with an increase in β until it reaches 0.07 and then the performances start to drop sharply. When β is small, the generated rules are highly specific and precise, so they cover only a small number of positive examples. Therefore, many of the positive examples not covered by these rules are assigned a score of 0 and are consequently ranked low. When β increases, rules become more general and cover more positive examples and the ranks of these newly covered positive examples move up, so the performance increases accordingly. When β becomes too large, however, rules tend to get much more general and much less precise, so that many negative examples are also covered. This causes the performance to drop. The three score combining methods perform similarly. It can also be seen that the reported AUC values barely beat the random performance. This shows the difficulty of the problem. Many positive examples cannot be covered by any rules with a recall larger than 1% and a precision larger than 60%.

Figure 6 shows the LAUC values. In contrast to the AUC, the best performances are achieved with smaller β values. This is what we had initially expected because a small β value results in highly specific and precise rules. As such, not many negative examples were covered by these rules. When β gets larger, however, rules cover more negative examples because they are more general and less precise. These covered negative examples are ranked high, so the performances on top ranked examples degrade. The three score combining methods seem to perform about the

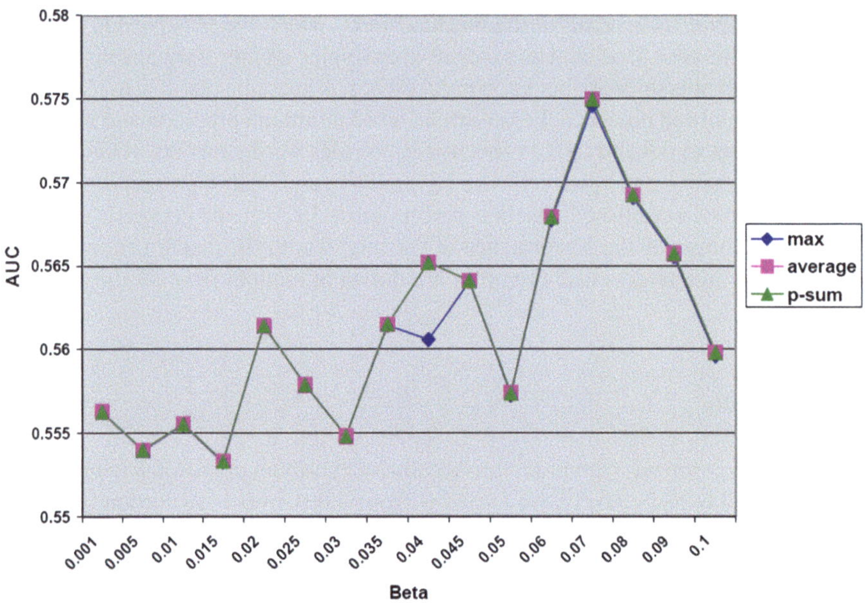

Fig. 5 AUC values reported for the first experiment

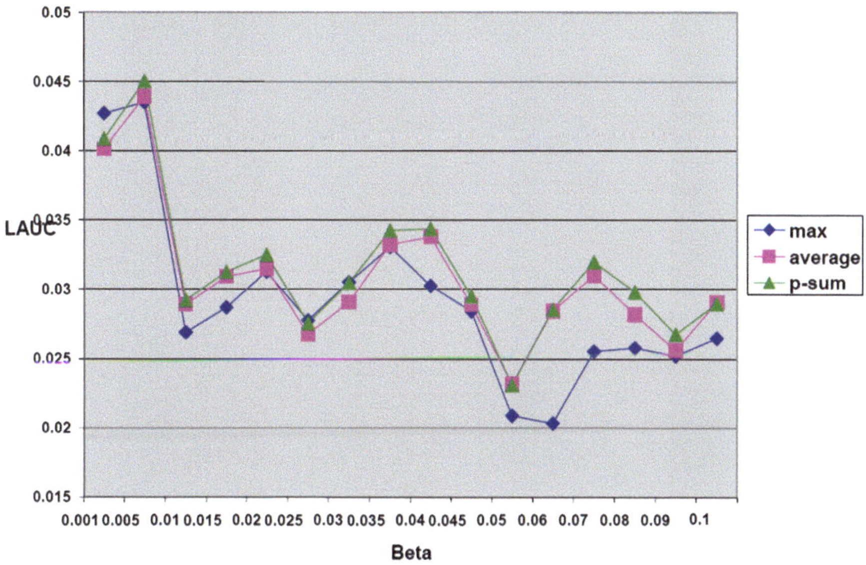

Fig. 6 LAUC values reported for the first experiment

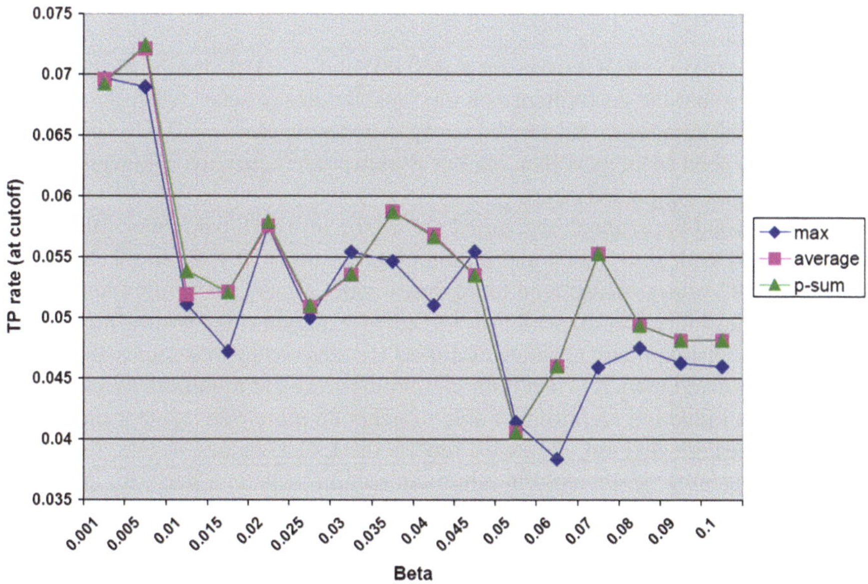

Fig. 7 True positive rates at cutoff point of 1% false positive rate reported for the first experiment

same for smaller β values and more specific rules. For larger β values, however, *P-Sum* and *Average* are better than *Max*, with *P-Sum* working slightly better than *Average*. This is because the chance of specific rules overlapping is smaller than that of general rules. Figure 7 displays the true positive rate at the cutoff point of 1% false positive rate. They are about 5–7 times better than random.

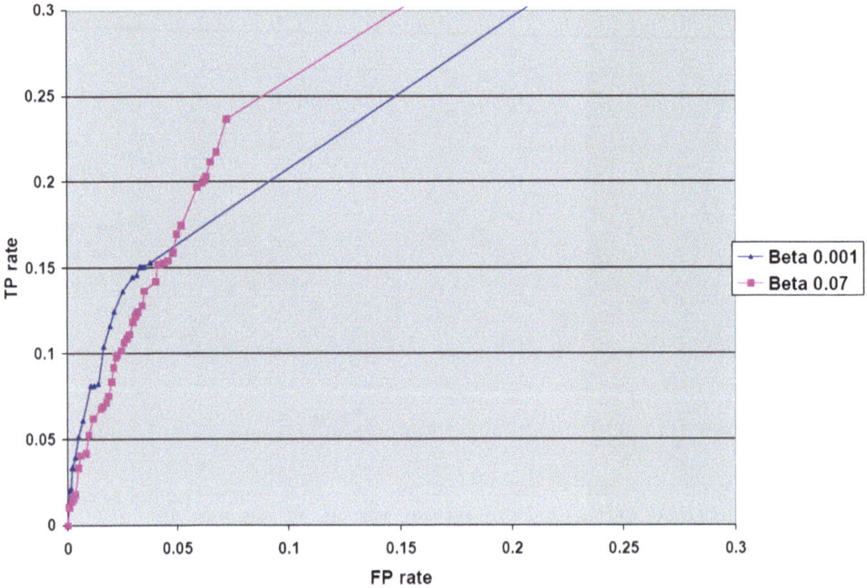

Fig. 8 ROC curves for $\beta = 0.001$ and $\beta = 0.07$ plotted for the first experiment

Figure 8 shows the ROC curves for $\beta = 0.001$ and $\beta = 0.07$ for the *Max* method. It can be seen that the performance on top ranked cases is better when $\beta = 0.001$, while the performance on all ranked cases is better when $\beta = 0.07$. The results of this study tend to suggest that smaller β values are better for achieving a high performance on top ranked cases.

In the second experiment, we varied the minimum recall parameter value from 0.005 to 0.04 with $\beta = 0.001$ and minimum precision = 0.4. Figure 9 shows the reported AUC values, which tend to go down with an increase in minimum recall. Since the rule induction algorithm is forced to generate rules with a high recall and a low precision, many negative examples are covered by these rules and correspondingly ranked higher. When both minimum recall and minimum precision are low, the rule induction algorithm is able to generate more rules so that more positive examples are covered. Figure 10 reports the LAUC values. Again, following the same reasoning, an increase in minimum recall results in a decrease in performance. Here, *P-Sum* and *Max* tend to do better than *Average* for lower minimum recall values.

In the third experiment, we varied the minimum recall parameter value from 0.005 to 0.04 with $\beta = 0.001$ and minimum precision = 0.7. Figure 11 shows the reported AUC values, which tend to go down with an increase in minimum recall for reasons similar to those discussed above for the second experiment. The difference is that when the minimum recall is increased, the number of rules generated goes down quickly. Actually, when the minimum recall is larger than 0.03, no rule is generated, so the reported AUC values are the same as those of a random selection.

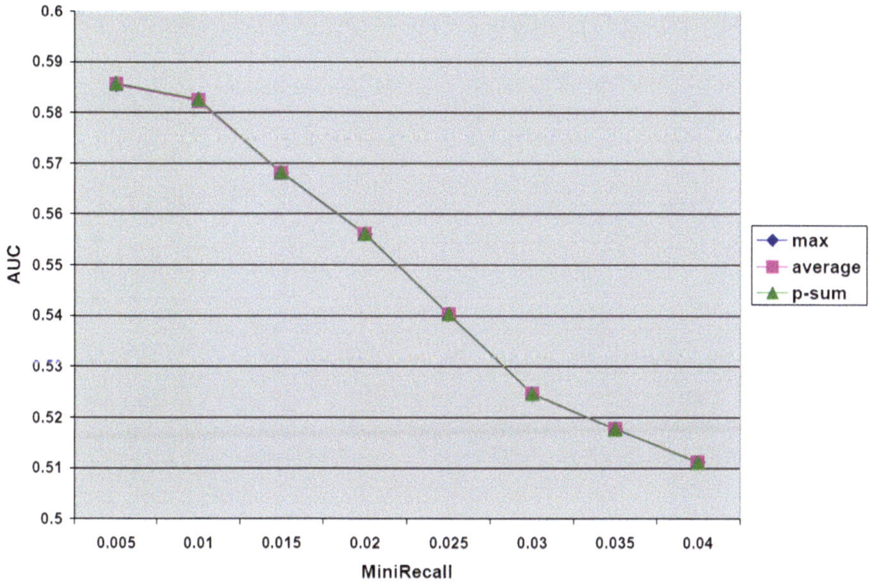

Fig. 9 AUC values reported for the second experiment

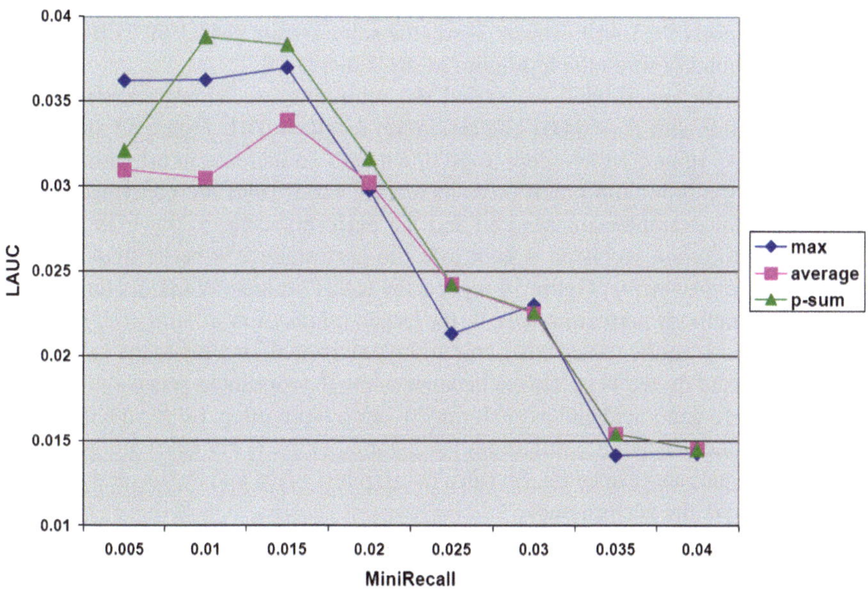

Fig. 10 LAUC values reported for the second experiment

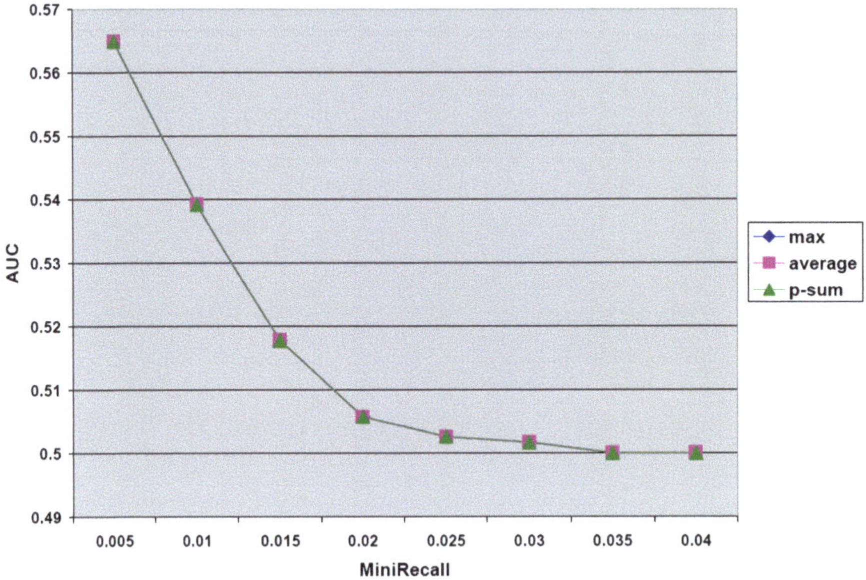

Fig. 11 AUC values reported for the third experiment

Figure 12 reports the LAUC values. Here, the same reason causes the performance to decrease quickly when the minimum recall is increased.

In the fourth experiment, we varied the minimum precision parameter value from 0.2 to 0.9 with $\beta = 0.001$ and minimum recall $= 0.01$. Figure 13 shows the reported AUC values, which seem to go down with an increase in minimum precision. When the minimum precision is increased, fewer rules are generated and so fewer positive examples are covered and the performance goes down as a result. When the minimum precision is kept low, the performance is better than those in all previous experiments. Figure 14 reports the LAUC values. While the three score combining methods perform similarly for larger values of minimum precision, the performances are quite varied when minimum precision is smaller. In the latter case, *P-Sum* seems to do the best. This is because when the minimum precision is small, many rules are generated and as such may overlap more often. So *P-Sum* improves the performance. When the minimum precision is large, fewer rules are generated and they are also more specific. So rules overlap less often and *P-Sum* and *Average* cannot improve the performance.

4.1.5 Summary

The experimental results clearly show that the performance measured by LAUC does not necessarily correlate with that measured by the AUC. They also show that the inductive bias impacts LAUC and AUC differently. More specific and more

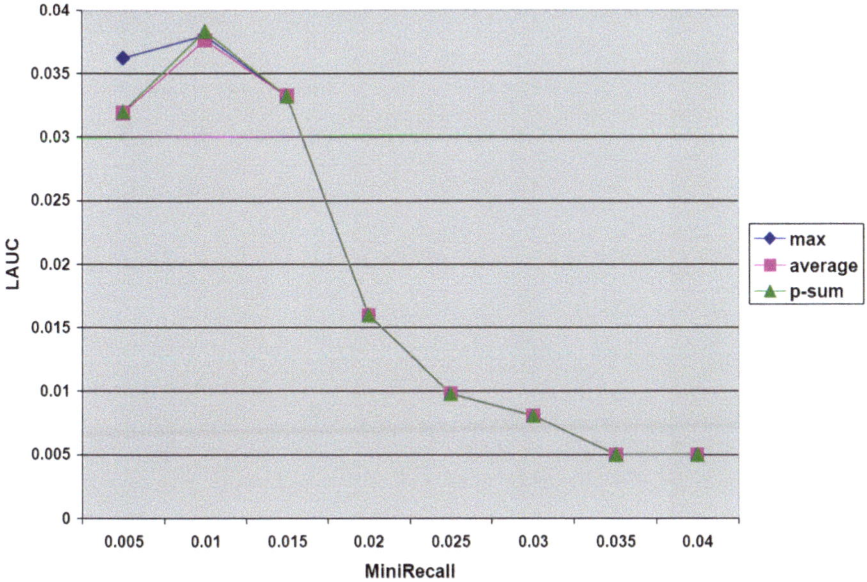

Fig. 12 LAUC values reported for the third experiment

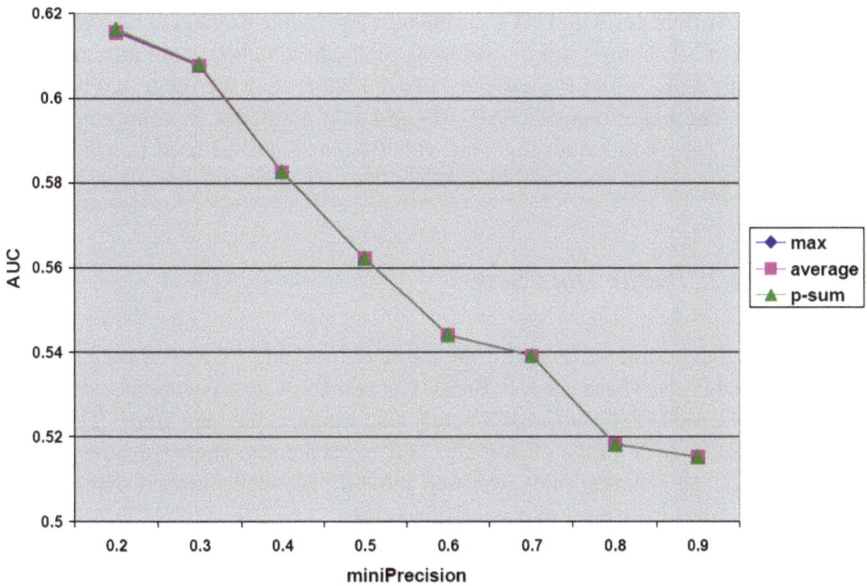

Fig. 13 AUC values reported for the fourth experiment

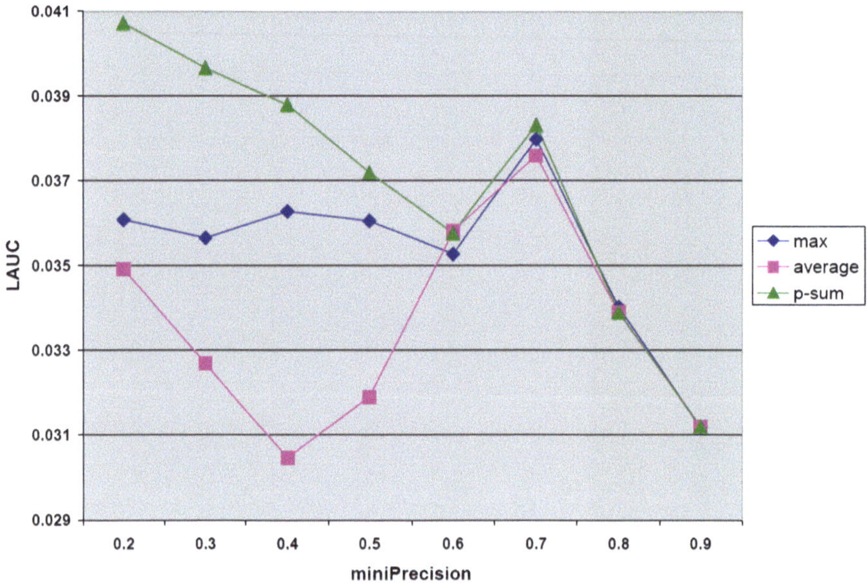

Fig. 14 LAUC values reported for the fourth experiment

precise rules work better with LAUC as the measure, while more general rules work better with AUC. The three score combining methods achieved about the same performance with AUC, while *P-Sum* generally did better than the other two methods with LAUC. The experimental results suggest that a smaller β, minimum recall and minimum precision values together with *P-Sum* should be used to achieve the optimal performance as measured by LAUC.

4.2 Study 2: Public Data Sets

We conducted experiments on six of the publicly available UCI Machine Learning Repository data sets. These are D1: Breast Cancer, D2: Chess (two-class scenario), D3: German Credit, D4: Japanese Credit, D5: Magic, and D6: Yeast. The sample sets present a wide range of domains and cover a comprehensive suite of data characteristics. To generate rules, we used the RIPPER algorithm [5] that induces classification rules from a set of preclassified cases.

The reported AUC values for each data set were calculated using a tenfold cross validation. Experiments were conducted with three groups of rules representing varying levels of simplifications for the generated rule sets (i.e., a larger number of more specific and precise rules or a smaller number of more general and overlapping rules). These variations were achieved by changing the RIPPER parameter settings. The training and test data sets were kept the same among these three groups of rule sets.

Six different scoring methods were used. The first three use probability estimation with three different methods, *Max*, *Average*, and *P-Sum*, for combining the scores of an example covered by multiple rules as describe in Sect. 3.4. These three methods are denoted as follows.

- AUC-FR0: *Max*
- AUC-FR1: *Average*
- AUC-FR2: *P-Sum*

The remaining group of methods utilizes the Laplace corrections, also with the three score combining methods. They are denoted as:

- AUC-LC0: *Max*
- AUC-LC1: *Average*
- AUC-LC2: *P-Sum*

We also use a measure that represents the average number of scores per example. It takes a value from 0 to 1 and is computed by dividing the number of unique scores divided by the number of test examples. For finer grained rankings, this number is generally closer to 1. In many applications, a finer grained ranking is preferred. This ranking-grain measure is used for the six different score computation methods as described above (*RankGrain-FR0, RankGrain-FR1, RankGrain-FR2, RankGrain-LC0, RankGrain-LC1,* and *RankGrain-LC2*).

Figure 15 depicts the summary of the reported AUC values for the six different methods. The reported values are the average on all six datasets. In the figure,

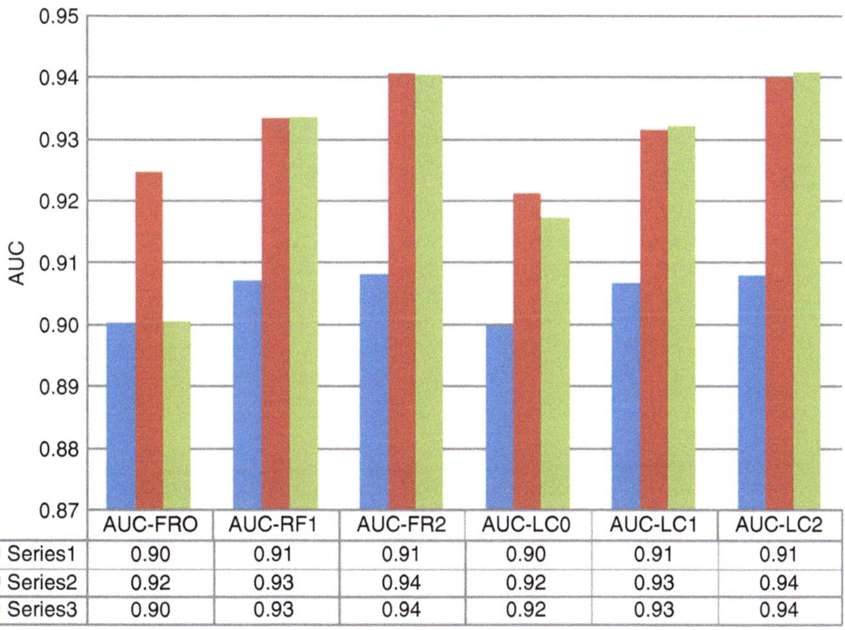

Fig. 15 Graphical summary of the reported AUC values

Series 1 is the rule set with the most general rules, Series 2 is the rule set with less general rules, and Series 3 is the rule set with the most specific rules. Except for FR0, rules of Series 2 and Series 3 outperformed rules of Series 1. Namely, specific rules outperformed general rules. It means specific rules are better for ranking than general rules, because specific rules are able to produce finer grained rankings than general rules. However, rules of Series 2 and Series 3 performed about the same. This suggests that highly specific rules may help with ranking. Highly specific rules tend to include more perfect rules, which could be better for ranking. It is also shown in the figure that *P-Sum* performed better than *Average*, which in turn did better than *Max*. *P-Sum* and *Average* produce finer grained rankings than *Max* and assume that examples covered by multiple rules should be ranked higher. Laplace correction achieved about the same performance as the simple probability estimation.

Table 1 reports the detailed AUC values for each of the six data sets. The R measurement represents the ratio of the number of perfect rules (rules that do not cover negative examples) to the total number of rules in a given rule set. Here, the average value of R is depicted (for each of the three runs on the six data sets). The R measurement is affected by using RIPPER's s parameter, which simplifies or specializes the generated rule set (i.e., trading off generality for accuracy by the degree

Table 1 Reported AUC values

Dataset	AUC-FR0	AUC-RF1	AUC-FR2	AUC-LC0	AUC-LC1	AUC-LC2
Series 1	$R = 0.19$					
D1	0.98	0.98	0.98	0.98	0.98	0.98
D2	0.77	0.79	0.79	0.77	0.78	0.79
D3	0.92	0.92	0.92	0.92	0.92	0.92
D4	0.82	0.84	0.84	0.82	0.84	0.84
D5	1.00	1.00	1.00	1.00	1.00	1.00
D6	0.91	0.91	0.91	0.91	0.91	0.91
Averages	0.90	0.91	0.91	0.90	0.91	0.91
Series2	$R = 0.25$					
D1	0.99	0.99	0.99	0.99	0.99	0.99
D2	0.84	0.85	0.88	0.83	0.85	0.87
D3	0.95	0.95	0.96	0.95	0.95	0.96
D4	0.85	0.88	0.89	0.84	0.87	0.89
D5	1.00	1.00	1.00	1.00	1.00	1.00
D6	0.92	0.93	0.93	0.92	0.93	0.93
Averages	0.92	0.93	0.94	0.92	0.93	0.94
Series 3	$R = 0.28$					
D1	0.92	0.99	0.99	0.95	0.99	1.00
D2	0.84	0.85	0.87	0.83	0.85	0.87
D3	0.96	0.96	0.96	0.95	0.95	0.96
D4	0.85	0.88	0.89	0.84	0.87	0.89
D5	0.90	1.00	1.00	1.00	1.00	1.00
D6	0.93	0.93	0.93	0.93	0.93	0.93
Averages	0.90	0.93	0.94	0.92	0.93	0.94

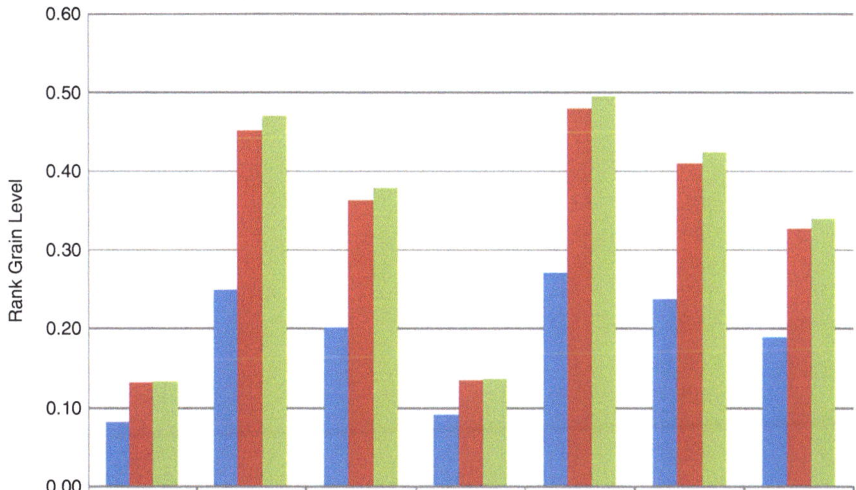

	RankGrain-FR0	RankGrain-FR1	RankGrain-FR2	RankGrain-LC0	RankGrain-LC1	RankGrain-LC2	AveRankGrain
Series1	0.08	0.25	0.20	0.09	0.27	0.24	0.19
Series2	0.13	0.45	0.36	0.13	0.48	0.41	0.33
Series3	0.13	0.47	0.38	0.14	0.50	0.42	0.34

Fig. 16 Graphical summary of the reported *RankGrain* values

of hypothesis simplification). When the parameter is set to favor the generation of more specific rules, then more accurate results can be achieved.

Figure 16 shows the summary of the *RankGrain* values graphically. Table 2 reports the *RankGrain* values. As seen in the figure, it seems that more specific rules result in finer grained rankings. The difference between Series 2 and Series 3 is small. *Average* and *P-Sum* produce more scores than *Max*. The number of different scores for *Max* cannot be larger than the number of rules and the number of scores for *Average* and *P-Sum* can never be smaller than *Max*. When there are many overlapping rules, *Average* and *P-Sum* may generate many more scores. It is interesting to see that *Average* produces more scores than *P-Sum*, which may be due to the number of perfect rules. As long as an example is covered by a perfect rule, *P-Sum* can assign it a perfect score of 1, no matter how many rules it is covered by. On the other hand, *Average* may give a different score to examples covered by some perfect rules. If all rules are perfect rules, both *Max* and *P-Sum* produce only two scores 0 and 1, but *Average* is still able to produce more scores. Laplace correction is able to produce slightly more scores than the simple probability estimation method because of the perfect rules. With Laplace correction, perfect rules may produce different scores depending on the number of examples covered.

The following are some of the conclusions drawn from the results:

- Specific rules produce higher AUC values than general rules.
- *P-Sum* generates the best AUC results in comparison with *Average* and *Max*.
- Simple probability estimation method performs about the same as Laplace correction.

Table 2 The reported *RankGrain* values

Dataset	RankGrain-FR0	RankGrain-FR1	RankGrain-FR2	RankGrain-LC0	RankGrain-LC1	RankGrain-LC2
Series 1	$R = 0.19$					
D1	0.11	0.31	0.24	0.11	0.31	0.31
D2	0.02	0.13	0.05	0.07	0.6	0.19
D3	0.13	0.29	0.29	0.13	0.29	0.29
D4	0.11	0.18	0.18	0.11	0.18	0.18
D5	0.03	0.37	0.24	0.03	0.37	0.25
D6	0.09	0.21	0.21	0.09	0.21	0.21
Averages	0.08	0.25	0.20	0.09	0.27	0.24
Series 2	$R = 0.25$					
D1	0.15	0.36	0.26	0.13	0.39	0.36
D2	0.02	0.13	0.05	0.07	0.27	0.19
D3	0.22	0.65	0.64	0.21	0.65	0.65
D4	0.19	0.45	0.44	0.18	0.45	0.45
D5	0.06	0.63	0.35	0.06	0.63	0.36
D6	0.15	0.49	0.44	0.14	0.49	0.45
Averages	0.13	0.45	0.36	0.13	0.48	0.41
Series 3	$R = 0.28$					
D1	0.14	0.40	0.28	0.13	0.42	0.37
D2	0.02	0.13	0.05	0.07	0.27	0.19
D3	0.23	0.68	0.66	0.23	0.68	0.68
D4	0.20	0.48	0.48	0.19	0.48	0.48
D5	0.06	0.63	0.35	0.06	0.63	0.35
D6	0.14	0.50	0.46	0.14	0.50	0.47
Averages	0.13	0.47	0.38	0.14	0.50	0.42

- Specific rules produce more scores than general rules.
- *Average* generates the best *RankGrain* results, while *Max* generates the worst *RankGrain* results.
- *RankGrain* values computed using the Laplace correction (*RankGrain-LCn*) are slightly better than the ones computed using class frequencies (*RankGrain-FRn*)
- The AUC value correlates with and increases with the number of scores.

5 Conclusions

In this chapter, we proposed a framework for ranking with rules. The framework introduces three types of rule-based ranking methods: post-analysis of rules, hybrid methods, and multiple rule set analysis. We also proposed three methods, *Max*, *Average*, and *Probabilistic Sum*, for combining the scores of an example covered by multiple rules. A successfully deployed data mining application in financial fraud detection was also discussed. The application aims at prioritizing human investigations of public companies suspected of engaging in fraudulent behavior based on their financial filings. Additional empirical studies were conducted using six of the UCI Machine Learning Repository data sets to evaluate the two simplest rule-based ranking methods: probability estimation with and without Laplace correction, as well as the three rule score combining methods. We investigated the impact of the inductive bias on the ranking performance. Experimental results clearly suggest that the inductive bias has an impact on the ranking performance. It is also shown that *Probabilistic Sum* and *Average* generally perform better than *Max*. It seems that a method that produces more scores usually outperforms a method that produces fewer scores. More experiments need to be conducted to verify this.

Future research will include an empirical validation of some of the other methods in the framework, specifically the ones that use geometric measures. Since with geometric methods more scores could be produced, it would be interesting to see whether such methods could improve the ranking performance. Also considered is a study comparing the performance of rule-based and decision tree-based ranking approaches.

References

1. I. Alvarez, S. Bernard, Ranking cases with decision trees: A geometric method that preserves intelligibility, in *Proceedings of the 19th International Joint Conference on AI (IJCAI-05)* (2005), pp. 635–640
2. L.W. Barsalou, Ideals, central tendency, and frequency of instantiation as determinants of graded structure in categories. J. Exp. Psychol. Learn. Mem. Cogn. **11**, 629–654 (1985)
3. E. Bauer, R. Kohavi, An empirical comparison of voting classification algorithms: bagging, boosting and variants. Mach. Learn. **36**, 105–142 (1999)
4. B.G. Buchanan, E.H. Shortliffe (eds.), *Rule-Based Expert Systems: The MYCIN Experiments of the Stanford Heuristic Programming Project* (Addison Wesley, Reading, MA, 1984)
5. W. Cohen, Text categorization and relational learning, in *Proceedings of the 12th International Conference on Machine Learning (ICML-95)* (1995), pp. 124–132
6. C. Cortes, M. Mohri, AUC Optimization vs. Error Rate Minimization, in *Advances in Neural Information Processing Systems (NIPS-03)* (MIT, 2003)
7. C. Ferri, P.A. Flach, J. Hernandez-Orallo, Improving the AUC of probabilistic estimation trees, in *Proceedings of the 14th European Conference on Machine Learning (ECML-03)* (Springer, 2003)
8. R. Kohavi, Scaling up the accuracy of Naive-Bayes classifiers: a decision-tree hybrid, in *Proceedings of the 2nd International Conference on Knowledge Discovery and Data Mining (KDD-96)* (1996)
9. C.X. Ling, R.J. Yan, Decision tree with better ranking, in *Proceedings of the 20th International Conference on Machine Learning (ICML-01)* (Morgan Kaufmann, 2003)
10. F.J. Provost, P. Domingos, Tree induction for probability-based ranking. Mach. Learn. **52**(30), 199–215 (2003)
11. P. Utgoff, Perceptron trees – a case study in hybrid concept representation, in *Proceedings of The Seventh National Conference on Artificial Intelligence (AAAI-88)* (1988), pp. 601–606
12. D.A. Waterman, *A Guide to Expert Systems* (Addison-Wesley Publishing, Reading, Mass, USA, 1985)
13. L.X. Wang, J.M. Mendel, Generating fuzzy rules by learning from examples. IEEE Trans. Syst. Man Cybern. **22**(6), 1414–1427 (1992)
14. J. Zhang, R.S. Michalski, An integration of rule induction and exemplar-based learning for graded concepts. Mach. Learn. **21**(3), 235–267 (1995)

Part III
Object Ranking

A Survey and Empirical Comparison of Object Ranking Methods

Toshihiro Kamishima, Hideto Kazawa, and Shotaro Akaho

Abstract Ordered lists of objects are widely used as representational forms. Such ordered objects include Web search results or bestseller lists. In spite of their importance, methods of processing orders have received little attention. However, research concerning orders has recently become common; in particular, researchers have developed various methods for the task of *Object Ranking* to acquire functions for object sorting from example orders. Here, we give a unified view of these methods and compare their merits and demerits.

1 Introduction

We survey methods for learning to estimate orders, and empirically compare these methods. The term *Order* indicates a sorted sequence of objects according to some property. For example, the responses from Web search engines are lists of pages sorted according to their relevance to queries. Bestseller lists, which are item-sequence sorted according to sales volume, are used on many E-commerce sites. In particular, several methods have been developed for learning functions used to sort objects from example orders. We call this task *Object Ranking* and emphasize its usefulness for sensory surveys,[1] information retrieval, and decision making. We give a unified view of the object ranking task that are independently proposed and discuss the connection with the other types of tasks dealing with orders. We then show experimental results to reveal the pros and cons of these object ranking methods.

---

[1]quantification of respondents' sensations or impressions

T. Kamishima (✉) and S. Akaho
National Institute of Advanced Industrial Science and Technology (AIST),
AIST Tsukuba Central 2, Umezono 1–1–1, Tsukuba, Ibaraki, 305–8568 Japan
e-mail: mail@kamishima.net⟨http://www.kamishima.net/⟩, s.akaho@aist.go.jp

H. Kazawa
Google Japan Inc., Cerulean Tower 6F, 26-1 Sakuragaoka-cho, Sibuya-ku, Tokyo
e-mail: kazawa@google.com

J. Fürnkranz and E. Hüllermeier (eds.), *Preference Learning*,
DOI 10.1007/978-3-642-14125-6_9,

We formalize the object ranking task in Sect. 2. We survey methods in Sect. 3. Experimental results on artificial data and on real data are shown in Sects. 4 and 5, respectively. We discuss and summarize the results in Sect. 6.

2 Orders and Object Ranking

This section shows basic notions regarding orders and formalizes the object ranking task.

2.1 Basics About Orders

We begin by defining basic notations regarding orders. An object or entity to be sorted is denoted by $\mathbf{x}_j$. The universal object set, $\mathcal{X}^*$, consists of all possible objects. Each object $\mathbf{x}_j$ is represented by the attribute value vector $\mathbf{x}_j = [x_{j1}, x_{j2}, \ldots, x_{jk}]^\top$, where k is the number of attributes. An order is denoted by $O = \mathbf{x}_{j_a} \succ \mathbf{x}_{j_b} \succ \cdots \succ \mathbf{x}_{j_c}$. Note that the subscript j of $\mathbf{x}_j$ does not mean "The jth object in this order," but that "The object is uniquely indexed by j in $\mathcal{X}^*$." An order $\mathbf{x}_1 \succ \mathbf{x}_2$ represents "$\mathbf{x}_1$ precedes $\mathbf{x}_2$." An object set $\mathcal{X}(O_i)$ or simply $\mathcal{X}_i$ is composed of all the objects in the order O_i. The length of O_i, i.e., $|\mathcal{X}_i|$, is shortly denoted by L_i. An order of all objects, i.e., O_i s.t. $\mathcal{X}(O_i) = \mathcal{X}^*$, is called a complete order; otherwise, the order is incomplete. Rank, $r(O_i, \mathbf{x}_j)$ or simply r_{ij}, is the cardinal number that indicates the position of the object $\mathbf{x}_j$ in the order O_i. For example, for $O_i = \mathbf{x}_1 \succ \mathbf{x}_3 \succ \mathbf{x}_2$, $r(O_i, \mathbf{x}_2)$ or r_{i2} is 3. For two orders, O_1 and O_2, consider an object pair $\mathbf{x}_a$ and $\mathbf{x}_b$ such that $\mathbf{x}_a, \mathbf{x}_b \in \mathcal{X}_1 \cap \mathcal{X}_2, a \neq b$. These two orders are concordant w.r.t. $\mathbf{x}_a$ and $\mathbf{x}_b$ if the two objects are placed in the same order, i.e., $(r_{1a} - r_{1b})(r_{2a} - r_{2b}) \geq 0$; otherwise, they are discordant. Further, O_1 and O_2 are concordant if O_1 and O_2 are concordant w.r.t. all object pairs such that $\mathbf{x}_a, \mathbf{x}_b \in \mathcal{X}_1 \cap \mathcal{X}_2, a \neq b$.

We then describe the distance between two orders, O_1 and O_2, composed of the same sets of objects, i.e., $\mathcal{X}(O_1) = \mathcal{X}(O_2) \equiv \mathcal{X}$. Various kinds of distance for orders have been proposed [1]. *Spearman distance* $d_S(O_1, O_2)$ is widely used. It is defined as the sum of the squared differences between ranks:

$$d_S(O_1, O_2) = \sum_{\mathbf{x}_j \in \mathcal{X}} \left(r_{1j} - r_{2j}\right)^2. \tag{1}$$

By normalizing the range to be $[-1, 1]$, *Spearman's rank correlation* ρ is derived.

$$\rho = 1 - \frac{6 d_S(O_1, O_2)}{L^3 - L}, \tag{2}$$

where L is the length of orders, i.e., $L = |\mathcal{X}|$. This exactly equals the correlation coefficient between ranks of objects. The *Kendall distance* $d_K(O_1, O_2)$ is another widely used distance. Consider a set of object pairs, $\{(\mathbf{x}_a, \mathbf{x}_b) \in \mathcal{X} \times \mathcal{X}\}$, $a \neq b$, $\mathbf{x}_a, \mathbf{x}_b \in \mathcal{X}$, including either $(\mathbf{x}_a, \mathbf{x}_b)$ or $(\mathbf{x}_b, \mathbf{x}_a)$. The Kendall distance is defined as the number of discordant pairs between O_1 and O_2 w.r.t. $\mathbf{x}_a$ and $\mathbf{x}_b$. Formally,

$$d_K(O_1, O_2) = \frac{1}{2}\left(M - \sum_{\{(\mathbf{x}_a, \mathbf{x}_b)\}} \operatorname{sgn}((r_{1a} - r_{1b})(r_{2a} - r_{2b}))\right), \tag{3}$$

where $\operatorname{sgn}(x)$ is a sign function that takes 1 if $x > 0$, 0 if $x = 0$, and -1 otherwise. $M = (L-1)L/2$ is the number of all object pairs. By normalizing the range to be $[-1, 1]$, *Kendall's rank correlation* τ is derived.

$$\begin{aligned} \tau &= 1 - \frac{2d_K(O_1, O_2)}{M} \\ &= \frac{1}{M} \sum_{\{(\mathbf{x}_a, \mathbf{x}_b)\}} \operatorname{sgn}\big((r_{1a} - r_{1b})(r_{2a} - r_{2b})\big). \end{aligned} \tag{4}$$

The computational costs for deriving ρ and τ are $O(L \log L)$ and $O(L^2)$, respectively. The values of τ and ρ are highly correlated, because the difference between two criteria is bounded by Daniels' inequality [2]:

$$-1 \leq \frac{3(L+2)}{L-2}\tau - \frac{2(L+1)}{L-2}\rho \leq 1. \tag{5}$$

Another inequality between d_K and d_S is Durbin–Stuart's inequality:

$$d_S \geq \frac{4}{3} d_K \left(1 + \frac{d_K}{L}\right).$$

In [1], you can find further description about other types of distance, such as Spearman's footrule, Cayley distance, and Ulam distance, and their characteristics.

Note that d_K is a metric, but d_S is not due to the violation of the triangular inequality condition. If two or more objects are tied, we give the same *midrank* to these objects [1]. For example, consider an order $\mathbf{x}_5 \succ \mathbf{x}_2 \sim \mathbf{x}_3$ ("$\sim$" denotes tie in rank), in which $\mathbf{x}_2$ and $\mathbf{x}_3$ are ranked at the 2nd or 3rd positions. In this case, the midrank 2.5 is assigned to both objects.

2.2 An Object Ranking Task

An *Object Ranking* task can be considered a regression or a fitting task whose target variables are orders. Further, input samples comprise not a set of vectors, but a

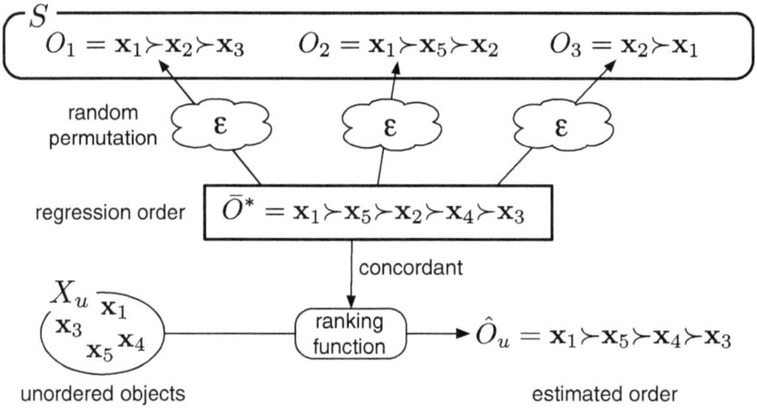

Fig. 1 The object ranking task

set of orders, $S = \{O_1, \ldots, O_N\}$, where N is the number of samples. The regression curve corresponds to a regression order. Analogous to the case of a regression function, a regression order is estimated so as to be concordant not only with given sample orders in S, but also with orders that will be generated in future. This task differs from a regression in two ways. First, since the target variables are orders, the modeling methods of regression orders and errors are needed. A regression order is modeled by a ranking function, ord(·), which represents a rule for sorting objects. Given an object set $\mathcal{X}_u$, a ranking function outputs the ideal order that consists of all objects in $\mathcal{X}_u$. Although errors of real values are modeled by an additive term of a random variable, errors in orders are modeled by a random permutation $\varepsilon(\cdot)$. That is to say, a sample order O_i is generated by $\varepsilon(\mathrm{ord}(\mathcal{X}_i))$. Second, since sample orders are generally incomplete, there may be objects not observed in given samples (e.g., x_4 in Fig. 1). Such objects should be ranked under the assumption that the neighboring objects in the attribute space would be close in rank. Object ranking is also different from classification, because no ordinal relations should be defined between classes.

We say that a ranking function is *absolute* if outputs of the function are concordant with each other; otherwise, it is *relative*. That is to say, for any $\mathcal{X}_u \subseteq \mathcal{X}^*$, while an absolute ranking function outputs orders that are concordant with orders a regression order that consists of all objects in $\mathcal{X}^*$, a relative ranking function does not. Being absolute is also equivalent to the condition 3, "The independence of irrelevant alternatives," of the Arrow's impossibility theorem [3]. Concretely, given unordered sets $\{\mathbf{x}_1, \mathbf{x}_2, \mathbf{x}_3\}$ and $\{\mathbf{x}_1, \mathbf{x}_2, \mathbf{x}_4\}$, an absolute ranking function outputs orders that are concordant w.r.t. $\mathbf{x}_1$ and $\mathbf{x}_2$ regardless of the existence of objects $\mathbf{x}_3$ or $\mathbf{x}_4$. However, a relative ranking function differently sorts $\mathbf{x}_1$ and $\mathbf{x}_2$ due to the existence of the other objects, e.g., $\mathbf{x}_1 \succ \mathbf{x}_2 \succ \mathbf{x}_3$ and $\mathbf{x}_2 \succ \mathbf{x}_4 \succ \mathbf{x}_1$. An absolute ranking function would be preferable in applications such as filtering or recommendation. For example, if one prefers an apple to an orange, he/she will always rank an apple higher than an orange when sorting any set of fruits according to degree of preference. One example application appropriate for relative ranking can be found in [4]. When summarizing multi-documents, after extracting important sentences, these sentences are ordered. In this case, appropriate order of sentences would be affected by the other extracted sentences.

Object ranking is closely related to a notion of a *Center of Orders* [1]; given sample orders S, center $\bar{O}$ is defined as the order that minimizes the sum of the distances $\sum_{O_i \in S} d(O_i, P(\bar{O}, \mathcal{X}_i))$. Note that $P(\bar{O}, \mathcal{X}_i)$ denotes the projection of $\bar{O}$ on the set $\mathcal{X}_i$, which is the order that consists of the objects in $\mathcal{X}_i$ and is concordant with $\bar{O}$. This notion is also referred by the other terms, such as aggregated ranking [5], in machine learning or information retrieval disciplines. We here adopt the term "center of orders" in statistics because this would be the first given denotation. This center differs from the above regression order in the points that concordance only with *given* samples is considered and that no description of objects, e.g., attribute values, is employed. The computation of centers is generally NP-hard, except for the cases such as employing Spearman's distance.

Object ranking is also related to *Ordinal Regression* [6, 7], which is a regression whose type of dependent variables is ordered categorical. Ordered categorical variables can take one of a finite set of predefined values, like categorical variables, and order these values additionally; for example, a domain of a variable is {"good","fair","poor"}. Ordered categories and orders are different in two points. First, while orders provide purely relative information, ordered categorical values additionally include absolute information. For example, while the category "good" means absolutely good, $\mathbf{x}_1 \succ \mathbf{x}_2$ means that $\mathbf{x}_1$ is relatively better than $\mathbf{x}_2$. Second, the number of grades that can be represented by ordered categorical variables is limited. Consider that there are four objects. Because at least two objects must be categorized into one of the three categories, {"good","fair","poor"}, the grades of these two objects are indistinguishable. However, orders can represent the differences of grades between any two objects. Because a task of object ranking is more complicated than ordinal regression, object ranking methods can be applicable to solve ordinal regression, but the converse is not true. But, generally speaking, since it is not efficient to solve too much complicated tasks, these two tasks should be carefully differentiated. For example, the computational cost of SVMs for object ranking (see Table 3) is about the square of $O(N^2\bar{L}^4)$, where $\bar{L}$ is the mean length of sample orders. The SVM specially designed for ordinal regression [8] demands less computational cost $O(N^2|\mathcal{Y}|^2)$, where $|\mathcal{Y}|$ is the number of grades.

3 Object Ranking Methods

We present five object ranking methods. Their abbreviations are given in parentheses in the section titles.

3.1 Cohen's method (Cohen)

Cohen's method [9] is designed to find the order $\hat{O}_u$ that maximizes

$$\sum_{\mathbf{x}_a \succ \mathbf{x}_b \in \hat{O}_u} \mathrm{P}[\mathbf{x}_a \succ \mathbf{x}_b | \mathbf{x}_a, \mathbf{x}_b], \tag{6}$$

where $\mathrm{P}[\mathbf{x}_a \succ \mathbf{x}_b | \mathbf{x}_a, \mathbf{x}_b]$ is the conditional probability given the attribute vectors of $\mathbf{x}_a$ and $\mathbf{x}_b$, and $\mathbf{x}_a \succ \mathbf{x}_b \in \hat{O}_u$ denotes all the ordered pairs concordant with $\hat{O}_u$. Note that O_u consists of all the objects in a given set of unordered objects, $\mathcal{X}_u$. Unfortunately, because the maximization of (6) is known as a linear ordering problem [10], which is NP-hard, it is not tractable to find the optimal solution if $|\mathcal{X}_u|$ is large. Cohen et al. hence proposed a greedy algorithm that sequentially chooses the most preceding object in Fig. 2. They proposed a more elaborate approximation method too, but any strict or approximation algorithms to solve the linear ordering problem [10] can be applied for this optimization.

$\mathrm{P}[\mathbf{x}_a \succ \mathbf{x}_b | \mathbf{x}_a, \mathbf{x}_b]$ is learned by Cohen et al.'s Hedge algorithm. The Hedge is an online algorithm, which is a variant of the Winnow [11]. Pairwise preference judgments are determined based on the linear combination of subordinate ordering functions. Given one preference feedback sample, a weight for an ordering function is increased according as the β parameter and the contribution of the ordering function to the concordance with the feedback. We set the β to 0.9; the attributes $\{\mathbf{x}_{jl}, -\mathbf{x}_{jl}\}_{l=1}^{k}$ are used as ordering functions in our experiments. To use the Hedge algorithm in offline mode, the objects in S are iteratively given as feedback, and iterations are repeated until the loss becomes stationary. Their Hedge algorithm is designed so that it takes only ordinal information of attributes into account and discards the numerical values themselves. Hence, our experimental condition in which objects are represented by nominal or numerical attributes are rather disadvantageous to this method.

$\hat{O}_u \leftarrow \emptyset$
for all $\mathbf{x} \in \mathcal{X}_u$ **do**
  $\mathrm{score}(x) \leftarrow \sum_{\mathbf{x}' \in \mathcal{X}_u} \mathrm{P}[\mathbf{x} \succ \mathbf{x}' | \mathbf{x}, \mathbf{x}']$
end for
while $\mathcal{X}_u \neq \emptyset$ **do**
  $\mathbf{x}_{top} \leftarrow \arg\max_{\mathbf{x}} \mathrm{score}(\mathbf{x})$
  $\hat{O}_u \leftarrow \hat{O}_u \succ \mathbf{x}_{top}, \quad \mathcal{X}_u \leftarrow \mathcal{X}_u - \{\mathbf{x}_{top}\}$
  for all $\mathbf{x} \in \mathcal{X}_u$ **do**
    $\mathrm{score}(\mathbf{x}) \leftarrow \mathrm{score}(\mathbf{x}) - \mathrm{P}[\mathbf{x} \succ \mathbf{x}_{top} | \mathbf{x}, \mathbf{x}_{top}]$
  end for
end while
return $\hat{O}_u$

Fig. 2 Cohen's greedy sorting algorithm

3.2 RankBoost (**RB**)

Freund et al. proposed *RankBoost* [12, 13], that is a boosting algorithm targeting orders. Inputs of the RankBoost are the feedback function $\Phi(\mathbf{x}_a, \mathbf{x}_b)$, which implies $\mathbf{x}_b \succ \mathbf{x}_a$ if $\Phi(\mathbf{x}_a, \mathbf{x}_b) > 0$, and a set of ranking features $f_l(\mathbf{x}_i)$, which conveys partial information about the target ordering. Given these inputs, RankBoost returns the final ranking $H(\mathbf{x}_i)$ that works as a score function. First, the initial distribution is calculated by $D_1(\mathbf{x}_a, \mathbf{x}_b) = \max(\Phi(\mathbf{x}_a, \mathbf{x}_b), 0)/Z_1$, where Z_1 is a normalization coefficient. Then, for each round $t = 1, \ldots, T$, the algorithm repeats the selection of weight α_t and weak learner $h_t(\mathbf{x})$, and the update of distribution by:

$$D_{t+1}(\mathbf{x}_a, \mathbf{x}_b) = \frac{1}{Z_t} D_t(\mathbf{x}_a, \mathbf{x}_b) \exp\bigl(\alpha_t (h_t(\mathbf{x}_a) - h_t(\mathbf{x}_b))\bigr).$$

Weak learners capture some information about target orders from ranking features, and output hypotheses such that $h_t(\mathbf{x}_b) > h_t(\mathbf{x}_a)$ implies $\mathbf{x}_b \succ \mathbf{x}_a$. α_t and h_t are selected so that the normalization factor Z_t is minimized. Once these weights and weak learners are acquired, unseen objects, $\mathbf{x} \in \mathcal{X}^*$, are sorted in descending order of $H(\mathbf{x}) = \sum_{t=1}^{T} \alpha_t h_t(\mathbf{x})$.

In our experiment, as ranking features, we adopt all terms that appear in n-order polynomials of object attributes after attribute values are transformed so as to range in [0, 1]. Let $\Phi(\mathbf{x}_a, \mathbf{x}_b)$ be $2\mathrm{P}[\mathbf{x}_b \succ \mathbf{x}_a] - 1$. Weak learners are set to ranking features themselves, i.e., $h(\mathbf{x}) = f_l(\mathbf{x})$. Selection method of α_t and h_t is the third option in Sect. 3.2 in [13]. The number of rounds, T is set to 100.

3.3 SVM-Based Methods: Order SVM (**OSVM**) and Herbrich's Method (**SVOR**)

We show two SVM-based methods: Order SVM and SVOR. In summary, the former is designed to discriminate whether or not a given object is ranked higher than jth, while the latter judges which of two objects precedes the other.

Since this paper concerns not categorical but ordinal rank, this method may appear to be a groundless attempt to discriminate high-ranked objects from low-ranked ones. However, we showed that the probability of finding an object sorted above a fixed rank is concordant with the true score function. Thus, if a classifier will discriminate the top j objects from the rest, its discrimination function must be concordant to some extent with probability and therefore with the true score function. This observation leads to the use of SVM as the estimator of a score function in [14].

We first show *Order SVM* [14]. To enhance the reliability of this estimation, we proposed training multiple SVMs with different threshold ranks and sorting unseen objects using the average of those SVMs. Its learning is formulated as the following optimization problem:

$$\min_{\mathbf{w},\mathbf{v}_t,b_t} \frac{1}{2}\|\mathbf{w}\|^2 + \frac{\lambda}{2}\sum_{t=1}^{L-1}\|\mathbf{v}_t\|^2 + C\sum_{t=1}^{L-1}\sum_{i=1}^{m}\sum_{j=1}^{L}\xi_i^j(t) \tag{7}$$

$$\text{s.t. } \mathrm{sgn}[j-t]((\mathbf{w}+\mathbf{v}_t)\cdot\mathbf{x}_i^j + b_t) \geq 1 - \xi_i^j(t),\ \xi_i^j(t) \geq 0, \quad \text{forall } i, j, t,$$

where $\mathbf{x}_i^j$ is the feature vector of the jth ranked object in the ith order, $\{\mathbf{x}_i^j\}_{i=1...m}^{j=1...L}$ are the training samples, and C and λ are hyperparameters. In our experiments, we set $C = 0.1$ and $\lambda = 1$ based on preparatory experiments on a small data set. The sgn[z] is 1 if $z \geq 0$; otherwise, -1. The SVM that discriminates the top t objects from the rest is $f_t(\mathbf{x}) = (\mathbf{w} + \mathbf{v}_t)\cdot\mathbf{x} + b_t$. Thus, the second regularizer $\sum_t \|\mathbf{v}_t\|^2$ makes all $f_t(\mathbf{x})$ agree on the predicted orders as much as possible. The order is predicted by sorting objects according to the score $\mathbf{w}\cdot\mathbf{x}$. The dual problem of (7) is similar to that of standard SVMs, and any kernel function can be used instead of the inner products between feature vectors [14]. This method performs discriminations whether an object is ranked higher than t for each $t = 1, \ldots, L-1$. The number of failures in these discriminations is equivalent to the absolute difference between object ranks in the estimated order and sample order. These differences are then summed up over all objects in orders. This sum can be considered as representing Spearman footrule [1], which is absolute distance between two orders. Therefore, this method aims at minimizing the sum of distances in Spearman footrule between an estimated order and each sample order.

We refer to the other SVM-based method as *Support Vector Ordinal Regression* (SVOR) [15] since its formulation is very similar to standard SVMs and the work on it appears to be inspired by that of past ordinal regression works [16]. This method was independently developed as *Ranking SVM* by Joachims [17] and was proposed in [18].

SVOR discriminates correctly ordered pairs from incorrectly ordered pairs. In contrast to the Cohen method in which precedences are independently learned from pairs and there is no guarantee that transitivity holds among the learned preferences, SVOR uses a single score function for learning and thus avoids the intractability problem of the sorting process shown in [9].

SVOR's learning is formulated as the following optimization problem:

$$\min_{\mathbf{w}} \frac{1}{2}\|\mathbf{w}\|^2 + C\sum_{i=1}^{m}\sum_{1\leq j<l\leq L}\xi_i^{jl} \tag{8}$$

$$\text{s.t. } \mathbf{w}\cdot(\mathbf{x}_i^j - \mathbf{x}_i^l) \geq 1 - \xi_i^{jl}, \quad \xi_i^{jl} \geq 0, \quad \text{for } ^\forall i, j < l,$$

where the same notations as OSVM are used for $\mathbf{x}_i^j$, m, L, and C (In our experiments, $C = 1$). SVOR tries to find the direction $\mathbf{w}$ along which sample objects are ordered so that the narrowest separation between samples is maximal. The prediction of orders is done by sorting objects according to the score $\mathbf{w}\cdot\mathbf{x}$. As in the case of OSVM, the dual problem of (8) can be written using only the inner products of $\mathbf{x}$;

thus, we can use any kernel function in SVOR as well. This method is designed so that for each object pair in orders, the discrimination whether one object precedes the other is performed. The number of failures in these discriminations is equivalent to Kendall distance. Therefore, this method aims at minimizing the sum of distances in Kendall distance between an estimated order and each sample order.

3.4 Expected Rank Regression (ERR)

We turn to our *Expected Rank Regression method* [19]. In this method, after expected ranks of objects are derived, the function to estimate these expected ranks is learned using a standard regression technique.

To derive expected ranks, assume that orders $O_i \in S$ are generated as follows: First, a complete sample order O_i^*, which cannot be observed and consists of all objects in $\mathcal{X}^*$, is generated. $|\mathcal{X}^*| - |O_i|$ objects are then selected uniformly at random, and these are eliminated from O_i^*; then, the O_i is observed. Under this assumption, a theoretical result in order statistics [20] shows that the conditional expectation of rank of the object $\mathbf{x}_j \in \mathcal{X}_i$ in the unobserved complete order O_i^* given the sample order O_i is

$$\mathrm{E}[r(O_i^*, \mathbf{x}_j)|O_i] \propto \frac{r(O_i, \mathbf{x}_j)}{|O_i| + 1}. \tag{9}$$

These expected ranks are calculated for all the objects in all the orders in a sample set S. Next, standard regression techniques are applied to derive weights of regression function, $f(\mathbf{x}_j)$, which predicts the above expectation of ranks. Samples for regression consist of the attribute vectors of objects, $\mathbf{x}_j$, and their corresponding expected ranks, $r(O_i, \mathbf{x}_j)/(|O_i| + 1)$; thus, the number of samples becomes $\sum_{O_i \in S} |\mathcal{X}(O_i)|$. Once weights of $f(\mathbf{x}_j)$ are learned, the order $\hat{O}_u$ can be estimated by sorting the objects $\mathbf{x}_j \in \mathcal{X}_u$ according to the corresponding values of $f(\mathbf{x}_j)$.

Below, we describe a rationale why this ERR works: We assume that sample orders are generated according to the Thurstone's model (case V) [21]. In this model, sample orders are generated by sorting objects $\mathbf{x} \in \mathcal{X}_u$ in the ascending order of the scores, $\mathbf{x}$, $f^*(\mathbf{x})$, that follow the following normal distribution:

$$f^*(\mathbf{x}) \sim \mathcal{N}(\mu(\mathbf{x}), \sigma^2), \tag{10}$$

where $\mu(\mathbf{x})$ and σ are mean and standard deviation, respectively. A complete regression order, $\bar{O}^*$, is determined by sorting all objects in $\mathcal{X}^*$ in the ascending order of the mean scores $\mu(\mathbf{x})$, and random permutation is caused by the additive noise in scores of objects; the above complete sample orders, $O_i^*, i = 1, \ldots, |S|$, are consequently generated. Next, we consider the probability that a pair of adjacent objects in $\bar{O}^*$ permutes. We assume that such probabilities are equal over all adjacent pairs, because there is no prior information that one pair is more frequently permuted than the other pair. This assumption makes the differences of the mean scores between these adjacent pairs, $|\mu(\mathbf{x}_i) - \mu(\mathbf{x}_j)|$, to be constant. Further, these mean can be replaced with rank in $\bar{O}^*$ without loss of generality, because an order derived based on scores is invariant for any linear transformation in scores. We then learn a function to predict rank $r(\hat{O}^*, \mathbf{x})$ for any object $\mathbf{x}$. Expectations of these ranks over random elimination of objects can be obtained by (9), and noise in scores follows normal distribution as in (10). Consequently, we can estimate ranks in $\bar{O}^*$ by applying standard regression to samples, $(\mathbf{x}_j, r(O_i, \mathbf{x}_j)/(|O_i| + 1))$.

In our experiments, n-order polynomials are adopted as a class of regression functions. No regularization terms were adopted in regression.

4 Experiments on Artificial Data

To reveal the characteristics of object ranking methods, we applied these methods to artificial data.

4.1 Experimental Conditions

Artificial data were generated in three steps. First, we generated two types of vectors: numerical (`num`) and binary (`bin`). Each numerical vector consists of $5(\equiv k)$ attributes, which follow the normal distribution, $N(0, 1)$. Binary vectors are composed of $15(\equiv k)$ attributes, and are randomly generated so that every object is represented by different value vectors. Note that, when the methods designed for numeric attributes, binary values are converted to the number, 0 or 1. Second, true regression orders are determined based on these attribute values. Objects are sorted according to the values of the function:

$$\text{utility}(\mathbf{x}_j) = \left(1 + \sum_{l=1}^{k} w_l x_{jl}\right)^{\text{dim}},$$

where w_l are random weights that follow the normal distribution, $\mathcal{N}(0, 1)$. We tested two settings: linear (dim $= 1$) and nonlinear (dim $= 2$ or 3), denoted by `li` and `nl`, respectively. Finally, N sample orders $O_i \in S$ were generated by randomly eliminating objects from the true regression orders.

As an error measure, we used Spearman's ρ (2) between the estimated order and the above true regression order. If ρ is 1, the two orders are completely concordant; if it is -1, one order is the reverse of the other. We will show the means of ρ over a tenfold cross validation for 10 different weight sets of utility(x_j). Regardless of the length of training orders, the size of the unordered set, $|\mathcal{X}_u|$, is set to 10, because errors cannot be precisely measured if orders are too short.

4.2 Experimental Results

As the basic experimental condition, we chose $N = 300$, $L_i = 5$, and $|\mathcal{X}^*| = 1{,}000$. Under this condition, the probability that one object in $\mathcal{X}_u$ is unobserved in the training samples seems rather low (25.8%). However, the probability that the object pairs in $\mathcal{X}_u$ become unobserved, which is intrinsic to ordinal relations, is fully high (99.5%). Therefore, we consider that this data set is well suited to evaluate the generalization ability. Note that algorithm parameter settings described in Sect. 3 are tuned for this basic condition under a noisy condition described later. By default, we used this basic condition in the following experiments. The other settings were: For Cohen, we adopt their Hedge and their greedy search algorithm. Note that we also applied the exhaustive search to derive the optimal orders that maximizes (6), but the results were rather inferior to that by the greedy search in Fig. 2. For RB, ranking features are all terms of a second-order polynomial. For SVOR and OSVM, Gaussian kernels with $\sigma = 1$ were used. For ERR, the second-order polynomials was used as class of regression functions.

Table 1a shows the means of ρ under this basic setting. Each row corresponds to each of the four data sets described in Sect. 4.1, and each column corresponds to each method in Sect. 3. The rank of each method is shown in brackets. Except for the difference between RB and SVOR of `nl/bin`, the difference between each method and the next-ranked one is statistically significant at the level of 1% when using a paired t-test and a Bonferroni multiple comparison.

Defects of the Cohen would be due to the fact that only the ordinal information of attributes is considered, as described in Sect. 3.1. The ERR methods were inferior in `bin` cases, but were superior in `num` cases. The performance with `nl/bin` data was unstable because the weights of a regression function have to be determined based on two points, 0 and 1. The SVM-based method could avoid this problem by adopting the regularization property. The two SVM-based methods, OSVM and SVOR, also bear a resemblance to each other. The RB was rather inferior for the `nl` case, but it could be improved by increasing the number of rounds T. For example, when we tried the number of iterations $T = 1{,}000$ for basic data under a noisy

Table 1 Basic Results: $|\mathcal{X}^*| = 1000$, $L_i = 10$, $N = 300$

(a) Under Noiseless Conditions

	Cohen	RB	SVOR	OSVM	ERR
li/num	0.860 [5]	0.959 [2]	0.914 [3]	0.886 [4]	0.982 [1]
li/bin	0.966 [2]	0.978 [1]	0.885 [4]	0.868 [5]	0.895 [3]
nl/num	0.682 [5]	0.763 [4]	0.911 [2]	0.878 [3]	0.935 [1]
nl/bin	0.786 [5]	0.875 [1]	0.866 [2]	0.842 [3]	0.830 [4]

(b) Under Noisy Conditions

	Cohen	RB	SVOR	OSVM	ERR
nl/num	0.652 [5]	0.719 [4]	0.818 [1]	0.797 [3]	0.813 [2]
nl/bin	0.764 [5]	0.842 [1]	0.817 [2]	0.809 [3]	0.796 [4]

condition in Table 1b, the ρ improved from 0.720 to 0.765. However, it was too slow compared with other methods, so we had to set $T = 100$ when performing our experiments.

Table 1a shows the results under a noiseless condition; that is to say, all the sample orders are perfectly concordant with the corresponding regression order. To test the robustness of the methods against the noise in the orders, we permuted two randomly selected pairs of adjacent objects in the original sample orders. By changing the number of times that objects are permuted, the noise level could be controlled. The order noise level is measured by the probability that the ρ between the original order and the permuted one is smaller than the ρ between the original order and a random one. This probability can be computed by using the statistical property of Spearman's ρ. We generated four types of data whose noise levels were 0%∼10%. Note that the 0% level noise is equivalent to the noiseless case.

Figure 3 shows the means of ρ in accordance with the order noise level for the `nl/num` data. In accordance with the increase of noise, the empirical ρ (between the estimated order and the permuted sample order) drastically became worse, whereas true ρ (between the estimated order and the noiseless order that cannot be observed in nonartificial data) did not decrease to a significant degree. For example, at the 10% noise level, the empirical ρ by the ERR for the `nl/num` data is 0.409, while the true ρ is 0.904. We then examined the robustness of these methods against noise in attribute values. For the numerical attributes, the α% level of noise is obtained by multiplying the true values by the random factors that follow $N(1, \alpha/100)$. Note that sample orders were calculated based on noiseless attribute values. Figure 4 shows the means of ρ in accordance with the level of attribute noise for the `nl/num` data. We generated five types of data whose α values were set to 0%∼160%. The results shown in Figs. 3 and 4 indicate a clear contrast. The SVM-based methods were robust against attribute noise, but not against order noise. Conversely, the other methods were robust against order noise, but not against attribute noise. This could be explained as follows: The SVM-based methods are sensitive to order noise because the exchanged ordered pairs tend to become support

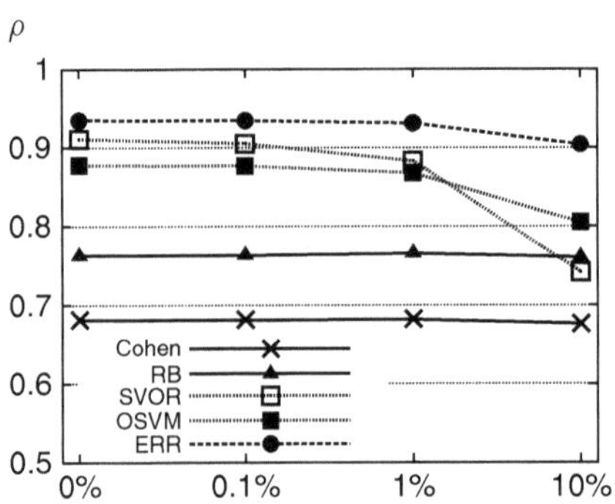

Fig. 3 Variation in the order noise

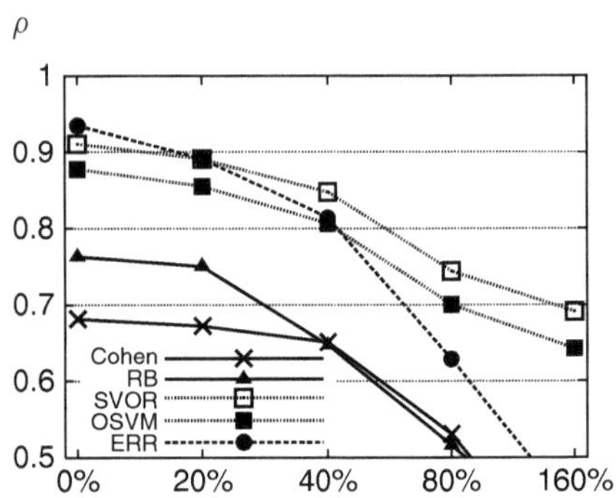

Fig. 4 Variation in the attribute noise

vectors, while perturbation of attribute values does not affect the support vectors as much. Inversely, the non-SVM-based methods can learn correctly if correct orders constitute the majority of the sample orders; thus, these methods are robust against order noise. However, any perturbation in attribute values affects their performance.

The noiseless setting of Table 1a is unrealistic, because real data generally include noise. We therefore investigated the behavior of object ranking methods under more realistic *noisy* conditions. According to the above results, the relative superiority of the prediction performance among the methods heavily depended on types of noise. That is to say, while the non-SVM-based methods were superior for data with more order noises, the SVM-based ones were superior for data with more attribute noises. Instead of comparing the relative superiority of methods, we investigated the patterns of the changes of the relative predictive performance in accordance with the variation of data properties. To check these, noise levels were selected so that ERR and SVOR gave roughly equal performance. In both the `nl/num` and the `nl/bin` data sets, order noise levels were set to 1%. While the attribute noise level of the `nl/num` was 40%, binary values were flipped with a probability of 1% for the `nl/bin`. We, however, used noiseless data when testing the variation in the predictive performance according to the length of sample orders (Fig. 5c), because the shorter sample orders were more seriously influenced by order noise. Along with fixing the algorithm parameter settings, we tested the changes of the prediction performance according to variation in the number of objects $|\mathcal{X}^*|$, the number of sample orders N, and the length of orders L_i.

Table 1b shows the results under this noisy condition for the basic data. Except between SVOR and ERR of `nl/num`, the difference between each method and the next-ranked one is statistically significant at the level of 1% when using a paired t-test and a Bonferroni multiple comparison. Figure 5 shows the means of ρ in accordance with the variations in the other properties of samples sets. The results for the `nl/num` and the `nl/bin` data are shown in each column. Rows (a), (b), and (c) show results when the number of objects $|\mathcal{X}^*|$, the number of sample orders N, and the length of orders L_i were varied, respectively.

In terms of Fig. 5a, the probability that an object in test orders has not been included in training samples decreases in accordance with the increase of $|\mathcal{X}^*|$; accordingly, a greater generalization ability is required. SVM-based methods were better if $|\mathcal{X}^*|$ was small, but their performance dropped for larger $|\mathcal{X}^*|$. Adoption of soft-margin parameter C tuning for $|\mathcal{X}^*|$ was required in order for the SVM-based methods to work well. The non-SVM-based methods results were rather flat. This would be because the number of model parameters to determine is fewer in these methods than in the SVM-based ones.

Turning to Fig. 5b and c, the Cohen method performed more poorly for the larger L_i. This would be because the paired comparison model used in the Cohen method assumes independence among ordered pairs. For small N or L_i, the performance of the SVM-based methods was inferior to those of the others. However, the performance was improved in accordance with the increase of N or L_i. This might be because the SVM-based methods are overfitted when the sample set is so small that the learned functions are not sparse. We also expected that this observation arises

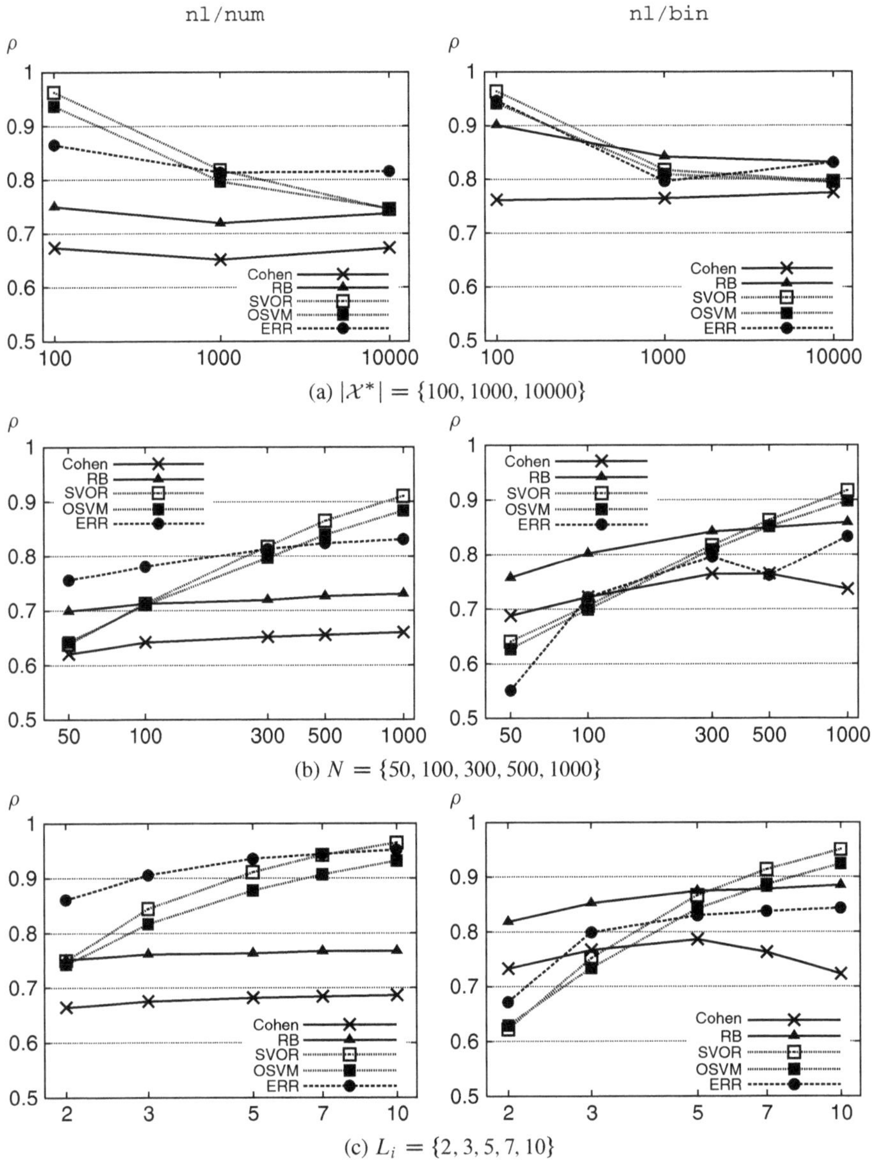

Fig. 5 Variation in the number of objects $|\mathcal{X}^*|$, the number of sample orders N, and the length of sample orders L_i

from the strength of model biases. Hence, we further checked the performance by changing the parameters of the methods, but we could not find a simple relation between the number of parameters to learn and the number of observed samples.

5 Experiments Using Real Data

We applied the methods described in Sect. 3 to real data from the following questionnaire surveys. The first data set was a survey of preferences in sushi (Japanese food), and is denoted as SUSHI[1] [19, 22]. In this data set, $N = 500$, $L_i = 10$, and $|\mathcal{X}^*| = 100$. Objects are represented by 12 binary and 4 numerical attributes. By using the k-o'means clustering method [23, 24], we generated two sample orders whose ordinal variances were broad and narrow. The probabilities that objects were selected in O_i were not uniform, as assumed in an ERR method. Objects were selected independently with probabilities that range from 3.2% to 0.13% from $\mathcal{X}^*$. The second data set was a questionnaire survey of news articles sorted according to their significance, and is denoted as NEWS. These news articles were obtained from "CD-Mainichi-Newspapers 2003." In this data set, $N = 4{,}000$, $L_i = 7$, and $|\mathcal{X}^*| = 11{,}872$. The variance among sample orders was slightly broader than the tight SUSHI data. Articles were represented by keyword and document frequencies, and these 22,297 elements were compressed to 20 attributes using latent semantic indexing (a contribution ratio is about 0.9). Additionally, we used 8 binary attributes to represent article categories. For both data sets, the N or L_i were varied by eliminating sample orders or objects. Orders became difficult to estimate as N and/or L_i decreased. Errors were measured by the empirical ρ between the sample order and the estimated one. The algorithm's parameters were set to the practical setting in Sect. 4.2. Ideally, these parameters should be tuned by using cross-validation, but some methods were too slow to perform such fine tuning. Therefore, these experiments reveal not the relative superiority among methods but the changes in performance along with the variation in N and/or L_i.

In Table 2, we show the means of ρ. The column labeled $N{:}L_i$ represents the number and the length of sample orders, and the letters "b" and "n" denote the types of variance, broad or narrow, respectively. In the SUSHI case, the differences among methods were less clear than those in artificial data. Although we expected that the SVM would work well for a tight data set, the variance in sample orders was less affected. We could say that this is due to the fitness of the SUSHI data to a linear model; we observed that the other method worked well when using a linear model. ERR showed good performance for large N or L_i, but poorer results for small N or L_i. This is due to too complicated model for regression, because the mean ρ increased to 0.249 by adopting a simpler linear regression model for 100:2:(b). Inversely, RB was better for small N or L_i. This is due to the setting of $T = 100$, the number of rounds. When $T = 300$, we observed that performance for large N or L_i improved, but it was depressed for small N or L_i because of overfitting. In the case of NEWS, sample orders were tight, but the correlation between sample orders and attributes were remarkably weak. Thus, all methods performed poorly. For such weak attributes, Cohen performed very poorly, even though we tuned β parameters. Again, ERR was better for large N or L_i, but was poorer for small N

[1] This data set can be downloaded from http://www.kamishima.net/sushi/

Table 2 Results on real data sets

	N:L_i	Cohen	RB	SVOR	OSVM	ERR
SUSHI	500:10(b)	0.364 [5]	0.384 [4]	0.393 [3]	0.400 [1]	0.397 [2]
	100:5(b)	0.354 [2]	0.356 [1]	0.284 [4]	0.315 [3]	0.271 [5]
	100:2(b)	0.337 [1]	0.281 [2]	0.115 [4]	0.208 [3]	0.010 [5]
	500:10(n)	0.543 [5]	0.583 [4]	0.719 [1]	0.708 [2]	0.705 [3]
	100:5(n)	0.548 [5]	0.612 [4]	0.646 [2]	0.655 [1]	0.617 [3]
	100:2(n)	0.577 [1]	0.542 [2]	0.522 [4]	0.540 [3]	0.421 [5]
NEWS	4000:7	−0.008 [5]	0.350 [3]	0.244 [4]	0.366 [2]	0.386 [1]
	1000:5	−0.009 [5]	0.340 [3]	0.362 [1]	0.353 [2]	0.312 [4]
	1000:2	−0.009 [5]	0.338 [3]	0.349 [1]	0.344 [2]	0.149 [4]

or L_i. However, this was due to the model complexity as in the above SUSHI case. In summary, the performances of four methods other than Cohen were roughly equal, when dealing with these real data.

6 Discussion and Conclusions

We first discuss the relation between object ranking methods and the probabilistic generative models for orders (see Appendix A). These models can be categorized as four types [1, 25]:

- *Thurstonian*: Objects are sorted according to the corresponding score.
- *Paired comparison*: Objects are compared in pairwise, and all objects are sorted based on these comparison results.
- *Distance-based*: Orders are generated with the probability that is determined based on the distance from a modal order.
- *Multistage*: Objects are sequentially arranged top to end.

Object ranking methods are commonly designed by incorporating a way to deal with attributes into these models. Error or loss in orders are designed based on these generative models. In Cohen, because the precedence between object pairs is first determined based on the learned model, we consider that this method adopts a paired comparison model. Regarding RB and SVM methods, a modal order is determined by sorting the outputs of a score function. Loss functions between a modal order and sample orders are defined based on the discordance between them. Therefore, these methods can be considered to be related to the distance-based model. While RB and SVOR adopt Kendall distance, OSVM adopts Spearman footrule. Finally, ERR is based on the Thurstonian model as described in Sect. 3.4.

We next summarize computational complexities of learning and sorting time in the first and second rows of Table 3. We assume that the number of ordered pairs and of objects in S are approximated by $N\bar{L}^2$ and $N\bar{L}$, respectively, where $\bar{L}$ is the mean length of the sample orders. The SVM's learning time is assumed to be

Table 3 Computational complexities

	Cohen	RB	SVOR	OSVM	ERR
Learn	$N\bar{L}^2k$	$N\bar{L}^2k$	$N^2\bar{L}^4k$	$N^2\bar{L}^4k$	$N\bar{L}k^2$
Sort	L^2	$L\log L$	$L\log L$	$L\log L$	$L\log L$

quadratic in the number of training samples. The learning time of Cohen's Hedge algorithm or the RB is linear in terms of the $N\bar{L}^2$, if the number of iteration T is fixed. However, if T is adaptively chosen according to $N\bar{L}^2$, their time complexity becomes superlinear. In terms of the number of attributes k, the SVM-based methods depend on the number of nonzero attribute values; thus, it is practically sublinear. Standard regression used in ERR can be computed faster, if attributes are sparse. Generally, in practical use, the learning time of the SVM-based methods is slow, that of Cohen and RB is intermediate, and that of ERR are much faster. In terms of time for sorting of $\mathbf{x}_j \in \mathcal{X}$, the Cohen greedy requires $O(L^2)$ while the others perform more quickly, $O(L\log L)$.

We finally summarize the pros and cons of each method. Our new ERR method was practically the fastest without sacrificing its prediction performance. This quickness make it to try many parameter settings in relatively short times. Even for the result where ERR were poor, e.g., SUSHI:100:2(b), we observed that it could be improved by re-tuning. In this method, the uniform distribution of the object observation is assumed, but SUSHI result demonstrated the robustness against the violation of this assumption to an extent. However, this method requires quadric time in terms of k, if attribute vectors are dense.

The most prominent merit of Cohen is to be an online method. For online learning purposes, the other methods cannot be used. Although the Cohen performed rather poorly in our experiments, this is because the Hedge algorithm is designed to take into account as its attributes only ordinal information. We observed that the performance could be improved by using other classifier, such as the naïve Bayes, instead of the Hedge, because this method was designed to deal with categorical or numerical attributes. Further, our experiments were absolute ranking task (see Sect. 2.2), but the Cohen acquires relative ranking function, because this method adopts a paired comparison model.

The unique property of the RB is rich options of weak learners. Because of this property, various types of attributes can be used. If objects are represented by vectors whose attribute types are mixtures of ordinal and numerical/categorical, the other algorithm cannot be used. Our experimental results of RB were rather inferior, but we observed that this could be considerably improved by adaptively increasing T. Due to too slow convergence, we had to stop iterations after the end of the drastic error drops at the beginning stage. However, it takes as same or more computation time as the SVM-based methods until complete convergence, and it should be also noted that too large T would cause overfitting.

Like a standard SVM, the SVOR and OSVM are advantageous if k is large. The our experimental results demonstrated that the two SVM-based methods and the others are robust against different types of noise. Hence, for data in which orders are permuted, the non-SVM-based methods are preferable, while for data whose attributes are disturbed, the SVM-based methods are preferable. The demerit of the SVM-based methods are slowness. The learning complexity of the two SVM-based methods is the same, but the OSVM is practically slower. However, it was more robust against order noise than SVOR.

A Probabilistic Generative Models for Orders

We briefly summarize the probabilistic generative models for orders. Readers that needs further information should refer a textbook [1] or a survey paper [25]. As described before, these models can be categorized into four types: Thurstonian, paired comparison, distance-based, and multistage. All these are generative model for complete orders, in which all objects are included. We sequentially introduce these four types.

As indicated by the name *Thurstonian*, this type of models were first proposed by Thurstone [21] and are also called by order statistics models. In this model, objects, $\mathbf{x}_i \in \mathcal{X}^*$, are sorted according to their corresponding scores. This score is a real value probabilistic function that is defined as

$$\mathrm{score}(\mathbf{x}_i) = \mu(\mathbf{x}_i) + \epsilon(\mathbf{x}_i),$$

where $\mu(\mathbf{x}_i)$ is the mean score for the object $\mathbf{x}_i$ and $\epsilon(\mathbf{x}_i)$ follows the distribution with zero mean. In original Thurstone's model, the normal distribution is used as $\epsilon(\mathbf{x}_i)$. Especially, the "case V", in which the variance of the normal distribution is constant over all objects, is widely used. In this case, the probability the object $\mathbf{x}_i$ precedes the $\mathbf{x}_j$ is simply computed by [26]:

$$\mathrm{P}[\mathbf{x}_i \succ \mathbf{x}_j] = \mathrm{P}[\mathrm{score}(\mathbf{x}_i) > \mathrm{score}(\mathbf{x}_j)] = \Phi\left(\frac{\mu(\mathbf{x}_i) - \mu(\mathbf{x}_j)}{\sqrt{2}\sigma}\right),$$

where $\Phi(\cdot)$ is the normal c.d.f. and σ^2 is its variance.

In a *paired comparison model*, which object precedes the other is determined for each pair of objects. Accordingly, $L(L-1)/2$ ordered pairs are generated, where L is the total number of objects, i.e. $L = |\mathcal{X}^*|$. If this set of ordered pairs is cyclic, i.e., $\mathbf{x}_i \succ \mathbf{x}_j$, $\mathbf{x}_j \succ \mathbf{x}_k$, and $\mathbf{x}_k \succ \mathbf{x}_i 0$, then this set is discarded, and all pairs are re-generated; otherwise, a total order is uniquely determined. The saturated model having $L(L-1)/2$ parameters, which represent the probabilities that one object precedes the other, is called by *Babington Smith model*. Babington Smith first showed the moments of this model in [27]. After that, some models with fewer parameters have been proposed. The next model having L parameters is called *Bradley–Terry model* [28]:

$$\mathrm{P}[\mathbf{x}_i \succ \mathbf{x}_j] = \frac{f(\mathbf{x}_i)}{f(\mathbf{x}_i) + f(\mathbf{x}_j)},$$

where $f(\cdot)$ is positive real function.

In *distance-based model*, orders, O, follows the following distribution

$$\mathrm{P}[O] = \frac{1}{Z(\lambda)} \exp\Bigl(-\lambda d(O_0, O)\Bigr),$$

where λ is a concentration parameter, which is nonnegative real value, the O_0 is a modal ranking, at which this distribution is peaked, $d(\cdot,\cdot)$ denotes the distance between orders, and $Z(\lambda)$ is a normalization factor. Especially, this model is called *Mallows ϕ-model* and *Mallows θ-model* if Kendall and Spearman distance are adopted, respectively. This is because these are the special cases of the *Mallows model* [29], which is a kind of a paired comparison model. If ith and jth ranked objects in a generated order are $\mathbf{x}_i$ and $\mathbf{x}_j$, respectively, Mallows model is defined as

$$\mathrm{P}[\mathbf{x}_i \succ \mathbf{x}_j] = \frac{\theta^{i-j}\phi^{-1}}{\theta^{i-j}\phi^{-1} + \theta^{j-i}\phi},$$

where θ and ϕ are positive real parameters. If $\theta = 1$ (resp. $\phi = 1$), this model becomes Mallows ϕ-model (resp. Mallows θ-model).

Finally, in *multistage model*, objects are sequentially arranged top to end. *Plackett–Luce model* [30], which is a kind of a multistage model, generates an order, $O = \mathbf{x}_{i_1} \succ \mathbf{x}_{i_2} \succ \cdots \mathbf{x}_{i_L}$, with the following procedure. The top ranked object, $\mathbf{x}_{i_1}$, is chosen with the probability:

$$\frac{f(\mathbf{x}_{i_1})}{\sum_{\mathbf{x}\in\mathcal{X}^*} f(\mathbf{x})},$$

the second ranked object, $\mathbf{x}_{i_2}$, is chosen with the probability:

$$\frac{f(\mathbf{x}_{i_2})}{\sum_{\mathbf{x}\in\mathcal{X}^*,\mathbf{x}\neq\mathbf{x}_{i_1}} f(\mathbf{x})},$$

and these procedures are iterated $L-1$ times. Orders generated by this model satisfies *Luce's choice axiom*: Roughly speaking, the probability that an object is top-ranked equals to the product of the probability that the object is top-ranked in any subset and the probability that the subset contains the object. More formally, $\mathrm{P}_{\mathcal{X}}[\mathbf{x}]$ denotes the probability that an object $\mathbf{x}$ is top ranked among the object set, $\mathcal{X}$, and $\mathrm{P}_{\mathcal{X}}[\mathcal{X}']$, $\mathcal{X}'\subset\mathcal{X}$, denotes the probability that the top object in $\mathcal{X}$ is contained in $\mathcal{X}'$. A set of choice probabilities is said to satisfy Luce's choice axiom if $\forall\mathcal{X}\subset\mathcal{X}^*$ with at least two objects, $\mathbf{x}_i$ and $\mathbf{x}_j$, satisfy:

- If $\mathrm{P}_{\{\mathbf{x}_i,\mathbf{x}_j\}}[\mathbf{x}_i] > 0 \;\forall\mathbf{x}_i,\mathbf{x}_j \in \mathcal{X}$ then $\forall\mathbf{x}_k \in \mathcal{X}' \subset \mathcal{X}, \mathrm{P}_{\mathcal{X}}[\mathbf{x}_k] = \mathrm{P}_{\mathcal{X}'}[\mathbf{x}_k]\mathrm{P}_{\mathcal{X}}[\mathcal{X}']$,
- If $\mathrm{P}_{\{\mathbf{x}_i,\mathbf{x}_j\}}[\mathbf{x}_i] = 0 \;\exists\mathbf{x}_i,\mathbf{x}_j \in \mathcal{X}$ then if $\mathbf{x}_j \in \mathcal{X}$, $\mathbf{x}_j \neq \mathbf{x}_i$, $\mathrm{P}_{\mathcal{X}}[\mathbf{x}_j] = \mathrm{P}_{\mathcal{X}\setminus\{\mathbf{x}_i\}}[\mathbf{x}_j]$.

This Plackett–Luce model is equivalent to the above Thurstonian model, if $\epsilon(\cdot)$ follows Gumbel distribution, whose c.d.f. is $1-\exp(-\exp(x))$.

Acknowledgements A part of this work is supported by the grant-in-aid 14658106 and 16700157 of the Japan society for the promotion of science. Thanks are due to the Mainichi Newspapers for permission to use the articles.

References

1. J.I. Marden, *Analyzing and Modeling Rank Data*, Vol. 64, *Monographs on Statistics and Applied Probability* (Chapman & Hall, 1995)
2. M. Kendall, J.D. Gibbons, *Rank Correlation Methods*, 5th edn. (Oxford University Press, 1990)
3. K.J. Arrow, *Social Choice and Individual Values*, 2nd edn. (Yale University Press, 1963)
4. D. Bollegala, N. Okazaki, M. Ishizuka, A machine learning approach to sentence ordering for multidocument summarization and its evaluation, in *Proceedings of the Natural Language Processing society of Japan* (2005)
5. C. Dwork, R. Kumar, M. Naor, D. Sivakumar, Rank aggregation methods for the Web, in *Proceedings of The 10th International Conference on World Wide Web* (2001), pp. 613–622
6. A. Agresti, *Categorical Data Analysis*, 2nd edn. (Wiley, 1996)
7. P. McCullagh, Regression models for ordinal data. J. Royal Stat. Soc. B **42**(2), 109–142 (1980)
8. A. Shashua, A. Levin, Ranking with large margin principle: Two approaches, in *Advances in Neural Information Processing Systems*, vol. 15 (2003), pp. 961–968
9. W.W. Cohen, R.E. Schapire, Y. Singer, Learning to order things. J. Artif. Intell. Res. **10**, 243–270 (1999)
10. M. Grötschel, M. Jünger, G. Reinelt, A cutting plane algorithm for the linear ordering problem. Oper. Res. **32**(6), 1195–1220 (1984)
11. N. Littlestone, Learning quickly when irrelevant attributes abound: A new linear-threshold algorithm. Mach. Learn. **2**, 285–318 (1988)
12. Y. Freund, R. Iyer, R.E. Schapire, Y. Singer, An efficient boosting algorithm for combining preferences, in *Proceedings of The 15th International Conference on Machine Learning* (1998), pp. 170–178
13. Y. Freund, R. Iyer, R.E. Schapire, Y. Singer, An efficient boosting algorithm for combining preferences. J. Mach. Learn. Res. **4**, 933–969 (2003)
14. H. Kazawa, T. Hirao, E. Maeda, Order SVM: a kernel method for order learning based on generalized order statistics. Syst. Comput. Jpn. **36**(1), 35–43 (2005)
15. R. Herbrich, T. Graepel, P. Bollmann-Sdorra, K. Obermayer, Learning preference relations for information retrieval, in *ICML-98 Workshop: Text Categorization and Machine Learning* (1998), pp. 80–84
16. R. Herbrich, T. Graepel, K. Obermayer, Support vector learning for ordinal regression, in *Proceedings of the 9th International Conference on Artificial Neural Networks* (1999), pp. 97–102
17. T. Joachims, Optimizing search engines using clickthrough data, in *Proceedings of The 8th International Conference on Knowledge Discovery and Data Mining* (2002), pp. 133–142
18. O. Luaces, G.F. Bayón, J.R. Quevedo, J. Díez, J.J. del Coz, A. Bahamonde, Analyzing sensory data using non-linear preference learning with feature subset selection, in *Proceedings of the 15th European Conference on Machine Learning* (2004), pp. 286–297 [LNAI 3201]
19. T. Kamishima, H. Kazawa, S. Akaho, Supervised ordering – an empirical survey, in *Proceedings of The 5th IEEE International Conference on Data Mining* (2005), pp. 673–676
20. B.C. Arnold, N. Balakrishnan, H.N. Nagaraja, *A First Course in Order Statistics* (Wiley, 1992)
21. L.L. Thurstone, A law of comparative judgment. Psychol. Rev. **34**, 273–286 (1927)
22. T. Kamishima, Nantonac collaborative filtering: Recommendation based on order responses, in *Proceedings of The 9th International Conference on Knowledge Discovery and Data Mining* (2003), pp. 583–588
23. T. Kamishima, J. Fujiki, Clustering orders, in *Proceedings of The 6th International Conference on Discovery Science* (2003), pp. 194–207 [LNAI 2843]

24. T. Kamishima, S. Akaho, Efficient clustering for orders, in *Mining Complex Data*, vol. 165, *Studies in Computational Intelligence*, ed. by D.A. Zighed, S. Tsumoto, Z.W. Ras, H. Hacid (Springer, 2009), pp. 261–280
25. D.E. Critchlow, M.A. Fligner, J.S. Verducci, Probability models on rankings. J. Math. Psychol. **35**, 294–318 (1991)
26. F. Mosteller, Remarks on the method of paired comparisons: I – the least squares solution assuming equal standard deviations and equal correlations. Psychometrika **16**(1), 3–9 (1951)
27. B. Babington Smith, Discussion on professor ross's paper. J. Royal Stat. Soc. B **12**, 53–56 (1950) (A.S.C. Ross, "*Philological Probability Problems*", pp. 19–41)
28. R.A. Bradley, M.E. Terry, Rank analysis of incomplete block designs – i. The method of paired comparisons. Biometrika **39**, 324–345 (1952)
29. C.L. Mallows, Non-null ranking models, I. Biometrika **44**, 114–130 (1957)
30. R.L. Plackett, The analysis of permutations. J. Royal Stat. Soc. C **24**(2), 193–202 (1975)

Dimension Reduction for Object Ranking

Toshihiro Kamishima and Shotaro Akaho

Abstract Ordered lists of objects are widely used as representational forms. Such ordered objects include Web search results and bestseller lists. Techniques for processing such ordinal data are being developed, particularly methods for an object ranking task: i.e., learning functions used to sort objects from sample orders. In this article, we propose two dimension reduction methods specifically designed to improve prediction performance in an object ranking task.

1 Introduction

Orders are sequences of objects sorted according to some property and are widely used to represent data. For example, responses from Web search engines are lists of pages sorted according to their relevance to queries. Bestseller lists, which are item sequences sorted according to sales volume, are used on many E-commerce sites. Processing techniques for orders have immediate practical value, and so research concerning orders has become very active in recent years. In particular, several methods are being developed for learning functions used to sort objects represented by attribute vectors from example orders. We call this task *Object Ranking* [1, 2] and emphasize its usefulness for sensory tests [2, 3],[1] information retrieval [4–8], and recommendation [9].

Several methods have been developed for the object ranking task. However, when the attribute vectors that describe objects are in very high-dimensional space, these object ranking methods are degraded in prediction performance. The main reason for this is that the number of model parameters to be learned grows in accordance

[1] measurement of respondents' sensations, feelings or impressions.

T. Kamishima (✉) and S. Akaho
National Institute of Advanced Industrial Science and Technology (AIST)
AIST Tsukuba Central 2, Umezono 1–1–1, Tsukuba, Ibaraki, 305–8568 Japan
e-mail: mail@kamishima.net ⟨http://www.kamishima.net/⟩, s.akaho@aist.go.jp

J. Fürnkranz and E. Hüllermeier (eds.), *Preference Learning*,
DOI 10.1007/978-3-642-14125-6_10, © Springer-Verlag Berlin Heidelberg 2010

with the increase of dimensionality; thus, the acquired functions might not perform well when sorting unseen objects due to overfitting.

Dimension reduction techniques are one obvious solution to the problems caused by high dimensionality. Dimension reduction is the task of mapping points originally in high-dimensional space to a lower dimensional subspace, while limiting the amount of lost information. Principal component analysis (PCA) is one of the typical techniques for dimension reduction. PCA is designed so that variations in original data are preserved as much as possible. It has been successfully used for other learning tasks but is less appropriate for an object ranking task. Since PCA is designed so as to preserve information regarding the objects themselves, useful information in terms of the target ordering might be lost by this approach. Therefore, in this paper, we propose *Rank Correlation Dimension Reduction (RCDR)* for dimension reduction in conjunction with object ranking. RCDR is designed to preserve information that is useful for mapping to the target ordering.

We propose our RCDR methods in Sect. 2. Experimental results are shown in Sect. 3. We discuss and summarize the results in Sect. 4.

2 Rank Correlation Dimension Reduction

In this section, we show a dimension reduction technique specially designed for object ranking methods. Note that the notations and the task of object ranking are described in our survey chapter of this book [1].

To obtain satisfactory results when using data mining or machine learning algorithms, it is important to apply preprocessing methods, such as feature selection, dealing with missing values, or dimension reduction. Appropriate preprocessing of data can improve prediction performance, and can occasionally reduce computational and/or memory costs. Some preprocessing techniques for mining or learning methods dealing with orders have been proposed. Bahamonde et al. [10] applied wrapper-type feature selection to an object ranking task. Slotta et al. [11] performed feature selection for classification of orders. In [12, 13], rank statistics were used for selecting informative genes from microarray data. To measure the similarities between orders, Kamishima and Akaho proposed a method to fill in missing objects in orders [14]. To our knowledge, however, dimension reduction techniques specially designed for an object ranking task have not yet been developed.

Similar to other types of learning tasks, such as classification or regression, dimension reduction techniques will be beneficial for object ranking tasks, in particular, if the number of attributes, K, is very large. With reduced dimensions, the generalization ability can be improved. Because the number of model parameters to be learned grows in accordance with K, the acquired functions might not perform well when sorting unseen objects due to overfitting. In particular, if there are many noninformative attributes or if complex models are used, the problem of overfitting will be alleviated by reducing dimensions.

To reduce the number of dimensions before performing object ranking, one might assume that reduction techniques used for other learning tasks can be used. However, this is not the case. Principal component analysis (PCA) is one of typical techniques for dimension reduction. PCA is designed so that information about data in original attribute vector space is preserved as much as possible. This approach is less appropriate for an object ranking task. Specifically, because an object ranking task must find a mapping from attribute vectors to the target ordering, it is not sufficient to preserve information only in source vectors. On the other hand, Diaconis' spectral analysis [15] for orders is another possibility. This is a technique to decompose distributions of orders into subcomponents. For example, first-order components represent the frequency that the object $\mathbf{x}_j$ is lth ranked, while second-order components represent the frequency that objects $\mathbf{x}_j$ and $\mathbf{x}_k$ are lth and mth ranked, respectively. However, our goal is not to find decomposition in an ordinal space, but to find a subspace in an attribute vector space.

From the above discussion, it should be clear that we had to develop reduction techniques that preserve information about mappings from attribute vectors to the target ordering. This is analogous to Fisher's discriminant analysis, which is a dimension reduction technique to preserve information about a mapping from an attribute vector to target classes.

Additionally, the computational cost for reducing dimensions should not be much higher than that for object ranking methods. Computational complexities of object ranking methods in the learning stage are summarized in Table 1. These columns show the complexities for Cohen's method [4], RankBoost [9], Order SVM [16], Support Vector Ordinal Regression (SVOR) [5] (also known as RankingSVM [6]), and Expected Rank Regression (ERR) [2]. You can find a summary of these methods in [1]. We assume that the number of ordered pairs and objects in S are approximated by $N\bar{L}^2$ and $N\bar{L}$, respectively ($\bar{L}$ is the mean length of the sample orders). The SVM's learning time is assumed to be quadratic in the number of training samples. The learning time of Cohen's method or the RankBoost is linear in terms of $N\bar{L}^2$, if the number of iterations is constant. However, in practical use, the number of iterations should be increased adaptively. In the experiment in [9], the number of iterations was linearly increased in accordance with the number of ordered pairs, $N\bar{L}^2$. Therefore, their time complexities approach $N^2\bar{L}^4k$. When dimension

Table 1 Computational complexities of object ranking algorithms

Cohen	RankBoost	SVOR	Order SVM	ERR
$N\bar{L}^2K$	$N\bar{L}^2K$	$N^2\bar{L}^4K$	$N^2\bar{L}^4K$	$N\bar{L}K^2$

Note: $\bar{L}$: the mean length of sample orders, N: the number of samples, and K: the dimension of attribute vectors. The number of ordered pairs and objects in S are approximated by $N\bar{L}^2$ and $N\bar{L}$, respectively. The SVM's learning time is assumed to be quadratic in the number of training samples. The learning complexities of Cohen's method or the RankBoost are as above if the number of iterations is constant. However, in practical use, because the number of iterations should be increased adaptively in accordance with the number of ordered pairs, their time complexities approach $N^2\bar{L}^4k$.

reduction methods require much higher computational costs than those in Table 1, the reduction of dimensions greatly lessens scalability.

Taking into account what is mentioned above, our dimension reduction methods should satisfy two requirements.

1. It must be designed so as to preserve information about mappings from object attributes to targeting orders.
2. The computational complexity for dimension reduction should not be much larger than that for object ranking algorithms.

To fill these requirements, we propose *Rank Correlation Dimension Reduction (RCDR)*. Given a basis that consists of l vectors, the next $l+1$ vector is selected so as to preserve as much information about target ordering as possible. By repeating this procedure, we obtain the final subspace.

First, we outline our RCDR method. Let $\mathbf{w}^{(l)}$ be the lth vector of a basis. The subspace spanned by the basis, $\{\mathbf{w}^{(1)},\ldots,\mathbf{w}^{(l)}\}$, is called the lth subspace. We represent this sub-space by the matrix, $W^{(l)} = [\mathbf{w}^{(1)},\ldots,\mathbf{w}^{(l)}]$. Let $W^{(l)\perp}$ be the complementary space of the $W^{(l)}$ that is spanned by $(K-l)$ vectors which are orthogonal to all vectors in the basis, $\{\mathbf{w}^{(1)},\ldots,\mathbf{w}^{(l)}\}$. We are given sample orders S and attribute vectors, $\{\mathbf{x}_j\}$, and the basis of the lth subspace. This condition is depicted in Fig. 1. The objects in the original K-dimensional spaces (marked by "∘" in Fig. 1) are projected to the complementary space $W^{(l)\perp}$ of the lth subspace. The projected objects (marked by "•" in Fig. 1) are denoted by $\mathbf{x}_j^{(l)}$, and $\mathbf{x}_j^{(0)} \equiv \mathbf{x}_j$. By this projection, we can eliminate information about the target ordering contained in the subspace $W^{(l)}$. For each $k = 1,\ldots,K$, objects are sorted in descending order of the kth attribute values of the objects projected to $W^{(l)\perp}$. In Fig. 1, examples of those orders are $\mathbf{x}_1 \succ \mathbf{x}_3 \succ \mathbf{x}_2$ in the first attribute and $\mathbf{x}_1 \succ \mathbf{x}_2 \succ \mathbf{x}_3$ in the second attribute. The rank correlations between each of these orders and each sample order are calculated. Then, the sum of these rank correlations are denoted by $R_k^{(l)}$ (strict definition will be given

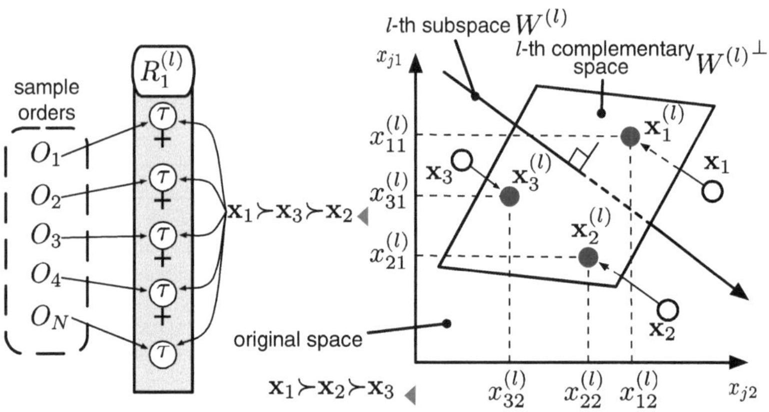

Fig. 1 An outline of rank correlation dimension reduction method

later). This $R_k^{(l)}$ represents the concordance between the target ordering and the kth attribute values of the objects projected on the lth complementary space. A new vector, $\mathbf{w}^{(l+1)}$, is chosen so that each element of this vector, $w_k^{(l+1)}$, is as proportional to the corresponding concordance, $R_k^{(l)}$, as possible.

Now, we formally describe our RCDR. Let $\mathbf{w}^{(l)} = [w_1^{(l)}, w_2^{(l)}, \ldots, w_K^{(l)}]^\top$ be the lth vector of a basis. These vectors are orthonormal to each other, i.e., $\mathbf{w}^{(l)\top}\mathbf{w}^{(m)} = 0$, $l \neq m$ and $\|\mathbf{w}^{(l)}\| = 1$. The dimension of the final subspace is denoted by K'. We are given a set of sample orders $S = \{O_1, \ldots, O_N\}$, the basis of the lth sub-space, $W^{(l)}$, and the objects $\{\mathbf{x}_j | \mathbf{x}_j \in \mathcal{X}_S\}$, $\mathcal{X}_S \equiv \cup_{O_i \in S} \mathcal{X}_i$. From these, we derive the $(l+1)$th vector, $\mathbf{w}^{(l+1)}$, as follows. First, we define $R_1^{(l)}, \ldots, R_K^{(l)}$ as the concordances between sample orders and the attribute values of the objects projected on the complementary space, $W^{(l)\perp}$. Let us focus on the sample order O_i and the kth attribute values of objects. Because the goal of an object ranking task is to estimate the orders of objects, the relative ordering of attribute values is more important than the attribute values themselves. We therefore sort the kth attribute values $x_{jk}^{(l)}$ of all objects $\mathbf{x}_j \in \mathcal{X}(O_i)$ in descending order, where $x_{jk}^{(l)}$ denotes the kth attribute value of the object, $\mathbf{x}_j^{(l)}$ projected on the lth complementary space. Note that the projected objects are represented as $[x_{j1}^{(l)}, \ldots, x_{jK}^{(l)}]^\top$ on the coordinates of the original space. The resultant order is denoted by $O(\mathcal{X}_i, x_{jk}^{(l)})$. Because both this $O(\mathcal{X}_i, x_{jk}^{(l)})$ and the sample order O_i consist of the same set of objects, the concordance between these two orders can be measured by Kendall's τ. Such rank correlations are calculated between the kth attribute values and each of sample orders in S, and these correlations are summed up:

$$R_k^{(l)} = \sum_{O_i \in S} \tau(O_i, O(\mathcal{X}_i, x_{jk}^{(l)})). \tag{1}$$

We use this sum as a measure of the concordance between the kth attribute values of objects and the target ordering. Next, to fill the first requirement of the RCDR, the $(l+1)$th vector is chosen so that the above concordance is preserved as much as possible. Let us consider the vector,

$$\mathbf{R}^{(l)} = [R_1^{(l)}, \ldots, R_K^{(l)}]^\top.$$

Because the elements of this vector are the concordances between attribute values and the target ordering, this vector would point in the direction that preserves information about the target ordering in the attribute space. Therefore, we choose the vector $\mathbf{w}^{(l+1)}$ so that it maximizes the cosine between $\mathbf{w}^{(l+1)}$ and $\mathbf{R}^{(l)}$ in the complementary space, $W^{(l)\perp}$. Further, the vector $\mathbf{R}^{(l)}$ is constant, and $\mathbf{w}^{(l+1)} = 1$; thus, the maximization of this cosine is equivalent to the $\mathbf{w}^{(l+1)}$. This optimization problem is formalized as follows:

Inputs: $S = \{O_1, \ldots, O_N\}$: a sample order set
$\mathbf{x}_j \in \mathcal{X}_S \equiv \cup_{O_i \in S} \mathcal{X}_i$: attribute value vectors
K': the dimension of sub-space

1: $\mathbf{x}_j^{(0)} \equiv \mathbf{x}_j$
2: **for all** l **in** $0, \ldots, (K'-1)$ **do**
3: compute $\mathbf{R}^{(l)}$ s.t. $R_k^{(l)} = \sum_{O_i \in S} \tau(O_i, O(\mathcal{X}_i, x_{jk}^{(l)}))$
4: **if** $l > 0$ **then**
5: $W^{(l)} = [\mathbf{w}^{(1)}, \ldots, \mathbf{w}^{(l)}]$
6: $\mathbf{R}^{(l)\perp} = (I - W^{(l)} W^{(l)\top}) \mathbf{R}^{(l)}$
7: **else**
8: $\mathbf{R}^{(l)\perp} = \mathbf{R}^{(l)}$
9: **end if**
10: $\mathbf{w}^{(l+1)} = \mathbf{R}^{(l)\perp} / \|\mathbf{R}^{(l)\perp}\|$
11: **for all** $\mathbf{x}_j$ **in** $\mathcal{X}_S$ **do**
12: $\mathbf{x}_j^{(l+1)} = \mathbf{x}_j^{(l)} - \mathbf{w}^{(l+1)} \mathbf{w}^{(l+1)\top} \mathbf{x}_j^{(l)}$
13: **end for**
14: **end for**
15: **return** $W^{(K')} = [\mathbf{w}^{(1)}, \ldots, \mathbf{w}^{(K')}]$

Fig. 2 Kendall rank correlation dimension reduction

$$\mathbf{w}^{(l+1)} = \arg\max_{\mathbf{w}} \mathbf{w}^\top \mathbf{R}^{(l)}, \tag{2}$$
$$\text{subject to:} \quad \|\mathbf{w}^{(l+1)}\| = 1, \mathbf{w}^{(l+1)^\top} \mathbf{w}^{(m)} = 0, \; m = 1, \ldots, l.$$

Note that one might think that $\mathbf{w}^{(l)}$ becomes a zero vector, if $l \geq 2$, but this is not the case. When the performing standard regression and Pearson's correlation is maximized, $\mathbf{w}^{(l)}$ would be a zero vector for $l \geq 2$. This is because zero Pearson's correlation implies such orthogonality in the attribute space. However, because rank correlation does not imply orthogonality, $\mathbf{w}^{(l)}$ is generally a non-zero vector even if $l \geq 2$.

Next, we solve (2). The derivation of $\mathbf{w}^{(l+1)}$ can be easily shown by the following procedure: Calculate the vector of the correlations sums, $\mathbf{R}^{(l)}$, project this vector to the lth complementary space, and normalize the projected vector. Once a new vector is derived, objects in the lth complementary space, $\mathbf{x}^{(l)}$, are mapped to the new complementary space, and iteratively the next vector can be computed. This algorithm is shown in Fig. 2. $\mathbf{R}^{(l)}$ is computed in line 3, projected to the current complementary space in lines 4–9, and normalized in line 10 so that its norm is one. In lines 11–13, the objects in the current complementary space are projected to the new complementary space. Because the concordance is measured by Kendall's τ, we call this method *Kendall RCDR*. The computational complexities of lines, 3, 4–9, 10, and 11–13 are $O(N\bar{L}^2K)$, $O(KK')$, $O(K)$, and $O(N\bar{L}K)$, respectively; thus, the complexity per one iteration is $O(N\bar{L}^2K)$ (generally $N\bar{L}^2 \gg K'$), and the total complexity is $O(N\bar{L}^2KK')$. As noted before, because the complexity of Cohen's method and RankBoost practically approaches $O(N^2\bar{L}^4K)$, our Kendall RCDR is strictly faster than object ranking methods except for ERR (see Table 1). In practical use, RCDR is not so slow than ERR, because $\bar{L}$ is generally small. To further save time complexity, we replace Kendall's τ in line 3 of the algorithm by Spearman's ρ, because ρ and τ are highly correlated. We call this method *Spearman RCDR*. Because its time complexity is $O(NKK'\bar{L}\log\bar{L})$, this method becomes faster than the ERR method if $K'\log\bar{L} < K$. Therefore, our RCDR methods satisfy the second requirement. Note that the Kendall RCDR is faster than the Spearman RCDR in the special case: $L_i = 2$, $O_i \in S$. Joachims et al. proposed a method to implicitly collect sample orders whose lengths are two [6]. The Kendall RCDR is useful in such cases.

3 Experiments

After showing a simple example of our RCDR methods, we describe the experimental results for real data sets.

3.1 A Preliminary Experiment

To show what is produced by our two RCDR methods, we present a simple example using artificial data. We give the ideal weight vector by $\mathbf{w}^* = [1, 1, 0.5, 0, 0]$, and set the dimensions of the original space as $K = 5$ and the number of objects as $|\mathcal{X}^*| =$ 1,000. For each object $\mathbf{x}_j \in \mathcal{X}^*$, the first to the fourth attribute values are randomly generated according to the normal distribution, $N(0, 1)$, while the fifth value is equal to the fourth. We generated 300 sample orders as follows: Five objects were selected uniformly at random from $\mathcal{X}^*$; then these objects were sorted in descending order of ${\mathbf{w}^*}^\top \mathbf{x}_j$. We applied Kendall RCDR, Spearman RCDR, and PCA to this data set. The first and second vectors are shown in the upper and lower parts of Table 2, respectively. In each row, we show vectors derived by Kendall RCDR, Spearman RCDR, and PCA. The first to the fifth columns show the elements of vectors. In the last column, the norm lengths of the sum vector of rank correlations per sample order, $\|\mathbf{R}^{(l)}\|/N$, are shown for the RCDR cases, and the contribution ratios are shown for the PCA cases.

Let us look at the first vector. The vectors derived by the two RCDR methods show resemblance. This indicates that one can use the faster RCDR method; concretely, Spearman RCDR is better except for the case $L_i = 2$. Because the fourth and the fifth elements of the $\mathbf{w}^*$ are zero, no information useful for the target ordering is represented in these axes. In our RCDR cases, the fourth and the fifth weights of vectors are almost zero; thus, these useless axes can be ignored. In the PCA case, the fourth weight is far from zero, because no information about the target ordering is taken into account. The PCA merely ignores axes that are correlated in attribute space, such as in the fifth element. Further, because variances in all dimensions are equal, the contribution ratio is not so large, even if the target ordering is decided by a linear function.

Table 2 Vectors of a Basis derived by our RCDRs and the PCA

Method	1	2	3	4	5	
			The first vector			
KRCDR	0.70	0.64	0.31	−0.06	−0.06	0.146
SRCDR	0.70	0.64	0.32	−0.06	−0.06	0.173
PCA	0.02	−0.74	0.54	−0.39	0.00	0.393
			The second vector			
KRCDR	−0.27	−0.17	0.93	−0.13	−0.13	0.007
SRCDR	−0.30	−0.15	0.94	−0.05	−0.05	0.007
PCA	−0.06	−0.18	0.39	0.90	0.00	0.213

Note: The first to fifth columns of each table show the components of vectors, $\mathbf{w}^{(1)}$ and $\mathbf{w}^{(2)}$. In the last columns, values of $\|\mathbf{R}^{(l)}\|/N$ are shown for the RCDRs and the contribution ratio is shown for the PCA.

We turn to the second component. In the RCDR cases, the correlation vector size $\|\mathbf{R}^{(2)}\|/N$ is much smaller than $\|\mathbf{R}^{(1)}\|/N$; this means that the second vector is far less informative than the first, because the target ordering is generated by a linear function in this example. In the PCA case, the contribution ratio indicates that useful information still remains in this vector. Note that it is not guaranteed that the $\|\mathbf{R}^{(l)}\|/N$ decreases in accordance with the increase of l, and vectors with bigger $\|\mathbf{R}^{(l)}\|/N$ do not always contribute to predicting the target ordering. However, we empirically observed that if $\|\mathbf{R}^{(l)}\|/N$ is very small, the corresponding vector is not informative. We believe that $\|\mathbf{R}^{(l)}\|/N$ can be used as an index for the importance of vectors.

3.2 Experiments on Real Data Sets

We applied the methods described in Sect. 2 to real data from questionnaire surveys.[2] The first data set was a survey of preferences in sushi (Japanese food), and is denoted by SUSHI. In this data set, $N = 500$, $L_i = 10$, and $|\mathcal{X}^*| = 100$. Objects are represented by 12 binary and 4 numerical attributes. The second data set was a questionnaire survey of news article titles sorted according to their significance, and is denoted by NEWS. These news articles were obtained from "CD-Mainichi-Newspapers 2003." In this data set, $N = 4{,}000$, $L_i = 7$, and $|\mathcal{X}^*| = 11{,}872$. The variance among sample orders was slightly broader than the SUSHI data. Titles were represented by 0-1 vectors indicating whether a specified keyword appears in the title. Among 18,381 keywords, we selected 595 keywords that were observed 30 or more times. Additionally, we used 8 binary attributes to represent article categories; thus, the number of attributes was 603 in total.

[2] http://www.kamishima.net/sushi/

To evaluate the usefulness of our dimension reduction methods, we applied the Support Vector Ordinal Regression (SVOR) [5] (also known as RankingSVM [6]), and Expected Rank Regression (ERR) [1,2]. to these two data sets. We used our original implementation for the ERR and the SVM^{light} [3] for the SVOR. We chose these two methods, because they differently behaved in the survey [2]. The SVOR was robust for the noise in attribute values, but not for the perturbations in sample orders. Contrarily, the ERR could resistant to ordinal noises, but not to the variation in attribute values. As a family of fitting functions, no kernel was used and a linear model was adopted.

Sample order sets were partitioned into testing and training sets. The ranking function was learned from training a sample order set with original attributes or reduced attributes. After learning, prediction performance was measured by the mean of ρ between an order in a testing set, O_t, and the corresponding estimated order, $\hat{O}_t$. The larger ρ was, the better the prediction performance was. The number of folds in cross-validation was ten for SUSHI and five for NEWS. Note that we reduced the number of folds in the NEWS experiment, because the size of the NEWS data set was larger and we had to save required computational time. In the left and right parts of the Figs. 3 and 4, we show the variation of mean ρ in accordance with the dimensions of the reduced space, K', for SUSHI and NEWS, respectively. For both data sets, N or L_i was varied by eliminating sample orders or objects; the results for these sets are shown in each subfigure. The charts that labeled, ERR and SVOR, show the results derived by the expected rank regression and the support vector ordinal regression, respectively. N and/or L_i increased from the subfigure (a) to (c); thus, orders became the most difficult to estimate in the sub-figure (a) case. The curves labeled by "KRCDR", "SRCDR", and "PCA" show the mean ρ derived by ERR after applying Kendall RCDR, Spearman RCDR, and PCA, respectively. The label "ORIG" indicates that no reduction method was used, and original attribute vectors were adopted.

From these figures, the following conclusions can be drawn. First, in terms of the variation according to the increase of the dimensions, both the ERR and the SVOR behaved very similarly. This showed that our RCDR technique is independent from the learning algorithms. Second, the two RCDR methods show resemblance; thus, the faster method can be used for dimension reduction. Third, both RCDRs performed better in prediction than PCA. The difference was particularly clear when the number of dimensions K' was small. This means that RCDR successfully preserved information useful for estimating target orders. Therefore, we can say that RCDR is more effective than PCA when carrying out an object ranking task. Fourth, our RCDR technique could improve the prediction performance. The curves labeled "SRCDR"/"KRCDR" were compared with those labeled "ORIG." The reduced vectors could lead to better prediction than the original vectors. We think that this is because the models used for ordering were simplified while useful information was preserved. This can be confirmed by the fact that the improvements were prominent

[3] http://svmlight.joachims.org/

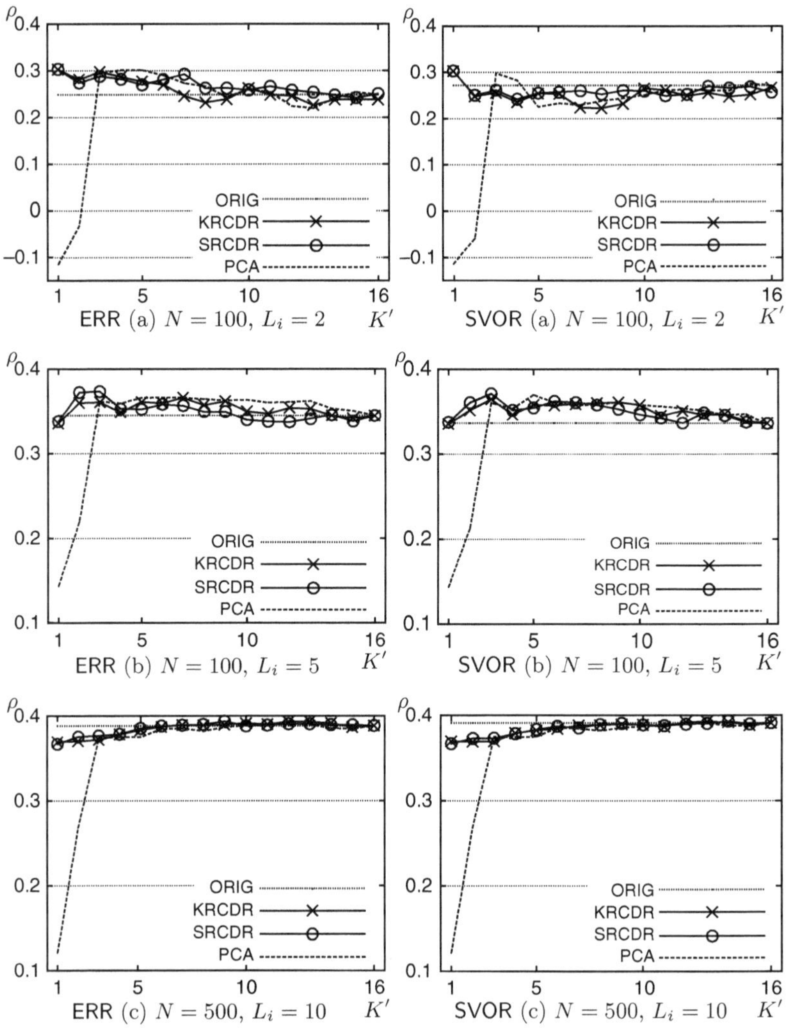

Note: The concordances between sample orders and estimated orders were measured by Spearman's ρ. These charts show variation of ρ in accordance with the number of dimensions K'. The charts that labeled, ERR and SVOR, show the results derived by the expected rank regression and the support vector ordinal regression, respectively. N is the size of a data set and L_i is the length of sample orders. The curves labeled "ORIG" show the result derived without application of dimension reduction. The curves labeled "KRCDR", "SRCDR", and "PCA" show the estimation results after reducing dimensions by the corresponding method.

Fig. 3 Comparison of dimension reduction methods on SUSHI data sets

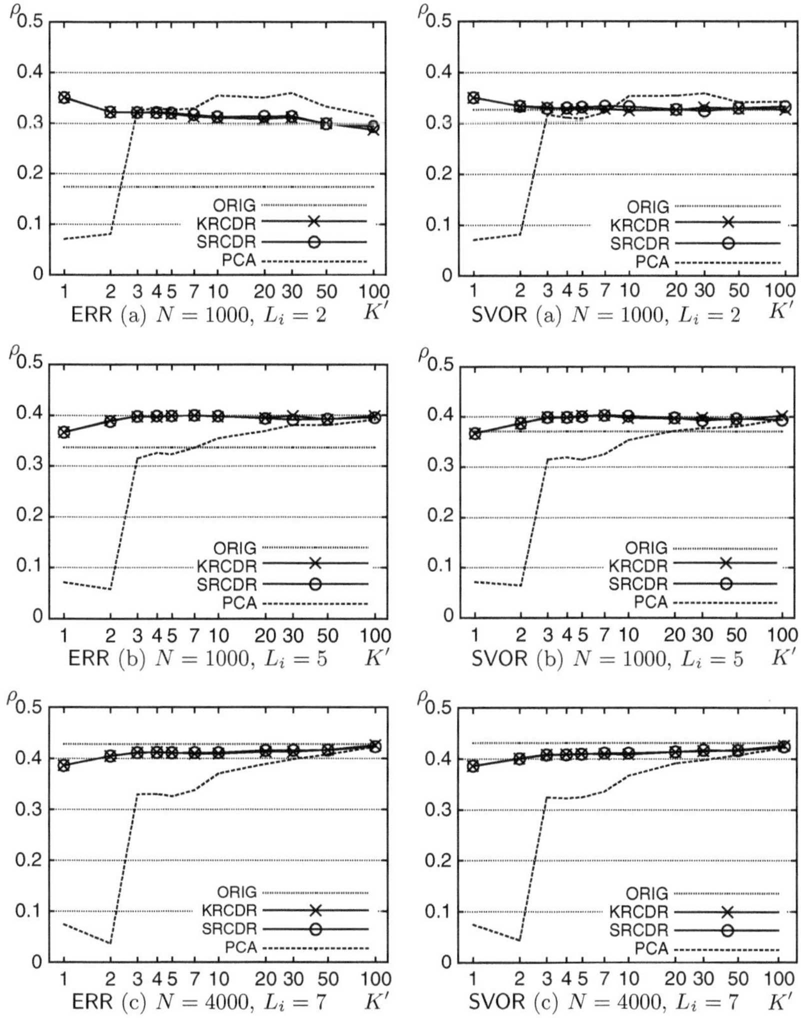

Note: The concordances between sample orders and estimated orders were measured by Spearman's ρ. These charts show variation of ρ in accordance with the number of dimensions K'. The charts that labeled, ERR and SVOR, show the results derived by the expected rank regression and the support vector ordinal regression, respectively. N is the size of a data set and L_i is the length of sample orders. The curves labeled "ORIG" show the result derived without application of dimension reduction. The curves labeled "KRCDR", "SRCDR", and "PCA" show the estimation results after reducing dimensions by the corresponding method.

Fig. 4 Comparison of dimension reduction methods on NEWS data sets

when N and/or L_i were small. The simpler model could produce better generalization ability for a limited number of samples. Therefore, our reduction technique is useful for improving prediction performance.

Finally, we can exploit the components of vectors for qualitative analysis. We obtained the first vector, $\mathbf{w}^{(1)}$, derived from the SUSHI, N=500, L_i=10 data set by applying our Kendall RCDR method. The components of the vectors, $w_1^{(1)}, \ldots, w_K^{(K)}$, were sorted in descending order of their absolute values, $|w_k^{(1)}|$. The top 5 components were as follows:

$w_{13}^{(1)}$	$= 0.5951$	the frequency the user eats
$w_{15}^{(1)}$	$= 0.4278$	how many restaurants supply the sushi
$w_{1}^{(1)}$	$= 0.4237$	red fish (e.g., fatty tuna)
$w_{14}^{(1)}$	$= 0.2822$	inexpensiveness
$w_{12}^{(1)}$	$= -0.2317$	lightness or non-oiliness in tasting

From these components, we can say that "users primarily prefer sushi that they frequently eat and that is supplied in many sushi restaurants."

4 Discussion and Conclusion

In this paper, we proposed a dimension reduction technique specialized for an object ranking task. The method was designed so as to preserve information about a relation from object attribute vectors to the target ordering. For this purpose, we developed Kendall RCDR and Spearman RCDR. We then applied these methods to real data sets. From the experimental results, we arrived at the following conclusions. First, the RCDR methods outperform PCA when carrying out an object ranking task. Second, by using the RCDR technique, performance in prediction can be improved, especially when training samples are not adequate. Finally, our two RCDR methods are comparable in prediction performance. Therefore, the faster method should be used; concretely, Spearman RCDR is better except for the condition where $L_i = 2$.

Intuitively speaking, in the lth iteration of the RCDR, the algorithm finds the vector that is most relevant to target ordering. After that, by mapping attribute vectors to the new subspace, components in attributes related to this vector are subtracted. At this time, it might be effective to subtract the explained component in the target ordering from sample orders. We will try such improvement by using a technique similar to Diaconis' spectral analysis [15].

Acknowledgements This work is supported by the grants-in-aid 14658106 and 16700157 of the Japan society for the promotion of science. Thanks are due to the Mainichi Newspapers for permission to use the articles. We would also like to thank Thorsten Joachims for providing the SVM^{light} software.

References

1. T. Kamishima, H. Kazawa, S. Akaho, A survey and empirical comparison of object ranking methods, in *Preference Learning*, ed. by J. Fürnkranz, E. Hüllermeier (Springer, 2010)
2. T. Kamishima, H. Kazawa, S. Akaho, Supervised ordering – an empirical survey, in *Proceedings of The 5th IEEE International Conference on Data Mining* (2005) pp. 673–676
3. O. Luaces, G.F. Bayón, J.R. Quevedo, J. Díez, J.J. del Coz, A. Bahamonde, Analyzing sensory data using non-linear preference learning with feature subset selection, in *Proceedings of the 15th European Conference on Machine Learning* (2004), pp. 286–297 [LNAI 3201]
4. W.W. Cohen, R.E. Schapire, Y. Singer, Learning to order things. J. Artif. Intell. Res. **10**, 243–270 (1999)
5. R. Herbrich, T. Graepel, P. Bollmann-Sdorra, K. Obermayer, Learning preference relations for information retrieval, in *ICML-98 Workshop: Text Categorization and Machine Learning* (1998), pp. 80–84
6. T. Joachims, Optimizing search engines using clickthrough data, in *Proceedings of The 8th International Conference on Knowledge Discovery and Data Mining* (2002), pp. 133–142
7. F. Radlinski, T. Joachims, Query chains: Learning to rank from implicit feedback, in *Proceedings of The 11th International Conference on Knowledge Discovery and Data Mining* (2005), pp. 239–248
8. H. Yu, SVM selective sampling for ranking with application to data retrieval, in *Proceedings of The 11th International Conference on Knowledge Discovery and Data Mining* (2005), pp. 354–363
9. Y. Freund, R. Iyer, R.E. Schapire, Y. Singer, An efficient boosting algorithm for combining preferences. J. Mach. Learn. Res. **4**, 933–969 (2003)
10. A. Bahamonde, G.F. Bayón, J.D.J.R. Quevedo, O. Luaces, J.J. del Coz, J. Alonso, F. Goyache, Feature subset selection for learning preferences: A case study, in *Proceedings of The 21st International Conference on Machine Learning* (2004), pp. 49–56
11. D.J. Slotta, J.P. Vergara, N. Ramakrishnan, L.S. Heath, Algorithms for feature selection in rank-order spaces. Technical Report TR-05-08, Computer Science, Virginia Tech. (2005)
12. M. Dettling, P. Bühlmann, Supervised clustering of genes. Genome Biol. **3**(12) (2002). doi:research0069.1–0069.15
13. L. Deng, J. Pei, J. Ma, D.L. Lee, A rank sum test method for informative gene discovery, in *Proceedings of The 10th International Conference on Knowledge Discovery and Data Mining* (2004), pp. 410–419
14. T. Kamishima, S. Akaho, Filling-in missing objects in orders, in *Proceedings of The 4th IEEE International Conference on Data Mining* (2004), pp. 423–426
15. P. Diaconis, A generalization of spectral analysis with application to ranked data. Ann. Stat. **17**(3), 949–979 (1989)
16. H. Kazawa, T. Hirao, E. Maeda, Order SVM: a kernel method for order learning based on generalized order statistics. Syst. Comput. Jpn. **36**(1), 35–43 (2005)

Learning of Rule Ensembles for Multiple Attribute Ranking Problems

Krzysztof Dembczyński, Wojciech Kotłowski, Roman Słowiński, and Marcin Szeląg

Abstract In this paper, we consider the multiple attribute ranking problem from a Machine Learning perspective. We propose two approaches to statistical learning of an ensemble of decision rules from decision examples provided by the Decision Maker in terms of pairwise comparisons of some objects. The first approach consists in learning a preference function defining a binary preference relation for a pair of objects. The result of application of this function on all pairs of objects to be ranked is then exploited using the Net Flow Score procedure, giving a linear ranking of objects. The second approach consists in learning a utility function for single objects. The utility function also gives a linear ranking of objects. In both approaches, the learning is based on the boosting technique. The presented approaches to Preference Learning share good properties of the decision rule preference model and have good performance in the massive-data learning problems. As Preference Learning and Multiple Attribute Decision Aiding share many concepts and methodological issues, in the introduction, we review some aspects bridging these two fields. To illustrate the two approaches proposed in this paper, we solve with them a toy example concerning the ranking of a set of cars evaluated by multiple attributes. Then, we perform a large data experiment on real data sets. The first data set concerns credit rating. Since recent research in the field of Preference Learning is motivated by the increasing role of modeling preferences in recommender systems and information retrieval, we chose two other massive data sets from this area – one comes from movie recommender system MovieLens, and the other concerns ranking of text documents from 20 Newsgroups data set.

K. Dembczyński (✉), W. Kotłowski, R. Słowiński, and M. Szeląg
Institute of Computing Science, Poznań University of Technology, 60-965 Poznań, Poland
e-mail: {kdembczynski,wkotlowski,rslowinski,mszelag}@cs.put.poznan.pl

R. Słowiński
Systems Research Institute, Polish Academy of Sciences
01-447 Warsaw, Poland

J. Fürnkranz and E. Hüllermeier (eds.), *Preference Learning*,
DOI 10.1007/978-3-642-14125-6_11,

1 Introduction to Multiple Attribute Decision Aiding

In this paper, we consider the multiple attribute ranking problem from a Machine Learning perspective. It is useful, however, to start with a short survey of the goals and methodologies known in the field of *Multiple Attribute Decision Aiding* (also called *Multiple Criteria Decision Aiding* or *Multiple Criteria Decision Making*). Although these methodologies are rarely based on statistical analysis of data, which is characteristic for *Machine Learning* methodologies, some goals, concepts, and models developed and investigated within Multiple Attribute Decision Aiding are useful in Machine Learning and, in particular, in the emerging subfield of Machine Learning, called *Preference Learning*. We will concentrate on the aspects bridging these two fields.

1.1 Multiple Attribute Decision Aiding

The aim of scientific decision aiding is to give the Decision Maker (DM) a recommendation concerning a set of *objects* (also called alternatives, solutions, acts, actions, options, candidates, ...) evaluated from multiple points of view considered relevant for the problem at hand and called *attributes* (also called features, variables, criteria, ...).

For example, a decision can concern:

1. diagnosis of pathologies for a set of patients, where patients are objects of the decision, and symptoms and results of medical tests are the attributes,
2. assignment of enterprises to classes of risk, where enterprises are objects of the decision, and financial ratio indices and other economic indicators, such as the market structure, the technology used by the enterprise, and the quality of management, are the attributes,
3. selection of a car to be bought from among a given set of cars, where cars are objects of the decision, and maximum speed, acceleration, price, fuel consumption, comfort, color, etc. are the attributes,
4. ordering of students applying for a scholarship, where students are objects of the decision, and scores in different disciplines are the attributes.

The following three main categories of decision problems are typically distinguished [48]:

- *classification*, when the decision aims at assigning objects to predefined classes,
- *choice*, when the decision aims at selecting the best objects,
- *ranking*, when the decision aims at ordering objects from the best to the worst.

Looking at the above examples, one can say that (1) and (2) are classification problems, (3) is a choice problem, and (4) is a ranking problem.

The above categorization can be refined by distinguishing three kinds of classification problems:

- *taxonomy*, when the value sets of attributes and the predefined classes are not preference ordered,
- *ordinal classification*, when the predefined classes are preference ordered, while the value sets of attributes are not, and
- *ordinal classification with monotonicity constraints* (also known as *multiple criteria sorting*), when both the value sets of attributes and the predefined classes are preference ordered [21].

The monotonicity constraints imposed on ordinal classification require that an improvement of an object's evaluation on any attribute should not deteriorate its class assignment. In the above examples, (1) is a taxonomy problem and (2) is an ordinal classification problem with monotonicity constraints.

An important step in Multiple Attribute Decision Aiding concerns construction or selection of attributes describing the objects. They are built on, or selected from among, elementary features of the objects. The aim is to set up a *consistent set of attributes*, that makes the pairwise comparison of all objects in the considered set meaningful. In other words, the consistent set of attributes should permit a meaningful distinction of objects, i.e., objects which are indiscernible with respect to a given set of attributes should be considered indifferent; if this was not the case, the given set of attributes would not be consistent.

1.2 Criteria: Attributes with Preference Ordered Scales

Very often, the description of objects by attributes is not neutral with respect to *preferences* of the DM, and then there is a need of taking into account that for the DM some values of attributes are more (or less) preferred than others. In such cases, in Multiple Attribute Decision Aiding, the value sets of these attributes are translated into a monotonic preference scale, which may be ordinal or cardinal [50]. Attributes with monotonic preference scales are called *criteria*.

When there is no relationship between value sets of attributes and DM's preferences, then, to distinguish such attributes from criteria, one calls them *regular attributes*. For example, in a decision regarding the selection of a car, its price is a criterion because, obviously, a low price is better than a high price. On the other hand, the color of a car is not a criterion but a regular attribute, because red is not intrinsically better than green. One can imagine, however, that also the color of a car could become a criterion if, for example, a DM would consider red better than green.

As the concept of criterion plays an important role in Multiple Attribute Decision Aiding, to make our survey more specific, we formalize this concept a little further.

Let $\boldsymbol{x}$ denote an object belonging to a universe of discourse $\mathcal{X}$. A *criterion* is a real-valued function g_h defined on $\mathcal{X}$, $g_h : \mathcal{X} \rightarrow \Re$, reflecting a worth of objects from a DM's point of view, such that to compare any two objects $\boldsymbol{x}, \boldsymbol{x}' \in \mathcal{X}$ from this point of view, it is sufficient to compare two values: $g_h(\boldsymbol{x})$ and $g_h(\boldsymbol{x}')$. Without loss of generality, $g_h(\boldsymbol{x}) \geq g_h(\boldsymbol{x}')$ means that "$\boldsymbol{x}$ is at least as good as $\boldsymbol{x}'$ with respect to criterion g_h", which is denoted by $\boldsymbol{x} \succeq_h \boldsymbol{x}'$. Therefore, it is supposed that $\succeq_h$ is a complete preorder, i.e., a strongly complete and transitive binary relation defined on $\mathcal{X}$ on the basis of evaluations $g_h(\cdot)$.

All points of view being relevant for a decision problem at hand form a *consistent set of criteria* $G = \{g_1, g_2, \ldots, g_m\}$. Comparing to the consistency condition of a set of attributes mentioned above, the consistency of the set of criteria involves one condition more, called condition of monotonicity. This condition requires that if for $\boldsymbol{x}, \boldsymbol{x}' \in \mathcal{X}$, object $\boldsymbol{x}$ is weakly preferred to object $\boldsymbol{x}'$ (denoted by $\boldsymbol{x} \succeq \boldsymbol{x}'$), then for another object $\boldsymbol{x}'' \in \mathcal{X}$, such that $g_h(\boldsymbol{x}'') \geq g_h(\boldsymbol{x})$, for all $h = 1, \ldots, m$, object $\boldsymbol{x}''$ should also be weakly preferred to object $\boldsymbol{x}'$ ($\boldsymbol{x}'' \succeq \boldsymbol{x}'$) [49].

Remark that while in Multiple Attribute Decision Aiding (or, more precisely, Multiple Criteria Decision Aiding) the construction of criteria with explicit monotonic preference scales is an important step in the procedure of decision aiding, in Preference Learning, the relationships between value sets of attributes and DM's preferences (if any) are discovered from data for a direct use in classification or ranking. This means that in Preference Learning, the monotonic preference scales converting regular attributes to criteria are neither used nor revealed explicitly.

1.3 Dominance Relation in the Set of Objects

For a given finite set of objects $X = \{\boldsymbol{x}_1, \boldsymbol{x}_2, \ldots, \boldsymbol{x}_o\} \subseteq \mathcal{X}$, and a set of criteria G, the only objective information that comes out from comparison of these objects on multiple criteria is a *dominance relation* $\rhd$ in set X. Object $\boldsymbol{x}_i$ (weakly) dominates object $\boldsymbol{x}_j$, which is denoted by $\boldsymbol{x}_i \rhd \boldsymbol{x}_j$, if and only if $g_h(\boldsymbol{x}_i) \geq g_h(\boldsymbol{x}_j)$ for each $h = 1, \ldots, m$. Object $\boldsymbol{x}_i$ is *nondominated* in set X (Pareto-optimal) if and only if there is no other object $\boldsymbol{x}_j \in X$ dominating $\boldsymbol{x}_i$. Therefore, the dominance relation $\rhd$ is a partial preorder, i.e., a reflexive and transitive binary relation defined on X on the basis of evaluations $g_h(\cdot)$, $h = 1, \ldots, m$.

To simplify notation, in the rest of the paper, we will identify $g_h(\boldsymbol{x})$ with x_h, whether it concerns the hth criterion or regular attribute ($h = 1, \ldots, m$). Thus, any object $\boldsymbol{x} = (x_1, x_2, \ldots, x_m) \in \mathcal{X} \subseteq \Re^m$.

Apart from trivial cases, the dominance relation $\rhd$ is rather poor and leaves many objects incomparable – these are all nondominated objects in set X. To enrich the dominance relation and make the objects in X more comparable, one needs additional information about value system of the DM, called *preference information*. This information permits to build a more or less explicit model of DM's preferences, called *preference model*. The preference model relates the decision to evaluations of the objects on the considered criteria. In other words, the preference model aggregates evaluations of objects on multiple criteria. It is inducing a *preference structure* in set X. A proper exploitation of this structure leads then to a *recommendation* in terms of sorting, or choice, or ranking of objects from set X.

In Preference Learning, the training data are the equivalent of preference information in Multiple Attribute Decision Aiding. Moreover, the aim of getting a preference model that permits the working out of a final recommendation is the same for both methodologies – roughly speaking, the difference resides in the statistical or nonstatistical way of processing the preference information.

It follows from above that the preference information and the preference model are two crucial components of both Multiple Attribute Decision Aiding and Preference Learning. Many of the methods existing in both fields differ by these two components only. Below, with respect to these two components, we review some recent trends in Multiple Attribute Decision Aiding.

1.4 Preference Information and Preference Model

As to the preference information, it depends on the adopted methodology: prices and interest rates for cost-benefit analysis, cost coefficients in objectives and technological coefficients in constraints for mathematical programming, a training set of decision examples for neural networks and machine learning, substitution rates for a value function of Multi-Attribute Utility Theory, pairwise comparisons of objects in terms of intensity of preference for the Analytic Hierarchy Process, attribute weights and several thresholds for ELECTRE methods, and so on (see the state-of-the-art survey [12]). This information has to be provided by the DM, possibly assisted by an analyst.

Very often this information is not easily definable. For example, this is the case of the price of many immaterial goods and of the interest rates in cost-benefit analysis, or the case of the coefficients of objectives and constraints in mathematical programming models. Moreover, the preference information given by the DM is often processed in a way which is not clear for her/him, such that (s)he cannot see what are the exact relations between the provided information and the final recommendation. Consequently, very often the decision aiding method is perceived by the DM as a *black box* whose result has to be accepted because the analyst's authority guarantees that the result is "right". In this context, the aspiration of the DM to find good reasons to make decision is frustrated and rises the need for a more transparent methodology in which the relation between the original information and the final recommendation is clearly shown. Such a transparent methodology searched for has been called *glass box* [28]. A part of such a methodology is certainly the preference information given in the form of a training set of decision examples.

The decision examples may either be provided by the DM on a set of real or hypothetical objects, or may come from observation of DM's past decisions. Such an approach follows the paradigm of inductive learning used in artificial intelligence [42], or robust ordinal regression becoming popular in operational research [32]. It is also concordant with the principle of posterior rationality postulated by March [41] since it emphasizes the discovery of DM's intentions as an interpretation of actions rather than as a priori position. This paradigm has been used to construct various preference models from decision examples, e.g., the general additive utility functions [13, 31], the outranking relations [32, 43], the monotonic decision trees [20], and the set of "*if* ..., *then* ..." decision rules [26].

Of particular interest is the last model based on decision rules – it has been introduced to decision analysis by Greco, Matarazzo, and Słowiński [22, 24, 54]. A popular saying attributed to Slovic is that "people make decisions and then search for rules that justify their choices". The rules explain the preferential attitude of the DM and enable understanding of the reasons of his/her past decisions. The recognition of the rules by the DM [40] justifies their use for decision support. So, the preference model in the form of rules derived from decision examples fulfills both explanation and recommendation goals of decision aiding.

For example, in case of a medical diagnosis problem, the decision rule approach requires as input information a set of examples of previous diagnoses, from which some diagnostic rules are induced, such as "if there is symptom α and the test result is β, then there is pathology γ". Each one of such rules is directly related to examples of diagnoses in the input information, where there is symptom α, test result β, and pathology γ.

Decision rules constituting the preference model have a special syntax which involves partial evaluation profiles and dominance relations on these profiles. The traditional preference models, which are the utility function and the outranking relation, can be represented in terms of equivalent decision rules. The clarity of the rule representation of preferences enables one to see the limits of these aggregation functions. Several studies [25, 27, 52] presented an axiomatic characterization of all three kinds of preference models in terms of conjoint measurement theory and in terms of a set of decision rules. The given axioms of "cancelation property" type are the weakest possible. In comparison to other studies on the characterization of preference models, these axioms do not require any preliminary assumptions about the scales of preferences of criteria. A side-result of these investigations is that the decision rule preference model is the most general among all known models.

1.5 Inconsistency of Preference Information and the Rough Set Concept

Preference information given in terms of decision examples is often inconsistent. For example, objects with the same description by the set of attributes may be assigned to different classes. This explains the interest in *rough set theory* proposed by Pawlak [45]. Rough set theory permits to structure the data set such that decision classes are represented by pairs of ordinary sets called *lower* and *upper approximations*. The differences between upper and lower approximations are called boundary sets, and their cardinalities indicate to what degree the data set is inconsistent. Moreover, rough set theory provides useful information about the role of particular attributes and their subsets in the approximation of decision classes. Induction of decision rules from data structured in this way permits to obtain certain or approximate decision rules [46, 51]. For example, in the above diagnostic context, cases where the presence of different pathologies is associated with the presence of the same symptoms and test results are inconsistent, and thus they are placed in the boundaries of the classes of pathologies; decision rules supported by these examples are approximate.

As the classical definition of rough sets is based on indiscernibility relation in the set of objects, it can handle only one kind of inconsistency of decision examples – the one related to indiscernibility of objects belonging to different decision classes. While this is sufficient for classification of taxonomy type, the classical rough set approach fails in case of ordinal classification with monotonicity constraints, where the value sets of attributes, as well as decision classes, are preference ordered. In this case, decision examples may be inconsistent in the sense of violation of the *dominance principle* which requires that an object $\boldsymbol{x}$ dominating object $\boldsymbol{x}'$ on all considered criteria (i.e., $\boldsymbol{x}$ having evaluations at least as good as $\boldsymbol{x}'$ on all considered criteria) should also dominate $\boldsymbol{x}'$ on the decision (i.e., $\boldsymbol{x}$ should be assigned to at least as good decision class as $\boldsymbol{x}'$). To deal with this kind of inconsistency, Greco, Matarazzo, and Słowiński generalized the classical rough set approach, so as to take into account preference orders and monotonic relationships between evaluations on criteria and assignment to decision classes. This generalization, called Dominance-based Rough Set Approach (DRSA), has been adapted to a large variety of decision problems [14, 22, 24, 28, 30, 54].

Moreover, DRSA has been adapted to handle granular (fuzzy) information [29], and incomplete information [7, 23].

The usefulness of DRSA goes beyond the frame of Multiple Attribute Decision Aiding. This is because the type of monotonic relationships handled by DRSA is also meaningful for problems where preferences are not considered but a kind of monotonicity relating ordered attribute values is meaningful for the analysis of data at hand. Indeed, monotonicity concerns, in general, mutual trends existing between different variables, such as distance and gravity in physics, or inflation rate and interest rate in economics. Whenever a relationship between different aspects of a phenomenon is discovered, this relationship can be represented by a monotonicity with respect to some specific measures or perception of the considered aspects, e.g., "the colder the weather, the higher the energy consumption" or "the more a tomato is red, the more it is ripe". The qualifiers, such as "cold weather", "high energy consumption", "red" and "ripe", may be expressed either in terms of some measurement units, or in terms of degrees of membership to fuzzy sets representing these concepts.

1.6 Statistical Learning from Preference Information

The above survey concerned those methodologies of Multiple Attribute Decision Aiding which have some links with Preference Learning, either through the type of decision problems being solved, or through the type of preference information being used, or, finally, through the type of preference model used to work out a recommendation. These methodologies are based on nonstatistical ways of processing preference information given in terms of decision examples. This type of preference information is characteristic for contemporary methods of Multiple Attribute Decision Aiding. Recently, we have been able to observe an increasing interest in statistical methods of processing this information, particularly in situations where the number of decision examples is very large. Statistical approach to learning preference models from decision examples is the core of Preference Learning, which is today one of the main topics in Machine Learning. This interest is motivated by new challenging applications related to Internet, in particular, recommender systems and information retrieval. In the first, the task is to recommend to the user a new item (like movie or book) that fits her/his preferences. The recommendation is computed on the base of the learning information describing the past behavior of the user. In the latter, the task is to sort (or rank) the documents retrieved by the search engine according to the user's preferences. There are several algorithms that are tailored for these kinds of problems. The most popular ones are based on *rank loss minimization*. These include variants of support vector machines [35] and boosting [16]. One should also note that there exist several other learning approaches in which preferences are modeled [15, 19, 47, 55]. Moreover, an interesting work has been done in the field of ordinal classification with monotonicity constraints [6, 8, 10, 38].

2 Multiple Attribute Ranking Problem

While in the classification problem the assignment of objects to decision classes is based on absolute evaluation of objects on attributes, in the ranking problem, the position of an object in the ranking depends on comparison of its evaluation on attributes with other objects. Thus, in case of multiple attribute ranking problems, the preference information given in terms of decision examples concerns *pairwise comparisons* of objects. The pairwise comparison of objects $\boldsymbol{x}, \boldsymbol{x}' \in \mathcal{X}$ (on particular criteria or comprehensively) states what is the intensity of preference of object $\boldsymbol{x}$ over object $\boldsymbol{x}'$. The intensity of preference can be expressed either on an ordinal scale or on a cardinal scale. In case of the ordinal scale, only the order of the intensity of preference matters; in case of the cardinal expression of the intensity, the interval scale permits to compare differences of the intensity, and the ratio scale permits, moreover, to express the ratio between the intensities. In the simplest case, the intensity of preference is expressed on a 3-value ordinal scale coded by $1, 0, -1$, which corresponds to preference ($\boldsymbol{x} \succ \boldsymbol{x}'$), indifference ($\boldsymbol{x} \sim \boldsymbol{x}'$), and inverse preference ($\boldsymbol{x}' \succ \boldsymbol{x}$) for a pair of objects, respectively. The weak preference, denoted by $\succeq$, groups the preference and the indifference, i.e., $\boldsymbol{x} \succeq \boldsymbol{x}' \Leftrightarrow \boldsymbol{x} \succ \boldsymbol{x}' \vee \boldsymbol{x} \sim \boldsymbol{x}'$.

Several approaches have been proposed to multiple attribute ranking based on preference information given in terms of decision examples. They mainly differ by the type of preference model constructed from examples of pairwise comparisons. Let us mention, for instance:

- the UTA^{GMS} and GRIP methods for the construction of necessary ranking (partial preorder) and possible ranking (strongly complete and negatively transitive relation) of objects from set $X \subseteq \mathcal{X}$, resulting from consideration of all general additive utility functions compatible with the preference information [13, 31],
- the construction of necessary and possible outranking (reflexive) relations in set $X \subseteq \mathcal{X}$, resulting from consideration of all outranking models compatible with the preference information [32, 43],
- the induction of a set of decision rules from a pairwise comparison table using the Variable Consistency DRSA [53], and exploitation of preference structure induced by application of the rules on set $X \times X$ ($X \subseteq \mathcal{X}$) using Weighted-Fuzzy Net Flow Score or Lexicographic-Fuzzy Net Flow Score [14].

The last approach based on decision rules is able to handle pairwise comparisons inconsistent with the dominance principle. In this context, the dominance principle says that if for two pairs of objects, $(\boldsymbol{x}, \boldsymbol{x}') \in \mathcal{X} \times \mathcal{X}$ and $(\boldsymbol{x}'', \boldsymbol{x}''') \in \mathcal{X} \times \mathcal{X}$, object $\boldsymbol{x}$ is preferred to object $\boldsymbol{x}'$ at least as much as object $\boldsymbol{x}''$ is preferred to object $\boldsymbol{x}'''$ on all considered criteria, then pair $(\boldsymbol{x}, \boldsymbol{x}')$ should be, comprehensively, weakly preferred to pair $(\boldsymbol{x}'', \boldsymbol{x}''')$, i.e., the comprehensive preference of $\boldsymbol{x}$ over $\boldsymbol{x}'$ should not be less intensive than that of $\boldsymbol{x}''$ over $\boldsymbol{x}'''$.

In this paper, we consider the multiple attribute ranking problem from a Machine Learning perspective. We propose two variants of a statistical approach to the learning of an *ensemble of decision rules* from decision examples. The learning is based on the boosting approach. The presented approach shares good properties of the decision rule preference model mentioned above and has good performance in the massive-data learning problems.

The decision examples, characterized below, boil down to a set of pairwise comparisons of objects. The objects used in decision examples constitute a *training set* denoted by $\mathcal{T} = \{\boldsymbol{x}_1, \boldsymbol{x}_2, \ldots, \boldsymbol{x}_n\}$. The objects to be ranked constitute set $X = \{\boldsymbol{x}_1, \boldsymbol{x}_2, \ldots, \boldsymbol{x}_o\}$. Of course, $\mathcal{T}, X \subseteq \mathcal{X}$ and, sometimes, $\mathcal{T} \subseteq X$.

We investigate two ways of learning of an ensemble of decision rules. They differ by the type of considered decision rules. The first type of decision rules specifies conditions on differences of evaluations of a pair of objects on particular attributes, which result in a comprehensive preference of one object over the other object. The second type of decision rules specifies conditions on evaluations of a single object on particular attributes, which result in a given increase or decrease of a comprehensive utility (called also score or value) of this object.

The ensemble composed of the first type of rules is applied on $X \times X$ and the resulting preference structure (represented in a graph form) is exploited using the Net Flow Score procedure to get the final ranking. The ensemble composed of the second type of rules is applied on X and the resulting comprehensive utility of individual objects induces the final ranking. In both cases, the final ranking is a complete preorder.

To ease the presentation, we will consider the two approaches as an approach to learning of an additive real-valued function:

$$f = \sum_{p=1}^{P} r_p, \tag{1}$$

called *rule ensemble*, where r_p is a marginal function corresponding to pth decision rule, and P is the number of decision rules used in the ensemble. Moreover,

- For the first type of decision rules, the argument of function f and of its marginal functions r_p is a pair of objects $\boldsymbol{x}, \boldsymbol{x}' \in \mathcal{X}$; if $f(\boldsymbol{x}, \boldsymbol{x}') \geq 0$, then one predicts $\boldsymbol{x} \succeq \boldsymbol{x}'$, and if $f(\boldsymbol{x}, \boldsymbol{x}') \leq 0$, then one predicts $\boldsymbol{x}' \succeq \boldsymbol{x}$,
- For the second type of decision rules, the argument of function f and of its marginal functions r_p is a single object $\boldsymbol{x} \in \mathcal{X}$; if $f(\boldsymbol{x}) \geq f(\boldsymbol{x}')$, then one predicts $\boldsymbol{x} \succeq \boldsymbol{x}'$, and if $f(\boldsymbol{x}) \leq f(\boldsymbol{x}')$, then one predicts $\boldsymbol{x}' \succeq \boldsymbol{x}$.

In the first case, f is called *preference function* and the resulting weak preference relation $\succeq$ is reflexive but, in general, nontransitive, and in the second case, f is called *utility function* (or scoring, or value function) and the resulting weak preference relation $\succeq$ is reflexive and transitive.

We also consider three types of preference information (decision examples) delivered by the DM, which result in preference relation $\succ$ for some pairs of objects from set $\mathcal{T}$:

- Pairwise comparisons of objects, i.e., $\boldsymbol{x}_i \succ \boldsymbol{x}_j$ for some $\boldsymbol{x}_i, \boldsymbol{x}_j \in \mathcal{T}$,
- Complete (linear) ranking of objects, i.e., $\boldsymbol{x}_1 \succ \boldsymbol{x}_2 \succ \cdots \succ \boldsymbol{x}_n$ for $\{\boldsymbol{x}_1, \ldots, \boldsymbol{x}_n\} \in \mathcal{T}$, which means $\boldsymbol{x}_i \succ \boldsymbol{x}_j$ for all $i < j$ and $i, j \in \{1, \ldots, n\}$,
- Rating of objects on a finite scale corresponding to additional attribute $y \in \mathcal{Y} = \{1, 2, \ldots, k\}$, such that $1 \prec 2 \prec \ldots \prec k$, i.e., for $y_i, y_j \in \mathcal{Y}$ assigned to $\boldsymbol{x}_i, \boldsymbol{x}_j \in \mathcal{T}$, respectively, if $y_i > y_j$, then $\boldsymbol{x}_i \succ \boldsymbol{x}_j$.

Preference information in terms of pairwise comparisons of objects is delivered, for example, in information retrieval. Let us consider the following example. Having a list of documents, the user has chosen the first, third, and seventh document. Given that the abstracts associated with the documents are sufficiently informative, this gives some indication of user's preferences. One can infer from this selection that the third document is more relevant (better) than the second document. Assuming that the user scanned the list from top to bottom, (s)he must have observed the second document before clicking on the third one, making a decision not to click on the second. Similarly, it is possible to infer that the seventh document is more relevant than the second, fourth, fifth, and sixth [37]. In the presence of such preference information, the learning problem is often referred to as *object ranking*.

Preference information in terms of a linear ranking means that the DM (teacher) has ordered all the training examples in a complete way, and no objects are incomparable. This kind of preference information is often encountered in Multiple Criteria Decision Aiding [36].

The rating of objects on a finite ordinal scale is typical to *ordinal classification*, also called *instance ranking* or *multipartite ranking*. In this case, the DM is assigning to each training object a rank from a finite scale. By comparing the ranks of objects pairwise, one gets the information on the preference relation in the training set. This is a common situation in recommender systems, which aim at ranking items to give a final recommendation. For example, in the movie recommender systems, such as Netflix [44], the users are often asked to rate movies on a five-value scale, e.g., one to five "stars". It is assumed that it is easier to obtain the preference information by assigning the ranks to movies, than by comparing movies directly pairwise.

The remainder of the paper is organized as follows. In the next section, we formulate the multiple attribute ranking problem in a machine learning setting. Section 4 describes the decision rule approach to the ranking problem. Section 5 presents the algorithm for learning of an ensemble of decision rules. A toy example that illustrates the presented methodology is analyzed in Sect. 6. An experiment on three data sets: credit rating, MovieLens, and 20 Newsgroups is described in Sect. 7. The last section concludes the paper.

3 Two Approaches to Multiple Attribute Ranking Learning

From machine learning perspective, the multiple attribute ranking problem can be seen as a problem of learning:

(a) preference function $f(\cdot,\cdot)$, and the resulting weak preference relation $\succeq$, which is to be exploited such as to get a linear ranking over set X,
(b) utility function $f(\cdot)$, which results directly in a linear ranking over set X.

From both approaches, one gets function f, which can be used by a ranking method to rank set X of objects. We investigate two ranking methods corresponding to the above approaches: the *PrefRules* algorithm based on the preference function, and the *RankRules* algorithm based on the utility function.

Let us consider two objects $\boldsymbol{x}, \boldsymbol{x}' \in \mathcal{X}$ described by m attributes, $\boldsymbol{x} = (x_1, x_2, \ldots, x_m)$, $\boldsymbol{x}' = (x'_1, x'_2, \ldots, x'_m)$. Since in machine learning one assumes that data are generated according to some probability distribution, the pairwise comparison of these objects is a random variable $s : \mathcal{X} \times \mathcal{X} \rightarrow \{1, -1\}$, indicating the preference or the inverse preference between $\boldsymbol{x}$ and $\boldsymbol{x}'$, respectively, i.e.:

◇ $s(\boldsymbol{x}, \boldsymbol{x}') = 1 \Leftrightarrow \boldsymbol{x} \succ \boldsymbol{x}'$,
◇ $s(\boldsymbol{x}, \boldsymbol{x}') = -1 \Leftrightarrow \boldsymbol{x}' \succ \boldsymbol{x}$.

We assume that the indifference between these two objects, $\boldsymbol{x} \sim \boldsymbol{x}'$, takes place when $\Pr(s(\boldsymbol{x}, \boldsymbol{x}') = 1 | \boldsymbol{x}, \boldsymbol{x}') = \Pr(s(\boldsymbol{x}, \boldsymbol{x}') = -1 | \boldsymbol{x}, \boldsymbol{x}')$.

In approach (a), the goal is to predict the value of $s(\boldsymbol{x}, \boldsymbol{x}')$. Let $f : \mathcal{X} \times \mathcal{X} \rightarrow \Re$ be a preference function that represents a prediction strategy, such that

◇ If $f(\boldsymbol{x}, \boldsymbol{x}') > 0$, then one predicts $\boldsymbol{x} \succ \boldsymbol{x}'$,
◇ If $f(\boldsymbol{x}, \boldsymbol{x}') < 0$, then one predicts $\boldsymbol{x}' \succ \boldsymbol{x}$,
◇ If $f(\boldsymbol{x}, \boldsymbol{x}') = 0$, then one predicts $\boldsymbol{x} \sim \boldsymbol{x}'$.

We can measure the quality of preference function $f(\boldsymbol{x}, \boldsymbol{x}')$ by the expected value of the so-called *rank loss*, defined as:

$$L\left(s(\boldsymbol{x}, \boldsymbol{x}'), f(\boldsymbol{x}, \boldsymbol{x}')\right) = \begin{cases} 0, & \text{if } s(\boldsymbol{x}, \boldsymbol{x}') \times f(\boldsymbol{x}, \boldsymbol{x}') > 0, \\ \frac{1}{2}, & \text{if } s(\boldsymbol{x}, \boldsymbol{x}') \times f(\boldsymbol{x}, \boldsymbol{x}') = 0, \\ 1, & \text{if } s(\boldsymbol{x}, \boldsymbol{x}') \times f(\boldsymbol{x}, \boldsymbol{x}') < 0. \end{cases} \tag{2}$$

Let us remark that this formulation of the ranking problem resembles binary classification problem in which the sign of $s(\boldsymbol{x}, \boldsymbol{x}')$ is to be guessed for a pair of objects $(\boldsymbol{x}, \boldsymbol{x}')$. For this reason, the optimal preference function corresponds to the so-called Bayes optimal decision, given by:

$$f^* = \arg\min_f \mathbb{E}\left[L\left(s(\boldsymbol{x}, \boldsymbol{x}'), f(\boldsymbol{x}, \boldsymbol{x}')\right)\right], \tag{3}$$

where $\mathbb{E}$ stands for the expectation with respect to the probability distribution generating the data. Thus, for a given pair $(\boldsymbol{x}, \boldsymbol{x}') \in \mathcal{X} \times \mathcal{X}$, the optimal preference function has the form:

$$f^*(\boldsymbol{x}, \boldsymbol{x}') = \begin{cases} > 0, & \text{if } \Pr(s(\boldsymbol{x}, \boldsymbol{x}') = 1|\boldsymbol{x}, \boldsymbol{x}') > \Pr(s(\boldsymbol{x}, \boldsymbol{x}') = -1|\boldsymbol{x}, \boldsymbol{x}'), \\ 0, & \text{if } \Pr(s(\boldsymbol{x}, \boldsymbol{x}') = 1|\boldsymbol{x}, \boldsymbol{x}') = \Pr(s(\boldsymbol{x}, \boldsymbol{x}') = -1|\boldsymbol{x}, \boldsymbol{x}'), \\ < 0, & \text{if } \Pr(s(\boldsymbol{x}, \boldsymbol{x}') = 1|\boldsymbol{x}, \boldsymbol{x}') < \Pr(s(\boldsymbol{x}, \boldsymbol{x}') = -1|\boldsymbol{x}, \boldsymbol{x}'). \end{cases} \tag{4}$$

The preference function f induces a weak preference relation $\succeq$ in set $\mathcal{X}$, which is reflexive but not necessarily transitive. As our goal is to get a linear ranking over X (reflexive and transitive), we have to exploit properly the preference structure induced by preference function f on set X. We propose to apply to this end the *Net Flow Score* (NFS) procedure characterized in [2, 4, 57]. The net flow score of each object $\boldsymbol{x} \in X$ can be calculated using the following formula:

$$NFS(\boldsymbol{x}) = \sum_{\boldsymbol{x}' \in X,\, \boldsymbol{x}' \neq \boldsymbol{x}} \Big[\mathrm{sgn}(f(\boldsymbol{x}, \boldsymbol{x}')) - \mathrm{sgn}(f(\boldsymbol{x}', \boldsymbol{x}))\Big]. \tag{5}$$

Obviously, $NFS(\boldsymbol{x})$ ranks all objects $\boldsymbol{x} \in X$ from the best to the worst.

In approach (b), the aim is to learn function $f : \mathcal{X} \to \Re$ that corresponds to the utility of a single object. To predict the value of $s(\boldsymbol{x}, \boldsymbol{x}')$, we will use the difference of utilities for objects $\boldsymbol{x}, \boldsymbol{x}' \in \mathcal{X}$, i.e., $f(\boldsymbol{x}) - f(\boldsymbol{x}')$, such that

- ◇ If $f(\boldsymbol{x}) - f(\boldsymbol{x}') > 0$, then one predicts $\boldsymbol{x} \succ \boldsymbol{x}'$,
- ◇ If $f(\boldsymbol{x}) - f(\boldsymbol{x}') < 0$, then one predicts $\boldsymbol{x}' \succ \boldsymbol{x}$,
- ◇ If $f(\boldsymbol{x}) - f(\boldsymbol{x}') = 0$, then one predicts $\boldsymbol{x} \sim \boldsymbol{x}'$.

Thus, in approach (b), $f(\boldsymbol{x}) - f(\boldsymbol{x}')$ plays the role of $f(\boldsymbol{x}, \boldsymbol{x}')$ in approach (a). In this case, the rank loss takes the following form:

$$L\left(s(\boldsymbol{x},\boldsymbol{x}'), f(\boldsymbol{x}) - f(\boldsymbol{x}')\right) = \begin{cases} 0, & \text{if } s(\boldsymbol{x},\boldsymbol{x}') \times (f(\boldsymbol{x}) - f(\boldsymbol{x}')) > 0, \\ \frac{1}{2}, & \text{if } s(\boldsymbol{x},\boldsymbol{x}') \times (f(\boldsymbol{x}) - f(\boldsymbol{x}')) = 0, \\ 1, & \text{if } s(\boldsymbol{x},\boldsymbol{x}') \times (f(\boldsymbol{x}) - f(\boldsymbol{x}')) < 0. \end{cases} \quad (6)$$

To distinguish the rank loss in approaches (a) and (b), we will call (2) *preference-based rank loss*, and (6) *utility-based rank loss*.

It is worth noting that the utility-based rank loss is closely related to the so-called AUC (Area Under the "Receiver Operator Characteristic" Curve) criterion, becoming a popular measure for evaluating the performance of binary classifiers. Binary classification can be treated as a special case of rating on a finite ordinal scale. It is often called *bipartite ranking*. In this case, one has two possible ranks, usually coded by $y \in \{1, -1\}$, indicating a positive and negative class, respectively. One can define the random variable $s(\boldsymbol{x},\boldsymbol{x}')$ as, $s(\boldsymbol{x},\boldsymbol{x}') = 1$, if $y > y'$, and $s(\boldsymbol{x},\boldsymbol{x}') = -1$, if $y < y'$, where *label* is the rank assigned to $\boldsymbol{x}$, and *label*$'$ is the rank assigned to $\boldsymbol{x}'$. The AUC criterion is defined as:

$$\mathrm{AUC}(f) = \Pr(f(\boldsymbol{x}) > f(\boldsymbol{x}')|y = 1, y' = -1) + \frac{1}{2}\Pr(f(\boldsymbol{x}) = f(\boldsymbol{x}')|y = 1, y' = -1),$$

where f is a real function returned by the classifier. One can observe that maximization of the AUC criterion boils down to the minimization of the rank loss, since:

$$\mathrm{AUC}(f) = 1 - \frac{\mathbb{E}[L(s(\boldsymbol{x},\boldsymbol{x}'), f(\boldsymbol{x}) - f(\boldsymbol{x}'))]}{2\Pr(y = 1, y' = -1))}.$$

In the following, we present approaches (a) and (b) to learning of function f in a unified framework. The learning is performed on the training set $\mathcal{T}$. For set $\mathcal{T} \times \mathcal{T}$ we know the value of $s(\boldsymbol{x}_i, \boldsymbol{x}_j)$ for all or some $i, j = 1, \ldots, n$, which represents the preference relation $\succ$ in set $\mathcal{T}$ following from preference information (decision examples) delivered by the DM. The goal is to minimize the corresponding loss function on the training set, the so-called *Empirical risk*:

$$R_{\mathrm{emp}} = \sum_{ij} L_{ij},$$

in which L_{ij} denotes $L\left(s(\boldsymbol{x}_i,\boldsymbol{x}_j), f(\boldsymbol{x}_i,\boldsymbol{x}_j)\right)$ or $L\left(s(\boldsymbol{x}_i,\boldsymbol{x}_j), f(\boldsymbol{x}_i) - f(\boldsymbol{x}_j)\right)$, depending on approach (a) or (b).

4 Decision Rules

Decision rule is a simple logical pattern having the form:

"if [*conjunction of elementary conditions*], then [*decision*]".

The "conjunction of elementary conditions" is called *condition part* or *premise*, and the "decision" is called *decision part* or *conclusion* of the rule. Depending on approach (a) or (b) to multiple attribute ranking learning, we consider two types of decision rules distinguished by the syntax of condition and decision parts:

(a) "If $(x_{i1} - x'_{i1} \geq \delta_{i1})$ and ... $(x_{il'} - x'_{il'} \geq \delta_{il'})$ and $(x_{j1} - x'_{j1} \leq \delta_{j1})$ and ... $(x_{jl''} - x'_{jl''} \leq \delta_{jl''})$, then $\boldsymbol{x} \succ \boldsymbol{x}'$ with credibility α",

(b) "If $(x_{i1} \geq \delta_{i1})$ and ... $(x_{il'} \geq \delta_{il'})$ and $(x_{j1} \leq \delta_{j1})$ and ... $(x_{jl''} \leq \delta_{jl''})$, then the utility of $\boldsymbol{x}$ increases by α",

where $\delta_{(..)} \in \Re$, and sets of attribute indices $\{i1, \ldots, il'\} \subseteq \{1, \ldots, m\}$ and $\{j1, \ldots, jl''\} \subseteq \{1, \ldots, m\}$ are not necessarily disjoint. Elementary conditions on attributes with indices $i1, \ldots, il'$ are called "at least" conditions, and elementary conditions on attributes with indices $j1, \ldots, jl''$ are called "at most" conditions. For a rule of type (a), when an attribute has an ordinal scale coded in $\Re$, then in the elementary conditions we allow $\delta_{(..)} = 0$ only, since we cannot say anything about the intensity of difference between two ordinal values. Thus, we take into account the sign of the difference only. For both types of rules, in the case of nominal attributes, we perform binarization of each such attribute by replacing it with a set of new binary attributes. These binary attributes are then treated as ordinal ones.

Let us remark that the negative value of α inverts the decision, i.e., in rules of type (a), the decision becomes $\boldsymbol{x}' \succ \boldsymbol{x}$ with credibility $|\alpha|$, and in rules of type (b), the utility of $\boldsymbol{x}$ decreases by $|\alpha|$.

The main advantage of decision rule representation is simplicity and human-interpretable form that can model interactions between attributes present in the condition part of the rule. The condition part of each above decision rule defines an axis-parallel subregion in the attribute difference space (rule of type (a)) or in the original attribute space $\mathcal{X}$ (rule of type (b)). In both cases, the subregion is constrained by positive and negative dominance cones corresponding to "at least" and "at most" elementary conditions, respectively (see [22, 24, 26, 54]).

As an example of decision rule of type (a), we present a rule from the toy example further discussed in Sect. 6:

if the road holding difference between cars $\boldsymbol{x}$ and $\boldsymbol{x}'$ is ≥ 0
and the cost difference between cars $\boldsymbol{x}$ and $\boldsymbol{x}'$ is ≤ 3348.0,
then car $\boldsymbol{x}$ is preferred over car $\boldsymbol{x}'$ with credibility 0.48.

From the same toy example, we present a decision rule of type (b):

if acceleration of car $\boldsymbol{x}$ is ≤ 28.6,
then the utility of car $\boldsymbol{x}$ increases by 0.520.

From the machine learning perspective, decision rule of type (a) or (b) can be treated as a simple classifier that gives a constant response for objects satisfying the condition part, and abstains from the response for all other objects.

For each decision rule of type (a), we define function $\Phi : \mathcal{X} \times \mathcal{X} \rightarrow \{0, 1\}$, such that $\Phi(\boldsymbol{x}, \boldsymbol{x}') = 1$ if pair of objects $(\boldsymbol{x}, \boldsymbol{x}') \in \mathcal{X} \times \mathcal{X}$ matches the condition part of the rule, and $\Phi(\boldsymbol{x}, \boldsymbol{x}') = 0$, otherwise.

Similarly, for each decision rule of type (b), we define function $\Phi : \mathcal{X} \rightarrow \{0, 1\}$, such that $\Phi(\boldsymbol{x}) = 1$ if object $\boldsymbol{x} \in \mathcal{X}$ matches the condition part of the rule, and $\Phi(\boldsymbol{x}) = 0$, otherwise.

The contribution of a single rule to a rule ensemble is defined by marginal function r_p appearing in (1), where p is the rule's identifier. Depending on approach (a) or (b), marginal function r_p is defined as:

$$r_p(\boldsymbol{x}, \boldsymbol{x}') = \alpha_p \Phi_p(\boldsymbol{x}, \boldsymbol{x}'),$$
$$\text{or } r_p(\boldsymbol{x}) = \alpha_p \Phi_p(\boldsymbol{x}),$$

where $\alpha_p \in \Re$ is the so-called response of the rule, assigned to the region defined by its condition part. Let us remark that α_p is equal to α of the pth rule and that absolute value of α_p indicates the weight of the rule.

While in preference learning it is not necessary to know *a priori* the preference order in the value sets of attributes, one can use this knowledge when available. For attributes which are criteria, it is then necessary to impose only "at least" or only "at most" elementary conditions, depending on the positive or negative sign of response α_p, respectively. This allows to constrain the search space of elementary conditions.

In the next section, we present a simple algorithm for learning an ensemble of decision rules that is based on the boosting approach.

5 Learning of an Ensemble of Decision Rules

The rule ensemble is constructed by the minimization of the empirical risk. We consider two formulations of the problem. The first one corresponds to approach (a), in which the preference function is built and is referred to as *PrefRules*. The second one corresponds to approach (b), in which the utility function is built and is referred to as *RankRules*. We will treat these two formulations jointly as long as it does not lead to any confusion.

The minimization of the rank loss over the training set is a hard optimization task, since the rank loss (in both approaches) is neither smooth nor convex. Instead, we use a convex and upper-bounding surrogate of the rank loss based on the exponential function:

$$L(y, z) = \mathrm{e}^{-yz}. \tag{7}$$

Such a loss function is typically used in the boosting learning machines [16, 17]. In the remainder, we will refer to the rank loss based on (7) as *exponential rank loss*.

In approach (a), the exponential rank loss is then defined as:

$$L\left(s(\boldsymbol{x}, \boldsymbol{x}'), f(\boldsymbol{x}, \boldsymbol{x}')\right) = \mathrm{e}^{-s(x,x') \times f(x,x')},$$

while in approach (b), it is defined as:

$$L\left(s(\boldsymbol{x}, \boldsymbol{x}'), f(\boldsymbol{x}) - f(\boldsymbol{x}')\right) = \mathrm{e}^{-s(x,x') \times (f(x) - f(x'))}.$$

Following the boosting approach, we apply an iterative procedure in which rules are added to the ensemble one by one, greedily minimizing the exponential rank loss over the training set. In the pth iteration, the marginal function r_p corresponding to the pth rule is given by:

$$r_p = \arg\min_r R_{\text{emp}}(f_{p-1} + r) = \arg\min_{\Phi,\alpha} R_{\text{emp}}(f_{p-1} + \alpha\Phi),$$

where f_{p-1} is the rule ensemble after $p-1$ iterations.

In approach (a), the Empirical risk in the pth iteration is then formulated as follows:

$$R_{\text{emp}}(f_{p-1} + \alpha\Phi) = \sum_{ij} \mathrm{e}^{-s(x_i,x_j)\times(f_{p-1}(x_i,x_j)+\alpha\Phi(x_i,x_j))}.$$

For approach (b), we obtain analogous formulation:

$$R_{\text{emp}}(f_{p-1} + \alpha\Phi) = \sum_{ij} \mathrm{e}^{-s(x_i,x_j)\times(f_{p-1}(x_i)-f_{p-1}(x_j)+\alpha(\Phi(x_i)-\Phi(x_j)))}.$$

Let us note that the value of $\mathrm{e}^{-s(x_i,x_j)\times f_{p-1}(x_i,x_j)}$ in approach (a), and the value of $\mathrm{e}^{-s(x_i,x_j)\times(f_{p-1}(x_i)-f_{p-1}(x_j))}$ in approach (b), is constant in the pth iteration for each pair of training objects $(\boldsymbol{x}_i, \boldsymbol{x}_j)$. This value can be treated as a weight of the pair of training objects. To simplify the notation, these weights are denoted by $w_{ij}^{(p)}$, $i, j = 1, \ldots, n$.

Let us also note that in approach (a), the only pairs of training objects influencing the empirical risk are those for which $\Phi(\boldsymbol{x}_i, \boldsymbol{x}_j) = 1$. We will then say that such pairs are covered by the rule. In approach (b), in turn, the only pairs of training objects influencing the empirical risk are those for which $\Phi(\boldsymbol{x}_i) \neq \Phi(\boldsymbol{x}_j)$. In this case also, we will say that these pairs are covered by the rule, despite the fact that only one object of each such pair matches the condition part of the rule.

To treat approaches (a) and (b) in a similar way, we transform both formulations to:

$$R_{\text{emp}}(f_{p-1} + \alpha\Phi) = \mathrm{e}^{-\alpha} W_> + \mathrm{e}^{\alpha} W_< + W_0.$$

In the above formulation, W_0 is a sum of weights over the pairs not covered by the rule, $W_>$ is a sum of weights over the pairs covered by the rule for which the first object is preferred over the second one, and $W_<$ is a sum of weights over the pairs covered by the rule for which the second object is preferred over the first one.

More formally, in approach (a) we define these sums of weights as:

$$W_> = \sum_{s(x,x')>0} \Phi(\boldsymbol{x}_i, \boldsymbol{x}_j) w_{ij}^{(p)}, \quad W_< = \sum_{s(x,x')<0} \Phi(\boldsymbol{x}_i, \boldsymbol{x}_j) w_{ij}^{(p)}, \quad W_0 = \sum_{\Phi(x_i,x_j)=0} w_{ij}^{(p)}.$$

Analogous definitions for approach (b) are:

$$W_> = \sum_{\substack{s(x,x')>0,\\ \Phi(x_i)\neq\Phi(x_j)}} w_{ij}^{(p)}, \quad W_< = \sum_{\substack{s(x,x')<0,\\ \Phi(x_i)\neq\Phi(x_j)}} w_{ij}^{(p)}, \quad W_0 = \sum_{\Phi(x_i)=\Phi(x_j)} w_{ij}^{(p)}.$$

Let us also denote $W = W_> + W_< + W_0$.

Now, we can state the minimization problem for approaches (a) and (b) by one mathematical formulation:

$$r_p = \arg\min_{\Phi,\alpha} (\mathrm{e}^{-\alpha} W_> + \mathrm{e}^{\alpha} W_< + W_0). \tag{8}$$

For given Φ_p, the problem of finding α_p is straightforward. After simple calculations, one obtains a closed-form solution:

$$\alpha_p = \frac{1}{2} \ln \frac{W_>}{W_<}. \tag{9}$$

To find Φ_p, we derive *impurity measure* $\mathcal{L}_p(\Phi)$ from (8) in such a way that the minimization problem no longer depends on α. Putting optimal value of α_p given by (9) into (8) results in the following impurity measure:

$$\mathcal{L}_p(\Phi) = 2\sqrt{W_> W_<} + W_0.$$

This can be further simplified by using short multiplication formulas and replacing W_0 by $W - W_> - W_<$ (remark that W is constant in a given iteration). Finally, we obtain:

$$\mathcal{L}_p(\Phi) = -\left|\sqrt{W_>} + \sqrt{W_<}\right|. \tag{10}$$

One can notice that the form of (10) resembles the measure used in SLIPPER [5] to generate rules in the binary classification case. A wide discussion on different rule impurity measures obtained in the boosting framework can be found in [9].

For given $\mathcal{L}_p(\Phi)$, the procedure constructing Φ_p resembles the way in which decision trees are built. The algorithm constructs only one path from the root to the leaf. At the beginning, Φ_p is empty and in each subsequent step an elementary condition minimizing $\mathcal{L}_p(\Phi)$ is added to Φ_p. This procedure ends if $\mathcal{L}_p(\Phi)$ cannot be decreased anymore. Let us underline that minimal value of $\mathcal{L}_p(\Phi)$ is a natural stop condition of the procedure and no other parameters have to be specified. This is due to the fact that the impurity measure represents a natural trade-off between misclassification and coverage of the rule.

The learning algorithm has three parameters that control its performance. The first one is the number P of rules to be generated. This is the main stop condition, but one can also consider other conditions, like the minimum value of rank loss computed over the training set. In this case, one can use the exponential rank loss or just the rank loss defined in the original way.

The second parameter, $\nu \in (0, 1]$, is responsible for shrinking [34] a newly generated rule with marginal function $r_p(\boldsymbol{x}) = \alpha_p \Phi_p(\boldsymbol{x})$ toward rules already present in the ensemble:

$$f_p(\boldsymbol{x}) = f_{p-1}(\boldsymbol{x}) + \nu \times r_p(\boldsymbol{x}).$$

This shrinkage parameter can be regarded as controlling the learning rate. For small ν, we obtain a more accurate but less interpretable ensemble, since we need to generate more rules.

The third parameter, $\zeta \in (0, 1]$, determines the fraction of training objects used in the procedure for finding Φ_p. This fraction of training objects is drawn without replacement, similarly as in [18]. Such an approach leads to a set of rules that are more diversified and less correlated. The diversification usually improves the performance of the ensemble. Moreover, finding Φ_p on a subsample reduces the computational cost. However, we pay once again the price of the interpretability.

At the end of this section, let us discuss some properties of the rule ensembles obtained in approaches (a) and (b).

In general, the rules of type (b) can be represented by the rules defined on pairs of objects (such as in approach (a)), but the inverse is not true. In approach (a), however, where elementary conditions used in the rules concern differences of attribute values for two objects, we assume that a given difference has the same intensity whatever the individual evaluations of the objects are. In other words, we assume that the difference between two consecutive values of each attribute has a constant intensity of preference in the entire scale of the attribute.

In approach (a), it is possible that just one rule can rank linearly all the objects if there exist strong monotonic relationships between attributes and the preference relation $\succ$ in the training set $\mathcal{T}$. This is not the case for the rules of type (b). In this case, we approximate the utility function by rules corresponding to piecewise-constant regions. Usually, we need a high number of rules to obtain a good approximation of the utility function. Otherwise, the resulting ranking would consist of many ties.

The clear advantage of approach (b) is lower computational complexity, since rules are defined in the original attribute space. In other words, the number of elementary conditions to be checked for a given attribute in rule construction procedure scales linearly with n. In approach (a), rules are defined in the attribute difference space, which results in checking the number of elementary conditions of order n^2.

6 A Toy Example

In this section, we present decision rules and linear rankings generated for the Thierry's choice problem [3], in the case of preference function learning and utility function learning. This problem concerns a ranking of 14 cars, described by means of five criteria (attributes): cost, acceleration, pick up, brakes, and road holding

Table 1 Data of the Thierry's choice problem [3]

#	Car	Cost	Acceleration	Pick up	Brakes	Road holding
x_1	Fiat Tipo	18,342	30.7	37.2	2.33	3
x_2	Alfa 33	15,335	30.2	41.6	2	2.5
x_3	Nissan Sunny	16,973	29	34.9	2.66	2.5
x_4	Mazda 323	15,460	30.4	35.8	1.66	1.5
x_5	Mitsubishi Colt	15,131	29.7	35.6	1.66	1.75
x_6	Toyota Corolla	13,841	30.8	36.5	1.33	2
x_7	Honda Civic	18,971	28	35.6	2.33	2
x_8	Opel Astra	18,319	28.9	35.3	1.66	2
x_9	Ford Escort	19,800	29.4	34.7	2	1.75
x_{10}	Renault 19	16,966	30	37.7	2.33	3.25
x_{11}	Peugeot 309 16V	17,537	28.3	34.8	2.33	2.75
x_{12}	Peugeot 309	15,980	29.6	35.3	2.33	2.75
x_{13}	Mitsubishi Galant	17,219	30.2	36.9	1.66	1.25
x_{14}	Renault 21	21,334	28.9	36.7	2	2.25

(interested reader can refer to [3] for the precise description of the meaning of these criteria). Evaluations of all 14 cars are presented in Table 1.

In the previous sections we assumed, without loss of generality, that the DM's preferences increase with increasing value of given criterion g_h, $h \in \{1, \ldots, m\}$. This is the case for criteria brakes and road holding. However, we can also consider criteria for which the DM's preferences increase with decreasing value of the criterion, which is the case for criteria cost, acceleration, and pick up. The only difference is that, for such a criterion, car $\boldsymbol{x}$ is at least as good as car $\boldsymbol{x}'$ if $g_h(\boldsymbol{x}) \leq g_h(\boldsymbol{x}')$.

Preference information is given by the DM in terms of a linear ranking of five cars: $\boldsymbol{x}_{11}$, $\boldsymbol{x}_3$, $\boldsymbol{x}_{13}$, $\boldsymbol{x}_9$, $\boldsymbol{x}_{14}$. Thus, we can consider 20 ordered pairs of cars. If the DM ranks car $\boldsymbol{x}_i$ higher than car $\boldsymbol{x}_j$, then car $\boldsymbol{x}_i$ is preferred to car $\boldsymbol{x}_j$. In this case, we have $s(\boldsymbol{x}_i, \boldsymbol{x}_j) = 1$. Analogously, if the DM ranks car $\boldsymbol{x}_i$ lower than car $\boldsymbol{x}_j$, then inverse preference occurs, which results in $s(\boldsymbol{x}_i, \boldsymbol{x}_j) = -1$. Thus, from the given ranking we get, for example, $s(\boldsymbol{x}_{11}, \boldsymbol{x}_{13}) = 1$ and $s(\boldsymbol{x}_9, \boldsymbol{x}_3) = -1$.

Below, we present a list of rules forming an ensemble generated for the problem of learning preference function $f(\cdot,\cdot)$ (approach (a)). The rules were iteratively added to the ensemble until preference-based rank loss, calculated according to (2), was reduced to zero for each pair of training objects. They involve elementary conditions in the attribute difference space (the difference is always calculated between the first and the second car in a pair). Decision of each rule concerns pairs covered by that rule. It can state whether the first car in a pair is preferred over the second one, or the second car in a pair is preferred over the first one. Moreover, in parentheses we give two characteristics of each rule, which are, respectively, weight of the rule and the ratio of covered positive (i.e., supporting rule's decision) and negative (i.e., opposing rule's decision) pairs of cars.

#	Rule
r_1:	*if* the cost difference between cars $\boldsymbol{x}$ and $\boldsymbol{x}'$ is $\leq 1{,}049.0$ *and* the brakes difference between cars $\boldsymbol{x}$ and $\boldsymbol{x}'$ is ≥ -0.5 *then* car $\boldsymbol{x}$ is preferred over car $\boldsymbol{x}'$ (1.46) (10:1)
r_2:	*if* the cost difference between cars $\boldsymbol{x}$ and $\boldsymbol{x}'$ is $\geq -1{,}049.0$ *and* the brakes difference between cars $\boldsymbol{x}$ and $\boldsymbol{x}'$ is ≤ 0.5 *then* car $\boldsymbol{x}'$ is preferred over car $\boldsymbol{x}$ (1.66) (10:1)
r_3:	*if* the road holding difference between cars $\boldsymbol{x}$ and $\boldsymbol{x}'$ is ≥ 0.0 *and* the cost difference between cars $\boldsymbol{x}$ and $\boldsymbol{x}'$ is $\leq 3{,}348.0$ *then* car $\boldsymbol{x}$ is preferred over car $\boldsymbol{x}'$ (0.48) (7:2)

The rules are compact, but at the cost of considering all pairwise comparisons that gives quadratic complexity with respect to the number of training objects. It can be noticed that rules r_1 and r_2 model indifference thresholds on criteria cost and brakes. Precisely, the indifference threshold for criterion cost is equal to 1,049, while the indifference threshold for criterion brakes is equal to 0.5. Positive difference on criterion cost up to 1,049 (i.e., in disfavor of the first car in a pair) and negative difference on criterion brakes up to −0.5 (i.e., in disfavor of the first car in a pair, again) did not prevent from deciding that the first car is preferred over the second one. Analogously, negative difference on criterion cost not lower than −1,049 (i.e., in favor of the first car in a pair) and positive difference on criterion brakes not greater than 0.5 (i.e., in favor of the first car in a pair, again) did not prevent from deciding that the second car is preferred over the first one. Different weights of rules r_1 and r_2 are an artifact of the minimization procedure.

The final ranking of cars is built using the NFS procedure, which operates on a preference structure resulting from application of the above three rules to the set of all pairs of cars. For the five cars included in the linear ranking given by the DM, the final ranking is the following:

Rank	Car	*NFS*
1	x_{11}	8
2	x_3	4
3	x_{13}	0
4	x_9	-4
5	x_{14}	-8

This ranking is exactly the same as the input linear ranking. Moreover, let us analyze how the net flow score is calculated for a single car. The following table shows which rules cover pairs involving car $\boldsymbol{x}_{11}$, values of preference function f for these pairs, and corresponding values of function $\text{sgn}(f)$.

Pair of cars	Covering rules	$f = \sum_i r_i$	sgn(f)
$(\boldsymbol{x}_{11}, \boldsymbol{x}_3)$	r_1, r_2, r_3	0.28	1
$(\boldsymbol{x}_{11}, \boldsymbol{x}_9)$	r_1, r_3	1.94	1
$(\boldsymbol{x}_{11}, \boldsymbol{x}_{13})$	r_1, r_3	1.94	1
$(\boldsymbol{x}_{11}, \boldsymbol{x}_{14})$	r_1, r_3	1.94	1
$(\boldsymbol{x}_3, \boldsymbol{x}_{11})$	r_1, r_2	−0.2	−1
$(\boldsymbol{x}_9, \boldsymbol{x}_{11})$	r_2	−1.66	−1
$(\boldsymbol{x}_{13}, \boldsymbol{x}_{11})$	r_2	−1.66	−1
$(\boldsymbol{x}_{14}, \boldsymbol{x}_{11})$	r_2	−1.66	−1

We compute the net flow score for $\boldsymbol{x}_{11}$ according to (5):

$$\text{NFS}(\boldsymbol{x}_{11}) = \sum_{x_i \in \{x_3, x_9, x_{13}, x_{14}\}} \Big[\text{sgn}(f(\boldsymbol{x}_{11}, \boldsymbol{x}_i)) - \text{sgn}(f(\boldsymbol{x}_i, \boldsymbol{x}_{11}))\Big].$$

Since for all pairs $(\boldsymbol{x}_{11}, \boldsymbol{x}_i)$, $\text{sgn}(f(\boldsymbol{x}_{11}, \boldsymbol{x}_i)) = 1$ and for all pairs $(\boldsymbol{x}_i, \boldsymbol{x}_{11})$, $\text{sgn}(f(\boldsymbol{x}_i, \boldsymbol{x}_{11})) = -1$, we obtain $NFS(\boldsymbol{x}_{11}) = 8$.

For the entire set of 14 cars, we obtained the following final ranking:

Rank	Cars	NFS
1	$\boldsymbol{x}_{12}$	21
2	$\boldsymbol{x}_2, \boldsymbol{x}_{10}$	19
3	$\boldsymbol{x}_3$	9
4	$\boldsymbol{x}_6, \boldsymbol{x}_{11}$	6
5	$\boldsymbol{x}_1$	5
6	$\boldsymbol{x}_5$	0
7	$\boldsymbol{x}_4$	−4
8	$\boldsymbol{x}_7$	−6
9	$\boldsymbol{x}_{13}$	−12
10	$\boldsymbol{x}_8$	−18
11	$\boldsymbol{x}_9$	−21
12	$\boldsymbol{x}_{14}$	−24

It should be noted that the NFS procedure inverted the ranks of cars $\boldsymbol{x}_{11}$ and $\boldsymbol{x}_3$ – in the above ranking car $\boldsymbol{x}_3$ is ranked higher than car $\boldsymbol{x}_{11}$. Such inversion was possible because cars $\boldsymbol{x}_3$ and $\boldsymbol{x}_{11}$ are incomparable. In fact, car $\boldsymbol{x}_3$ is better on criteria cost and brakes, while car $\boldsymbol{x}_{11}$ is better on criteria acceleration, pick up and road holding. In general, although NFS is a popular and intuitive ranking procedure with good properties, it does not fully respect preference information given by the DM. This is, however, also the case for other ranking procedures different from NFS [4, 57], especially when the preference structure induced by the preference function $f(\cdot, \cdot)$ is intransitive and contains cycles, which is the case in our toy example.

Now, let us consider the problem of learning utility function $f(\cdot)$ (approach (*b*)). The list of rules generated for this problem is the following:

#	Rule
1.	*if* cost of car $\boldsymbol{x}$ is $\leq$ 18,668.5 *then* the utility of car $\boldsymbol{x}$ increases by 0.973 (3)
2.	*if* brakes of car $\boldsymbol{x}$ is $\leq$ 2.165 *then* the utility of car $\boldsymbol{x}$ decreases by 0.753 (3)
3.	*if* acceleration of car $\boldsymbol{x}$ is $\leq$ 28.6 *then* the utility of car $\boldsymbol{x}$ increases by 0.520 (1)
4.	*if* cost of car $\boldsymbol{x}$ is $\leq$ 20,567.0 *then* the utility of car $\boldsymbol{x}$ increases by 0.489 (4)

As before, the rules were iteratively added to the ensemble until utility-based rank loss, calculated according to (6), was reduced to zero for each pair of training objects. In parentheses, we show the number of cars covered by each rule. The rules involve elementary conditions in the space of original attributes. Decision of each rule defines how the utility of a covered car is increased or decreased. Obviously, the final utility of a car is a sum of utilities given to it by different covering rules.

In the case of the utility function learning, we do not need to apply NFS procedure since utility function already induces a linear ranking of cars. Final ranking of the 5 cars for which the DM expressed her/his preferences is calculated as follows:

Rank	Car	r_1	r_2	r_3	r_4	$f = \sum_i r_i$
1	$\boldsymbol{x}_{11}$	0.973		0.520	0.489	1.982
2	$\boldsymbol{x}_3$	0.973			0.489	1.462
3	$\boldsymbol{x}_{13}$	0.973	−0.753		0.489	0.709
4	$\boldsymbol{x}_9$		−0.753		0.489	−0.264
5	$\boldsymbol{x}_{14}$		−0.753			−0.753

For the entire set of 14 cars, utility function induces the following final ranking:

Rank	Car	r_1	r_2	r_3	r_4	$f = \sum_i r_i$
1	$\boldsymbol{x}_{11}$	0.973		0.520	0.489	1.982
2	$\boldsymbol{x}_1, \boldsymbol{x}_3, \boldsymbol{x}_{10}, \boldsymbol{x}_{12}$	0.973			0.489	1.462
3	$\boldsymbol{x}_7$			0.520	0.489	1.009
4	$\boldsymbol{x}_2, \boldsymbol{x}_4, \boldsymbol{x}_5, \boldsymbol{x}_6, \boldsymbol{x}_8, \boldsymbol{x}_{13}$	0.973	−0.753		0.489	0.709
5	$\boldsymbol{x}_9$		−0.753		0.489	−0.264
6	$\boldsymbol{x}_{14}$		−0.753			−0.753

As it can be seen in the above two tables, preference information given by the DM in terms of a linear ranking of 5 cars have been preserved in both final rankings obtained in approach (b). However, we can notice that the final ranking of all 14 cars contains a lot of ties. This is due to the fact that the utility function is here approximated by piecewise-constant regions (rules). To make a finer distinction, we would need more rules.

7 Massive Data Experiment

To verify how the proposed approaches behave in practice, we performed computational experiments on real data sets. The first data set concerns credit rating [11]. Moreover, since the recent research in the field of preference learning is motivated by an increasing role of modeling preferences in the recommender systems and information retrieval, we chose two other massive data sets. The first one comes from movie recommender system MovieLens [33], while the second one concerns ranking of text documents from 20 Newsgroups data set [39].

Before we move into details of each data set, let us describe how the performance of the algorithms is assessed in the experiments. For each data set, a separate testing set X is given and the algorithms are then applied on this set. The output of each algorithm is the final ranking of the objects from X. For any pair of the objects $(\boldsymbol{x}, \boldsymbol{x}')$, the error occurs whenever $s(\boldsymbol{x}, \boldsymbol{x}') = -1$ and $\boldsymbol{x}$ is ranked higher than $\boldsymbol{x}'$, or whenever $s(\boldsymbol{x}, \boldsymbol{x}') = 1$ and $\boldsymbol{x}$ is ranked lower than $\boldsymbol{x}'$ in the final ranking. In the case of tie in the final ranking, we count a half of the error. Then, the performance of the algorithm is the total number of errors over all pairs from $X \times X$ normalized by the number of all pairs $(\boldsymbol{x}, \boldsymbol{x}')$ such that $s(\boldsymbol{x}, \boldsymbol{x}') = 1$ or $s(\boldsymbol{x}, \boldsymbol{x}') = -1$. Thus, the performance measure varies between 0 (no errors) and 1 (all preference relations are reversed).

Note that we do not assess the performance of the intermediate steps, such as preference function $f(\boldsymbol{x}, \boldsymbol{x}')$. Only the final ranking (based on $f(\boldsymbol{x})$ for *RankRules* and on $NFS(\boldsymbol{x})$ for *PrefRules*) is compared with the testing data.

All the algorithms used in this experiment were implemented using the Weka environment [58].

7.1 Credit Rating Data

The credit rating data set is described in [11]. It is based on the rating of 1328 firms issued by Qui Credit Assessment Ltd., a UK credit rating agency. On the basis of this rating, the firms are classified into five groups of risk: "Secure", "Stable", "Normal", "Unstable", and "High risk". The firms are also described by 10 numerical attributes, such as "solvency ratio" or "return on total assets". The risk groups constitute the preference information: each time a given firm $\boldsymbol{x}$ is classified into a group higher

Table 2 Results of the *RankRules* and *PrefRules* on the credit rating data

# Rules	RankRules		PrefRules	
	Error	Time [s]	Error	Time [s]
10	0.095	0.0878	0.054	175.5468
20	0.077	0.1686	0.054	318.4124
50	0.066	0.4472	0.054	602.303
80	0.063	0.7549	0.057	911.6686
100	0.058	0.9688	0.052	1067.0126

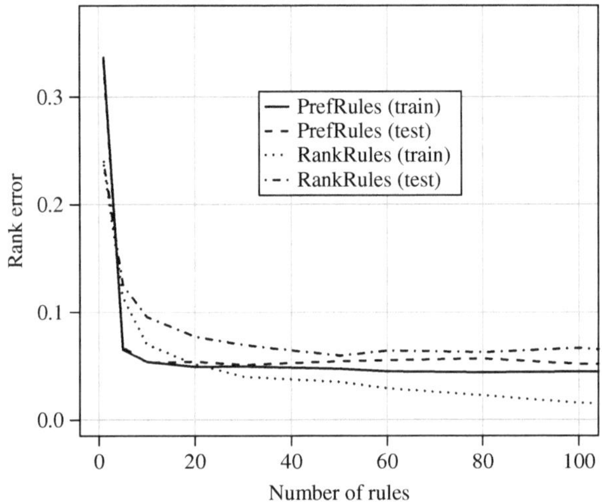

Fig. 1 Performance of *RankRules* and *PrefRules* on the credit rating data set as a function of the number of rules varying from 0 to 100

than that of firm x', we consider that x is preferred over x' (i.e., $s(x,x') = 1$). Similarly, if x is classified into a lower group than that of x', we have $s(x,x') = -1$.

We ran *RankRules* and *PrefRules*, setting all the algorithms' parameters to default values (those values were chosen on the basis of the past experiments with the algorithms, and were not optimized to any extent on the data sets described in this paper). Only the number of rules in the ensemble varied between 10 and 100. The results are given in Table 2. The error rate was calculated as described above, by randomly drawing 5 times testing sets containing 30% of objects. We also visualize the error rate as a function of the number of rules – see Fig. 1.

One can see that both algorithms perform similarly, with a slight advantage to *PrefRules*. On the other hand, *PrefRules* is much slower, as it works on the pairs of objects (constructing preference function) instead of single objects, which is the case of *RankRules* (constructing utility function). Thus, *PrefRules* scales quadratically with the size of the data. However, *PrefRules* needs much less rules to achieve its best performance – even 10 rules is enough – while *RankRules* works best with 100 rules. Therefore, in the next experiments, we decided to run *PrefRules* with smaller ensemble sizes. Note that *RankRules* is able to decrease the training error more than

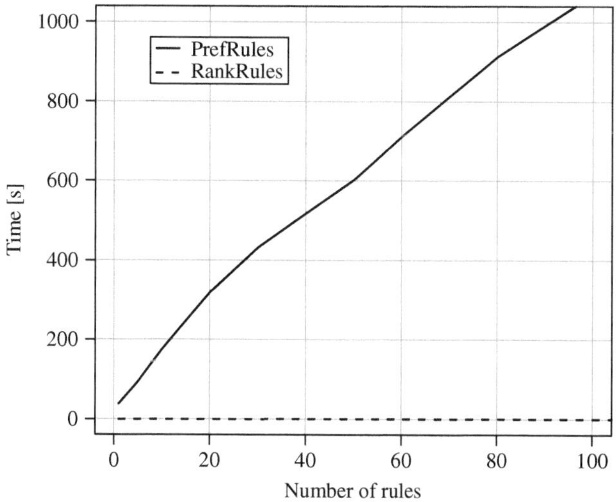

Fig. 2 Computational time on the credit rating data set as a function of the number of rules varying from 0 to 100

PrefRules. This is due to the fact that *PrefRules* uses differences of attribute values in the condition part of decision rules. Therefore, the difference between, e.g., 10 and 11 is the same as between 1 and 2, so the rules cannot model a variable intensity of preference in the scales of attributes.

We also show in Fig. 2 how computational time increases with increasing number of rules. The dependence is roughly linear. Notice that the time of training *RankRules* is negligible comparing to that of *PrefRules*, since the former scales linearly with the number of objects, while the latter scales quadratically.

7.2 Recommender System

The first massive data set concerns movie recommender system MovieLens in which users are asked to rate movies on an ordinal scale, assigning to each movie from one to five "stars". We used "100 K" version of the data set, consisting of 100,000 ratings on 1,682 movies given by 943 users. The goal is to sort movies unseen by the user according to his/her preferences. The task is mapped into our framework in the following way. Let us fix a user and let $\boldsymbol{x}$ and $\boldsymbol{x}'$ be two movies. Then $s(\boldsymbol{x},\boldsymbol{x}') = 1$ if the user ranks movie $\boldsymbol{x}$ higher than movie $\boldsymbol{x}'$; $s(\boldsymbol{x},\boldsymbol{x}') = -1$ if movie $\boldsymbol{x}$ is ranked lower than movie $\boldsymbol{x}'$.

We treated the problem of recommendation of movies as a machine learning problem in the following way. We chose 200 most popular users (i.e., users who rated the greatest number of movies) and removed the remaining users from the data set. Then, for each user, we considered ratings of this user as ranks while the

Table 3 Results of the *RankRules* and *PrefRules* on MovieLens

# Algorithm	Error	Time [s]
RankRules	0.29	2,823.691
PrefRules	0.31	99,865.923

ratings of the other 199 users formed the description of movies in terms of attributes. In other words, each movie is an object, described by 199 ordinal attributes and a decision attribute associated with a given user. In such a data table, a lot of values are missing since none of the users ranks all of the movies. We do not fill these missing values, since decision rules can easily handle them by treating a missing value on an attribute as a value which never satisfies a rule condition associated with the corresponding attribute.

The above procedure leads to 200 separate problems, one for each user. The performance of the algorithms is averaged over all those problems. For each problem, the data is separated into the training and testing set in a way recommended in the MovieLens data (for every user, exactly 10 ratings are selected to the testing set, while all the other ratings constitute the training set). The number of rules for *PrefRules* was set to 100, while for *RankRules*, it was set to 500. The results are presented in Table 3.

The difference in the performance of both algorithms is small and statistically insignificant at significance level $\alpha = 0.05$ (value of Student's t statistic calculated by treating difference in performance on each data set as a separate observation equals 1.397). On the other hand, the computational time for *PrefRules* is much higher than for *RankRules*, which makes *RankRules* more preferable for this kind of problems. The main advantage of *PrefRules* is a more concise model – it needs much less rules than *RankRules*.

7.3 Text Ranking

The 20 Newsgroups data set is a collection of approximately 20,000 text messages, partitioned evenly across 20 different newsgroups. We used the 18,828 version of the data set [1], where there are no duplicates and each message contains only "From" and "Subject" headers. We considered five binary problems:

1. class `comp.sys.ibm.pc.hardware` versus `comp.sys.mac.hardware`,
2. class `alt.atheism` versus `soc.religion.christian`,
3. class `rec.sport.baseball` versus `rec.sport.hockey`,
4. class `talk.politics.misc` versus `talk.politics.mideast`,
5. class `rec.autos` versus `rec.motorcycles`.

In each problem, we chose similar categories to make the problem harder. Similarly to previous data sets, we treated ordered classification as ranking by choosing any of the two considered classes as more preferred. Thus, document $\boldsymbol{x}$ is preferred over document $\boldsymbol{x}'$ if $\boldsymbol{x}$ belongs to a more preferred class, while $\boldsymbol{x}'$ – to a less preferred one.

Table 4 Results of the *RankRules*, *PrefRules* and SVM on the 20 Newsgroups data

# Data set	*RankRules*		*PrefRules*		SVM	
	Error	Time [s]	Error	Time [s]	Error	Time [s]
Atheism vs. Christian	0.008	10.5	0.017	227.2	0.047	2.7
Pc vs. Mac	0.033	9.4	0.057	192.4	0.115	3.1
Autos vs. Motorcycles	0.016	9.8	0.023	202.1	0.043	1.9
Baseball vs. Hockey	0.007	9.7	0.017	209.9	0.029	1.5
Misc vs. Mideast	0.011	9.7	0.024	208.3	0.036	1.9

To represent messages in a bag-of-words representation, for each binary problem we performed the feature selection as follows. First, we split message texts into tokens and removed stopwords. Second, we did stemming using Porter's stemmer. Third, we removed terms that occurred in the considered data set less than 20 times. Fourth, from the remaining terms we selected 500 terms with the highest frequency of occurrence. Finally, we removed terms that occurred in less than 10 messages. Thus, as a final result, each message was represented in a feature space, where attribute value of 1 corresponded to the presence of a given term, and attribute value of 0 – to its absence (i.e., we ignored term frequencies).

To evaluate the performance of the algorithms, we split the data sets into the training and testing part. We used both *RankRules* and *PrefRules*, along with a state-of-the-art Support Vector Machine [56] classifier (with linear kernel and all the other parameters set to default values), which serves as a baseline for the comparison. The results are given in Table 4. *RankRules* clearly outperforms *PrefRules* on every data set, both in terms of error rate and training time, and is a true winner of this experiment. On the other hand, both of our algorithms outperform SVM, proving their usefulness in the text mining problems. The differences in performance are statistically significant at the significance level $\alpha = 0.05$ – treating difference in results for each data set as a separate observation. Student's t statistic equals 4.05 when comparing *RankRules* with *PrefRules*, and 3.03 when comparing *PrefRules* with SVM (both above the critical value 2.57, using Bonferroni correction).

8 Conclusions

Preference Learning and Multiple Attribute Decision Aiding share many concepts and methodological issues. In this paper, we have considered a ranking problem, aiming to propose a methodology having good properties from both learning and decision aiding perspectives.

The proposed methodology is based on learning of an ensemble of decision rules from decision examples given by the DM in terms of pairwise comparisons of some objects. Decision rule models are of particular interest in Multiple Attribute Decision Aiding, since decision aiding procedures using these models are "glass box" procedures providing a clear justification of recommended decisions in the language of the DM. Moreover, the decision rule model is the most general among all known preference models. Decision rules can also be efficiently learned in the boosting framework. As shown in the experiment on a large data set, the rule-based learning algorithms are competitive to SVM.

We proposed two approaches to ranking learning. In the first one (referred to as approach (a)), the rule ensemble is represented by a preference function defining a binary preference relation on the set of objects. The result of application of this function on all pairs of objects to be ranked is then exploited using the Net Flow Score procedure, giving a linear ranking of objects. In the second approach (referred to as approach (b)), the rule ensemble is represented by a utility function for single objects. This utility function directly induces a linear ranking of objects.

The rules of type (a) concern direct comparison of two objects, while the rules of type (b) determine a change of utility for single objects. This is why the former rules are often treated as more intuitive in solving multiple attribute ranking problems. However, the ensembles based on the latter rules perform better, as shown in the massive data experiments – their generation is much faster and their accuracy is not significantly worse; rather, it is better.

Acknowledgements The authors wish to acknowledge financial support from the Ministry of Science and Higher Education, grant N N519 314435.

References

1. http://people.csail.mit.edu/jrennie/20Newsgroups/20news-18828.tar.gz
2. D. Bouyssou, Ranking methods based on valued preference relations: A characterization of the net flow method. Eur. J. Oper. Res. **60**, 61–67 (1992)
3. D. Bouyssou, T. Marchant, M. Pirlot, P. Perny, A. Tsoukiás, P. Vincke (eds.), *Evaluation and Decision Models: A critical perspective*, Chap. 6.1 (Kluwer, 2000), pp. 91–101
4. D. Bouyssou, P. Vincke, Ranking alternatives on the basis of preference relations: A progress report with special emphasis on outranking relations. J. Multi-Crit. Dec. Anal. **6**, 77–85 (1997)
5. W.W. Cohen, Y. Singer, A simple, fast, and effective rule learner, in *Proceedings of National Conference on Artificial Intelligence (AAAI 1999)* (AAAI/MIT, Orlando, USA, 1999), pp. 335–342
6. K. Dembczyński, S. Greco, W. Kotłowski, R. Słowiński, Statistical model for rough set approach to multicriteria classification, in *Knowledge Discovery in Databases (PKDD 2007)*, vol. 4702 *Lecture Notes in Artificial Intelligence*, ed. by J.N. Kok, J. Koronacki, R.L. de Mántaras, S. Matwin, D. Mladenič, A. Skowron (Springer, Warsaw, Poland, 2007), pp. 164–175
7. K. Dembczyński, S. Greco, R. Słowiński, Rough set approach to multiple criteria classification with imprecise evaluations and assignments. Eur. J. Oper. Res. **198**(2), 626–636 (2009)
8. K. Dembczyński, W. Kotłowski, R. Słowiński, Ensemble of decision rules for ordinal classification with monotonicity constraints, in *Rough Sets and Knowledge Technology 2008 (RSKT 2008)*, vol. 5009, *Lecture Notes in Artificial Intelligence*, ed. by G. Wang, T. Li, J.W. Grzymała-Busse, D. Miao, A. Skowron, Y. Yao (Springer, Chengdu, China, 2008), pp. 260–267
9. K. Dembczyński, W. Kotłowski, R. Słowiński, A general framework for learning an ensemble of decision rules, in *Proceedings of the LeGo '08: From Local Patterns to Global Models, ECML/PKDD 2008 Workshop*, ed. by J. Fürnkranz, A. Knobbe (Antwerp, Belgium, 2008), pp. 17–36

10. K. Dembczyński, W. Kotłowski, R. Słowiński, Ordinal classification with decision rules, in *Mining Complex Data (MCD 2007)*, vol. 4944, *Lecture Notes in Artificial Intelligence*, ed. by Z.W. Raś, S. Tsumoto, D. Zighed (Springer, 2008), pp. 169–181
11. M. Doumpos, F. Pasiouras, Developing and testing models for replicating credit ratings: A multicriteria approach. Comput. Econ. **25**, 327–341 (2005)
12. J. Figueira, S. Greco, M. Ehrgott (eds.), *Multiple Criteria Decision Analysis: State of the Art Surveys* (Springer, Berlin, 2005)
13. J. Figueira, S. Greco, R. Słowiński, Building a set of additive value functions representing a reference preorder and intensities of preference: Grip method. Eur. J. Oper. Res. **195**(2), 460–486 (2009)
14. P. Fortemps, S. Greco, R. Słowiński, Multicriteria decision support using rules that represent rough-graded preference relations. Eur. J. Oper. Res. **188**(1), 206–223 (2008)
15. E. Frank, M. Hall, A simple approach to ordinal classification. in *Proceedings of the European Conference on Machine Learning (ECML 2001)*, vol. 2167, *Lecture Notes in Artificial Intelligence*, ed. by L. De Raedt, P.A. Flach (Springer, Freiburg, Germany, 2001), pp. 145–157
16. Y. Freund, R. Iyer, R.E. Schapire, Y. Singer, An efficient boosting algorithm for combining preferences. J. Mach. Learn. Res. **6**(4), 933–969 (2003)
17. Y. Freund, R.E. Schapire, A decision-theoretic generalization of on-line learning and an application to boosting. J. Comput. Syst. Sci. **55**(1), 119–139 (1997)
18. J.H. Friedman, B.E. Popescu, Importance sampled learning ensembles. Research report, Department of Statistics, Stanford University, 2003
19. J. Fürnkranz, E. Hüllermeier, Pairwise preference learning and ranking, in *Proceedings of the European Conference on Machine Learning (ECML 2003)*, vol. 2837 *Lecture Notes in Artificial Intelligence*, ed. by N. Lavrac, D. Gamberger, L. Todorovski, H. Blockeel (Springer, Cavtat-Dubrovnik, Croatia, 2003), pp. 145–156
20. S. Giove, S. Greco, B. Matarazzo, R. Słowiński, Variable consistency monotonic decision trees, in *Rough Sets and Current Trends in Computing (RSCTC 2002)*, vol. 2475 *Lecture Notes in Artificial Intelligence*, ed. by J.J. Alpigini, J.F. Peters, A. Skowron, N. Zhong (Springer, Malvern, USA, 2002), pp. 247–254
21. S. Greco, B. Matarazzo, R. Słowiński, A new rough set approach to evaluation of bankruptcy risk, in *Operational Tools in the Management of Financial Risks*, ed. by C. Zopounidis (Kluwer, Dordrecht, 1998), pp. 121–136
22. S. Greco, B. Matarazzo, R. Słowiński, Rough approximation of a preference relation by dominance relations. Eur. J. Oper. Res. **117**, 63–83 (1999)
23. S. Greco, B. Matarazzo, R. Słowiński, Dealing with missing data in rough set analysis of multi-attribute and multi-criteria decision problems, in *Decision Making: Recent Developments and Worldwide Applications*, ed. by S.H. Zanakis, G. Doukidis, C. Zopounidis (Kluwer, Dordrecht, 2000), pp. 295–316
24. S. Greco, B. Matarazzo, R. Słowiński, Rough sets theory for multicriteria decision analysis. Eur. J. Oper. Res. **129**, 1–47 (2001)
25. S. Greco, B. Matarazzo, R. Słowiński, Axiomatic characterization of a general utility function and its particular cases in terms of conjoint measurement and rough-set decision rules. Eur. J. Oper. Res. **158**(2), 271–292 (2004)
26. S. Greco, B. Matarazzo, R. Słowiński, Decision rule approach, in *Multiple Criteria Decision Analysis: State of the Art Surveys*, Chap. 13, ed. by J. Figueira, S. Greco, M. Ehrgott (Springer, Berlin, 2005), pp. 507–561
27. S. Greco, B. Matarazzo, R. Słowiński, Preference representation by means of conjoint measurement and decision rule model, in *Aiding Decisions with Multiple Criteria - Essays in Honor of Bernard Roy*, ed. by D. Bouyssou, E. Jacquet-Lagrèze, P. Perny, R. Słowiński, D. Vanderpooten, P. Vincke (Kluwer, Boston, 2005), pp. 263–313
28. S. Greco, B. Matarazzo, R. Słowiński, Dominance-based rough set approach to interactive multiobjective optimization, in *Multiobjective Optimization: Interactive and Evolutionary Approaches*, Chap. 5, ed. by J. Branke, K. Deb, K. Miettinen, R. Słowiński (Springer, Berlin, 2008), pp. 121–156

29. S. Greco, B. Matarazzo, R. Słowiński, Granular computing for reasoning about ordered data: the dominance-based rough set approach, in *Handbook of Granular Computing*, Chap. 15, ed. by W. Pedrycz, A. Skowron, V. Kreinovich (Wiley, Chichester, 2008), pp. 347–373
30. S. Greco, B. Matarazzo, R. Słowiński, Dominance-based rough set approach to decision under uncertainty and time preference. Ann. Oper. Res. (2009). doi: 10.1007/s10479-009-0566-8
31. S. Greco, V. Mousseau, R. Słowiński, Ordinal regression revisited: multiple criteria ranking using a set of additive value functions. Eur. J. Oper. Res. **191**(2), 415–435 (2008)
32. S. Greco, R. Słowiński, J. Figueira, V. Mousseau, Robust ordinal regression, in *New Advances in Multiple Criteria Decision Analysis*, ed. by M. Ehrgott, J. Figueira, S. Greco, (Springer, Berlin, 2009)
33. http://www.grouplens.org/node/73.
34. T. Hastie, R. Tibshirani, J.H. Friedman, *Elements of Statistical Learning: Data Mining, Inference, and Prediction* (Springer, Berlin, 2003)
35. R. Herbrich, T. Graepel, K. Obermayer, Regression models for ordinal data: A machine learning approach. Technical report TR-99/03, TU Berlin, 1999
36. E. Jacquet-Lagrèze, Y. Siskos, Assessing a set of additive utility functions for multicriteria decision making: The UTA method. Eur. J. Oper. Res. **10**, 151–164 (1982)
37. T. Joachims, Optimizing search engines using clickthrough data, in *Proceedings of the International Conference on Knowledge Discovery and Data Mining (ACM SIGKDD 2002)* (ACM, Alberta, Canada, 2002), pp. 133–142
38. W. Kotłowski, K. Dembczyński, S. Greco, R. Słowiński, Stochastic dominance-based rough set model for ordinal classification. Inf. Sci. **178**(21), 4019–4037 (2008)
39. K. Lang, Learning to filter netnews, in *Proceedings of the International Conference on Machine Learning (ICML 1995)*, ed. by A. Prieditis, Stuart J. Russell (Morgan Kaufmann, Tahoe City, USA, 1995), pp. 331–339
40. P. Langley, H.A. Simon, Fielded applications of machine learning, in *Machine Learning and Data Mining*, ed. by Ryszard S. Michalski, I. Bratko, M. Kubat (Wiley, New York, 1998), pp. 113–129
41. J.G. March, Bounded rationality, ambiguity, and the engineering of choice, in *Decision Making, Descriptive, Normative and Prescriptive Interactions*, ed. by David E. Bell, H. Raiffa, A. Tversky (Cambridge University Press, New York, 1988), pp. 33–58
42. R.S. Michalski, A theory and methodology of inductive learning, in *Machine Learning: An Artificial Intelligence Approach*, ed. by R.S. Michalski, J.G. Carbonell, T.M. Mitchell (Tioga Publishing, Palo Alto, 1983), pp. 83–129
43. V. Mousseau, R. Słowiński, Inferring an ELECTRE TRI model from assignment examples. J. Global Optim. **12**(2), 157–174 (1998)
44. http://www.netflixprize.com
45. Z. Pawlak, *Rough Sets. Theoretical Aspects of Reasoning about Data* (Kluwer, Dordrecht, 1991)
46. Z. Pawlak, R. Słowiński, Rough set approach to multi-attribute decision analysis. Eur. J. Oper. Res. **4**(3), 443–459 (1994)
47. J. Rennie, N. Srebro, Loss functions for preference levels: Regression with discrete ordered labels, in *Proceedings of the IJCAI Multidisciplinary Workshop on Advances in Preference Handling* (Edinburgh, Scotland, 2005)
48. B. Roy, *Méthodologie Multicritère d'Aide à la Décision* (Economica, Paris, 1985)
49. B. Roy, D. Bouyssou, *Aide Multicritère à la Décision: Méthodes et Cas* (Economica, Paris, 1993)
50. P.J.H. Shoemaker, The expected utility model: its variants, purposes, evidence and limitations. J. Econ. Lit. **20**, 529–562 (1982)
51. R. Słowiński, Rough set learning of preferential attitude in multi-criteria decision making, in *Methodologies for Intelligent Systems*, vol. 689, *Lecture Notes in Artificial Intelligence*, ed. by J. Komorowski, Z. Raś (Springer, Berlin, 1993), pp. 642–651
52. R. Słowiński, S. Greco, B. Matarazzo, Axiomatization of utility, outranking and decision-rule preference models for multiple-criteria classification problems under partial inconsistency with the dominance principle. Control Cybern. **31**(4), 1005–1035 (2002)

53. R. Słowiński, S. Greco, B. Matarazzo, Mining decision-rule preference model from rough approximation of preference relation, in *Proceedings of the IEEE Annual International Conference on Computer Software and Applications (COMPSAC 2002)* (Oxford, UK, 2002), pp. 1129–1134
54. R. Słowiński, S. Greco, B. Matarazzo, Rough set based decision support. in *Introductory Tutorials on Optimization, Search and Decision Support Methodologies*, Chap. 16, ed. by E. Burke, G. Kendall (Springer, New York, 2005), pp. 457–524
55. N. Srebro, J. Rennie, T. Jaakkola, Maximum margin matrix factorizations, in *Advances in Neural Information Processing Systems 17 (NIPS 2004)* (MIT, 2005), pp. 1329–1336
56. V. Vapnik, *Statistical Learning Theory*, (Wiley-Interscience, 1998)
57. P. Vincke, Exploitation of a crisp relation in a ranking problem. Theory Decis. **32**, 221–240 (1992)
58. I.H. Witten, E. Frank, *Data Mining: Practical machine learning tools and techniques*, (2nd edn.) (Morgan Kaufmann, San Francisco, 2005)

Part IV
Preferences in Multi-Attribute Domains

Learning Lexicographic Preference Models

Fusun Yaman, Thomas J. Walsh, Michael L. Littman, and Marie desJardins

Abstract Lexicographic preference models (LPMs) are one of the simplest yet most commonly used preference representations. In this chapter, we formally define LPMs and present learning algorithms for mining these models from data. In particular, we study a greedy algorithm that produces a "best guess" LPM that is consistent with the observations and two voting-based algorithms that approximate the target using the votes of a *collection* of consistent LPMs. In addition to our theoretical analyses of these algorithms, we empirically evaluate their performance under different conditions. Our results show that voting algorithms outperform the greedy method when the data is noise-free. The dominance is more significant when the training data is scarce. However, the performance of the voting algorithms quickly decays with even a little noise, whereas the greedy algorithm is more robust. Inspired by this result, we adapt one of the voting methods to consider the amount of noise in an environment and empirically show that the modified voting algorithm performs as well as the greedy approach even with noisy observations. We also introduce an intuitive yet powerful learning bias to prune some of the possible LPMs. We demonstrate how this learning bias can be used with variable and model voting and show that the learning bias improves learning performance significantly, especially when the number of observations is small.

F. Yaman (✉)
BBN Technologies, 10 Moulton St. Cambridge, MA 02138, USA
e-mail: fusun@bbn.com

T.J. Walsh and M.L. Littman
Rutgers University, Department of Computer Science, Piscataway, NJ 08854, USA
e-mail: thomaswa@cs.rutgers.edu, mlittman@cs.rutgers.edu

M. desJardins
University of Maryland Baltimore County, Computer Science and Electrical Engineering Department, Baltimore, MD 21250, USA
e-mail: mariedj@cs.umbc.edu

J. Fürnkranz and E. Hüllermeier (eds.), *Preference Learning*,
DOI 10.1007/978-3-642-14125-6_12,

1 Introduction

Lexicographic preference models (LPMs) are one of the simplest preference representations. An LPM defines an order of importance on the variables that describe the objects in a domain and uses this order to make preference decisions. For example, when choosing between two flights (with the same price), a traveler may prefer a nonstop flight to a flight with one or more stops. Among the nonstop flights, the traveler prefers a morning flight to an afternoon flight. This set of preferences can be represented as an LPM. Similarly, consider the task of picking a movie. Typically, the most important feature one considers is the genre of the movie (e.g., action, sci-fi, or romance). Then among the movies in the preferred category, one chooses the one with the best reviews, if there are still many movies with similar reviews, one might examine to a feature such as the movie's director or lead actor.

Despite the simplicity of LPMs, several studies on human decision making [4,9,14] have experimentally demonstrated that when facing equal cost alternatives, humans often make decisions using lexicographic reasoning instead of more mathematically sophisticated methods, such as linear additive value maximization [6].

In this chapter, we formally define LPMs and present learning algorithms for mining these models from data. In particular, we study a greedy method [12] that can discover one of many LPMs that are consistent with preference observations and two voting-based methods [16] that sample from the set of consistent LPMs and use the votes of this ensemble to predict the preferred outcome. In addition to our theoretical analyses of these algorithms, we empirically evaluate their performance under different conditions.

It is highly likely that a pick-any-consistent model approach (like the greedy approach we study) will produce poor approximations of the target when there are few observations. Our experiments show that voting algorithms outperform the average and worst-case performance of the greedy algorithm when the training data is scarce but noise-free.

To further improve the performance of the learning algorithms when the number of observations is small, we investigate the effects of using background knowledge, to impose a bias on learning. We focus on an intuitive yet powerful learning bias that defines equivalence classes on the variables, indicating the most important set of variables, the second most important set, and so on. We show how this learning bias can be used with voting algorithms and show that the learning bias improves learning performance significantly, especially when the number of observations is small.

We also investigate the effect of imperfect data on the learning algorithms. We consider two kinds of imperfections: *faulty* observations (noise) and *hidden ties* (ties that are broken arbitrarily). Our empirical evaluation demonstrates that all of the algorithms we consider are robust in the presence of hidden ties. However, even a small number of faulty observations significantly reduce the performance of the voting algorithms. On the other hand, the greedy algorithm is resilient: that is, the performance decline is proportional to the amount of noise in the data. We take a lesson from this, and adapting one of the voting methods to consider the amount of noise in an environment we empirically show the resulting heuristic is on par with the greedy approach in the case of noisy observations.

In the rest of this chapter, we present LPMs, then describe the greedy learning algorithm and our voting-based methods. We then introduce the learning bias and show how we can generalize the voting methods to exploit such a bias. Finally, we present the results of our experiments, followed by related work and concluding remarks.

2 Lexicographic Decision Models

In this section, we introduce the lexicographic preference model (LPM). Here and throughout this chapter, we only consider binary variables whose domain is $\{0, 1\}$. Note that the representation can easily be generalized to monotonic preferences with ordinal variables, such that 1 corresponds to a preference on the values in increasing order, and 0 to a decreasing order. For clarity in introducing the algorithms, we assume that the preferred value of each variable is known, although this assumption can be relaxed by considering a duplicate set of variables which are compliments of the originals (see Yaman et al. [15] for details). Without loss of generality, we will assume that 1 is always preferred to 0.

An object z is represented as a feature vector which means,

$$z = (z_1, z_2, \ldots, z_m) \in \mathcal{Z} = \mathcal{Z}_1 \times \mathcal{Z}_2 \times \cdots \times \mathcal{Z}_m \,.$$

We use $\boldsymbol{Z}$ to denote the set of features (also called variables), i.e. $\boldsymbol{Z} = \{\mathcal{Z}_1 \ldots \mathcal{Z}_m\}$. We use the notation $z(\mathcal{Z}_i)$ to refer the value of $\mathcal{Z}_i$ in the object z.

A *lexicographic preference model* $\mathcal{M}$ on $\boldsymbol{Z}$ is a total order on a subset R of $\boldsymbol{Z}$. We denote this total order with $\prec_{\mathcal{M}}$. Any variable in R is *relevant* with respect to $\mathcal{M}$; similarly, any variable in $I = \boldsymbol{Z} - R$ is *irrelevant* with respect to $\mathcal{M}$. If z and z' are two objects, then the preferred object given $\mathcal{M}$ is determined as follows:

- Find the smallest variable $\mathcal{Z}_i$ in $\prec_{\mathcal{M}}$ such that $\mathcal{Z}_i$ has different values in z and z'. The object that has the value 1 for $\mathcal{Z}_i$ is the most preferred.
- If all relevant variables in $\mathcal{M}$ have the same value in z and obj', then the objects are equally preferred (a tie).

Example 1. Suppose that $\mathcal{Z}_1 \prec \mathcal{Z}_2 \prec \mathcal{Z}_3$ is the total order defined by an LPM $\mathcal{M}$, and consider objects $z_1 = (1, 0, 1, 1)$, $z_2 = (0, 1, 0, 0)$, $z_3 = (0, 0, 1, 1)$, and $z_4 = (0, 0, 1, 0)$. z_1 is preferred over z_2 because $z_1(\mathcal{Z}_1) = 1$, and $\mathcal{Z}_1$ is the most important variable in $\mathcal{M}$. z_2 is preferred over z_3 because $z_2(\mathcal{Z}_2) = 1$ and both objects have the same value for $\mathcal{Z}_1$. Finally, z_3 and z_4 are equally preferred because they have the same values for the relevant variables.

2.1 Observations

An *observation* $o = (z, z')$ is an ordered pair of objects, connoting that z is preferred to z', also denoted as $z \succ z'$. Note that this representation does not allow for the explicit marking of ties, which would indicate that z and z' have the same values for attributes that are relevant to the preference decision but may differ on irrelevant attributes. This omission stems from the assumption – which holds in many practical applications – that preference observations are gathered from the demonstration of an expert who breaks ties arbitrarily. Thus, in some observations, z and z' may actually be tied even though z is reported to be preferred to z'. We call such observation, *hidden ties*. For example, consider the objects z_3 and obj_4 and the LPM $\mathcal{M}$ introduced in Example 1. The observations (z_3, z_4) and (z_4, z_3) are hidden ties because z_3 and z_4 are equally preferred by $\mathcal{M}$. Note that this case is more general than one where ties are reported, so our algorithms, which all work in the presence of hidden ties, are still poised to take advantage of reported ties.

An observation $o = (z, z')$ is *faulty* if z' is actually preferred over z. Such observations, also called *noise*, can be produced due to various reasons such as hardware and software failures, or can simply be data collected from users who do not always apply consistent, hard-and-fast preference rules.

2.2 Learning an LPM Consistent with Observations

An LPM $\mathcal{M}$ is *consistent* with an observation (z, z') iff $\mathcal{M}$ implies that z is preferred to z' or that z and z' are equally preferred.

Example 2. Once again consider, the variables, $\mathcal{Z}_1$, $\mathcal{Z}_2$, $\mathcal{Z}_3$, $\mathcal{Z}_4$ and the objects $z_1 = (1,0,1,1)$, $z_2 = (0,1,0,0)$, $z_3 = (0,0,1,1)$, and $z_4 = (0,0,1,0)$. Given the set of observations $o_1 = (z_1, z_2)$, $o_2 = (z_1, z_3)$, $o_3 = (z_2, z_3)$, and $o_4 = (z_3, z_4)$, the LPMs $\mathcal{Z}_1 \prec \mathcal{Z}_2 \prec \mathcal{Z}_3$ and $\mathcal{Z}_1 \prec \mathcal{Z}_2 \prec \mathcal{Z}_3 \prec \mathcal{Z}_4$ are consistent with those observations.

The problem of learning an LPM is defined as follows: Given a set of observations, find an LPM $\mathcal{M}$ that is consistent with the observations.

Given a noise-free set of observations finding an LPM that is consistent with the observations, if one exists, can be done in low-order polynomial time (see Sect. 3 for an algorithm that constructs a consistent LPM). However, if the observations are noisy, or if the target preference model was not an LPM, then we need to rephrase the learning problem as: Find an LPM that does not violate more than a constant number of the observations. Schmitt et al. [12] have proven that the complexity of this problem is NP-complete.

3 Greedy Algorithm

Schmitt et al. [12] proposed a greedy algorithm that is guaranteed to find one of the LPMs that is consistent with the observations, if one exists.

Algorithm 1 is Schmitt at al.'s greedy variable-permutation algorithm, which we use as a performance baseline. The algorithm refers to a function *MISS*$(\mathcal{Z}_i, O)$,

Algorithm 1 *greedyPermutation*

Require: A set of variables $\boldsymbol{Z}$ and a set of observations O.
Ensure: An LPM that is consistent with O, if one exists.

```
for i = 1, ..., n do
    Arbitrarily pick one of Z_j ∈ Z such that
    MISS(Z_j, O) = min_{Z_k ∈ Z} MISS(Z_k, O)
    π(Z_j) := i, assign the rank i to Z_j
    Remove Z_j from Z
    Remove all observations (z, z') from O such that z(Z_j) ≠ z'(Z_j)
end for
Return the total order ≺ on Z such that Z_i ≺ Z_j iff π(Z_i) < π(Z_j)
```

Table 1 The observation set before each iteration of the for-loop in algorithm 1 and the number of misses each attribute would cause at each iteration

Iterations	O	$Miss(\mathcal{Z}_1, 0)$	$Miss(\mathcal{Z}_2, 0)$	$Miss(\mathcal{Z}_3, 0)$	$Miss(\mathcal{Z}_4, 0)$	$Miss(\mathcal{Z}_5, 0)$
1	$\{o_1, o_2, o_3\}$	2	0	0	3	2
2	$\{o_1, o_3\}$	1	–	0	2	2
3	$\{o_3\}$	0	–	–	1	1
4	$\{\}$	–	–	–	0	0
5	$\{\}$	–	–	–	–	0

which is defined as $|\{(z, z') \in O : z(\mathcal{Z}_i) < z'(\mathcal{Z}_i)\}|$; that is, the number of observations violated in O if the most important variable is selected as $\mathcal{Z}_i$. Basically, the algorithm greedily constructs a total order by choosing the variable at each step that causes the minimum number of inconsistencies with the observations. If multiple variables have the same minimum, then one of them is chosen arbitrarily. The algorithm runs in polynomial time, specifically $O(m^2 n)$, where m is the number of variables and n is the number of observations.

The following example demonstrates how Algorithm 1 works.

Example 3. Suppose $\boldsymbol{Z} = \{\mathcal{Z}_1, \mathcal{Z}_2, \mathcal{Z}_3, \mathcal{Z}_4, \mathcal{Z}_5\}$ and O contains $o_1 = ((0, 1, 1, 0, 0), (1, 1, 0, 1, 1))$, $o_2 = ((0, 1, 1, 0, 1), (1, 0, 0, 1, 0))$, and $o_3 = ((1, 0, 1, 0, 0), (0, 0, 1, 1, 1))$. Table 1 shows $Miss(\mathcal{Z}_k, O)$, the number of misses that each attribute would cause had it been selected as the most important attribute in each iteration of the for-loop in Algorithm 1 and what would O be before each iteration. At every row, the attribute with the bold-faced entry is selected for the ith rank. Before the first iteration, the set of observations is $O = \{o_1, o_2, o_3\}$. The algorithm produces the LPM $\mathcal{Z}_2 \prec \mathcal{Z}_3 \prec \mathcal{Z}_1 \prec \mathcal{Z}_4 \prec \mathcal{Z}_5$ and the set of observations O_i after the ith iteration are $O_1 = \{o_1, o_3\}$, $O_2 = \{o_3\}$, $O_3 = \{\}$, $O_4 = \{\}$, and $O_5 = \{\}$. The learned LPM is consistent with all of the observations because O_5 is empty.

Schmitt et al. [12] have proven that if there is no LPM that is consistent with all the observations O, then the LPM $\mathcal{M}$ that Algorithm 1 produces is guaranteed to be consistent with at least half of the observations in O. This result suggests that Algorithm 1 is able to tolerate noisy data to some extent. Schmitt et al. did not perform any studies on the effect of hidden ties in the data in terms of learning the target LPM correctly. In Sect. 6, we provide experimental evidence that observations with hidden ties produces better approximations of the target.

Another shortcoming of this greedy approach is that it only produces a single LPM, even though multiple LPMs may fit the data. In the next section, we show that learning algorithms can sample many of these models while preserving tractability and theoretical guarantees of convergence and sample complexity. Later, our empirical studies show these *voting* approaches greatly outperforming Algorithm 1 in many situations.

4 Voting Algorithms

This section presents a democratic approach for approximating the target LPM that produced a set of noise-free observations, possibly containing hidden ties. Instead of finding just one of the consistent LPMs, it reasons with a collection of LPMs that are consistent with the observations. Given two objects, such an approach prefers the one that a majority of its models prefer. A naive implementation of a voting algorithm would enumerate all LPMs that are consistent with a set of observations. However, since the number of models consistent with a set of observations can be exponential in the number of attributes, the naive implementation is infeasible.

In this section, we describe two methods – *variable voting* and *model voting* – that sample the set of consistent LPMs and use voting to predict the preferred object. The following subsections explain the variable voting and model voting methods and summarize some of our theoretical results.

4.1 Variable Voting

Variable voting uses a generalization of the LPM representation. Instead of a total order on the variables, variable voting reasons with a *partial* order ($\preceq$) to find the preferred object in a given pair. Among the variables that are different in both objects, the ones that have the smallest rank (and are hence the most salient) in the partial order vote to choose the preferred object. The object that has the most "1" values for the voting variables is declared to be the preferred one. If the votes are equal, then the objects are equally preferred.

Definition 1 (Variable Voting). Suppose that $\boldsymbol{Z}$ is a set of variables and $\preceq$ is a partial order on $\boldsymbol{Z}$. Given two objects, z and z', the variable voting process with respect to $\preceq$ for determining which of the two objects is preferred is:

- Define D, the set of variables that differ in z and z'.
- Define D^*, the set of variables in D that have the smallest rank among D with respect to $\preceq$.

Algorithm 2 *learnVariableRank*

Require: A set of $\boldsymbol{Z}$ of variables, and a set O of observations
Ensure: A partial order on $\boldsymbol{Z}$.
$\Pi(Z_i) = 1, \forall\ Z_i \in \boldsymbol{Z}$
while Π can change **do**
  for Every observation $(z, z') \in O$ **do**
    Let D be the variables that differ in z and z'
    $D^* = \{Z_i \in D | \forall Z_j \in D, \Pi(Z_i) \leq \Pi(Z_j)\}$
    V_z is the set of variables in D^* that are 1 in z
    $V_{z'}$ is the set of variables in D^* that are 1 in z'
    if $|V_{z'}| \geq |V_z|$ **then**
      for $Z_i \in V_{z'}$ such that $\Pi(Z_i) < |\boldsymbol{Z}|$ **do**
        $\Pi(Z_i) = \Pi(Z_i) + 1$
      end for
    end if
  end for
end while
Return partial order $\preceq$ on $\boldsymbol{Z}$ such that $Z_i \preceq Z_j$ iff $\Pi(Z_i) < \Pi(Z_j)$

- Define N_z as the number of variables in D^* that favor z (i.e., that have value 1 in z and 0 in z') and $N_{z'}$ as the number of variables in D^* that favor z'.
- If $N_z > N_{z'}$, then z is preferred. If $N_z < N_{z'}$, then z' is preferred. Otherwise, they are equally preferred.

Example 4. Suppose that $\preceq$ is the partial order $\{Z_2, Z_3\} \prec \{Z_1\} \prec \{Z_4, Z_5\}$. Consider objects $z = (0, 1, 1, 0, 0)$ and $z' = (0, 0, 1, 0, 1)$. D is $\{Z_2, Z_5\}$. D^* is $\{Z_2\}$ because Z_2 is the smallest ranking variable in D with respect to $\preceq$. Z_2 favors z because $z(Z_2) = 1$. Thus, variable voting with $\preceq$ prefers z over z'.

Algorithm 2 presents the algorithm *learnVariableRank*, which learns a partial order $\preceq$ on the variables from a set of observations such that variable voting with respect to $\preceq$ will correctly predict the preferred objects in the observations. Specifically, it finds partial orders that define equivalence classes on the set of variables. The algorithm maintains the minimum possible rank for every variable that does not violate an observation with respect to variable voting. Initially, all variables are considered equally important (rank of 1). The algorithm loops over the set of observations until the ranks converge. At every iteration and for every pair, variable voting predicts a winner. If it is correct, then the ranks stay the same. Otherwise, the ranks of the variables that voted for the wrong object are incremented, thus reducing their importance[1]. Finally, the algorithm builds a partial order $\preceq$ based on the ranks such that $x \preceq y$ if and only if x has a lower rank than y.

Example 5. Suppose, as in Example 3 , $\boldsymbol{Z} = \{Z_1, Z_2, Z_3, Z_4, Z_5\}$ and O consists of $((0, 1, 1, 0, 0), (1, 1, 0, 1, 1))$, $((0, 1, 1, 0, 1), (1, 0, 0, 1, 0))$ and $((1, 0, 1, 0, 0),$

[1] In our empirical results, we also update the ranks when the prediction was correct but not unanimous. This produces a heuristic speed-up without detracting from the worst case guarantees.

Table 2 The rank of the variables after each iteration of the for-loop in line 3 of the algorithm *learnVariableRank*

Observations	z_1	z_2	z_3	z_4	z_5
Initially	1	1	1	1	1
(0, 1, 1, 0, 0), (1, 1, 0, 1, 1)	2	1	1	2	2
(0, 1, 1, 0, 1), (1, 0, 0, 1, 0)	2	1	1	2	2
(1, 0, 1, 0, 0), (0, 0, 1, 1, 1)	2	1	1	3	3

$(0, 0, 1, 1, 1))$. Table 2 illustrates the ranks of every variable in $\boldsymbol{Z}$ after each iteration of the for-loop in line 3 of the algorithm *learnVariableRank*. The ranks of the variables stay the same during the second iteration of the while-loop; thus, the loop terminates. The partial order $\preceq$ based on ranks of the variables is the same as the order given in Example 4.

We next summarize our theoretical results about the algorithm *learnVariableRank*.

Correctness

Suppose that $\preceq$ is a partial order returned by *learnVariableRank*$(\boldsymbol{Z}, O)$. It can be shown that if the observations are noise-free, then any LPM $\mathcal{M}$ such that $\sqsubset_{\mathcal{M}}$ is a topological sort of $\preceq$ is consistent with O. Furthermore, *learnVariableRank* never increments the ranks of the relevant variables beyond their actual rank in the target LPM. The ranks of the irrelevant variables can be incremented as far as the number of variables.

Convergence

learnVariableRank has a mistake bound of $O(m^2)$, where m is the number of variables, because each mistake increases the sum of the potential ranks by at least 1 and the sum of the ranks that the target LPM induces is $O(m^2)$. This bound guarantees that given enough observations (as described in the background section), *learnVariableRank* will converge to a partial order $\preceq$ such that every topological sort of $\preceq$ has the same prefix as the total order induced by the target LPM. If all variables are relevant, then $\preceq$ will converge to the total order induced by the target LPM.

Complexity

A very loose upper bound on the time complexity of *learnVariableRank* is $O(m^3 n)$, where m is the number of variables and n is the number of observations. This bound holds because the while-loop on line 2 runs at most $O(m^2)$ times, and the for-loop in line 3 runs for n observations. The time complexity of one iteration of the for-loop is $O(m)$; therefore, the overall complexity is $O(m^3 n)$. We leave the investigation of tighter bounds and the average case analysis for future work.

4.2 Model Voting

The second method we present employs a Bayesian approach. This method randomly generates a sample set, S, of distinct LPMs that are consistent with the observations. When a pair of objects is presented, the preferred one is predicted using weighted voting. That is, each $\mathcal{M} \in S$ casts a vote for the object it prefers, and this vote is weighted according to its posterior probability $P(\mathcal{M}|S)$.

Definition 2 (Model Voting). Let U be the set of all LPMs, O be a set of observations, and $S \subset U$ be a set of LPMs that are consistent with O. Given two objects, z and z', model voting prefers z over z' with respect to S if:

$$\sum_{\mathcal{M} \in U} P(\mathcal{M}|S) V^{\mathcal{M}}_{(z \succ z')} > \sum_{\mathcal{M} \in U} P(\mathcal{M}|S) V^{\mathcal{M}}_{(z' \succ z)}, \qquad (1)$$

where $V^{\mathcal{M}}_{(z \succ z')}$ is 1 if z is preferred with respect to $\mathcal{M}$, and 0 otherwise. $V^{\mathcal{M}}_{(z' \succ z)}$ is defined analogously. $P(\mathcal{M}|S)$ is the posterior probability of $\mathcal{M}$ being the target LPM given S, calculated as discussed below.

We first assume that all LPMs are equally likely *a priori*. In this case, given a sample S of size k, the posterior probability of an LPM $\mathcal{M}$ will be $1/k$ if and only if $\mathcal{M} \in S$, and 0 otherwise. Note that if S is maximal, this case degenerates into the naive voting algorithm. However, it is generally not feasible to enumerate all consistent LPMs – in practice, the sample has to be small enough to be feasible and large enough to be representative.

In constructing S, we exploit the fact that many consistent LPMs share prefixes in the total order that they define on the variables. We wish to discover and compactly represent such LPMs. To this end, we introduce the idea of *aggregated LPMs*. An aggregated LPM, $(\mathcal{Z}_1, \mathcal{Z}_2 \ldots, \mathcal{Z}_k, *)$, represents a set of LPMs that define a total order with the prefix $\mathcal{Z}_1 \prec \mathcal{Z}_2 \prec \cdots \prec \mathcal{Z}_k$. Intuitively, an aggregated LPM states that any possible completion of the prefix is consistent with the observations. The algorithm *sampleModels* in Algorithm 3 implements a "smart sampling" approach by constructing an LPM that is consistent with the given observations, returning an aggregated LPM when possible. We start with an arbitrary consistent LPM (such as the empty set, which is always consistent) and add more variable orderings, extending the input LPM. We first identify the variables that can be used in extending the prefix – that is, all variables $\mathcal{Z}_i$ such that in every observation, either $\mathcal{Z}_i$ is 1 in the preferred object or is the same in both objects. We then select one of those variables randomly and extend the prefix. Finally, we remove the observations that are explained with this selection and continue with the rest of the observations. If, at any point, no observations remain, then we return the aggregated form of the prefix,

Algorithm 3 *sampleModels*

Require: Set of variables $\boldsymbol{Z}$, set of observations O, and *rulePrefix*, LPM to be extended.
Ensure: An LPM (possibly aggregated) consistent with O.

$candidates = \{Z_i : Z_i \notin rulePrefix \mid \forall(z, z') \in O, z(Z_i) = 1 \text{ or } z(Z_i) = z'(Z_i)\}$.
while $candidates \neq \emptyset$ **do**
  if $O = \emptyset$ **then**
    return $(rulePrefix, *)$
  end if
  Randomly remove a variable Z_j from *candidates*
  Remove any observation (z, z') from O such that $z(Z_j) \neq z'(Z_j)$
  Extend *rulePrefix*: $rulePrefix = (rulePrefix, Z_j)$
  Recompute *candidates*
end while
return *rulePrefix*

since every completion of the prefix will be consistent with the null observation. Running *sampleModels* several times and eliminating duplicates will produce a set of (possibly aggregated) LPMs.

Example 6. Consider the same set of observations O as in Example 5. Then, the LPMs that are consistent with O are as follows: (), (Z_2), (Z_2, Z_3), $(Z_2, Z_3, Z_1, *)$, (Z_3), $(Z_3, Z_1, *)$, (Z_3, Z_2), and $(Z_3, Z_2, Z_1, *)$. To illustrate the set of LPMs that an aggregate LPM represents, consider $(Z_2, Z_3, Z_1, *)$, which has a total of five extensions: (Z_2, Z_3, Z_1), (Z_2, Z_3, Z_1, Z_4), (Z_2, Z_3, Z_1, Z_5), $(Z_2, Z_3, Z_1, Z_4, Z_5)$, and $(Z_2, Z_3, Z_1, Z_5, Z_4)$. Every time the algorithm *sampleModels* runs, it will randomly generate one of the aggregated LPMs: $(Z_2, Z_3, Z_1, *)$, $(Z_3, Z_1, *)$, or $(Z_3, Z_2, Z_1, *)$. Note that the shorter models that are not produced by *sampleModels* are all subprefixes of the aggregated LPMs and it is easy to modify *sampleModels* to return those models as well.

An aggregate LPM in a sample saves us from having to enumerate all possible extensions of a prefix, but it also introduces complications in computing the weights (posteriors) of the LPMs, as well as their votes. For example, when comparing two objects z and z', some extensions of an aggregate LPM might vote for z and some for z'. Thus, we need to find the total number of LPMs that an aggregate LPM represents and determine what proportion of them favor z over z' (or vice versa), without enumerating all extensions. Suppose there are m variables and $\mathcal{M}$ is an aggregated LPM with a prefix of length k. Then, the number of extensions of $\mathcal{M}$ is denoted by $F_{\mathcal{M}}$ and is equal to f_{m-k}, where f_t is defined to be:

$$f_t = \sum_{i=0}^{t} \binom{t}{i} \times i! = \sum_{i=0}^{t} \frac{(t)!}{(t-i)!}. \tag{2}$$

Intuitively, f_t counts every possible permutation with at most t items. Note that f_t can be computed efficiently and that the number of all possible LPMs when there are m variables is given by f_m.

Consider a pair of objects, z and z'. We wish to determine how many extensions of an aggregate LPM $\mathcal{M} = (\mathcal{Z}_1, \mathcal{Z}_2, \ldots, \mathcal{Z}_k, *)$ would vote for one of the objects. We will call the variables $\mathcal{Z}_1 \ldots \mathcal{Z}_k$ the *prefix variables*. If z and z' have different values for at least one prefix variable, then all extensions will vote in accordance with the smallest such variable. Suppose that all prefix variables are tied and m is the set of all nonprefix variables. Then, m is composed of three disjoint sets a, b, and w, such that a is the set of variables that favor z, b is the set of variables that favor z', and w is the set of variables that are neutral (that is, that have the same value in z and z').

An extension $\mathcal{M}'$ of $\mathcal{M}$ will produce a tie iff all variables in a and b are irrelevant in $\mathcal{M}'$. The number of such extensions is $f_{|w|}$. The number of extensions that favor z over z' is directly proportional to $|a|/(|a| + |b|)$. The number of extensions of $\mathcal{M}$ that will vote for z over z' is denoted by $N^{\mathcal{M}}_{z \succ z'}$, which is given by:

$$N^{\mathcal{M}}_{z \succ z'} = \frac{|a|}{|b| + |a|} \times (f_{|m|} - f_{|w|}). \tag{3}$$

The number of extensions of $\mathcal{M}$ that will vote for z' over z is computed similarly. Note that the computation of $N^{\mathcal{M}}_{z \succ z'}$, $N^{\mathcal{M}}_{z' \succ z}$, and $F_{\mathcal{M}}$ can be done in linear time by caching the recurrent values.

Example 7. Suppose $\boldsymbol{Z}$ and O are as defined in Example 5. The first column of Table 3 lists all LPMs that are consistent with O. The second column gives the posterior probabilities of these models given the sample S_1, which is the set of all consistent LPMs. The third column is the posterior probability of the models given the sample $S_2 = \{(\mathcal{Z}_2, \mathcal{Z}_3, \mathcal{Z}_1, *), (\mathcal{Z}_3, \mathcal{Z}_1, *), (\mathcal{Z}_3, \mathcal{Z}_2, \mathcal{Z}_1, *)\}$. Given two objects $z = (0, 1, 1, 0, 0)$ and $z' = (0, 0, 1, 0, 1)$, the number of votes for each object based on each LPM is given in the last two columns. Note that the total number of votes for z and z' does not add up to the total number of extensions of $(\mathcal{Z}_3, \mathcal{Z}_1, *)$ because two of its extensions – $(\mathcal{Z}_3, \mathcal{Z}_1)$ and $(\mathcal{Z}_3, \mathcal{Z}_1, \mathcal{Z}_4)$ – prefer z and z' equally.

Algorithm 4 describes *modelVote*, which takes a sample of consistent LPMs and a pair of objects as input, and predicts the preferred object using the weighted votes of the LPMs in the sample.

Table 3 The posterior probabilities and number of votes of all LPMs in Example 7

LPMs	$P(L\|S_1)$	$P(L\|S_2)$	$N^{\mathcal{M}}_{z \succ z'}$	$N^{\mathcal{M}}_{z' \succ z}$
()	1/31	0	0	0
$(\mathcal{Z}_2)$	1/31	0	1	0
$(\mathcal{Z}_2, \mathcal{Z}_3)$	1/31	0	1	0
$(\mathcal{Z}_2, \mathcal{Z}_3, \mathcal{Z}_1, *)$	5/31	5/26	5	0
$(\mathcal{Z}_3)$	1/31	0	0	0
$(\mathcal{Z}_3, \mathcal{Z}_1, *)$	16/31	16/26	7	7
$(\mathcal{Z}_3, \mathcal{Z}_2)$	1/31	0	1	0
$(\mathcal{Z}_3, \mathcal{Z}_2, \mathcal{Z}_1, *)$	5/31	5/26	5	0

Algorithm 4 *modelVote*

Require: A set of LPMs, S, and two objects, z and z'
Ensure: Returns either one of z or z' or *tie*

Initialize *sampleSize* to the number of non-aggregated LPMs in S
for every aggregated LPM $\mathcal{M} \in S$ **do**
  $sampleSize += F_{\mathcal{M}}$
end for
$Vote(z) = 0$; $Vote(z') = 0$
for every LPM $\mathcal{M} \in S$ **do**
  if $\mathcal{M}$ is not an aggregate rule **then**
    winner is the object $\mathcal{M}$ prefers among z and z'
    Increment $Vote(winner)$ by $1/sampleSize$
  else
    if z and z' differ in at least one prefix variable of $\mathcal{M}$ **then**
      $\mathcal{M}^*$ is an extension of $\mathcal{M}$ referring only the prefix
      winner is the object $\mathcal{M}^*$ prefers among z and z'
      $Vote(winner) += F_{\mathcal{M}}/sampleSize$
    else
      $Vote(z) += N^L_{z \succ z'}/sampleSize$
      $Vote(z') += N^L_{z' \succ z}/sampleSize$
    end if
  end if
end for
if $Vote(z) = Vote(z')$ **then**
  Return a *tie*
else
  Return the object *obj* with the highest $Vote(obj)$
end if

Returning to Example 7, the reader can verify that model voting will prefer z over z'. Next, we present our theoretical results on the *sampleModels* and *modelVote* algorithms.

Complexity

The time complexity of *sampleModels* is bounded by $O(m^2n)$, where m is the number of variables and n is the number of observations: the while-loop in line 2 runs at most m times; at each iteration, we have to process every observation, each time performing computations that take $O(m)$ time. If we call *sampleModels* s times, then the total complexity of sampling is $O(sm^2n)$. For constant s, this bound is still polynomial. Similarly, the complexity of *modelVote* is $O(sm)$ because it considers each of the s rules in the sample, counting the votes of each rule, which can be done in $O(m)$ time.

Comparison to Variable Voting

The set of LPMs that is sampled via *learnVariableRank* is a subset of the LPMs that *sampleModels* can produce. The running example in the paper demonstrates that *sampleModels* can generate the LPM $(\mathcal{Z}_3, \mathcal{Z}_1, *)$; however, none of its extensions is consistent with the partial order that *learnVariableRank* returns.

5 Introducing Bias

In general, when there are not many training examples for a learning algorithm, the space of consistent LPMs is large. In such cases, it is usually not possible to find a good approximation of the target model. To overcome this problem, we can introduce a bias (domain knowledge), indicating that certain solutions should be favored over the others. In this section, we propose a bias in the form of equivalence classes over the set of attributes. These equivalence classes indicate the set of most important attributes, second most important attributes, and so on. For example, when buying a used car, most people consider the most important attributes of a car to be the mileage, the year, and the make of the car. The second most important set of attributes is the color, number of doors, and body type. Finally, perhaps the least important properties are the interior color and the wheel covers. Throughout this section, we assume that the preferred value of a variable is known or is given in the prescribed bias. We now formally define a learning bias and what it means for an LPM to be consistent with a learning bias.

Definition 3 (Learning Bias). A learning bias $\mathcal{B}$ for learning a lexicographic preference model on a set of variables $\boldsymbol{Z}$ is a total order on a partition of $\boldsymbol{Z}$. $\mathcal{B}$ has the form $E_1 \prec E_2 \prec \cdots \prec E_k$, where $\cup_i E_i = \boldsymbol{Z}$. Intuitively, $\mathcal{B}$ defines a partial order on $\boldsymbol{Z}$ such that for any two variables, $x \in E_i$ and $y \in E_j$, $x \prec y$ iff $E_i \prec E_j$. We denote this partial order by $\preceq_B$.

Definition 4. Suppose that $\boldsymbol{Z} = \{\mathcal{Z}_1, \ldots \mathcal{Z}_n\}$ is a set of variables, $\mathcal{B}$ a learning bias, and $\mathcal{M}$ an LPM. $\mathcal{M}$ is consistent with $\mathcal{B}$ iff the total order $\prec_{\mathcal{M}}$ is consistent with the partial order $\preceq_{\mathcal{B}}$.

Intuitively, an LPM that is *consistent* with a learning bias respects the variable orderings induced by the learning bias. The learning bias prunes the space of possible LPMs. The size of the partition determines the strength of the bias; for example, if there is a single variable per set, then the bias defines a specific LPM. In general, the number of LPMs that is consistent with a learning bias of the form $E_1 \prec E_2 \prec \cdots \prec E_k$ can be computed with the following recursive formula:

$$G([e_1, \ldots e_k,]) = f_{e_1} + e_1! \times (G([e_2, \ldots e_k]) - 1), \tag{4}$$

where $e_i = |E_i|$ and the base case for the recursion is $G([]) = 1$. The first term in the formula counts the number of possible LPMs using only the variables in E_1, which are the most important variables. The definition of consistency entails that a variable can appear in $\prec_{\mathcal{M}}$ iff all of the more important variables are already in $\prec_{\mathcal{M}}$, hence the term e_1. Note that the recursion on G is limited to the number of sets in the partition, which is bounded by the number of variables; therefore, it can also be computed in linear time by caching precomputed values of f.

To illustrate the power of a learning bias, consider a learning problem with nine variables. Without a bias, the total number of LPMs is 905,970. If a learning bias partitions the variables into three sets, each with three elements, then the number of LPMs consistent with the bias is only 646. A bias with four sets, where the first set has three variables and the rest have two, limits the number to 190.

We can easily generalize the *learnVariableRank* algorithm to utilize the learning bias by changing only the first line of *learnVariableRank*, which initializes the ranks of the variables. Given a bias of the form $S_1 \prec \cdots \prec S_k$, the generalized algorithm assigns the rank 1 (most important rank) to the variables in S_1, rank $|S_1| + 1$ to those in S_2, and so forth. This initialization ensures that an observation (z, z') is used for learning the order of variables in a class S_i only when A and B have the same values for all variables in classes $S_1 \ldots S_{i-1}$ and have different values for at least one variable in S_i.

The algorithm *modelVote* can also be generalized to use a learning bias $\mathcal{B}$. In the sample generation phase, we use *sampleModels* as presented earlier, and then eliminate all rules whose prefixes are not consistent with the bias. Note that even if the prefix of an aggregated LPM $\mathcal{M}$ is consistent with a bias, this may not be the case for every extension of $\mathcal{M}$. Thus, in the algorithm *modelVote*, we need to replace any references to $F_{\mathcal{M}}$ and $N^{\mathcal{M}}_{z \succ z'}$ (or $N^{\mathcal{M}}_{z' \succ z}$) with $F^{\mathcal{B}}_{\mathcal{M}}$ and $N^{\mathcal{M},\mathcal{B}}_{z \succ z'}$ (or $N^{\mathcal{M},\mathcal{B}}_{z' \succ z}$), respectively, where:

- $F^{\mathcal{B}}_{\mathcal{M}}$ is the number of extensions of $\mathcal{M}$ that are consistent with $\mathcal{B}$, and
- $N^{\mathcal{M},\mathcal{B}}_{z \succ z'}$ is the number of extensions of $\mathcal{M}$ that are consistent with $\mathcal{B}$ and prefer z. ($N^{\mathcal{M},\mathcal{B}}_{z' \succ z}$ is analogous.)

Suppose that $\mathcal{B}$ is a learning bias $E_1 \prec \cdots \prec E_m$. Let Y denote the prefix variables of an aggregate LPM $\mathcal{M}$ and let E_k be the first set such that at least one variable in E_k is not in Y. Then, $F^{\mathcal{B}}_{\mathcal{M}} = G([|E_k - Y|, |E_{k+1} - Y|, \ldots |E_m - Y|])$.

When counting the number of extensions of $\mathcal{M}$ that are consistent with $\mathcal{B}$ and prefer z, we again need to examine the case where the prefix variables equally prefer the objects. Suppose Y is as defined as above and D_i denotes the set difference between E_i and Y. Let D_j be the first non-empty set and D_k be the first set such that at least one variable in D_k has different values in the two objects. Obviously, only the variables in D_k will influence the prediction of the preferred object. If

- $d_i = |D_i|$, the cardinality of D_i, and
- a is the set of variables in D_k that favor z, b is the set of variables in D_k that favor z', and w is the set of variables in D_k that are neutral,

then $N^{\mathcal{M},\mathcal{B}}_{z \succ z'}$, the number of extensions of $\mathcal{M}$ that are consistent with $\mathcal{B}$ and prefer z, can be computed as follows:

$$N^{\mathcal{M},\mathcal{B}}_{z \succ z'} = \frac{|a|}{|a| + |b|} \times (F^{\mathcal{B}}_{\mathcal{M}} - G([d_j \ldots d_{k-1}, |w|])). \quad (5)$$

6 Experiments

In this section, we explain our experimental methodology and discuss the results of our empirical evaluations. We define the *prediction performance* of an algorithm P with respect to a set of test observations T as:

$$performance(P, T) = \frac{Correct(P, T) + 0.5 \times Tie(P, T)}{|T|}, \tag{6}$$

where $Correct(P, T)$ is the number of observations in T that are predicted correctly by P (including any prediction for $t \in T$ where t is actually a tie) and $Tie(P, T)$ is the number of observations in T that P predicted as a tie when one object should actually have been preferred over the other. Note that an LPM returned by *greedyPermutation* never returns a tie. In contrast, variable voting with respect to a partial order in which every variable is equally important will only return ties, so the overall performance will be 0.5, which is no better than randomly selecting the preferred objects. We will use MV, VV, and G to denote the model voting, variable voting, and the greedy approximations of an LPM.

Given sets of training and test observations, (O, T), we measure the *average* and *worst* performances of VV, MV, and G. When combined with *learnVariableRank*, VV is a deterministic algorithm, so the average and worst performances of VV are the same. However, this is not the case for MV with sampling, because *sampleModels* is randomized. Even for the same training and test data (O, T), the performance of MV can vary. To mitigate this effect, we ran MV 10 times for each (O, T) pair, and called *sampleModels* S times on each run (thus, the sample size is at most S), recording the average and worst of its performance. The greedy algorithm G is also randomized (in line 2, one variable is picked arbitrarily), so we ran G 200 times for every (O, T), recording its average and worst performance.

For our experiments, the control variables are R, the number of relevant variables in the target LPM; I, the number of irrelevant variables; N_O, the number of training observations; and N_T, the number of test observations. For MV experiments, the sample size (S) is also a control parameter. For fixed values of R and I, an LPM $\mathcal{M}$ is randomly generated. (If a bias B is given, then $\mathcal{M}$ is also consistent with B.) We randomly generated N_O and N_T pairs of objects, each with $I + R$ variables. Finally, we labeled the preferred objects according to $\mathcal{M}$. In all of the figures, the data points are averages over 20 different pairs of training and test sets (O, T).

6.1 Comparison of MV, VV and G

Figure 1a shows the average performance of G, MV with sample size $S = 200$, and VV for $R = 15$, $I = 0$, and $N_T = 20$, as N_O ranges from 2 to 20. Figure 1(b) shows the worst performance for each algorithm. The average performance of VV and MV is better than the average performance of G, and the

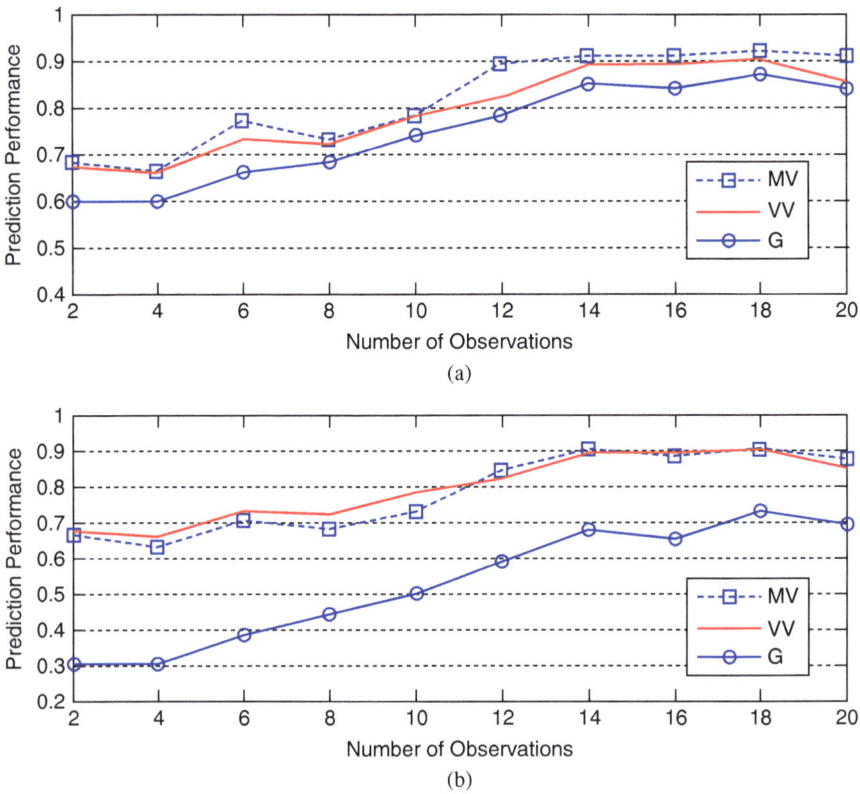

Fig. 1 (**a**) Average prediction performance and (**b**) worst prediction performance of the greedy algorithm, variable voting, and model voting

difference is significant at every data point. Also, note that the worst-case performance of G after seeing two observations is around 0.3, which suggests a very poor approximation of the target. VV and MV's worst-case performances are much better than the worst-case performances of G, justifying the additional complexity of the algorithms MV and VV.

6.2 *Effect of Hidden Ties*

Figure 2 shows the prediction performance of VV and G (average and worst case) for problems with 10 relevant variables ($R = 10$) and five irrelevant variables ($I = 5$). The number of observations (N_o) is always 50, but the number of these observations that are hidden ties varies from 0 to 45 (x-axis).

In general, as the number of hidden ties increase in the observations, the prediction performance degrades. This was expected because the number of useful

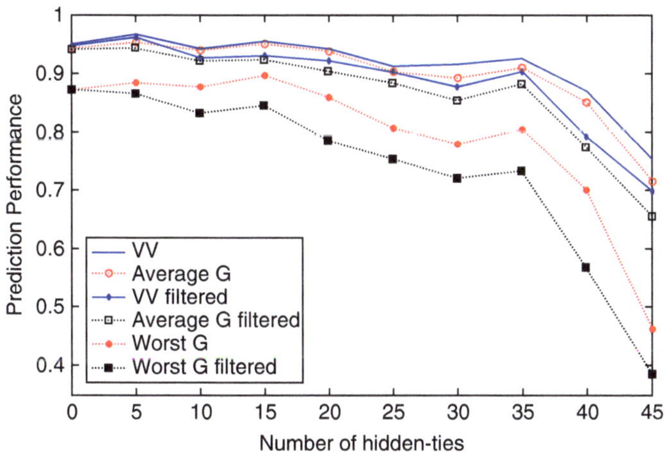

Fig. 2 The prediction performance of VV and G for varying number of hidden ties in 50 observations. The curves for filtered VV and G are obtained by eliminating the hidden ties from the observations

observations that can help the algorithms learn the ranking on the relevant attributes decreases as the number of hidden ties increases. The performance of MV on the same data sets was very similar to VV and has thus been omitted for clarity.

A more interesting result, however, is that the existence of hidden ties in the data actually *improves* the performance of both algorithms, over a smaller dataset with the hidden tie observations omitted (the "filtered" versions in Fig. 2). Our explanation for this phenomenon is while hidden ties do not provide useful information for learning the order on the relevant attributes, their existence helps the algorithms identify the irrelevant attributes (because hidden ties can increase the number of mistakes that would be caused only by the irrelevant attributes) and push them further up in the ranking (decreasing their importance), thus allowing the other observations to clarify the ordering of the identified relevant attributes.

6.3 Effect of Noise

Figure 3 shows the average prediction performance of MV and G for problems with 10 relevant variables ($R = 10$) and five irrelevant variables ($I = 5$). The total number of observations (N_o) is always 50 but the number of these observations that are faulty varies from 0 to 45 (x-axis). Figure 4 shows the worst performance for the same setting.

The results show that both the average and the worst performance of MV (which operates under the assumption of noise-free data) are significantly compromised by even small amounts of noise, which it interprets as a refutation of the correct model (as well as many of the "almost correct" models). The asymptotic performance at

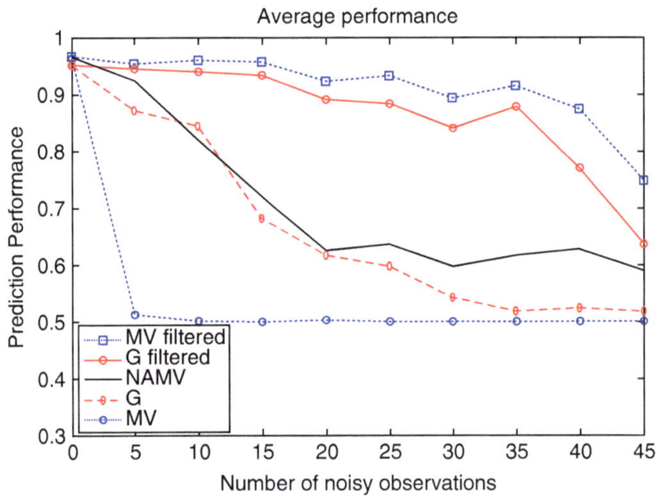

Fig. 3 The average prediction performance of model voting, greedy, and *NAMV* as the number of noisy observations increases

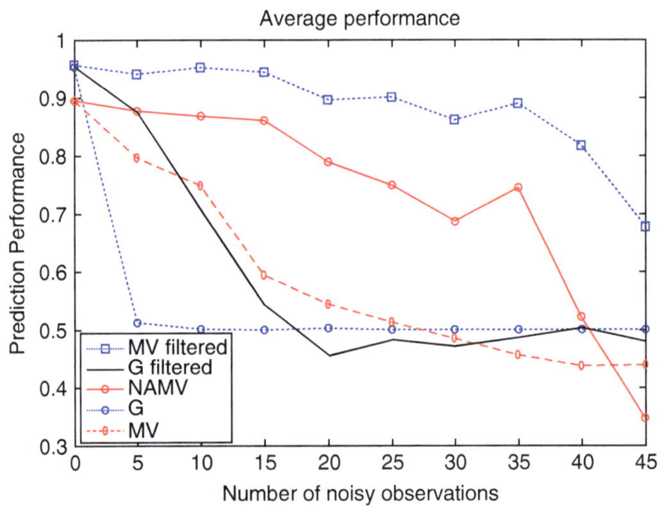

Fig. 4 The worst prediction performance of model voting, greedy, and *NAMV* as the number of noisy observations increases

0.5 reflects the fact that this noise causes MV to eliminate all the models from its version space, causing it to predict a tie for every testing observation (essentially making it a random selection algorithm). We omitted the results for VV from the figure since VV's behavior closely resembled that of MV.

The performance of G decays far more gracefully than MV or VV because G allows for some of the observations to be discarded. Although inferring an LPM from noisy data is NP-complete, the moderate empirical success of the Greedy algorithm gives us an intuition as to how a heuristic solution with voting can be developed. Specifically, the Greedy approach iteratively constructed an LPM where each added attribute violated the fewest number of observations. We borrow this intuition in the following heuristic extension of *modelVote*, which is provided with the expected number of noisy observations ϵ in a data set. The new algorithm, Noise-Aware Model Vote (*NAMV*), changes *sampleModels* to only consider a variable as a *candidate* if adding it will not violate more (in total with the other variables) than ϵ observations. Notice that this remains a stochastic LPM construction and can still consider many more LPMs than the greedy approach. In Figs. 3 and 4, we compare the average and worst performance of this approach to the other algorithms in this paper, including "filtered" versions where the noisy data was omitted (which provided as baselines). Notice that unlike the original *modelVote*, which was confounded by even a small amount of noise, the gentle decay of *NAMV* mirrors the greedy approach's robustness to noise and performs comparably on this data set. Asymptotically, if all of the observations are noisy, *NAMV* still performs better than *modelVote* in the average case, because it does not eliminate all of the possible models, so instead of defaulting to random selection, it favors the test observation with the most 1s.

6.4 Effect of Bias on Performance

Figure 5 shows the positive effect of learning bias on the performance of voting algorithms for $R = 10$, $I = 0$, and $N_T = 20$, as N_O ranges from 2 to 20. In addition, this experiment aims to show that bias does not undermine the advantage that voting algorithms held over the greedy algorithm in the unbiased case. To this end, we have trivially generalized G to produce LPMs that are consistent with a given bias. The data points are averages over 20 different pairs of training and test sets (O, T). We have arbitrarily picked two biases: $B_1 : \{\mathcal{Z}_1, \mathcal{Z}_2, \mathcal{Z}_3, \mathcal{Z}_4, \mathcal{Z}_5\} \prec \{\mathcal{Z}_6, \mathcal{Z}_7, \mathcal{Z}_8, \mathcal{Z}_9, \mathcal{Z}_{10}\}$ and $B_2 : \{\mathcal{Z}_1, \mathcal{Z}_2, \mathcal{Z}_3\} \prec \{\mathcal{Z}_4, \mathcal{Z}_5\} \prec \{\mathcal{Z}_6, \mathcal{Z}_7, \mathcal{Z}_8\} \prec \{\mathcal{Z}_9, \mathcal{Z}_{10}\}$. For our experiments, we have randomly generated LPMs that are consistent with those biases. The performance of VV improved greatly with the introduction of both learning biases. B_2 is a stronger bias than B_1, and therefore prunes the space of consistent LPMs more than B_1. As a result, the performance gain due to B_2 is greater than that due to B_1. The difference between the bias and nonbias curves is statistically significant except at the last point. Note that the biases are particularly effective when the number of training observations is small. The worst-case performances of G with biases B_1 and B_2 are also shown in Fig. 5. For both, the worst-case performance of G is significantly lower than the performance of VV with the corresponding bias.

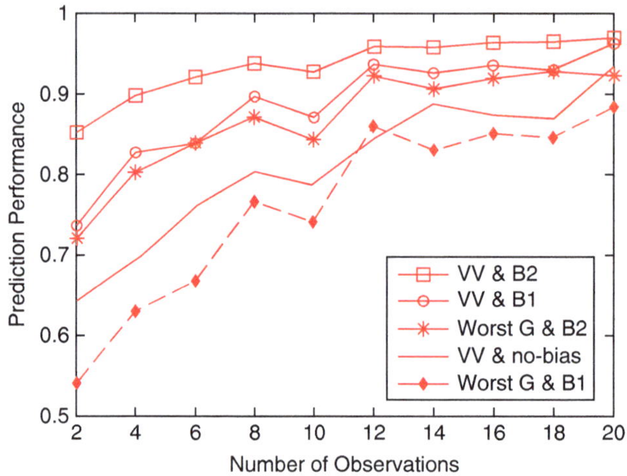

Fig. 5 The effect of bias on VV and G, using two arbitrarily selected biases, where $B2$ is a stronger bias than $B1$

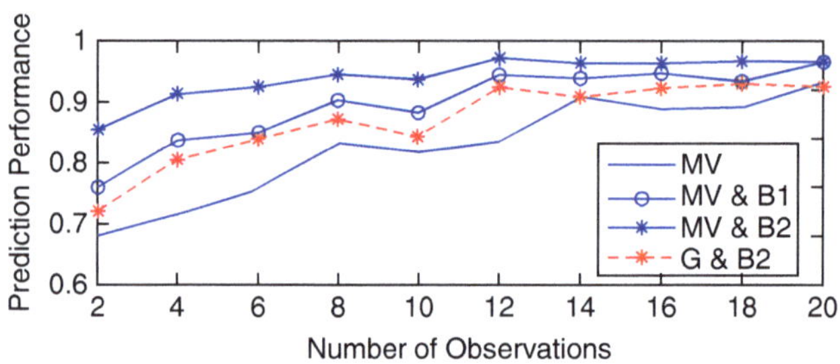

Fig. 6 The effect of bias on average MV performance, using two arbitrarily selected biases, where $B2$ is a stronger bias than $B1$

Using the same experimental scenario, we obtained very similar results with MV, as seen in Fig. 6. In summary, the worst-case performance of greedy algorithm with bias B_2 outperforms the average performance of MV without any bias. However, even with a weaker bias such as B_1, the average performance of MV is better than G with B_2.

7 Related Work

Lexicographic orders and other preference models have been utilized in several research areas, including multicriteria optimization [1], linear programming [5], and game theory [10]. The lexicographic model and its applications have been surveyed by Fishburn [8].

The most relevant existing work for learning and/or approximating LPMs are the approaches presented by Schmitt and Martignon [12], which was summarized in Sect. 3, and by Dombi et al. [7]. Dombi et al. showed that if there are n variables, all of which are relevant, then $O(n \log n)$ queries to an oracle suffice to learn an LPM. Furthermore, it is possible to learn any LPM with $O(n^2)$ observations if all pairs differ in only two variables. They proposed an algorithm that can find the unique LPM induced by the observations. In case of noise or hidden ties due to irrelevant attributes, the algorithm does not return an answer.

In general, preferences and ranking are similar. The ranking problem as described by Cohen et al. [3] is similar to the problem of learning an LPM. However, that line of work poses learning as an optimization problem, with the goal of finding the ranking that maximally agrees with the given preference function. Our work generally assumes noise-free data, for which an optimization approach is not needed.

Torrey et al. [13] employ an inductive logic programming approach to learn multiattribute ranking rules. In principle, these rules can represent lexicographic preference models.

A more complex preference representation is CP-Nets [2]. CP-nets can model conditional preferences under a *ceteris paribus* (all else being equal) assumption. However, despite its complexity, this representation will not necessarily capture lexicographic preference models, and is therefore not directly applicable to the problem we have considered.

Another analogy, identified in Schmitt and Martignon [12], is between LPMs and decision lists [11]. Specifically, it was shown that LPMs are a special case of 2-decision lists, but that the algorithms for learning these two classes of models are not directly applicable to each other.

8 Conclusions and Future Work

In this chapter, we presented democratic approximation methods for learning lexicographic preference models (LPMs) given a set of preference observations. Instead of committing to just one of the consistent LPMs, we maintain a set of models and predict based on the majority of votes. We described two such methods: *variable voting* and *model voting*.

Variable voting implicitly samples possible LPMs by constructing a partial order on the variables such that any linearization of the partial order will correspond to a consistent LPM. Model voting, on the other hand, explicitly samples the space of consistent LPMs but allows aggregation of LPMs to achieve compact representation of LPMs that share the same prefix. We showed that both methods can be implemented in polynomial time and exhibit much better worst- and average-case performance than the existing methods in case of noise-free data. In addition, we have defined a learning bias that can improve performance when the number of observations is small and incorporated this bias into the voting-based methods, significantly improving their empirical performance.

Future directions of this work allow for a number of extensions and further theoretical investigations. We will improve the voting algorithms by continuing to investigate heuristics such as *NAMV* that make them more robust against noise. While the problem of learning LPMs from noisy data is NP-complete, the superior performance of the voting algorithms over the greedy method in the noise-free case indicates that it may be possible to identify and characterize other restricted problem settings in which heuristic extensions such as *NAMV* would significantly outperform the greedy approach.

Acknowledgements This work was supported by the Defense Advanced Research Projects Agency and the U. S. Air Force through BBN Technologies Corp. under contract number FA8650-06-C-7606. Approved for Public Release, Distribution Unlimited.

References

1. D.P. Bertsekas, J.N. Tsitsiklis, *Parallel and distributed computation: numerical methods* (Athena Scientific, 1997)
2. C. Boutilier, R.I. Brafman, C. Domshlak, H.H. Hoos, D. Poole, CP-nets: A tool for representing and reasoning with conditional ceteris paribus preference statements. J. Artif. Intell. Res. **21**, 135–191 (2004)
3. W.W. Cohen, R.E. Schapire, Y. Singer, Learning to order things. J. Artif. Intell. Res. **10**, 243–270 (1999)
4. A.M. Colman, J.A. Stirk, Singleton bias and lexicographic preferences among equally valued alternatives. J. Econ. Behav. Organ. **40**(4), 337–351 (1999)
5. G.B. Dantzig, A. Orden, P. Wolfe, The generalized simplex method for minimizing a linear form under linear inequality restraints. Pac. J. Math. **5**, 183–195 (1955)
6. R.M. Dawes, The robust beauty of improper linear models in decision making. Am. Psychol. **34**, 571–582 (1979)
7. J. Dombi, C. Imreh, N. Vincze, Learning lexicographic orders. Eur. J. Oper. Res. **183**(2), 748–756 (2007)
8. P.C. Fishburn, Lexicographic orders, utilities and decision rules: A survey. Manage. Sci. **20**(11), 1442–1471 (1974)
9. J. Kevin Ford, N. Schmitt, S.L. Schechtman, B.M. Hults, M.L. Doherty, Process tracing methods: contributions, problems and neglected research issues. Organ. Behav. Hum. Decis. Process. **43**, 75–117 (1989)
10. A. Quesada, Negative results in the theory of games with lexicographic utilities. Econ. Bull. **3**(20), 1–7 (2003)
11. R.L. Rivest, Learning decision lists. Mach. Learn. **2**(3), 229–246 (1987)
12. M. Schmitt, L. Martignon, On the complexity of learning lexicographic strategies. J. Mach. Learn. Res. **7**, 55–83 (2006)
13. L. Torrey, J. Shavlik, T. Walker, R. Maclin, Relational macros for transfer in reinforcement learning, in *Proceedings of the Seventeenth Conference on Inductive Logic Programming* (Corvallis, Oregon, 2007)
14. M.R.M. Westenberg, P. Koele, Multi-attribute evaluation processes: methodological and conceptual issues. Acta Psychol. **87**, 65–84 (1994)
15. F. Yaman, T.J Walsh, M. Littman, M. desJardins, Democratic approximation of lexicographic preference models, 2009. Submitted to Artificial Intelligence
16. F. Yaman, T.J. Walsh, M.L. Littman, M. desJardins, Democratic approximation of lexicographic preference models, in *Proceedings of the Twenty-Fifth International Conference (ICML 2008)* (Helsinki, Finland, 2008), pp. 1200–1207

Learning Ordinal Preferences on Multiattribute Domains: The Case of CP-nets*

Yann Chevaleyre, Frédéric Koriche, Jérôme Lang, Jérôme Mengin, and Bruno Zanuttini

Abstract A recurrent issue in decision making is to extract a preference structure by observing the user's behavior in different situations. In this paper, we investigate the problem of learning ordinal preference orderings over discrete multiattribute, or combinatorial, domains. Specifically, we focus on the learnability issue of conditional preference networks, or *CP-nets*, that have recently emerged as a popular graphical language for representing ordinal preferences in a concise and intuitive manner. This paper provides results in both passive and active learning. In the passive setting, the learner aims at finding a CP-net compatible with a supplied set of examples, while in the active setting the learner searches for the cheapest interaction policy with the user for acquiring the target CP-net.

1 Introduction

Suppose we observe a user expressing her preferences about airplane tickets. Namely, she prefers an Aeroflot flight landing at Heathrow to a KLM flight landing at Gatwick, while she prefers an Aeroflot flight landing at Heathrow to a KLM flight

*Partially supported by the ANR projects CANAR (ANR-06-BLAN-0383-02) and PHAC (ANR-05-BLAN-0384-01)

Y. Chevaleyre (✉) and J. Lang
LAMSADE, Université Paris-Dauphine, CNRS, France
e-mail: lang@lamsade.dauphine.fr, yann.chevaleyre@lamsade.dauphine.fr

F. Koriche
LIRMM, Université Montpellier II, CNRS, France
e-mail: frederic.koriche@lirmm.fr

J. Mengin
IRIT, Université Paul Sabatier, CNRS, France
e-mail: jerome.mengin@irit.fr

B. Zanuttini
GREYC, Université de Caen Basse-Normandie, CNRS, ENSICAEN, France
e-mail: bruno.zanuttini@info.unicaen.fr

J. Fürnkranz and E. Hüllermeier (eds.), *Preference Learning*,
DOI 10.1007/978-3-642-14125-6_13,

landing at Heathrow. An intuitively correct hypothesis that explains her behavior is that she prefers Aeroflot to KLM unconditionally, and Heathrow to Gatwick, again unconditionally. Such an hypothesis allows for predicting that she will prefer an Aeroflot flight landing at Heathrow to anything else, and an Aeroflot flight landing at Gatwick to a KLM flight landing at Gatwick. Yet, this hypothesis is not able to predict whether she will prefer an Aeroflot flight landing at Gatwick or a KLM flight landing at Heathrow. Now, if we observe later that she prefers a KLM flight landing at Gatwick to a KLM flight landing at Heathrow, the current hypothesis must be updated. A new possible hypothesis, among others, could be that she prefers Aeroflot to KLM, and Heathrow to Gatwick when flying on Aeroflot and *vice versa* when flying on KLM.

The learning problem underlying this scenario is to extract a preference structure by observing the user's behavior in situations involving a choice among several alternatives. Each alternative can be specified by many attributes, such as the flight company, the airport location, the arrival and departure time, the number of transits, and so on. As a result, the space of possible situations has a *combinatorial* structure. Furthermore, the preferences induced by the user's behavior are intrinsically related to *conditional preferential independence*, a key notion in multiattribute decision theory [20]. Indeed, the initial hypothesis is unconditional in the sense that the preference over the values of each attribute is independent of the values of other attributes. By contrast, in the final hypothesis, the user's preference between airports is conditioned by the airline company.

Preferences over combinatorial domains have been investigated in detail by researchers in multiattribute Decision Theory (DT) and Artificial Intelligence (AI). In multiattribute DT, researchers have focused on *modeling* preferences, that is, giving axiomatic characterizations of classes of preference relations or utility functions, while researchers in AI have concentrated on the development of languages for *representing* preferences that are computationally efficient; such languages have to express preferences as succinctly as possible, and to come with fast algorithms for finding optimal alternatives.

These classes of models and languages can be partitioned by examining the mathematical nature of the preferences they consider. Namely, a distinction is made between *ordinal* preferences that consist in ranking the alternatives, and *numerical* preferences consisting of utility functions mapping each alternative to some number. Learning or eliciting numerical preferences has received a great deal of attentions in the literature [7, 9, 10, 16, 18]. A related stream of work is preference elicitation in the context of combinatorial auctions [26]; what has to be learnt is the valuation function of every buyer, which associates with every combination of goods the maximum value that she is ready to pay for it. Recently, there has been a growing interest for learning ordinal preferences using numerical models. Many standard machine learning methods, such as neural networks [8] or support vector machines [14], have been adapted to this framework, often called *learning to rank instances* by the machine learning community.

In contrast, learning preferences using *ordinal* models has received much less attention. In fact, the most studied model is the *lexicographic preference model* that provides ordering relations between examples described as pairwise comparisons between tuples of values. Dombi et al. [13] propose a learning algorithm that elicits a lexicographic preference model by guiding the user through a sequence of queries involving test examples. Although it is possible to determine in polynomial time whether there exists a lexicographical model compatible with a set of such examples, Schmitt and Martignon [27] show that the corresponding optimization problem of minimizing preference disagreement is NP-hard, and can even not be approximated in polynomial time to within a constant factor. They also give the Vapnik–Chervonenkis dimension of lexicographical preference relations: it is equal to the number of attributes. Finally, Yaman et al. [32] do not commit to a single lexicographic preference relation but approximate the target using the votes of a *collection* of consistent lexicographic preference relations. In a nutshell, learning lexicographic preference relations proves not to be so hard, but this comes with a price, namely, the induced hypothesis is highly restrictive.

In this paper, we examine a different class of ordinal preference models, where the hypotheses we make bear on the preferential dependence structure. As emphasized in the above scenario, a key point when dealing with ordinal preferences on combinatorial domains is the dependence structure between attributes. To this point, conditional preference networks, also known as *CP-nets*, are a graphical language for representing preferences based on conditional preferential independence [5]. Informally, a CP-net consists in a collection of attributes pointing to a (possibly empty) set of parents, and a set of conditional tables associated with each attribute, expressing the local preference on the values of the attribute given all possible combinations of values of its parents. The transitive closure of these local preferences is a partial order over the set of alternatives, which can be extended into several total orders. CP-nets and their generalizations are probably the most popular compact representation language for ordinal preferences in multiattribute domains.

While many facets of CP-nets have been studied into detail, such as consistency, dominance checking, and optimization (constrained and unconstrained), the problem of learning CP-nets from examples have only rarely, and very recently, been addressed. One exception is [12], who proposed an algorithm that, given a set of examples, outputs a CP-net which implies them (see Sect. 5 for more details). Although not directly concerned with CP-nets, a related work is [25] which proposes to learn preference theories in the sense of [15].

The aim of this position paper is to examine the problem of learning CP-nets according to several dimensions that naturally emerges in preference learning. The first dimension is to consider whether the user's preferences are representable, or not, by a CP-net. If the target preference relation can be described by a CP-net, the goal is to identify this network. Alternatively, if the target preference relation is not representable by a CP-net, we can only hope finding an approximation of it. Among the candidate approximations, some of them are particularly relevant from a reasoning viewpoint. From this perspective, we shall concentrate on finding CP-nets for which the target relation is a "completion" of the hypothesized relation. Orthogonally, a second dimension in CP-net learning is to consider whether the learning process is merely *passive*, by simply observing the user's behavior in given situations, or *active*, by allowing the learner to test the user's behavior in some carefully chosen situations. In many contexts, it is relevant to identify, or at least approximate, a CP-net by mixing both active and passive learning. Consider, for instance, a system helping a user to find a flat from a large database, such as in [29, 30]. A flat is described by attributes such as price, location, size, etc. The system can start to extract a pool of preferences by observing the user's behavior, and then use queries to converge on the ideal hypothesis while minimizing the number of interactions.

In the different learning models that arise from these dimensions, we will investigate the learnability issues in terms of the worst-case number of resources required to converge toward the desired CP-net, where resources refer both to the running time, the sample complexity in passive learning, and the query complexity in active learning. Section 2 provides the necessary background about CP-nets. In Sect. 3, we extend the paradigm of concept learning to preference learning and introduce two frameworks, one for the problem of learning partial orderings that are representable by a CP-net, and the other for the problem of learning linear orderings that are not representable by a CP-net. Our learnability results lie in the next three sections. Namely, Sect. 4 focuses on the VC-dimension and the approximate fingerprint property for classes of CP-nets. Section 5 addresses passive learning of CP-nets. Section 6 considers active learning of CP-nets. Finally, Sect. 7 briefly discusses issues for further work.

Given that this is a position paper, the proofs of the results are only sketched, or even omitted. They can be found in [11, 21, 23].

2 Conditional Preference Networks

Throughout this paper, we shall assume a finite list $V = \langle X_1, \ldots, X_n \rangle$ of *attributes*, with their associated finite *domains* $D = \langle D_1, \ldots, D_n \rangle$. An attribute X_i is *binary* if D_i has two elements, which by convention we note x_i and $\overline{x}_i$. By $\mathcal{V} = \times_{X_i \in V} D_i$, we denote the set of all complete assignments, called *outcomes*.

For any nonempty subset X of V, we let $\mathcal{X} = \times_{X_i \in X} D_i$. Elements of $\mathcal{X}$ are called X-assignments and denoted using vectorial notation, *e.g.*, $\boldsymbol{x}$. For any disjoint subsets X and Y of V, the concatenation of assignments $\boldsymbol{x} \in \mathcal{X}$ and $\boldsymbol{y} \in \mathcal{Y}$, denoted $\boldsymbol{x}\boldsymbol{y}$, is the $(X \cup Y)$-assignment which assigns to attributes in X (resp. Y) the value assigned by $\boldsymbol{x}$ (resp. $\boldsymbol{y}$).

A *preference relation* is a reflexive and transitive binary relation $\succeq$ over $\mathcal{V}$. A *complete preference relation* is a preference relation $\succeq$ that is connected, that is, for every $\boldsymbol{x}, \boldsymbol{y} \in \mathcal{X}$ we have either $\boldsymbol{x} \succeq \boldsymbol{y}$ or $\boldsymbol{y} \succeq \boldsymbol{x}$. A *strict preference relation* $\succ$ is an irreflexive and transitive (thus asymmetric) binary relation over $\mathcal{V}$. A *linear preference relation* is a strict preference relation that is connected. From a preference relation $\succeq$, we define a strict preference relation in the usual way: $\boldsymbol{x} \succ \boldsymbol{y}$ iff $\boldsymbol{x} \succeq \boldsymbol{y}$ and not $(\boldsymbol{y} \succeq \boldsymbol{x})$.

Preferences between outcomes that differ in the value of one attribute only, all other attributes being equal (or *ceteris paribus*) are often easy to assert, and to understand. CP-nets [5] are a graphical language for representing such preferences. Informally, a CP-net is composed of a directed graph representing the preferential dependencies between attributes, and a set of conditional preference tables expressing, for each attribute, the local preference on the values of its domain given all possible combinations of values of its parents.

Let us call a *swap* any pair of outcomes $(\boldsymbol{x}, \boldsymbol{y})$ that differ in the value of one attribute only, and let us then call *swapped attribute* the attribute that has different values in $\boldsymbol{x}$ and $\boldsymbol{y}$.

Definition 1. Suppose that V is partitioned into the subsets X, Y, and Z. Let $\succ$ be a linear preference relation over $\mathcal{V}$. Then, we say that X is *preferentially independent* of Y given Z (w.r.t. $\succ$) if for all $\boldsymbol{x}_1, \boldsymbol{x}_2 \in \mathcal{X}$, $\boldsymbol{y}_1, \boldsymbol{y}_2 \in \mathcal{Y}$, $\boldsymbol{z} \in \mathcal{Z}$,

$$\boldsymbol{x}_1\boldsymbol{y}_1\boldsymbol{z} \succ \boldsymbol{x}_2\boldsymbol{y}_1\boldsymbol{z} \quad \text{if and only if} \quad \boldsymbol{x}_1\boldsymbol{y}_2\boldsymbol{z} \succ \boldsymbol{x}_2\boldsymbol{y}_2\boldsymbol{z}$$

Example 1. Consider three binary attributes X_1, X_2, and X_3 and suppose that the four swaps on X_2 are ordered as follows:

$$\begin{aligned} x_1x_2x_3 &\succ x_1\overline{x}_2x_3 \\ \overline{x}_1\overline{x}_2x_3 &\succ \overline{x}_1x_2x_3 \\ x_1x_2\overline{x}_3 &\succ x_1\overline{x}_2x_3 \\ \overline{x}_1\overline{x}_2\overline{x}_3 &\succ \overline{x}_1x_2\overline{x}_3 \end{aligned}$$

We can see that, irrespective of the value of X_3, if x_1 is the case, then x_2 is preferred to $\overline{x}_2$, whereas if $\overline{x}_1$ is the case, then $\overline{x}_2$ is preferred to x_2. This ordering on the X_2-swaps with two *conditional preferences*: $x_1 : x_2 \succ \overline{x}_2$ and $\overline{x}_1 : \overline{x}_2 > x_2$. We remark that X_2, given X_1, is preferentially independent of X_3.

Definition 2. A *CP-net* N over $V = \{X_1, \ldots, X_n\}$ consists in a directed graph over V, and a set of preference tables $\text{CPT}(X_i)$ associated with each $X_i \in V$. For attribute X_i, we denote by $\text{Pa}(X_i)$ the set of parents of X_i in the graph of N, and by $\text{NonPa}(X_i)$ the set $V \backslash (\{X_i\} \cup \text{Pa}(X_i))$.

Each conditional preference table $\text{CPT}(X_i)$ is a list of rows, also called *entries* or *rules*, of the form $\boldsymbol{u} : x_{i_1} \succ \cdots \succ x_{i_m}$, where $\boldsymbol{u}$ is an instantiation of $\text{Pa}(X_i)$ and $x_{i_1} \succ \cdots \succ x_{i_m}$ is a linear ordering of the domain D_i (with $m = |D_i|$). It indicates that $\boldsymbol{u}\boldsymbol{z}x_{i_j} \succ \boldsymbol{u}\boldsymbol{z}x_{i_{j+1}}$ for every possible instantiation $\boldsymbol{z}$ of $\text{NonPa}(X_i)$.

Example 2. A CP-net over the binary attributes X_1, X_2, and X_3 is:

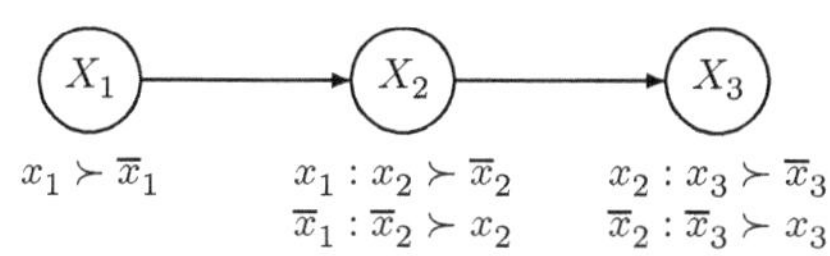

where an edge from X to Y means "X is a parent of Y". The associated ordering of the swaps is:

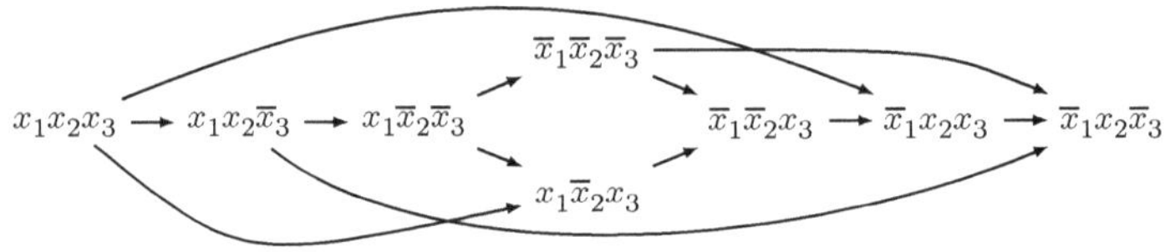

where $\boldsymbol{x} \to \boldsymbol{y}$ means "$\boldsymbol{x}$ is preferred to $\boldsymbol{y}$".

The *size* of a CP-net N, denoted by $|N|$, is the number of entries in the preference tables of N:

$$|N| = \sum_{X_i \in V} \prod_{X_j \in \mathrm{Pa}(X_i)} |\mathcal{X}_j|$$

which in the case of binary CP-nets boils down to $|N| = \sum_{X_i \in V} 2^{|\mathrm{Pa}(X_i)|}$.

Although a CP-net only specifies an ordering of all swaps, we are naturally interested in the transitive closure of this ordering; for a CP-net N, we write $\succ_N$ for this transitive closure. Note that this relation $\succ_N$ may not be total, and it may not be a irreflexive since it may contain cycles. Yet, we know from [5] that if the graph of N is acyclic, then $\succ_N$ is a strict preference relation (i.e. contains no cycles), and we say that N is *consistent*. Otherwise, we say that N is inconsistent. Note that even if N is consistent, $\succ_N$ may still not be connected; it can then be completed in a number of linear preference relations. If $\succ$ is one of them, we say that $\succ$ is a *completion* of N, or that it is *compatible* with N.

It is important to keep in mind that several CP-nets can induce the same preference relation: for example, if a CP-net N over the binary attributes X_1 and X_2 contains the table $x_2 : x_1 \succ \overline{x}_1$ and $\overline{x}_2 : x_1 \succ \overline{x}_1$, then the CP-net N' in which X_1 has no parent and the table $x_1 \succ \overline{x}_1$ is equivalent to N. In general, for every consistent CP-net N there is a unique CP-net N' equivalent to N that is minimal in the number of parents/table entries for each variable.

It is easy to verify that every linear preference relation is compatible with exactly one (minimal) CP-net. For instance, $x_1x_2 \succ \overline{x}_1\overline{x}_2 \succ x_1\overline{x}_2 \succ \overline{x}_1x_2$ is compatible with the CP-net

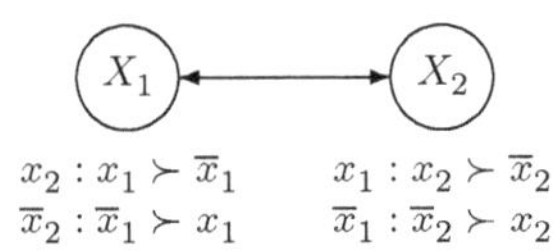

The following property will be frequently used in the remaining sections.

Proposition 1. *([5]) Let N be an acyclic CP-net and $\boldsymbol{x}$, $\boldsymbol{y}$ two outcomes. Then $\boldsymbol{x} \succ_N \boldsymbol{y}$ iff there is a sequence of swaps $(\boldsymbol{x}^0, \boldsymbol{x}^1), (\boldsymbol{x}^1, \boldsymbol{x}^2), \ldots, (\boldsymbol{x}^{k-1}, \boldsymbol{x}^k)$ such that $\boldsymbol{x}^0 = \boldsymbol{x}$, $\boldsymbol{x}^k = \boldsymbol{y}$, and for every $0 \leq i < k$, $\boldsymbol{x}^i \succ_N \boldsymbol{x}^{i+1}$, that is, if X_{j_i} is the attribute swapped between $\boldsymbol{x}^i$ and $\boldsymbol{x}^{i+1}$, and if $\boldsymbol{u}$ is the vector of values commonly assigned by $\boldsymbol{x}$ and $\boldsymbol{y}$ to the parents of X_{j_i}, then N contains $\boldsymbol{u} : x^i_{j_i} \succ x^{i+1}_{j_i}$.*

Although a CP-net is usually defined as above, most of the results presented here extend to *possibly incomplete CP-nets*. Such a CP-net is one in which each conditional preference table $CPT(X_i)$ contains *at most* one conditional preference rule per instantiation of $\text{Pa}(X_i)$ (instead of *exactly one*). A particular case is when some of the tables are empty. The semantics of an incomplete CP-net is still given by the transitive closure of the dominance relation on swaps induced by the rules. An important difference is that in an incomplete CP-net, not all swaps are comparable.

The rationale for considering incomplete CP-nets can be understood with the following example. A user may well know that she prefers traveling by bus rather than in the subway in Paris, and *vice-versa* in London, but be unable to state her preference into a city which she has never been, say Madrid. This does not mean that she would not have a preference, rather that she does not know it (so far). In this case, a variable encoding the transportation means (with values *subway* and *bus*) would have the variable encoding the city as its parent, with values *Paris*, *London*, and *Madrid*, but would contain only two rules.

3 Learning CP-nets: Learning What?

The problem of concept learning is to extrapolate from a collection of examples, each labeled as either positive or negative by some unknown target concept, a representation of this concept that accurately labels future, unlabeled examples. Most concept learning algorithms operate over some *concept class*, which captures the set of concepts that the learner can potentially generate over all possible sets of training examples.

In the setting suggested by our framework, a *concept* is a strict preference relation $\succ$ over $\mathcal{V}$. We say that a concept $\succ$ is *representable* by a CP-net N if the induced ordering $\succ_N$ coincides with $\succ$, that is, $\succ = \succ_N$. For example, the preference relation $\succ$ defined by $\{ab \succ a\overline{b}, ab \succ \overline{a}b, \overline{a}\overline{b} \succ a\overline{b}, \overline{a}\overline{b} \succ \overline{a}b\}$ is representable by the CP-net N specified by $\{a : b \succ \overline{b}, \overline{a} : \overline{b} \succ b, b : a \succ \overline{a}, \overline{b} : \overline{a} \succ a\}$. A *representation class* is a collection $\mathcal{N}$ of consistent CP-nets, and the *concept class* $\mathcal{C}_\mathcal{N}$ defined over $\mathcal{N}$ is the set of all preference relations $\succ$ that are representable by a CP-net N in $\mathcal{N}$.

Since we consider consistent CP-nets only, any target concept $\succ$ in a class $\mathcal{C}_\mathcal{N}$ can be represented by a unique minimal CP-net N. So, with a slight abuse of language, we shall simply say that $\mathcal{C}_\mathcal{N}$ is the class of all CP-nets in $\mathcal{N}$. For instance, the concept class $\mathcal{C}_{\text{ACY}}$ is the class of all acyclic CP-nets; and $\mathcal{C}_{\text{TREE}}$ is the class of tree-structured CP-nets.

With these notions in hand, we assume that the user has in mind a target preference ordering $\succ$, and the learner has at its disposal a predetermined and known class of CP-nets $\mathcal{C}_\mathcal{N}$. In this study, we shall consider two different types of target concepts.

(a) The target concept is a preference relation $\succ$ that belongs to the learner's concept class $\mathcal{C}_{\mathcal{N}}$. In other words, there is a CP-net N in $\mathcal{N}$, such that $\succ_N$ coincides with $\succ$. In this context, the goal of the learner is to find N.

(b) The target concept is a preference relation $\succ$ that does *not necessarily* belong to the learner's concept class $\mathcal{C}_{\mathcal{N}}$. For example, we can easily observe that the linear ordering $\succ$ defined by $\{ab \succ \overline{a}\overline{b} \succ a\overline{b} \succ \overline{a}b\}$ cannot be represented by any CP-net. Still, we shall make the assumption that $\succ$ is a completion of some representation in $\mathcal{N}$. Specifically, we say that $\succ$ is a *completion* of a CP-net N if $\boldsymbol{x} \succ_N \boldsymbol{y}$ implies $\boldsymbol{x} \succ \boldsymbol{y}$ for any pair of outcomes $(\boldsymbol{x}, \boldsymbol{y})$. For example, the above ordering $\succ$ is a completion of $N = \{a : b \succ \overline{b}, \overline{a} : \overline{b} \succ b, b : a \succ \overline{a}, \overline{b} : \overline{a} \succ a\}$. In this "agnostic" setting, the goal of the learner is to find a CP-net N of which $\succ$ is a completion.

Note that the distinction between (a) and (b) is not on the set of objects we want to learn, but on their interpretation, which has crucial consequences on how examples are interpreted.

3.1 Learning a Preference Relation Induced by a CP-net

Let us start with context (a) where the target concept is representable by a CP-net N of some given representation class $\mathcal{N}$. Here, an instance, or *example*, is a pair $(\boldsymbol{x}, \boldsymbol{y})$ of outcomes, and an *instance class* is a set $\mathcal{E}$ of examples. Given a target concept $\succ_N$, and an example $(\boldsymbol{x}, \boldsymbol{y})$, we say that $(\boldsymbol{x}, \boldsymbol{y})$ is *positive* for $\succ_N$ if $\boldsymbol{x} \succ_N \boldsymbol{y}$, that is, $\boldsymbol{x}$ dominates $\boldsymbol{y}$ according to $\succ_N$. Dually, $(\boldsymbol{x}, \boldsymbol{y})$ is *negative* for $\succ_N$ if $\boldsymbol{x} \not\succ_N \boldsymbol{y}$. It is important to keep in mind that, in general, if the pair $(\boldsymbol{x}, \boldsymbol{y})$ is a negative example of $\succ_N$, then the reverse pair $(\boldsymbol{y}, \boldsymbol{x})$ is *not* necessarily a positive example of $\succ_N$.

In this context, our learning problem can be seen as a standard concept learning problem: given a set $\mathcal{T}$ of positive and negative training examples, we want to find a CP-net that "implies" all positive examples and no negative example.

Definition 3. Let N be a CP-net over $\mathcal{V}$. An example $(\boldsymbol{x}, \boldsymbol{y})$ is *entailed by* N if $\boldsymbol{x} \succ_N \boldsymbol{y}$. A set of examples $\mathcal{T}$ is *implicatively consistent with* N, or *implied* by N, if

- all positive examples in $\mathcal{T}$ are entailed by N;
- no negative example in $\mathcal{T}$ is entailed by N.

Finally, we shall say that a training set $\mathcal{T}$ is *implicatively compatible* if it is implied by at least one CP-net.

3.2 Learning a CP-net Which Approximates the User's Preferences

Now, we turn to context (b) and make the assumption that the user's preference relation $\succ$ is a linear order, in general not exactly representable by any CP-net. In this framework, we start by examining an appropriate notion of consistency between a CP-net and a training set. Consider the following example:

Example 3. We have two binary attributes X_1 and X_2 (with domains $\{x_1, \overline{x}_1\}$ and $\{x_2, \overline{x}_2\}$), and the set of positive examples

$$\mathcal{T} = \{(x_1x_2\,,\, x_1\overline{x}_2),\ (x_1\overline{x}_2\,,\, \overline{x}_1x_2),\ (\overline{x}_1x_2\,,\, \overline{x}_1\overline{x}_2)\}$$

What do we expect to learn from the above set of examples $\mathcal{T}$? The transitive closure of $\mathcal{T}$ is the complete preference relation $x_1x_2 \succ x_1\overline{x}_2 \succ \overline{x}_1x_2 \succ \overline{x}_1\overline{x}_2$. This preference relation is *separable* (the agent unconditionally prefers x_1 to $\overline{x}_1$ and x_2 to $\overline{x}_2$). The fact that $x_1\overline{x}_2$ is preferred to $\overline{x}_1x_2$ simply means that when asked to choose between X_1 and X_2, the agent prefers to give up X_2 (think of X_1 meaning "getting rich" and X_2 meaning "beautiful weather tomorrow"). Intuitively, since $\mathcal{T}$ is separable, we expect to output a structure N that contains $x_1 \succ \overline{x}_1$ and $x_2 \succ \overline{x}_2$. However, no CP-net implies $\mathcal{T}$, whatever the dependencies. The structure N induces a *partial* preference relation in which $x_1\overline{x}_2$ and $\overline{x}_1x_2$ are incomparable. More generally, no *ceteris paribus* structure can "explain" that $x_1 \succ \overline{x}_1$ is "more important" than $x_2 \succ \overline{x}_2$ (i.e., with no intermediate alternative). Therefore, if we look for a structure *implying* all the examples, we will simply output "failure". On the other hand, if we look for a separable CP structure that is simply *contingent* with the examples, i.e., that does not imply the contrary of the examples, we will output N.

The explanation is that when an agent expresses a CP-net, the preference relation induced by this CP-net *is not meant to be the whole agent's preference relation, but a subset (or a lower approximation) of it.* In other terms, when an agent expresses the CP-net N, she simply expresses that she prefers x_1 to $\overline{x}_1$ *ceteris paribus* (i.e., for a fixed value of X_2) and similarly for the preference $x_2 \succ \overline{x}_2$; the fact that $x_1\overline{x}_2$ and $\overline{x}_1x_2$ are incomparable in N surely does not mean that the user really sees them incomparable, but, more technically, that CP-nets are not expressive enough for representing the missing preference $x_1\overline{x}_2 \succ \overline{x}_1x_2$.[1]

Therefore, in such cases we do not look for a CP-net which implies the examples. Rather, we look for one whose preference relation is consistent with the examples. A first way of understanding consistency is to require that the learnt CP-net N be such that the examples are consistent with at least one preference relation extending N. Yet, there are cases where it may even be too strong to require that one of the completions of $\succ_N$ contain all the examples, in particular, if they come from multiple users (given that we want to learn the generic preferences of a group of users), or a single user in different contexts:

Example 4. Suppose that we learn that all users in a group unconditionally prefer x_1 to $\overline{x}_1$ and x_2 to $\overline{x}_2$, whereas their preferences between $x_1\overline{x}_2$ and $\overline{x}_1x_2$ may differ (think as x_1 and x_2 as, respectively, "being invited to a fine dinner" and "receiving a

[1] If we want to do this, we have to resort to a more expressive language such as TCP-nets [6] or conditional preference theories [31].

\$50 award"): then $\mathcal{T} \supseteq \{(\overline{x}_1 x_2\,,\, x_1\overline{x}_2),\ (x_1\overline{x}_2\,,\, \overline{x}_1 x_2)\}$. $\mathcal{T}$ is clearly inconsistent, so there cannot be any preference structure whose ordering can be completed into a linear preference relation that contains $\mathcal{T}$. However, if $\mathcal{N} = \{x_1 \succ \overline{x}_1, x_2 \succ \overline{x}_2\}$, then each example in $\mathcal{T}$ is (individually) contained in at least one completion of $\succ_N$.

Such considerations lead us to define two new notions of compatibility of a CP-net with a set of examples. Note that because the target concept is a linear order, $(\boldsymbol{x}, \boldsymbol{y})$ is a negative example if and only if $(\boldsymbol{y}, \boldsymbol{x})$ is a positive one. For this reason, we can make the assumption that all examples in the learner's training set $\mathcal{T}$ are *positive*, with the implicit knowledge that the reverse $(\boldsymbol{y}, \boldsymbol{x})$ of any pair $(\boldsymbol{x}, \boldsymbol{y})$ in $\mathcal{T}$ is negative.

Definition 4. Let N be a CP-net over $\mathcal{V}$. An example $(\boldsymbol{x}, \boldsymbol{y})$ is *consistent by completion with* N if there is a completion $\succ$ of $\succ_N$ such that $\boldsymbol{x} \succ \boldsymbol{y}$. Furthermore, we will say that a set of examples $\mathcal{T}$ is:

- *strongly consistent by completion with* N if there is a completion $\succ$ of $\succ_N$ such that for all $(\boldsymbol{x}, \boldsymbol{y}) \in \mathcal{T}$, $\boldsymbol{x} \succ \boldsymbol{y}$;
- *weakly consistent by completion with* N if every example $(\boldsymbol{x}, \boldsymbol{y}) \in \mathcal{T}$ is individually consistent by completion with N.

Finally, we will say that $\mathcal{T}$ is *strongly / weakly compatible* if it is strongly / weakly consistent by completion with at least one CP-net. Clearly, strong compatibility implies weak compatibility. Moreover, since an example $(\boldsymbol{x}, \boldsymbol{y})$ is consistent by completion with a CP-net N if and only if $(\boldsymbol{y}, \boldsymbol{x})$ is not implied by N, implicative compatibility implies strong compatibility.

As it stands, it turns out to be significantly more difficult to search for a CP-net strongly or weakly consistent with a set of examples than to search for a CP-net implicatively consistent with it. Therefore, we mainly focus on implicative compatibility; strong and weak compatibility will only be discussed in the context of *separable* CP-nets, that is, CP-nets where all variables are independent.

4 Learnability of CP-nets

In this section, we investigate the theoretical limits concerning the learnability issue of CP-nets. In essence, the Vapnik–Chervonenkis dimension of a class gives upper bounds on the difficulty to learn, in terms of numbers of examples, while the approximate fingerprint property gives lower bounds.

4.1 Vapnik–Chervonenkis Dimension

The *Vapnik–Chervonenkis (VC) dimension* of a class of concepts is a fundamental complexity measure used in theoretical machine learning. Intuitively, the VC-dimension of $\mathcal{C}$ is the maximum number of informative examples which can be received by the learner, where an "informative" example is an observation that helps the learner reducing the number of consistent hypotheses.

Formally, let $\mathcal{C}$ be a concept class defined over some representation class $\mathcal{R}$. A set of instances $\mathcal{T}$ is said to be *shattered* by $\mathcal{C}$ if, whatever the partition of $\mathcal{T}$ into $\mathcal{T}^+ \cup \mathcal{T}^-$, there is a concept $N \in \mathcal{C}$ which admits all instances in $\mathcal{T}^+$ as positive examples and all instances in $\mathcal{T}^-$ as negative examples. The *Vapnik–Chervonenkis dimension of* $\mathcal{C}$, denoted $VC(\mathcal{C})$, is the maximum size of a set of examples $\mathcal{T}$ which is shattered by $\mathcal{C}$. The intuition is that for such a set $\mathcal{T}$, as long as the learner ignores the label of at least one example, at least two concepts in $\mathcal{C}$ are consistent with the labels it has seen so far.

When the learner has access only to a certain kind of example (e.g., swap examples), it makes sense to adapt the notion of VC-dimension. So if $\mathcal{E}$ is a class of instances, we write $VC_{\mathcal{E}}(\mathcal{C})$ for the VC-dimension of $\mathcal{C}$ *with respect to* $\mathcal{E}$, that is, the maximum size of a $\mathcal{T} \subseteq \mathcal{E}$ which is shattered by $\mathcal{C}$. Clearly, if $\mathcal{E} \subseteq \mathcal{E}'$, then $VC_{\mathcal{E}}(\mathcal{C}) \leq VC_{\mathcal{E}'}(\mathcal{C})$.

Observe that, as there are 2^m partitions of a set of m examples into positive and negative examples, each of which must be captured by a different concept, $VC(\mathcal{C}) \leq \log_2 |\mathcal{C}|$ always holds (whatever the class of examples).

We now give the VC-dimension of the class of all CP-nets which have a fixed graph. The intuition here is that since the parents of each variable are known, the quantity of information needed to characterize a CP-net is exactly 1 per possible rule in the CP-net, namely, one pair of outcomes which dictates the conclusion of the rule.

Call *subgraph* of G a graph with the same vertices as G but whose set of edges is included in that of G.

Proposition 2. *Let G be a graph, and let $\mathcal{C}_G$ be the class of all concepts which are representable by a binary complete CP-net whose graph is G or a subgraph of G. Then the VC-dimension of $\mathcal{C}_G$ with respect to swap examples is exactly the number of conditional preference rules in any such CP-net. If CP-nets are possibly incomplete, then the VC-dimension (w.r.t. swap examples) is still the number of rules in any complete CP-net on (a subgraph of) G.*

This property can help us finding an upper bound of the VC-dimension of acyclic CP-nets. Intuitively, the number of acyclic graphs with indegree at most k is lower bounded by $|\mathcal{C}_G|$ for a convenient graph with k roots and $n-k$ vertices with indegree exactly k, and upper bounded by $(n-1)^{n(k+1)}$ (for each vertex, choose at most k parents). Since any binary-valued CP-net built over an acyclic graph of degree at most k allows at most $n2^k$ entries, there are at most $(n-1)^{n(k+1)}2^{n2^k}$ binary-valued acyclic CP-nets with degree at most k. We therefore obtain the following result.

Corollary 1. *Let $k \in o(n)$, and let $\mathcal{C}^k_{\text{ACY}}$ be the class of all concepts which are representable by a possibly incomplete binary CP-net whose graph is acyclic and with indegree at most k. Then, the VC-dimension of $\mathcal{C}^k_{\text{ACY}}$ with respect to swap or arbitrary examples is in $\tilde{\Theta}(n2^k)$.*

Finally, without any restriction over the degree, it can be shown that the VC-dimension grows as $\Theta(2^n)$, which is still much below the VC-dimension of the class of all possible CP-nets.

Corollary 2. *Let $\mathcal{C}_{\text{ACY}}$ be the class of all concepts that are representable by a possibly incomplete binary CP-net whose graph is acyclic. Then, the VC-dimension of $\mathcal{C}_{\text{ACY}}$ with respect to swap examples is in $\Theta(2^n)$.*

4.2 Approximate Fingerprints

Approximate fingerprints are a powerful tool for obtaining nonlearnability results in active learning. Intuitively, a class of concepts $\mathcal{C}$ has the approximate fingerprint property if there is a subset $\mathcal{C}^*$ of $\mathcal{C}$ such that for any concept $N \in \mathcal{C}$, there is an example with which N is consistent, but with which only a superpolynomially small fraction of the concepts in $\mathcal{C}^*$ are also consistent.

This property can be used to show that in an interactive learning setting, a hypothesis $\widehat{N} \in \mathcal{C}$ may fail on an example which only gives clues about superpolynomially few candidate hypotheses. Hence, if the learner only gets information from such failures, in the worst case it *necessarily* makes an exponential number of errors before correctly identifying the target concept. We refer the reader to [2] for formal details.

As for the VC-dimension, the definition of approximate fingerprints can be restricted to instance classes $\mathcal{E}$. Observe that if $\mathcal{E} \subseteq \mathcal{E}'$ and $\mathcal{C}$ has the approximate fingerprint property with respect to $\mathcal{E}$, then it also has this property with respect to $\mathcal{E}'$.

Proposition 3 ([21]). *Let $\mathcal{C}_{\text{ACY}}$ be the class of all concepts which are representable by a binary complete CP-net whose graph is acyclic. Then $\mathcal{C}_{\text{ACY}}$ has the approximate fingerprint property with respect to swap examples.*

Proposition 4 ([21]). *Let $\mathcal{C}_{\text{TREE}}$ be the class of all concepts which are representable by a binary complete CP-net whose graph is a tree. Then $\mathcal{C}_{\text{TREE}}$ has the approximate fingerprint property with respect to arbitrary examples.*

5 Passive Learning of CP-nets

In this section, we investigate *passive* learning of CP-nets. In this setting, the only information about the target concept available to the learner is a set of examples. We shall concentrate here on the widely studied *Probably Approximately Correct* (PAC) learning model introduced by [28]. The intent of this model is to obtain with high probability a representation that is a good approximation of the target concept. To formalize the notion of a good approximation, we need to assume that there is some fixed, but unknown, probability distribution $\mathcal{D}$ defined on the example space $\mathcal{E}$, from which the available examples were drawn. In our case, $\mathcal{D}$ would define a probability over each instance $(\boldsymbol{x}, \boldsymbol{y})$. Given a target concept $\succ$, we then define the *error* of a hypothesized CP-net $\widehat{N}$ as the probability that $\succ$ and $\succ_{\widehat{N}}$ disagree on an example:

$$error(\widehat{N}) = \mathbf{Pr}_{(x,y)\sim\mathcal{D}}\left[\left(\boldsymbol{x} \succ \boldsymbol{y} \text{ and } \boldsymbol{x} \not\succ_{\widehat{N}} \boldsymbol{y}\right) \text{ or } \left(\boldsymbol{x} \not\succ \boldsymbol{y} \text{ and } \boldsymbol{x} \succ_{\widehat{N}} \boldsymbol{y}\right)\right]$$

How does one generate a *good* approximation? In the PAC model, one does this by looking at an example set, in which each example $(\boldsymbol{x}, \boldsymbol{y})$ has been drawn independently at random from the distribution $\mathcal{D}$, and labeled with "+" (positive) if $\boldsymbol{x} \succ \boldsymbol{y}$ and with "–" (negative) if $\boldsymbol{x} \not\succ \boldsymbol{y}$.

Thus, in the PAC setting, training and testing use the same distribution, and there is no noise in either phase. A learning algorithm is then a computational procedure that takes a sample of the target concept $\succ$, consisting of a sequence of independent random examples of $\succ$, and returns a hypothesis. We can define PAC learnability of CP-nets as follows.

Definition 5 (PAC learning). A concept class $\mathcal{C}_{\mathcal{N}}$ is *PAC learnable* by an example class $\mathcal{E}$ if there is a polynomial time learning algorithm A and a polynomial $p(\cdot,\cdot,\cdot)$ such that for any target concept $\succ$ in $\mathcal{C}_{\mathcal{N}}$ over n variables, any probability distribution $\mathcal{D}$ over $\mathcal{E}$, and any parameters $\delta, \epsilon \in (0, 1)$, if the algorithm A is given at least $p(n, \frac{1}{\epsilon}, \frac{1}{\delta})$ independent random examples of $\succ$ drawn according to $\mathcal{D}$, then with probability at least $1 - \delta$, A returns a hypothesis $\widehat{N} \in \mathcal{N}$ with $error(\widehat{N}) \leq \epsilon$. The smallest such polynomial p is called the *sample complexity* of the learning algorithm A.

The intent of this definition is that the learning algorithm must process the examples in polynomial time, and must be able to produce a good approximation of the target concept with high probability using only a reasonable number of training examples.

It is important to emphasize that our learnability results are defined over specific instance classes. In particular, if $\mathcal{E}$ is the class of all swap instances, then any distribution $\mathcal{D}$ over $\mathcal{E}$ will assign a zero probability to any "non-swap" instance. This restriction has deep consequences on the predictive power of CP-nets. Namely, even if a positive learnability result with swap instances guarantees that the hypothesized CP-net $\widehat{N}$ is expected to correctly classify "swap" instances drawn independently at random according to the distribution $\mathcal{D}$, such a result does *not* ensure that the learner will correctly classify arbitrary outcome pairs. Indeed, even if the probability of making a mistake on swaps is low, the probability of making a mistake on an arbitrary instance $(\boldsymbol{x}, \boldsymbol{y})$ may increase along an improving sequence from $\boldsymbol{y}$ to $\boldsymbol{x}$.

Many positive learnability results in the PAC model are obtained by showing that (1) there is an efficient algorithm capable of finding a hypothesized representation that is *consistent by implication* with a given sample of the target concept (called a *consistent algorithm*), and (2) the sample complexity of any such algorithm is polynomial.

The sample complexity of a consistent learning algorithm is usually measured using the VC-dimension of the concept class $\mathcal{C}_{\mathcal{N}}$. Indeed, it is shown in [4] that the sample complexity of a consistent learning algorithm is at most

$$\frac{1}{\epsilon(1-\sqrt{\epsilon})}\left(2\mathrm{VC}(\mathcal{C}_{\mathcal{N}})\ln\frac{6}{\epsilon}+\ln\frac{2}{\delta}\right) \tag{1}$$

A preliminary work on passive learning of CP-nets is [12]. They give an algorithm which, given a set of positive examples, outputs a CP-net that implies them, under some conditions. It is not entirely clear yet which class of CP-nets is learned by this algorithm.

5.1 PAC Learning of Acyclic CP-nets

We first investigate PAC learnability of various classes of acyclic CP-nets when the examples provided to the learner are swaps. Recall that for such examples, the dominance test with acyclic CP-nets is linear-time solvable (simple lookup in the conditional preference table), contrary to the general case.

We first show that even for the restricted class of concepts which are representable by a CP-net whose graph is a chain, the consistency problem is NP-complete. It follows directly (unless P = NP) that this class is not PAC-learnable with the very weak restriction that the produced hypothesis classifies correctly the examples received.

Proposition 5. *Deciding whether there exists a binary complete CP-net whose graph is a chain and which implies a given set of swaps is NP-complete. The result holds even if all examples are positive.*

A proof based on a reduction from the Hamiltonian path problem can be found in [11]. Another interesting class of CP-nets is that of acyclic *singlyconnected* CP-nets [5], that is, those acyclic CP-nets in whose graph each pair of vertices is connected by at most one directed path. Unfortunately, again we have a negative result for PAC learnability of such CP-nets.

Proposition 6. *Deciding whether there exists a binary complete CP-net whose graph is acyclic singlyconnected and which implies a given set of swaps is NP-complete. The result holds even if all examples are positive.*

This result [11] can be proven with a reduction from propositional satisfiability (SAT). We conjecture that a similar negative result holds for more general classes of acyclic CP-nets.

We now turn to positive results with swaps. Observe that if $\mathcal{C}_{\mathcal{N}'}$ is PAC learnable with swaps and $\mathcal{C}_{\mathcal{N}} \subseteq \mathcal{C}_{\mathcal{N}'}$, then $\mathcal{C}_{\mathcal{N}}$ is not necessarily PAC learnable with swaps as well. So our positive results do not contradict the negative ones. The main result is that tree CP-nets are PAC-learnable from swaps.

Proposition 7. *The class of all concepts which are representable by a (possibly incomplete) tree binary CP-net is PAC-learnable from swap examples.*

In addition, learning the structure of such a CP-net can be reduced to finding a spanning tree in a directed graph [11]. Since in general there may be several CP-nets which imply a given set of examples, it is interesting to impose some restrictions, e.g., on the degree of the forest (maximum number of children of a node). The next result states that the class of CP-nets whose graph is a forest with degree at most k is *improperly* PAC-learnable (in quasi-polynomial time) [11]. That is, it is "PAC-learnable", but the hypothesis may be in a larger representation class than the target concept.

Proposition 8. *There is a quasi-polynomial time algorithm which, given a set of swaps $\mathcal{T}$ over n variables implied by a (possibly incomplete) binary-valued CP-net whose graph is a forest of degree k, computes a binary-valued CP-net which implies $\mathcal{T}$ and whose graph is a forest of degree at most $k + \log n$.*

Finally, we give a more general result about tree CP-nets with a bounded number of tables on *arbitrary* examples. By Cayley's formula, we know that there are k^{k-1} rooted trees with k vertices. Each root is labeled by an unconditional rule of the form $p \succ \overline{p}$, and all other nodes are labeled by conditional rules of the form $p' : p \succ \overline{p}$, where p and p' are literals. There are $2n$ tables with no condition per rule and $6n(n-1)$ (possibly incomplete) tables with one condition per rule. It follows that the number C_k of CP-trees with at most k tables is bounded by $\sum_{i=0}^{k} 2n$ $(6kn(n-1)+1)^{k-1}$, which is indeed polynomial in k. So the VC-dimension of such CP-trees is polynomial in n.

Based on this result, we can use a simple consistent algorithm specified as follows. Start with the hypothesis set $\mathcal{N}$ of all CP-trees with at most k tables. For each example $(\boldsymbol{x}, \boldsymbol{y})$ in $\mathcal{T}$, remove any hypothesis N in $\mathcal{N}$ that is inconsistent with $(\boldsymbol{x}, \boldsymbol{y})$, that is, any hypothesis N for which the dominance test over $(\boldsymbol{x}, \boldsymbol{y})$ disagrees with its label. If the resulting set $\mathcal{N}$ is empty then $\mathcal{T}$ is not consistent with N. Otherwise, pick an arbitrary tree from $\mathcal{N}$. Because the dominance test is quadratic in the number of variables for binary-valued CP-trees, the running time is polynomial in C_k.

Proposition 9. *The class $\mathcal{C}^k_{\text{TREE}}$ of all concepts representable by a (possibly incomplete) binary-valued CP-tree with at most k tables is PAC learnable from arbitrary examples.*

5.2 *PAC Learning of Separable CP-nets*

We now consider the task of learning a CP-net of the simplest form: the variables are independent of each other. With binary variables, this means that if the possible values for variable X are x and $\overline{x}$, and the target CP-net is complete, then the preference table for X contains either $x \succ \overline{x}$ or $\overline{x} \succ x$.

In this case, checking if a given CP-net N entails $\boldsymbol{x} \succ_N \boldsymbol{y}$ for an arbitrary example is easy. Let $\mathrm{Diff}(\boldsymbol{x},\boldsymbol{y}) = \{x_i \mid (\boldsymbol{x})_i = x_i \text{ and } (\boldsymbol{y})_i = \overline{x_i}\} \cup \{\overline{x_i} \mid (\boldsymbol{x})_i = \overline{x_i} \text{ and } (\boldsymbol{y})_i = x_i\}$. Then $\boldsymbol{x} \succ_N \boldsymbol{y}$ if and only if N contains $x_i \succ \overline{x}_i$ for every $x_i \in \mathrm{Diff}(\boldsymbol{x},\boldsymbol{y})$ and $\overline{x}_i \succ x_i$ for every $\overline{x}_i \in \mathrm{Diff}(\boldsymbol{x},\boldsymbol{y})$ (this is a corollary of Theorems 7 and 8 by [5]).

Now, with each example $(\boldsymbol{x},\boldsymbol{y})$ we associate the clause $C^-_{x,y}$ that contains $\neg x_i$ iff $x_i \in \mathrm{Diff}(\boldsymbol{x},\boldsymbol{y})$ and x_i iff $\overline{x}_i \in \mathrm{Diff}(\boldsymbol{x},\boldsymbol{y})$. The intended meaning of the literal $\neg x_i$ is that $\overline{x}_i$ is preferred to x_i, whereas the meaning of the literal x_i is that x_i is preferred to $\overline{x}_i$; hence, the meaning of the clause $C^-_{x,y}$ is that $\boldsymbol{x} \not\succ_N \boldsymbol{y}$ for every separable CP-net N in which at least one of these local preferences is true, by virtue of the lemma above. For instance, if $\boldsymbol{x} = \overline{x}_1 x_2 x_3 \overline{x}_4$ and $\boldsymbol{y} = x_1 \overline{x}_2 x_3 x_4$, then $\mathrm{Diff}(\boldsymbol{x},\boldsymbol{y}) = \{\overline{x}_1, x_2, \overline{x}_4\}$ and $C^-_{x,y} = x_1 \vee \neg x_2 \vee x_4$. This clause expresses that x_1 is preferred to $\overline{x}_1$, or $\overline{x}_2$ is preferred to x_2, or x_4 is preferred to $\overline{x}_4$.

With each positive example $(\boldsymbol{x},\boldsymbol{y})$ we can also associate the cube (conjunction of literals) $C^+_{x,y} \equiv \neg C^-_{x,y}$. Given a set of training examples $\mathcal{T}$, let $\Gamma_{\mathcal{T}} = \bigwedge\{C^*_e \mid e \in \mathcal{T}\}$, where $C^*_{x,y} = C^+_{x,y}$ if $(\boldsymbol{x},\boldsymbol{y})$ is a positive example, and $C^*_{x,y} = C^-_{x,y}$ if the example is negative. Clearly, $\Gamma_{\mathcal{T}}$ is equivalent to a set of clauses.

Now consider the following one-to-one correspondence between truth assignments M over $\{x_1,\ldots,x_n\}$ and separable CP-nets N_M over $\mathcal{V}$: N_M contains the preference $x_i \succ \overline{x}_i$ for every i such that $M \models x_i$ and the preference $\overline{x}_i \succ x_i$ for every i such that $M \models \neg x_i$. For instance, if $M(x_1) = M(x_4) = \top$ and $M(x_2) = M(x_3) = \bot$, N_M contains the preference tables $\{x_1 \succ \overline{x}_1,\ \overline{x}_2 \succ x_2,\ \overline{x}_3 \succ x_3,\ x_4 \succ \overline{x}_4\}$. Then for an interpretation M, it is easily seen that $M \models C^-_{x,y}$ if and only if N_M does not imply $(\boldsymbol{x},\boldsymbol{y})$ or, equivalently, that $M \models C^+_{x,y}$ if and only if N_M implies $(\boldsymbol{x},\boldsymbol{y})$.

It follows that a set of examples $\mathcal{T}$ is implicatively consistent with N_M for a given model M if and only if $M \models \Gamma_{\mathcal{T}}$. Therefore, searching for a CP-net, which implies a given set of examples, amounts to searching for a model of the corresponding set of clauses, the size of which grows polynomially with the size of the set of examples.

This technique easily extends to nonbinary variables: we can use a propositional variable x_i^{kl} for every pair of distinct values $\{x_i^k, x_i^l\}$ for every variable X_i, where the intended meaning of x_i^{kl} is $x_i^k \succ_N x_i^l$, and add clauses to represent the transitivity of the relation $\succ_N$; there is a polynomial number of them (details can be found in [22]).

The one-to-one correspondence given above is a reduction from our learning problem to satisfiability. It is actually possible to find a reduction in the opposite direction (see [22]), from which we get the following result.

Proposition 10. *[22] Deciding whether there is a (binary or nonbinary) complete separable CP-net, that implies a given set of arbitrary examples is NP-complete. The result holds even if all examples are negative.*

It follows directly that this class is not PAC-learnable with the very weak restriction that the produced hypothesis classifies correctly the examples received. However,

Proposition 11. *[22] Deciding whether there is a binary-valued, complete, separable CP-net which implies a given set of positive examples can be done in polynomial time.*

Observe that this result can be extended to PAC-learnability of (possibly incomplete) separable CP-nets from positive examples, in the setting of one-sided errors [28]. This is because there is always a unique minimal (in terms of rules) incomplete separable CP-net which implies a given set of positive examples.

Now, as soon as $\mathcal{T}$ becomes large with respect to the number of attributes n, the chances that $\mathcal{T}$ is implicatively consistent with a separable CP-net become low. In this case, we may want to determine a separable CP-net that is implicatively consistent with as many examples of $\mathcal{T}$ as possible. This problem amounts to solving a MAXSAT problem, when each example corresponds to exactly one clause of $\Gamma_{\mathcal{T}}$, that is when we have no positive example: the separable CP-net that best fits a set of positive examples corresponds to the interpretation maximizing the number of clauses from $\Gamma_{\mathcal{T}}$ satisfied. In this case, we can reuse algorithms for MAXSAT for computing a separable CP-net that best fits a set of positive examples, as well as polynomial approximation schemes. This extends to nonbinary variables, with the difference that the clauses representing transitivity of the local preference tables are protected.

Finally, again using the same kind of translation, we easily get the following results.

Proposition 12. *If all variables are binary and all examples in $\mathcal{T}$ differ at most on two variables, then deciding whether there exists a separable CP-net, which implies that $\mathcal{T}$ can be done in polynomial time; however, the corresponding optimization problem remains* NP*-hard.*

5.3 Learning a Complete Preference Relation

We close this section by providing results about the learning context (b) specified in Sect. 3: the target concept is a linear order, not necessarily representable by a CP-net. Our goal is to find a CP-net that would be a good representation for this relation. Recall that since the target is a linear order, we only need to consider *positive* examples.

In the rest of this section, we investigate in turn the problems of finding a CP-net that is weakly consistent with a given set of examples, then strongly consistent with it. We focus on the problem of finding *separable* CP-nets. A set of examples is said to be *weakly separable* (resp. *strongly separable*) if there exists a separable CP-net with which it is weakly (resp. strongly) consistent.

We start by showing how the search for a separable CP-net that is weakly consistent with a set of examples can be rewritten as an instance of propositional satisfiability (SAT). Recall from Sect. 5.2 that an example $(\boldsymbol{x}, \boldsymbol{y})$ can be translated into a clause $C^-_{\boldsymbol{y},\boldsymbol{x}}$, the models of which correspond to separable CP-nets that are consistent with $(\boldsymbol{x}, \boldsymbol{y})$. Given a set of examples $\mathcal{T}$, let $\Phi_{\mathcal{T}} = \{C^{-}_{\boldsymbol{y},\boldsymbol{x}} \mid (\boldsymbol{x}, \boldsymbol{y}) \in \mathcal{T}\}$. Then, given an interpretation M, a set of examples $\mathcal{T}$ is weakly consistent with N_M if and only if $M \models \Phi_{\mathcal{T}}$. As a consequence, $\mathcal{T}$ is weakly separable if and only if $\Phi_{\mathcal{T}}$ is satisfiable.

Example 5. Consider three binary attributes A, B, C, and the set of examples

$$\mathcal{T} = \{(abc\,,\, \overline{a}bc),\ (\overline{a}\overline{b}c\,,\, ab\overline{c}),\ (\overline{a}b\overline{c}\,,\, a\overline{b}c),\ (\overline{a}bc\,,\, \overline{a}\overline{b}\overline{c})\}$$

$\Phi_{\mathcal{T}}$ has a unique model, corresponding to the separable CP-net $N = \{a \succ \overline{a}, b \succ \overline{b}, c \succ \overline{c}\}$. Therefore, N is the unique separable CP-net weakly consistent with N, and $\mathcal{T}$ is weakly separable.

As in Sect. 5.2, a similar translation can be used with nonbinary variables, and algorithms for solving MAXSAT can be used to search for a CP-net that is weakly consistent with as many examples as possible.

Proposition 13. *Deciding whether a set of examples over (binary or nonbinary) attributes is weakly separable is* NP*-complete.*

Now, let us turn to the notion of strong compatibility. Characterizing such a property is less easy. Indeed, the difference between weak and strong compatibility is that while in weak compatibility we look for a separable CP-net which is consistent with each individual example in $\mathcal{T}$, in strong compatibility we look for a separable CP-net which is consistent with the whole set of examples $\mathcal{T}$.

Example 5, continued *$\mathcal{T}$ is not strongly consistent with N, because $\mathcal{T} \cup \succ_N$ has the following cycle:*

$$a\overline{b}c \succ_N \overline{a}\overline{b}c \succ_{\mathcal{T}} ab\overline{c} \succ_N \overline{a}b\overline{c} \succ_{\mathcal{T}} a\overline{b}c$$

Since $\mathcal{T}$ is not strongly consistent with any separable CP-net other than N (because N is the unique one with which $\mathcal{T}$ is weakly compatible), $\mathcal{T}$ is not strongly separable.[2]

Note that all outcomes in the cycle on the example above appear in $\mathcal{T}$. More generally, if we denote by $\mathcal{O}(\mathcal{T})$ the set of outcomes that appear in $\mathcal{T}$, it can be proved that $\mathcal{T}$ is strongly consistent with N if and only if the restriction of $\succ_N \cup \mathcal{T}$ to $\mathcal{O}(\mathcal{T})$ is acyclic. Since this restriction has at most $2\,|\mathcal{T}|$ vertices, checking if it possesses a cycle can be done in polynomial time. Thus, checking whether $\mathcal{T}$ is strongly consistent with a given CP-net N is in P, and we have the following result:

Proposition 14. *[23] Checking whether $\mathcal{T}$ is strongly separable is* NP*-complete.*

Note that although weak and strong separability have the same complexity, weak separability enjoys the nice property that there is a simple solution-preserving

---

[2] Note that $\mathcal{T}$ is both weakly separable and does not contain any cycles as it was the case for Example 4, yet is not strongly separable.

translation into SAT (the models of $\Phi_{\mathcal{T}}$ correspond bijectively to the CP-nets that are weakly consistent with $\mathcal{T}$), which allows weak separability to be computed in practice using algorithms for SAT[3]. Contrastingly, to compute a separable CP-net strongly consistent with $\mathcal{T}$, we can generate structures N weakly consistent with $\mathcal{T}$, and test for acyclicity of $\succ_N \cup \mathcal{T}$ using graph algorithms.

6 Active Learning of CP-nets

In this section, we investigate the learnability issues of CP-nets in the paradigm of active learning. Recall that in the standard PAC learning model, examples are drawn at random according to an unknown but fixed distribution. This model of learning is merely passive in the sense that the learner has no control over the selection of examples. One can increase the flexibility of this model by allowing the learner to ask about particular examples, that is, the learner makes *membership queries* [1]. This capability appears to increase the power of polynomial-time learning algorithms. For instance, it is known that propositional Horn formulas are PAC-learnable with membership queries [3], but the results of [19] show that without membership queries, Horn formulas are no easier to learn than general CNF or DNF formulas.

In the setting of active preference learning, we assume that the user has in mind a target preference structure $\succ$, but does not know how to represent this structure into a CP-net. However, the user is disposed to help the learner by answering membership queries of the form "does $\boldsymbol{x}$ dominates $\boldsymbol{y}$?", where $\boldsymbol{x}$ and $\boldsymbol{y}$ are outcomes chosen by the learner. A membership query for a target concept $\succ$ is a map MQ that takes as input a pair of outcomes $(\boldsymbol{x}, \boldsymbol{y})$ and returns as output `yes` if $\boldsymbol{x} \succ \boldsymbol{y}$, and `no` if $\boldsymbol{x} \not\succ \boldsymbol{y}$.

From a practical perspective, one must take into account the fact that outcomes are typically *not* comparable with an equivalent cost. As observed in [17], users can meaningfully compare outcomes if they differ only on very few attributes. To this end, we define the *width* of MQ$(\boldsymbol{x}, \boldsymbol{y})$ to be the number of variables on which $\boldsymbol{x}$ and $\boldsymbol{y}$ differ. A membership query of width 1 is called a *swap membership query*.

Based on these considerations, a minimal requirement behind active learning is to ask as few membership queries as possible. An additional desiderata for minimizing the cognitive effort spent by the user in answering preference queries is to restrict to swap membership queries.

Definition 6 (PAC learning with membership queries). A concept class $\mathcal{C}_{\mathcal{N}}$ is *PAC learnable with swap membership queries* over an instance class $\mathcal{E}$ if there is a polynomial time learning algorithm A and two polynomials $p(\cdot,\cdot,\cdot)$ and $q(\cdot)$ such that for any target concept $\succ$ in $\mathcal{C}_{\mathcal{N}}$, any probability distribution $\mathcal{D}$ over $\mathcal{I}$, and any

[3] Such a translation exists for strong separability (which we do not give here), but unfortunately, the set of clauses generated uses $\mathcal{O}(n^2)$ variables (where n is the set of examples), which limits its practical applicability.

parameters $\delta, \epsilon \in (0, 1)$, after receiving $p(n, \frac{1}{\epsilon}, \frac{1}{\delta})$ random examples of $\succ$ drawn independently according to $\mathcal{D}$, and asking $q(n)$ swap membership queries, then with probability at least $1 - \delta$, A returns a hypothesis $\widehat{N} \in \mathcal{N}$ with $error(\widehat{N}) \leq \epsilon$. The smallest such polynomial q is called the *query complexity* of the learning algorithm A.

6.1 Active Learning with Swap Examples

We now investigate the problem of active preference learning, where the target concept can be represented by an *acyclic* CP-net, and the questions are restricted to swap membership queries.

In this setting, we can build an online algorithm for learning actively acyclic CP-nets. Recall that online learning proceeds into trials. Initially, the learner chooses a hypothesis $\hat{N}$. During each trial, the learner first receives an example $(\boldsymbol{x}, \boldsymbol{y})$, next predicts the label "+" or "–" of this instance according to its current hypothesis, and then receives the correct label from the user. If the prediction was incorrect, then the learner is charged one mistake.

The basic idea underlying our online learning algorithm is to start from the empty CP-net $\hat{N} = \varnothing$ and, during each trial, iteratively revise $\hat{N}$ by maintaining two invariants. The first invariant is that each rule (or entry) in the learner's hypothesis $\hat{N}$ is subsumed by a rule in the minimal representation of the target CP-net N. In other words, for each rule $\hat{\boldsymbol{u}} : x \succ \overline{x}$ in $\hat{N}$, there is a rule $\boldsymbol{u} : x \succ \overline{x}$ in the minimal representation of N, with $\hat{\boldsymbol{u}} \subseteq \boldsymbol{u}$. The second invariant is that each such rule $r = \hat{\boldsymbol{u}} : x \succ \overline{x}$ in $\hat{N}$ is *supported* by an instance $(\boldsymbol{x}_r, \boldsymbol{y}_r)$ of N, that is, $\boldsymbol{x}_r \succ_N \boldsymbol{y}_r$, $\boldsymbol{x}_r$ and $\boldsymbol{y}_r$ satisfy $\hat{\boldsymbol{u}}$, $\boldsymbol{x}_r$ satisfies x, and $\boldsymbol{y}_r$ satisfies $\overline{x}$.

Technically, the algorithm proceeds as follows. On seeing an example $(\boldsymbol{x}, \boldsymbol{y})$, if the learner predicts this instance as negative, while it is positive, then it expands its CP-net with a new rule $\boldsymbol{u} : x \succ \overline{x}$, where the support $(\boldsymbol{x}, \boldsymbol{y})$ is stored. Here, x is the literal in $\boldsymbol{x}$ whose value differs from $\boldsymbol{y}$, and $\boldsymbol{u}$ is the projection of the outcome $\boldsymbol{x}$ onto the current parent set of the variable X in $\hat{N}$. Dually, if the learner predicts $(\boldsymbol{x}, \boldsymbol{y})$ as positive, while it is negative, then it expands the condition $\boldsymbol{u}$ of the misclassifying rule $\boldsymbol{u} : x \succ \overline{x}$ with a new parent. Using the support $(\boldsymbol{x}_r, \boldsymbol{y}_r)$ of this rule, a new parent can be found by asking at most $n - 1$ membership queries. To this end, we simply need to incrementally transform $\boldsymbol{x}$ into $\boldsymbol{x}_r$ by iteratively flipping those literals that differ from $\boldsymbol{x}$ and $\boldsymbol{x}_r$, until we find the first literal x_j for which the label of $(\boldsymbol{x}_j[x], \boldsymbol{x}_j[\overline{x}])$ is positive; this literal is kept as a new parent of X. In fact, only $\log_2 n$ membership queries are needed to find this parent, by performing a binary search over this sequence of transformations. A detailed implementation of this algorithm is given in [21].

Example 6. Assume so far the learner has only learnt the rule $x_1 : x_3 > \overline{x}_3$, supported by $x_1 x_2 x_3 x_4 x_5 \succ_N x_1 x_2 \overline{x}_3 x_4 x_5$.

Now assume it receives the positive example $(\boldsymbol{x}, \boldsymbol{y})$ with $\boldsymbol{x} = \overline{x}_1 x_2 \overline{x}_3 x_4 x_5$, $\boldsymbol{y} = \overline{x}_1 x_2 x_3 x_4 x_5$. Then it learns the rule $\overline{x}_1 : \overline{x}_3 > x_3$, and stores $(\boldsymbol{x}, \boldsymbol{y})$ as its support.

Finally, assume it receives the positive example $x_1 \overline{x}_2 \overline{x}_3 \overline{x}_4 \overline{x}_5 \succ_N x_1 \overline{x}_2 x_3 \overline{x}_4 \overline{x}_5$, which contradicts its first rule. Then it searches for a new parent of x_3. Using the support of this rule and the new example, it asks the membership query $x_1 x_2 \overline{x}_3 \overline{x}_4 \overline{x}_5 \succ_N x_1 x_2 x_3 \overline{x}_4 \overline{x}_5$. Assuming the answer is `yes`, it goes on with the membership query $x_1 x_2 \overline{x}_3 x_4 \overline{x}_5 \succ_N x_1 x_2 x_3 x_4 \overline{x}_5$. Assuming this one answers `no`, the learner deduces that x_4 is a parent of x_3, hence it updates the previously learnt rules according to their support, resulting in rules $x_1 x_4 : x_3 > \overline{x}_3$ and $\overline{x}_1 x_4 : \overline{x}_3 > x_3$, and creates a new rule $x_1 \overline{x}_4 : \overline{x}_3 > x_3$ so as to cover the new example.

By using a well-known conversion from online learning to PAC-learning [24], we derive the following result.

Proposition 15. *Acyclic (possibly incomplete) binary CP-nets are PAC-learnable with swap membership queries, over the instance class of swaps. There is an online learning algorithm A for this class, such that for any target concept $\succ$ of description size s, the algorithm makes at most s mistakes and uses at most s $\log_2 n$ membership queries.*

6.2 Active Learning with Unrestricted Examples

When the instances supplied to the learner are unrestricted, even predicting their label is a difficult task, because dominance testing is NP-hard for acyclic CP-nets. As observed in the previous section, an important class of concepts for which dominance testing can be accomplished in polynomial time is the class of *tree CP-nets*. In this section, we briefly discuss an online algorithm for learning tree CP-nets.

The algorithm can be specified as follows. Initially, the learner starts from the hypothesis $\hat{N} = \varnothing$ and iteratively expands $\hat{N}$ until it finds the target representation N. An invariant of the algorithm is that $\hat{N}$ is always included in N so the learner can only make mistakes on positive examples $(\boldsymbol{x}, \boldsymbol{y})$. In such cases, the algorithm considers in turn each variable on which $\boldsymbol{x}$ and $\boldsymbol{y}$ differ, and builds its CP-table and that of all its ascendants in the tree. It stops whenever such a variable already has a CP-table in its current hypothesis. Again, to find a parent for each candidate variable, the algorithm can use a binary search strategy.

Proposition 16. *Tree (possibly incomplete) binary CP-nets are PAC-learnable with swap membership queries, over arbitrary examples: there is an online learning algorithm A for this class, such that for any target concept $\succ$ with k nodes, the algorithm makes at most k mistakes and uses $O(k \log_2 n)$ membership queries.*

We conclude this section by emphasizing that some classes of CP-nets are *teachable*, that is, learnable by asking a polynomial number of membership queries, without the need of observing any sample. Thus after making those queries, the learner is guaranteed to correctly predict any instance.

We focus on the class of acyclic CP-nets whose graph has indegree at most k. The learner proceeds by using a levelwise generate-and-test procedure and membership queries for uncovering the set of parents of each variable. Clearly, this approach is acceptable only for small bounded degrees, such as tree CP-nets.

Proposition 17. *The class of concepts representable by a binary-valued acyclic CP-net whose graph has degree at most k is teachable: there is an algorithm which outputs such a CP-net using $O(kn^{k+1}2^{k-1})$ membership queries and no other queries or examples, where n is the number of variables and k is the degree of the target concept.*

7 Conclusion and Open Problems

In this paper, we have addressed many issues related to learning CP-nets. We have argued that the first important problem is whether the CP-net that we aim at learning is such that the user's preference relation coincides with its induced preference relation, or is an lower approximation of the user's complete preference relation. Then we gave a few theoretical results on the learnability of CP-nets, and considered two different learning frameworks: passive learning (from a set of examples), and active learning (by queries).

We hope that our preliminary results in the learnability issue of CP-net open the door to new theoretical results and practical learning algorithms. First of all, we do not have a general method (other than brute-force search) for computing a CP-net that is weakly or strongly consistent with a set of examples in the nonseparable case, nor do we have algorithms for outputting a CP-net that realizes an optimal tradeoff between simplicity and accuracy.

We have already emphasized the lack of expressivity in CP-nets. Although CP-nets are a representation language wellsuited to expressing preferential (in)dependencies, they do not allow one, for instance, to express statements of relative importance between variables, as lexicographic orders do. We may desire to learn preferences that combine both aspects (preferential dependencies and relative importance). For this, it is necessary to study preference learning with more expressive languages, such as TCP-nets [6] or (even more general) conditional preference theories [31].

References

1. D. Angluin, Queries and concept learning. Mach. Learn. **2**(4), 319–342 (1988)
2. D. Angluin, Negative results for equivalence queries. Mach. Learn. **5**, 121–150 (1990)
3. D. Angluin, M. Frazier, L. Pitt, Learning conjunctions of Horn clauses. Mach. Learn. **9**, 147–164 (1992)

4. M. Anthony, N. Biggs, *Computational Learning Theory* (Cambridge University Press, 1992)
5. C. Boutilier, R. Brafman, C. Domshlak, H. Hoos, D. Poole, CP-nets: a tool for representing and reasoning with conditional ceteris paribus preference statements. J. Artif. Intell. Res. **21**, 135–191 (2004)
6. R. Brafman, C. Domshlak, S. Shimony, On graphical modeling of preference and importance. J. Artif. Intell. Res. **25**, 389–424 (2006)
7. D. Braziunas, C. Boutilier, Local utility elicitation in GAI models, in *Proceedings of the 21st Conference on Uncertainty in Artificial Intelligence (UAI)* (2005), pp. 42–49
8. C.J.C. Burges, T. Shaked, E. Renshaw, A. Lazier, M. Deeds, N. Hamilton, G. Hullender, Learning to rank using gradient descent, in *Proceedings of the 22nd International Conference on Machine Learning (ICML)* (2005)
9. U. Chajewska, L. Getoor, J. Norman, Y. Shahar, Utility elicitation as a classification problem, in *Proceedings of the 14th Conference on Uncertainty in Artificial Intelligence (UAI)* (1998), pp. 79–88
10. U. Chajewska, D. Koller, R. Parr, Making rational decisions using adaptive utility elicitation, in *Proceedings of the 17th National Conference on Artificial Intelligence (AAAI)* (2000), pp. 363–369
11. Y. Chevaleyre, A short note on passive learning of CP-nets. Rapport de recherche, Lamsade, mars 2009
12. Y. Dimopoulos, L. Michael, F. Athienitou, Ceteris paribus preference elicitation with predictive guarantees, in *Proceedings of the 21st International Joint Conference on Artificial Intelligence (IJCAI 2009)* (2009)
13. J. Dombi, C. Imreh, N. Vincze, Learning lexicographic orders. Eur. J. Oper. Res. **183**, 748–756 (2007)
14. C. Domshlak, T. Joachims, Efficient and non-parametric reasoning over user preferences. User Model. User-adapt. Interact. (UMUAI) **17**(1-2), 41–69 (2007)
15. J. Doyle, Y. Shoham, M. Wellman, A logic of relative desire (preliminary report), in *Proceedings of the 6th International Symposium on Methodologies for Intelligent Systems (ISMIS)* (Springer, 1991), pp. 16–31
16. Ch. Gonzales, P. Perny, GAI networks for utility elicitation, in *Principles of Knowledge Representation and Reasoning: Proceedings of the 9th International Conference (KR)* (2004), pp. 224–234
17. P. Green, V. Srinivasan, Conjoint analysis in consumer research: Issues and outlook. J. Consumer Res. **5**(2), 103–123 (1978)
18. V. Ha, P. Haddawy, Problem-focused incremental elicitation of multi-attribute utility models, in *Proceedings of the 13th Conference on Uncertainty in Artificial Intelligence (UAI)* (1997), pp. 215–222
19. M.J. Kearns, M. Li, L. Pitt, L.G. Valiant, On the learnability of boolean formulae, in *Proceedings of the Nineteenth Annual ACM Symposium on Theory of Computing* (1987), pp. 285–295
20. R. Keeney, H. Raiffa, *Decision with Multiple Objectives: Preferences and Value Trade-offs* (Wiley, 1976)
21. F. Koriche, B. Zanuttini, Learning conditional preference networks with queries, in *Proceedings of the 21st International Joint Conference on Artificial Intelligence (IJCAI 2009)* (2009)
22. J. Lang, J. Mengin, The complexity of learning *ceteris paribus* separable preferences, in *Proceedings of the 21st International Joint Conference on Artificial Intelligence (IJCAI 2009)* (2009)
23. J. Lang, J. Mengin, The complexity of learning separable ceteris paribus preferences. Rapport de recherche RR-2009-3-FR, IRIT, Université Paul Sabatier, Toulouse, mars 2009
24. N. Littlestone, From on-line to batch learning, in *Proceedings of the Second Annual Workshop on Computational Learning Theory* (Morgan Kaufmann, 1989), pp. 269–284
25. M. Sachdev, On learning of ceteris paribus preference theories. Master's thesis, Graduate Faculty of North Carolina State University, 2007

26. T. Sandholm, C. Boutilier, *Preference Elicitation in Combinatorial Auctions*, Chap. 10, in *Combinatorial Auctions*, ed. by Cramton, Shoham, and Steinberg (MIT, 2006)
27. M. Schmitt, L. Martignon, On the complexity of learning lexicographic strategies. J. Mach. Learn. Res. **7**, 55–83 (2006)
28. L.G. Valiant, A theory of the learnable. Commun. ACM **27**(11), 1134–1142 (1984)
29. P. Viappiani, B. Faltings, P. Pu, Evaluating preference-based search tools: a tale fo two approaches, in *Proceedings of the 21st National Conference on Artificial Intelligence (AAAI)* (2006), pp. 205–210
30. P. Viappiani, B. Faltings, P. Pu, Preference-based search using example-critiquing with suggestions. J. Artif. Intell. Res. **27**, 465–503 (2006)
31. N. Wilson, Extending CP-nets with stronger conditional preference statements, in *Proceedings of the 19th National Conference on Artificial Intelligence (AAAI)* (2004), pp. 735–741
32. F. Yaman, Th. Walsh, M. Littman, M. desJardins, Democratic approximation of lexicographic preference models, in *Proceedings of the 35th International Conference in Machine Learning (ICML)* (2008), pp. 1200–1207

Choice-Based Conjoint Analysis: Classification vs. Discrete Choice Models*

Joachim Giesen, Klaus Mueller, Bilyana Taneva, and Peter Zolliker

Abstract Conjoint analysis is a family of techniques that originated in psychology and later became popular in market research. The main objective of conjoint analysis is to measure an individual's or a population's preferences on a class of options that can be described by parameters and their levels. We consider preference data obtained in choice-based conjoint analysis studies, where one observes test persons' choices on small subsets of the options. There are many ways to analyze choice-based conjoint analysis data. Here we discuss the intuition behind a classification based approach, and compare this approach to one based on statistical assumptions (discrete choice models) and to a regression approach. Our comparison on real and synthetic data indicates that the classification approach outperforms the discrete choice models.

1 Introduction

Conjoint analysis is a popular family of techniques mostly used in market research to assess consumers' preferences, see [6] for an overview and recent developments. Preferences are assessed on a set of options that are specified by multiple parameters and their levels. In general conjoint analysis comprises two tasks: (a) preference data assessment, and (b) analysis of the assessed data. Common to all conjoint

---

*Joachim Giesen and Peter Zolliker have been partially supported by the Hasler Stiftung (Proj. #2200).

J. Giesen (✉)
Friedrich-Schiller Universität Jena, Germany

K. Mueller
Stony Brook University, USA

B. Taneva
Max-Planck Institut für Informatik, Saarbrücken, Germany

P. Zolliker
EMPA Dübendorf, Switzerland

J. Fürnkranz and E. Hüllermeier (eds.), *Preference Learning*,
DOI 10.1007/978-3-642-14125-6_14,

analysis methods is that preferences are estimated from conjoint measurements, i.e., measurements taken on all parameters simultaneously.

Choice-based conjoint analysis is a sub-family of conjoint analysis techniques named after the employed data assessment/measurement method, namely a sequence of choice experiments. In a choice experiment a test person is confronted with a small number of options sampled from a parameterized space, and has to choose his preferred option. The measurement is then just the observation of the test person's choice. Choice-based conjoint analysis techniques can differ in the analysis stage. Common to all methods is that they aim to put all the options on a common scale computed from the assessed choice data. Most straightforward is to compute an ordinal scale, i.e., a ranking of all the options, but it is more popular to compute an interval scale where a numerical value, i.e., a scale value, is assigned to every option. The interpretation is, that an option that gets assigned a larger scale value is more preferred. Differences of scale values have a meaning, but there is no natural zero. That is, an interval scale is invariant under translation and re-scaling by a positive factor. Note, that here we are not dealing with an instance of an object ranking problem, see the survey by Kamishima et al. in this volume [7], but a scaling problem, i.e., our dependent variables (scale values) are measured on an interval scale in contrast to dependent variables (ranks) on an ordinal scale.

The purpose of our paper is to provide intuition based on geometric duality why the analysis approach of Evgeniou et al. [3] that uses ideas from maximum margin classification aka support vector machines performs well (though applicability of the kernel trick, which mostly contributed to the popularity of support vector machines, seems not so important for conjoint analysis), and to compare this approach empirically to one based on statistical assumptions on measured and synthetic data. The latter approach can be seen as an extension of the popular discrete choice methods, see for example [10], to the case of conjoint measurements. We also study a geometrically inspired regression approach based on computing the largest ball that can be inscribed into a (constraint) polytope. All approaches compute an interval scale from choice data, and can be used to compute the scale for either a population of test persons from choice data assessed on the population, or for an individual solely from his choice data.

2 Notation

Formally, the options in the choice experiments are elements in the Cartesian product $A = A_1 \times \ldots \times A_n$ of parameter sets A_i, which in general can be either discrete or continuous – here we assume that they are finite. The choice data are of the form $a \succ b$, where $a = (a_1, \ldots, a_n), b = (b_1, \ldots, b_n) \in A$ and a was preferred over b by some test person in some choice experiment. Our goal is to compute an interval scale $v : A \to \mathbb{R}$ on A from a set of choice data.

Often it is assumed that the scale v is linear, i.e., that it can be decomposed as

$$v(a) = v\big((a_1, \ldots, a_n)\big) = \sum_{i=1}^{n} v_i(a_i),$$

where $v_i : A_i \to \mathbb{R}$ are also interval scales. In the case of continuous parameters A_i the linearity of the scale is (essentially) justified when the parameters are preferentially independent, i.e., the order $\times_{i \in I} A_i$ for any non-empty subset I of $\{1, \dots, n\}$ is independent of the choice of parameter levels for the remaining parameters $A_j, j \in \{1, \dots, n\} \setminus I$ (see [8] for details). For finite parameter sets linearity still implies preferential independence, but the reverse is in general not true anymore. Nevertheless, in practice linearity is almost always assumed. Also the two methods that we are going to discuss here both assume linearity of the scale.[1] The discrete choice models approach first estimates the scales v_i from the choice data individually first and then combines them in a second step. Note that the choice data are obtained from conjoint measurements, i.e., choices among options in A and not in A_i. The classification/regression (maximum margin/largest inscribed ball) approaches estimate the scales v_i simultaneously from the choice data. Note that both approaches have to estimate the same number of parameters, namely all the values $v_i(a), a \in A_i, i = 1, \dots, n$.

3 Conjoint Analysis as Classification Problem

The naive approach to choice-based conjoint analysis would be to compute scale values $v_i(a) \in \mathbb{R}, a \in A_i, i = 1, \dots, n$ that satisfy constraints of the form

$$\sum_{i=1}^{n} v_i(a_i) - v_i(b_i) > 0,$$

whenever $a = (a_1, \dots, a_n)$ was preferred over $b = (b_1, \dots, b_n)$ by some test person in a choice experiment. The geometric interpretation of this approach is picking a point $v \in \mathbb{R}^m$, where $m = \sum_{i=1}^n m_i$ and $m_i = |A_i|$, in the *feasible region*, i.e., the subset of $\mathbb{R}^m$ that satisfies all choice constraints. A choice experiment is defined by the characteristic vectors $\chi_a \in \{0, 1\}^m$, whose i'th component is 1 if the corresponding parameter level is present in option a, and 0 otherwise. The feasible region can now be re-written as

$$v^t(\chi_a - \chi_b) > 0, \text{ if } a \succ b \text{ in a choice experiment,}$$

or equivalently $v^t n_{ab} > 0$, where $n_{ab} = (\chi_a - \chi_b)$. Note that the feasible region is a cone whose apex is the origin. Let H_{ab} be the hyperplane $\{v \in \mathbb{R}^m \mid v^t n_{ab} = 0\}$ with normal n_{ab}, and let

$$H_{ab}^{+} = \{v \in \mathbb{R}^m \mid v^t n_{ab} > 0\} \qquad \text{and} \qquad H_{ab}^{-} = \{v \in \mathbb{R}^m \mid v^t n_{ab} < 0\}$$

[1] The linearity assumption can be mitigated by combining dependent parameters into a single one, see [5] for a practical example.

be the two open halfspaces bounded by H_{ab}. Note that $H_{ab}^{+} = H_{ba}^{-}$. If a was preferred over b in a choice experiment, then we have a constraint of the form $v \in H_{ab}^{+}$, otherwise, if b was preferred over a in a choice experiment, then we have $v \in H_{ab}^{-}$. That is, we can assign a label $+$, or $-$, respectively to the hyperplane H_{ab} depending on the outcome of a choice experiment for this hyperplane. Since the label attached to the hyperplane H_{ba} is just the opposite of the label attached to H_{ab} we can restrict ourselves to one of the two hyperplanes for every pair $a \neq b \in A$, e.g., by fixing an arbitrary order on the elements of A, and only considering hyperplanes H_{ab}, where a comes before b in this order. The feasible region can be written as the intersection of the halfspaces

$$H_{ab}^{+} \text{ if } a \succ b \text{ in a choice experiment, and } H_{ab}^{-} \text{ if } b \succ a \text{ in a choice experiment.}$$

The feasible region contains many points that all encode a ranking of the options in A that complies with the choices, provided the region is not empty. Since we have only combinatorial information, namely choices, there is no way to distinguish between the points in the feasible region. Among all the *feasible points* in the feasible region we want to choose one with good generalization properties, namely one that allows to predict the outcome of further choice experiments. The situation is similar to linear classification: for data with binary labels that are linearly separable, i.e., there exists a hyperplane that has the points with different labels on different sides, we want to choose a separating hyperplane with good generalization properties when it is used as class boundary. Regularization theory [2] provides theoretical arguments that the margin maximizing hyperplane has good generalization properties, i.e., the hyperplane that maximizes the distance to the convex hulls of the data points of the respective classes. We will discuss now that results from binary classification can be transferred to our conjoint analysis scenario by *geometric duality*.

The duality transform that we want to consider here maps points $h \in \mathbb{R}^{m+1}$ to non-vertical hyperplanes H in $\mathbb{R}^{m+1}$ and vice versa. The non-vertical hyperplane dual to $h = (h_1, \ldots, h_{m+1})$ is given as

$$H = \left\{ v \in \mathbb{R}^{m+1} \,\middle|\, v_{m+1} = \sum_{i=1}^{m} h_i v_i - h_{m+1} \right\}.$$

The important observation about the duality transform is that relative positions of points with respect to hyperplanes are maintained: let

$$H^{+} = \left\{ v \in \mathbb{R}^{m+1} \,\middle|\, v_{m+1} \geq \sum_{i=1}^{m} h_i v_i - h_{m+1} \right\},$$

$$H^{-} = \left\{ v \in \mathbb{R}^{m+1} \,\middle|\, v_{m+1} \leq \sum_{i=1}^{m} h_i v_i - h_{m+1} \right\},$$

then we have

$$p \in \begin{cases} H^+ \\ H^- \\ H \end{cases} \qquad \text{if and only if} \qquad h \in \begin{cases} P^+ \\ P^- \\ P \end{cases}$$

where we denote hyperplanes with capital letters and their dual points with the same lowercase letter, see also Fig. 1. Unfortunately, our hyperplanes H_{ab} can be vertical. We circumvent this difficulty by augmenting the vectors n_{ab} with a $(m+1)$'th coordinate with entry $\epsilon \in [0,1]$. For small values of ϵ the augmented hyperplane is almost vertical with respect to the $(m+1)$'th dimension, and the augmented constraints (halfspaces) $(n_{ab}, \epsilon)^t (v, v_{m+1}) \geq \epsilon$ are almost equivalent to $n_{ab}^t v \geq 0$ (analogously for "$\leq$") when ϵ and $|v_{m+1}|$ are small. The augmented hyperplanes have the following non-vertical formulation,

$$\left\{ v \in \mathbb{R}^{m+1} \,\middle|\, v_{m+1} = -\frac{1}{\epsilon} n_{ab}^t v + 1 \right\},$$

and their dual points are $-\frac{1}{\epsilon}(n_{ab}, \epsilon)$. If we attach the same labels (depending on the outcome of the choice experiment) to the dual points as we attach to the hyperplanes, then we can formulate a standard linear binary classification problem, i.e., finding the maximum margin hyperplane that separates the labeled points. Let such a separating hyperplane be given by a non-zero normal vector $v \in \mathbb{R}^{m+1}$ and an offset $w/\|v\| \in \mathbb{R}$. Hence we have for all $a, b \in A$ that have been compared in a choice experiment,

$$y_{ab} \left(-\frac{1}{\epsilon} v^t (n_{ab}, \epsilon) - w \right) > 0,$$

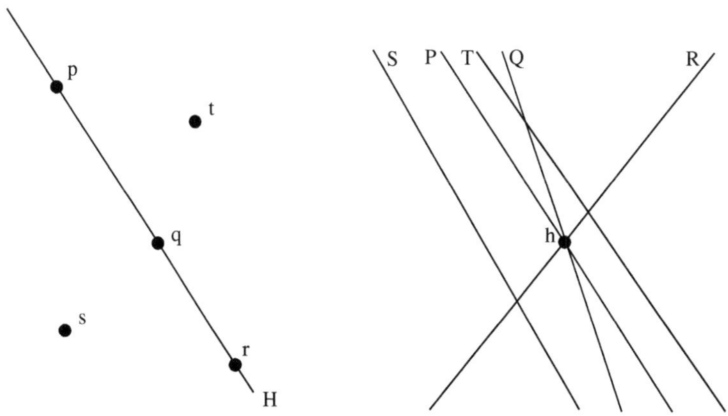

Fig. 1 Duality of non-vertical hyperplanes and points preserves relative positions. Dual points are labeled by lowercase letters, and dual hyperplanes by capital letters

where $y_{ab} \in \{\pm 1\}$ is the label of the dual point of the hyperplane corresponding choice experiment comparing a with b. Let

$$d(\epsilon) = \min_{a,b \in A: a \text{ has been compared with } b} \left| y_{ab} \left(-\frac{1}{\epsilon} v^t (n_{ab}, \epsilon) - w \right) \right| > 0.$$

Thus we have

$$y_{ab} \left(-\frac{1}{\epsilon} v^t (n_{ab}, \epsilon) - w \right) \geq d(\epsilon),$$

or by setting $v_i = -v_i$ for $i = 1, \ldots, m$,

$$y_{ab} v^t n_{ab} - \epsilon y_{ab} v_{m+1} \geq \epsilon d(\epsilon) + \epsilon y_{ab} w. \quad (1)$$

The maximum margin hyperplane problem now reads as

$$\begin{aligned} &\min_{v, v_{m+1}, w} \frac{1}{2} \left(\|v\|^2 + v_{m+1}^2 \right) \\ &\text{s.t.} \quad y_{ab} v^t n_{ab} - \epsilon y_{ab} v_{m+1} \geq \epsilon d(\epsilon) + \epsilon y_{ab} w \\ &\qquad \text{if } a \succ b \text{ in a choice experiment.} \end{aligned}$$

As discussed in [4] taking the limit $\epsilon \to 0$ gives the following optimization problem for the vector v of scale values whose entries are indexed by the $a \in A_i, i = 1, \ldots, n$:

$$\begin{aligned} &\min_v \sum_{i=1}^{n} \sum_{a \in A_i} v_i(a)^2 \\ &\text{s.t.} \quad \sum_{j=1}^{n} (v_i(a_j) - v_i(b_j)) \geq 1, \\ &\qquad \text{if } a = (a_1, \ldots, a_n) \succ b = (b_1, \ldots, b_n) \text{ in a choice experiment.} \end{aligned}$$

Note that contradictory information, i.e., choices of the form $a \succ b$ and $b \succ a$ renders the feasible region empty. In practice we will have to deal with contradictory information, especially when we assess preferences on a population, but also individuals can be inconsistent in their choices. It is essentially the contradictory information that makes the problem interesting and justifies the computation of an interval scale instead of an ordinal scale (i.e., a ranking or an enumeration of all partial rankings compliant with the choices) from choice information. The choice information now can no longer be considered purely combinatorial since also the frequency of $a \succ b$ for all comparisons of a and b will be important. To avoid an empty feasible region we introduce a non-negative slack variable z_j for every choice, i.e., $v_a - v_b + z_j \geq 0, z_j \geq 0$ if a was preferred over b in the j'th choice experiment. Now the feasible region will always be non-empty and it is natural to aim for minimal total slack, i.e., $\sum_{j=1}^{k} z_j$ if we have information from k choice experiments. That is, we end up with the following (soft margin) optimization problem to compute the scale values (using the standard trade-off between model complexity and quality of fit on the observed data with trade-off parameter $c > 0$):

$$\min_{v,z_j} \sum_{i=1}^{n} \sum_{a \in A_i} v_i(a)^2 + c \sum_{j=1}^{m} z_j$$

$$\text{s.t.} \quad \sum_{j=1}^{n} (v_i(a_j) - v_i(b_j)) + z_l \geq 1,$$

$$\text{if } (a_1, \ldots, a_n) \succ b = (b_1, \ldots, b_n) \text{ in the } l\text{'th choice experiment.}$$

$$z_l \geq 0, l = 1, \ldots, m$$

This is exactly the optimization problem suggested by Evgeniou et al. [3] to compute scale values.

4 Largest Inscribed Ball

We will compare the classification approach from the previous section to a geometrically inspired regression approach, namely computing the largest ball inscribed into the feasible region defined by the constraints

$$\sum_{i=1}^{n} v_i(a_i) - v_i(b_i) \geq 0, \text{ if } (a_1, \ldots, a_n) \succ (b_1, \ldots, b_n) \text{ in a choice experiment.}$$

We want to estimate the $v_i(a)$ for $i = 1, \ldots, n$ and all $a \in A_i$. That is, we want to estimate the entries of a vector v with $m = \sum_{i=1}^{n} m_i$ components, where $m_i = |A_i|$. As discussed earlier a choice experiment is defined by the characteristic vectors $\chi_a \in \{0, 1\}^m$, whose i'th component is 1 if the corresponding parameter level is present in the option a, and 0 otherwise. The feasible region can be characterized by

$$v^t(\chi_a - \chi_b) \geq 0, \text{ if } a \succ b \text{ in a choice experiment,}$$

or equivalently $v^t \bar{n}_{ab} \geq 0$, where $\bar{n}_{ab} = (\chi_a - \chi_b)/\|\chi_a - \chi_b\|$.

The distance of a point $v \in \mathbb{R}^m$ to the hyperplane (subspace) $\{v \in \mathbb{R}^m \mid v^t \bar{n}_{ab} = 0\}$ is given by $v^t \bar{n}_{ab}$. The largest inscribed ball problem now becomes (when using the standard trade-off between model complexity and quality of fit on the observed data)

$$\max_{v,r,z} \; r + c\sum_{j=1}^{k} z_j$$

$$\text{s.t.} \quad v^t \bar{n}_{ab} \geq r - z_k, \text{ if } a \succ b \text{ in the } j\text{'th choice experiment.}$$
$$z_j \geq 0, \; j = 1, \ldots, k$$

where r is the radius of the ball and $c > 0$ is the trade-off parameter. This is a linear program, in contrast to the classification approach based on maximizing the margin which results in a convex quadratic program.

The largest inscribed ball approach does not work directly. To see this observe that the line given by $v_1 = v_2 = \ldots = v_m =$ constant is always in the feasible region. If the feasible region contains only this line (which often is the case), then the optimal solution of our problem would be on this line. A solution $v_1 = v_2 = \ldots = v_m =$ constant however does not give us meaningful scale values. To make deviations from the line $v_i =$ constant possible we add a small constant $\epsilon > 0$ to the left hand side of all the comparison constraints. In our experiments we chose $\epsilon = 0.1$.

5 Discrete Choice Models

Finally, we want to discuss a statistically motivated approach to analyze choice-based conjoint analysis data. Discrete choice models deal with the special case of a single parameter, i.e., in a sense the non-conjoint case. Let the finite set A denote this parameter set. Choice data are now of the form $a \succ b$ with $a, b \in A$ and the goal is to compute $v : A \to \mathbb{R}$ or equivalently $\{v_a = v(a) \mid a \in A\}$. Discrete choice models make the assumption that the observed choices are outcomes of random trials: confronted with the two options $a, b \in A$ a test person assigns values $u_a = v_a + \epsilon_a$ and $u_b = v_b + \epsilon_b$, respectively, to the options, where (the error terms) ϵ_a and ϵ_b are drawn independently from the same distribution, and chooses the option with larger value. Hence the probability p_{ab} that a is chosen over b is given as

$$p_{ab} = Pr[u_a \geq u_b] = Pr[v_a + \epsilon_a \geq v_b + \epsilon_b] = Pr[v_a - v_b \geq \epsilon_b - \epsilon_a]$$

Discrete choice models can essentially be distinguished by the choice of distribution for the ϵ_a. Popular choices are normal distributions (Thurstone's (probit) model [9]) or extreme value distributions (Bradley-Terry's (logit) model [1]), see also [10]. The values v_a can be computed for both models either via the difference $v_a - v_b$ from the probability p_{ab} which can be estimated by the frequency f_{ab} that a was preferred over b in the choice experiments, or computationally more involved by a maximum likelihood estimator. Here we introduce a least squares approach using the frequency estimates for the p_{ab}.

5.1 Thurstone's Model (probit)

In Thurstone's model [9] the error terms ϵ_a are drawn from a normal distribution $N(0, \sigma^2)$ with expectation 0 and variance σ^2. Hence the difference $\epsilon_b - \epsilon_a$ is also normally distributed with expectation 0 and variance $2\sigma^2$ and hence

$$\begin{aligned} p_{ab} &= Pr[u_a \geq u_b] = Pr[\epsilon_b - \epsilon_a \leq v_a - v_b] \\ &= \frac{1}{\sqrt{4\pi\sigma^2}} \int_{-\infty}^{v_a - v_b} \mathrm{e}^{-\frac{x^2}{4\sigma^2}} dx = \Phi\left(\frac{v_a - v_b}{\sqrt{2}\sigma}\right), \end{aligned}$$

where Φ is the cumulative distribution function of the standard normal distribution

$$\Phi(x) = \frac{1}{\sqrt{2\pi}} \int_{-\infty}^{x} \mathrm{e}^{-y^2/2} \mathrm{d}y.$$

This is equivalent to

$$v_a - v_b = \sqrt{2}\sigma\Phi^{-1}(p_{ab}).$$

Using the frequency f_{ab} that a was preferred over b (number of times a was preferred over b divided by the number that a and b have been compared) we set

$$v_{ab} = \sqrt{2}\sigma\Phi^{-1}(f_{ab}).$$

5.2 Bradley-Terry's Model (logit)

In Bradley-Terry's model [1] the error terms ϵ_a are drawn from a standard Gumbel distribution, i.e., the distribution with location parameter $\mu = 0$ and scale parameter $\beta = 1$. Since the difference of two independent Gumbel distributed random variables is logistically distributed we have

$$\begin{aligned} p_{ab} &= Pr[u_a \geq u_b] = Pr[\epsilon_b - \epsilon_a \leq v_a - v_b] \\ &= \frac{1}{1 + \mathrm{e}^{-(v_a - v_b)}} = \frac{\mathrm{e}^{v_a - v_b}}{1 + \mathrm{e}^{v_a - v_b}} = \frac{\mathrm{e}^{v_a}}{\mathrm{e}^{v_a} + \mathrm{e}^{v_b}}. \end{aligned}$$

This implies

$$\frac{\mathrm{e}^{v_a}}{\mathrm{e}^{v_b}} = \frac{p_{ab}}{1 - p_{ab}},$$

which is equivalent to

$$v_a - v_b = \ln\left(\frac{p_{ab}}{1 - p_{ab}}\right).$$

Analogously to what we did for Thurstone's model we set

$$v_{ab} = \ln\left(\frac{f_{ab}}{1 - f_{ab}}\right).$$

5.3 Computing Scale Values

From both Thurstone's and Bradley-Terry's model we get an estimate v_{ab} for the difference of the scale values v_a and v_b. Our goal is to estimate the v_a's (and not only their differences). This can be done by computing v_a's that best approximate the v_{ab}'s (all equally weighted) in a least squares sense. That is, we want to minimize the residual

$$r(v_a | a \in A) = \sum_{a,b \in A; b \neq a}^{n} (v_a - v_b - v_{ab})^2.$$

A necessary condition for the minimum of the residual is that all partial derivatives vanish, which gives

$$\frac{\partial r}{\partial v_a} = 2 \sum_{b \in A; b \neq a} (v_a - v_b - v_{ab}) = 0.$$

Hence

$$|A| v_a = \sum_{b \in A} v_b + \sum_{b \in A; b \neq a} v_{ab}.$$

Since we aim for an interval scale we can assume that $\sum_{b \in A} v_b = 0$. Then the values that minimize the residual are given as

$$v_a = \frac{1}{|A|} \sum_{b \in A; b \neq a} v_{ab}.$$

We can specialize this now to the discrete choice models and get for Thurstone's model

$$v_a = \frac{\sqrt{2}\sigma}{|A|} \sum_{b \in A; b \neq a} \Phi^{-1}(f_{ab}),$$

and for Bradley-Terry's model

$$v_a = \frac{1}{|A|} \sum_{b \in A; b \neq a} \ln\left(\frac{f_{ab}}{1 - f_{ab}}\right).$$

5.4 Multi-Parameter (conjoint) Case

Now we turn to the multi-parameter case where the options are elements in $A = A_1 \times \ldots \times A_n$. We assume a linear model and describe a compositional approach to compute the scales for the parameters A_i. In a first step we compute scales v_i using a discrete choice model for the one parameter case, and then in a second step compute *re-scale* values w_i to make the scales v_i comparable. Our final scale for A is then given as $v = \sum_{i=1}^{n} w_i v_i$, i.e.,

$$v\big((a_1, \ldots, a_n)\big) = \sum_{i=1}^{n} w_i v_i(a_i).$$

To compute the scales v_i we make one further assumption: if $a = (a_1, \ldots, a_n) \in A$ is preferred over $b = (b_1, \ldots, b_n) \in A$ in a choice experiment we interpret this as a_i is preferred over b_i to compute the frequencies $f_{a_i b_i}$. If the parameter levels in the choice experiments are all chosen independently at random, then the frequencies $f_{a_i b_i}$ should converge (in the limit of infinitely many choice experiments) to the frequencies that one obtains in experiments involving only a single parameter A_i.

To compute the re-scale values w_i we use again a maximum margin approach (similarly to the one discussed before in Sect. 3). The approach makes the same trade-off between controlling the model complexity (maximizing the margin) and accuracy of the model (penalizing outliers). The trade-off is controlled by a parameter $c > 0$ and we assume that we have data from k choice experiments available.

$$\min_{w_i, z_j} \sum_{i=1}^{n} w_i^2 + c \sum_{j=1}^{k} z_j$$

$$\begin{aligned} \text{s.t.} \quad & \sum_{i=1}^{n} w_i \big(v_i(a_i) - v_i(b_i)\big) + z_j \geq 1, \\ & \quad \text{if } (a_1, \ldots, a_n) \succ (b_1, \ldots, b_n) \text{ in the } j\text{'th choice experiment.} \\ & z_j \geq 0,\ j = 1, \ldots, k \end{aligned}$$

6 Cross Validation

The classification and regression approaches (but also our re-scaling approach) have a free parameter $c > 0$ that controls the trade-off between model complexity and model accuracy. The standard way to choose this parameter is via k-fold cross validation. For k-fold cross-validation the set of choice data is partitioned into k partitions (aka strata) of equal size. Then $k - 1$ of the strata are used to compute the scale values, which can be validated on the left out stratum. For the validation we use the scale value to predict outcome in the choice experiments in the left-out stratum. Given $v(a)$ and $v(b)$ for $a, b \in A$ such that a and b have been compared in the left-out stratum, to predict the outcome one can either predict the option with the higher scale value, or one can make a randomized prediction, e.g., by using the Bradley-Terry model: predict a with probability $e^{v(a)}/(e^{v(a)} + e^{v(b)})$. The validation score in both cases is the percentage of correct predictions. For simplicity we decided to use the percentage of correct predictions.

7 Data Sets

We compared the different approaches to analyze choice-based conjoint analysis data on two different types of data sets: (a) data that we assessed in a larger user study to measure the perceived quality for a visualization task [5], and (b) synthetic data that we generated from a statistical model of test persons. Let us describe the visualization study first.

7.1 Visualization Study

The purpose of volume visualization is to turn 3D volume data into images that allow a user to gain as much insight into the data as possible. A prominent example of a volume visualization application is MRI (magnetic resonance imaging). Turning volume data into images is a highly parameterized process. Among the many parameters there are for example:

1. The choice of color scheme: often there is no natural color scheme for the data, but even when it exists it need not best suited to provide insight.
2. The viewpoint: an image is a 2D projection of the 3D data, but not all such projections are equally valuable in providing insights.
3. Other parameters like image resolution or shading schemes.

In our study [5] we were considering six parameters (with two to six levels each) for two data sets (foot and engine) giving rise to 2250 (foot) or 2700 (engine) options, respectively. Note that options here are images, i.e., different renderings of the data sets.

On these data sets we were measuring preferences by either asking for the better liked image (aesthetics), or for the image that shows more detail (detail). That is, in total we conducted four studies (foot-detail, foot-aesthetics, engine-detail, and engine-aesthetics). We had 317 test persons for the two details question studies and 366 test persons for the aesthetics studies, respectively. In each study the test persons were shown two images from the same category, i.e., either foot or engine, rendered with different parameter settings and asked which of the two images they prefer

Fig. 2 Data set foot: Which rendering do you like (left or right)?

Fig. 3 Data set engine: Which rendering shows more detail (left or right)?

(with respect to either the aesthetics or the details question). Hence in each choice experiment there were only two options, see Figs. 2 and 3 for examples.

In [5] we evaluated the choice data using the Thurstone discrete choice model for the whole population of test persons. There we used a different method to re-scale the values from the first stage than described here. The method we used is based on the normal distribution assumption and thus not as general as the method described here.

7.2 *Synthetic Data*

We were also interested to see how well the different methods perform when we only use information provided by a single person. Unfortunately the information provided individually by the test persons in the visualization studies is very sparse (only 20 comparisons per person). Thus we also generated synthetic data as follows:

1. We simulated a study with five parameters and five levels each.
2. We generated 200 synthetic test persons represented by a scale value for every parameter level. The scale values for the levels of each parameter were chosen from normal distributions with mean $-2, -1, 0, 1$ and 2, respectively. The normal distributions always had the same standard deviation, which we choose to be

2, 5 or 8 (for three different studies). Varying the standard deviation was used to model different degrees of heterogeneity in the population.

3. The synthetic test persons provided answers to 200 binary choice problems following the Bradley-Terry model. That is, given two options a and b and test person dependent scale values $v(a)$ and $v(b)$, respectively, the test person prefers a over b with probability $p_{ab} = e^{v(a)}/(e^{v(a)} + e^{v(b)})$. To simulate the choices we generated random numbers uniformly in [0, 1] and compared them to the p_{ab}'s.

8 Results and Discussion

As pointed out in Sect. 5 when assigning scale values to parameter levels in the discrete choice approach, we estimate the probability that level a is preferred over level b by the relative frequency f_{ab} that a was preferred over b. For sparse data (only very few comparisons per test person) the frequency matrix can also be sparse even in the sense of missing entries, i.e., levels a and b that never have been compared. To deal with sparseness we exploit a "transitivity of preferences" assumption. Whenever $a \succ b$ and $b \succ c$ we interpret this also as a (weak) vote for $a \succ c$. We implemented this idea as follows: we initialized f_{ab} with either the measured relative frequency, or when this is not available with $1/2$. Then we updated f_{ab} iteratively until convergence using the following formula (with some constant $c \in (0, 1)$, we obtained good results for $c = 0.3$):

$$f_{ab} = (1-c) f_{ab} + \frac{c}{n-2} \sum_{d \neq a,b} \frac{f_{ad} f_{db}}{f_{ad} f_{db} + f_{da} f_{bd}}$$

That is, we smoothed the frequencies using all the information available.

8.1 *Visualization Studies*

Let us start with a summary of the performance of our analysis methods on the data from the four visualization studies. The summary is given in Table 1.

Table 1 Average percentage of correct predictions for the four visualization studies. Shown is the mean for $k = 10$ strata and the estimated standard deviation in parentheses. See also Fig. 4

	Engine-aesthetics	Engine-detail	Foot-aesthetics	Foot-detail
Thurstone	0.7535(8)	0.8265(8)	0.6640(10)	0.7388(5)
Bradley-Terry	0.7536(9)	0.8267(5)	0.6640(1)	0.7387(10)
Classification	0.7529(9)	0.8401(20)	0.6638(10)	0.7411(10)
Largest ball	0.7530(9)	0.8414(7)	0.6638(16)	0.7405(10)

Table 2 Average percentage of correct predictions for the four visualization studies for individual test persons. For the prediction only data provided by the individual test person were used. See also Fig. 4

	Engine-aesthetics	Engine-detail	Foot-aesthetics	Foot-detail
Thurstone	0.636(2)	0.670(2)	0.597(2)	0.596(3)
Bradley-Terry	0.640(3)	0.675(2)	0.598(3)	0.599(3)
Classification	0.618(3)	0.651(2)	0.585(2)	0.589(3)
Largest ball	0.616(5)	0.631(4)	0.578(3)	0.580(4)

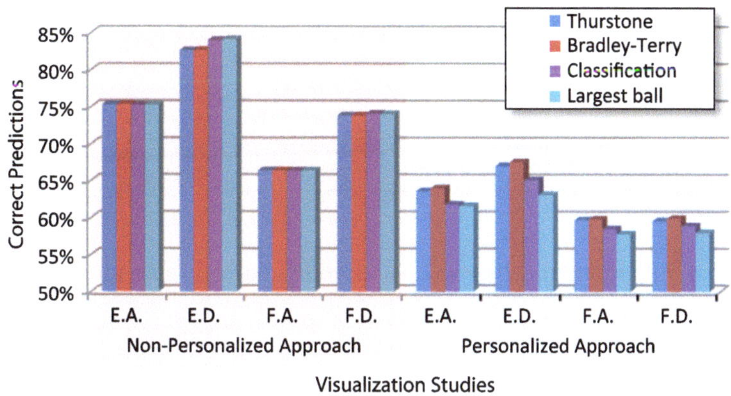

Fig. 4 Summarizing Tables 1 and 2

We consider the Thurstone and Bradley-Terry discrete choice models, the classification approach, and the largest inscribed ball regression approach. Here we report k-fold cross validation values, i.e., the percentage of correct predictions (on the left out strata). We were using 20 random partitions into $k = 10$ strata (also for everything that follows) and report the mean and estimated standard deviation of the percentage of correct predictions (i.e., for every correct prediction percentage that we report we had 200 data points).

For the data that we report in Table 2 we consider only data provided by a test person to compute personal scale values for this person. The presented results are the mean percentage of correct predictions on the left out strata also averaged over all test persons that participated in the study. The standard deviation is computed with respect to the left out strata and the different test persons.

While the classification/regression approaches performed slightly but not statistically significant better than the discrete choice approaches in the non-personalized analysis they perform significantly worse if one uses the approach for personalized prediction on the same data sets. We also generated synthetic data, because we wanted to study the behavior of the different approaches when we have more data per test person available. Note that in the visualization studies we were restricted to only very few choice experiments per test person since these test persons volunteered to take part in the studies during an exhibition at the computer science department of ETH Zürich. Our conjecture was that the classification/regression

Table 3 Average percentage of correct predictions for the four visualization studies analyzed with the classification approach using the values $c = 100, 10, 1$ and 0.01 for the trade-off parameter

	Engine-aesthetics	Engine-detail	Foot-aesthetics	Foot-detail
100	0.7525(6)	0.8401(20)	0.6635(10)	0.7402(10)
10	0.7529(10)	0.8396(20)	0.6636(10)	0.7401(10)
1	0.7529(10)	0.8341(20)	0.6638(10)	0.7411(10)
0.01	0.7405(10)	0.8313(10)	0.6585(10)	0.7167(10)

Table 4 Comparison of the average percentages of correct predictions on the artificial data with 40 choice experiments per person. Shown are prediction percentages for the classification and the Bradley-Terry discrete choice model approaches (non-personalized and personalized)

Standard Deviation	Classification	Pers.	Bradley-Terry	Pers.
2	0.716(1)	0.765(3)	0.723(1)	0.749(4)
5	0.606(1)	0.783(3)	0.619(1)	0.745(2)
8	0.574(1)	0.784(3)	0.565(2)	0.749(4)

approaches outperform the discrete choice models once we have more data per test person.

Another interesting observation regarding the visualization studies is that we were not able to detect a statistically significant difference between Thurstone's model and Bradley-Terry's model. We expected Bradley-Terry's model to perform slightly better, because of the more realistic fat tail assumption of the underlying distribution. Actually, it performs slightly better in the personalized analysis, but the advantage is not statistically relevant.

Finally, in Table 3 we report on the dependence of the classification approach on the regularization parameter $c > 0$ that controls the trade-off between model complexity and training error. Here we report only on a non-personalized analysis, since the behavior in the other settings is similar.

Interestingly, the classification approach depends only marginally on the choice of the regularization parameter c. All the results that we report here are at least in the range of the other approaches.

8.2 Synthetic Data

As we have mentioned earlier our main motivation to generate synthetic data was to study the effect of the number of choice experiments per test person on the performance of the different approaches.

To check the validity of our model from which we generated the artificial data we first show in Table 4 the correct prediction percentages for the classification approach and the Bradley-Terry discrete choice model for varying standard deviation (which is one of the parameters we can control when generating the synthetic data).

Table 5 Average percentage of correct predictions for the synthetic data set using 20 to 200 choice experiments per (artificial) test person. Shown are results for the classification (maximum margin), Bradley-Terry, and maximum inscribed ball regression approach. See also the figure below

# Comparisons	Classification	Non-pers.	Bradley-Terry	Non-pers.	Largest ball	Non-pers.
20	0.703(4)	0.717(2)	0.692(4)	0.714(2)	0.658(3)	0.730(1)
40	0.765(3)	0.7164(8)	0.749(4)	0.723(1)	0.754(5)	0.7296(8)
60	0.812(2)	0.7231(6)	0.767(2)	0.7147(5)	0.786(3)	0.7234(4)
80	0.839(2)	0.7192(5)	0.788(1)	0.7169(5)	0.821(4)	0.7197(4)
100	0.847(2)	0.7246(4)	0.805(2)	0.7198(5)	0.839(2)	0.7173(3)
120	0.860(1)	0.7302(3)	0.815(1)	0.7143(5)	0.851(1)	0.7220(3)
140	0.869(1)	0.7306(2)	0.824(1)	0.7295(3)	0.866(1)	0.7207(3)
160	0.8773(9)	0.7291(3)	0.836(1)	0.7259(4)	0.8741(7)	0.7361(3)
180	0.8785(8)	0.7188(3)	0.841(1)	0.7242(3)	0.8806(7)	0.7298(3)
200	0.8841(8)	0.7239(2)	0.850(1)	0.7226(2)	0.8832(4)	0.7357(2)

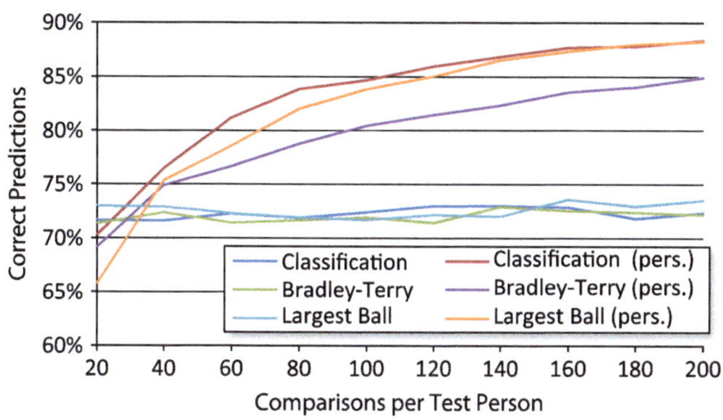

As expected it becomes more difficult to predict the outcome of a choice experiment on the population level when we increase the standard deviation (which is meant to model population heterogeneity), whereas in the personalized setting (where we consider only data provided by a test person to compute personal scale values for this test person) the prediction accuracy does not depend on the variance. The heterogeneity is actually large enough that it pays off (in terms of prediction accuracy) to personalize even for standard deviation 2.

In Table 5 we summarize the dependence on the number of choice experiments (comparisons) for the different approaches in the personalized setting. The results for the personalized setting show that – as we expected – the percentage of correct predictions hardly improves with growing number of choice experiments per test person. At the same time – which is also expected – the variance of correct prediction percentages goes down. But we do not only observe that the percentage of correct prediction increases with growing number of choice experiments per test persons, but also that the classification/regression approaches (which includes the largest inscribed ball approach) outperform the Bradley-Terry model (which essentially behaves the same as the Thurstone model). Thus here we observe statistically significant what we already conjectured for the visualization studies, namely, that classification/regression outperforms the other methods once enough data are available. Actually, the results show that the amount of data need not be very large before classification/regression outperforms the other approaches.

Let us also briefly comment on the results in the non-personalized setting, where the percentage of correct predictions hardly improves with increasing number of comparisons per person. This means that if our goal is to compute scale values for a population of respondents, it can be enough for the test persons to participate in few choice experiments, e.g., 20 for our model (but probably more for conjoint studies with more parameters). This is good news for studies in which the respondents cannot (or are not willing to) participate in many choice experiments.

9 Conclusion

We compared two discrete choice approaches, namely Thurstone's model and Bradley-Terry's model with a classification and a regression approach for choice-based conjoint data analysis. We introduced a new regression approach based on inscribing the largest ball into a feasible region, and provided some intuition why the classification approach should perform well. At least our personalized results on a synthetic data set suggest that the classification/regression approaches outperform the discrete choice models – provided there is enough data per test person available.

Our main interest is in using conjoint analysis techniques to measure users' preferences for visualization and imaging algorithms. In the conjoint studies that we performed to this end we typically only got test persons to participate in a small number of choice experiments (about 20 choice experiments per test person – which takes roughly three minutes). In this range of numbers of choice experiments discrete choice models even seem to have a small advantage over classification/regression (at least in non-personalized analysis). In the future we plan to conduct more user studies to figure out the best analysis approach for varying numbers of choice experiments and objectives (e.g., non-personalized vs. personalized).

References

1. R.A. Bradley, M.E. Terry, Rank analysis of incomplete block designs, i. the method of paired comparisons. Biometrika **39**, 324–345 (1952)
2. T. Evgeniou, M. Pontil, T. Poggio, Regularization networks and support vector machines. Adv. Comput. Math. **13**(1), 1–50 (2000)
3. T. Evgeniou, C. Boussios, G. Zacharia, Generalized robust conjoint estimation. Mark. Sci. **24**(3), 415–429 (2005)
4. J. Giesen, Choice based conjoint analysis as the geometric dual of a classification problem. Manuscript, 2009

5. J. Giesen, K. Mueller, E. Schuberth, L. Wang, P. Zolliker, Conjoint analysis to measure the perceived quality in volume rendering. IEEE Trans. Vis. Comput. Graph. **13**(6), 1664–1671 (2007)
6. A. Gustafsson, A. Herrmann, F. Huber, Conjoint analysis as an instrument of market research practice, in *Conjoint Measurement. Methods and Applications*, ed. by A. Gustafsson, A. Herrmann, F. Huber (Springer, Berlin, 2000), pp. 5–45
7. T. Kamishima, H. Kazawa, S. Akaho, A survey and empirical comparison of object ranking methods, in *Preference Learning*, ed. by J. Fürnkranz, E. Hüllermeier (Springer, 2010)
8. R.L. Keeney, H. Raiffa, *Decisions with Multiple Objectives: Preferences and Value Trade-Offs* (Cambridge University Press, Cambridge, 1993)
9. L.L. Thurstone, A law of comparative judgement. Psychol. Rev. **34**, 273–286 (1927)
10. K. Train, *Discrete Choice Methods with Simulation* (Cambridge University Press, 2003)

Learning Aggregation Operators for Preference Modeling

Vicenç Torra

Abstract Aggregation operators are useful tools for modeling preferences. Such operators include weighted means, OWA and WOWA operators, as well as some fuzzy integrals, e.g. Choquet and Sugeno integrals. To apply these operators in an effective way, their parameters have to be properly defined. In this chapter, we review some of the existing tools for learning these parameters from examples.

1 Introduction

Several approaches have been proposed in the literature to model preferences [6] on a set of alternatives. One of them is to use utility functions. Formally, they are functions that evaluate each alternative according to a predefined scale (e.g., in the [0,1] interval). In general, the larger the evaluation is, the more preferred the alternative.

Real decision making faces an important problem due to the fact that alternative selection needs to take into account different criteria, instead of a single one. This type of problems is known as multicriteria decision making [12, 13, 21]. The difficulty is due to the fact that relevant criteria in decision making are not positively correlated but, usually, in contradiction. For example, in the case of selection of a home, the most preferred home with respect to size, location, and public transport is usually not the best option with respect to the afforded price. Therefore, the houses we prefer the most with respect to price are not the houses we prefer the most with respect to location or size.

One approach for multicriteria decision making is to build an overall criterion (a combination of the basic criteria) and then select the preferred alternative according to this overall criterion. This overall criterion represents a trade-off between the different alternatives. Aggregation operators [19, 20] are appropriate tools for building

V. Torra
IIIA, Artificial Intelligence Research Institute, CSIC, Spanish National Research Council
Campus UAB s/n 08193 Bellaterra (Catalonia, Spain)
e-mail: vtorra@iiia.csic.es

J. Fürnkranz and E. Hüllermeier (eds.), *Preference Learning*,
DOI 10.1007/978-3-642-14125-6_15,

this overall criterion, as they are the usual way of computing a trade-off of a set of values.

At present a large number of aggregation operators exist. Most of them depend on a parameter. For example, the weighted mean depends on a weighting vector. Similarly, other operators have other types of parameters. Then, a relevant issue when using these operators for expressing the preferences on the alternatives is to select an appropriate aggregation operator and to select the appropriate parameters.

In this chapter, we present an overview of some aggregation operators together with their parameters. Then, we discuss how to learn preferences, from the examples. This corresponds to learning the parameters of the aggregation operators. The area of learning preferences has attracted much interest recently, and several alternative approaches have been considered in the literature (see e.g. [8]).

The structure of the paper is as follows. In Sect. 2, we formalize the multicriteria decision-making problem, in terms of alternatives and criteria. We describe the use of aggregation operators to construct an overall criterion over the set of alternatives. Then, in Sect. 3, we will review some of the most used aggregation operators. Finally, in Sect. 4, we will review some of the approaches to learn the preferences from examples. The paper finishes with a summary.

2 Multicriteria Decision Making: A Formalization

Our approach to decision making assumes that we have to select an alternative over a finite set of them. These alternatives can be either a set of objects in a standard decision making problem (e.g., cars or houses), or other objects in machine learning, e.g. cases (in a case-based reasoning system [1]) or examples. We assume that we have M alternatives and we will denote the set of all alternatives by $A = \{x_1, \ldots, x_M\}$.

As stated above, we assume that the set of alternatives is finite. This is the typical case in multicriteria decision making. In contrast, multiobjective decision making usually considers the case of infinite set of alternatives. This is the case of considering the alternatives to be the values of a continuous variable (e.g., the alternative is the price we have to assign to a commodity).

Then, let us consider the evaluation of the alternatives in terms of a set of N criteria. We will represent these criteria by $C = \{c_1, \ldots, c_N\}$. That is, each alternative x_j is evaluated with respect to each criteria c_i. As stated in the introduction, one approach is to consider that alternatives are evaluated using utility functions. In this case, this means to have a function for each criteria. Let u_{c_i} be the function of criteria c_i in C.

Utility functions evaluate each of the alternatives, in a given domain D. It is usual to consider D to be a subset of the real line, although other domains D are conceivable (e.g., ordered scales, fuzzy sets). We will consider the standard approach and consider D a subset of the real line, and for the sake of simplicity, we consider D to be the unit interval. Taking all this together, $u_{c_i} : A \to [0, 1]$.

Then, $u_{c_i}(x_j)$ evaluates alternative x_j with respect to criteria c_i, and, thus, expresses our preference of x_j with respect to c_i. This is so because the larger the $u_{c_i}(x_j)$, the larger our preference. Naturally, $u_{c_i}(x_j) = 1$ is our ideal.

Alternative selection consists on determining the *preferred* elements $x \subset A$. This is usually done computing an overall preference or an overall criterion that synthesizes all the criteria and then selecting the elements x in A with the best evaluation. In the case described above where preferences are expressed through utility functions, this consists of building a criterion $c_{\mathbb{C}}$ that synthesises ($\mathbb{C}$ombines) the criteria c_i, and building its corresponding utility function $u_{\mathbb{C}}$. This aggregated, or combined, utility function is defined for each x_j as the combination of the values $u_{c_i}(x_j)$.

Formally, let $\mathbb{C}$ be a function that combines N values (i.e., $\mathbb{C} : [0,1]^N \rightarrow [0,1]$); then, we define $u_{\mathbb{C}}(a_j)$ as follows:

$$u_{\mathbb{C}}(x_j) = \mathbb{C}(u_{c_1}(x_j), \ldots, u_{c_N}(x_j)).$$

When $\mathbb{C}$ depends on a parameter P, we will use

$$u_{\mathbb{C}_P}(x_j) = \mathbb{C}_P(u_{c_1}(x_j), \ldots, u_{c_N}(x_j)).$$

The arithmetic mean and the weighted mean are typical examples of functions $\mathbb{C}$. Naturally, the former has no parameter while the latter uses a weighting vector p.

For the sake of simplicity and convenience, as we can just consider aggregation for a single x_j without considering the rest of the x_j, we will ignore often in our notation the element x_j. Then, we will use one of the following two expressions:

$$\begin{aligned} u_{\mathbb{C}_P} &= \mathbb{C}_P(a_1, \ldots, a_N) \\ u_{\mathbb{C}_P} &= \mathbb{C}_P(u(c_1), \ldots, u(c_N)). \end{aligned}$$

So, for a given x_j, $u_{c_i}(x_j)$ will be denoted, when no confusion arises, either by a_i or $u(c_i)$.

3 Aggregation Operators

There are a large number of aggregation operators [19, 20]. Of them, the most well known are the arithmetic mean and the weighted mean. In this section, we review a few other operators. We focus on the operators that belong to the family of the Choquet integral. This family encompasses both arithmetic and weighted mean, in the sense that the Choquet integral, with some particular parameters, reduces to these two operators.

In this chapter, we use the term of aggregation operators or aggregation functions as functions that satisfy the following properties:

Unanimity or idempotency: $\mathbb{C}(a, \ldots, a) = a$ for all a
Monotonicity: $\mathbb{C}(a_1, \ldots, a_N) \geq \mathbb{C}(a'_1, \ldots, a'_N)$ when $a_i \geq a'_i$

Some researchers (e.g., [3]) replace the unanimity condition by a looser condition on unanimity on the boundaries of the [0,1] interval. That is, they require unanimity only on 0 and 1:

Unanimity or idempotency at 0: $\mathbb{C}(0, \ldots, 0) = 0$
Unanimity or idempotency at 1: $\mathbb{C}(1, \ldots, 1) = 1$

In this case, t-norms and t-conorms [10] are also considered aggregation operators. In this case, the aggregation operators satisfying unanimity for all a are known as mean operators.

In addition to these conditions, there is another one that is often required. This condition, known as the symmetry, is defined as follows:

Symmetry: For any permutation π on $\{1, \ldots, N\}$ it holds that

$$\mathbb{C}(a_1, \ldots, a_N) = \mathbb{C}(a_{\pi(1)}, \ldots, a_{\pi(N)}).$$

This condition means that none of the arguments have special significance when computing the result. There are several operators that satisfy this property. t-norms and t-conorms, which are some of the operators not satisfying, in general, unanimity, are some of them. We will not consider this condition as, in general, we want to be able to express that some of the arguments, some of the criteria, are more important or relevant than the others. So, we consider aggregation operators that satisfy monotonicity and unanimity for all a in [0,1]. From these two conditions, we can prove that all aggregation operators satisfy internality. Internality is defined as follows:

Internality: $\min_i a_i \leq \mathbb{C}(a_1, \ldots, a_N) \leq \max_i a_i$.

3.1 *Basic Aggregation Operators*

We begin our review with the definitions of the arithmetic mean, the weighted mean, and the OWA operator. As the latter two combine the data with respect to a weighting vector, we also give an explicit definition of this weighting vector.

Definition 1. A vector $v = (v_1 \ldots v_N)$ is a *weighting vector* of dimension N if and only if $v_i \in [0, 1]$ and $\sum_i v_i = 1$.

Definition 2. A mapping AM: $\mathbb{R}^N \rightarrow \mathbb{R}$ is an *arithmetic mean* of dimension N if $AM(a_1, \ldots, a_N) = (1/N) \sum_{i=1}^{N} a_i$.

Definition 3. Let $\mathbf{p}$ be a weighting vector of dimension N; then, a mapping WM: $\mathbb{R}^N \rightarrow \mathbb{R}$ is a *weighted mean* of dimension N if $WM_{\mathbf{p}}(a_1, \ldots, a_N) = \sum_{i=1}^{N} p_i a_i$.

Now, we present the OWA operator. The definition of this operator is similar to the one of the weighted mean in the sense that it is a linear combination of the data with respect to the weighting vector. Nevertheless, a permutation σ plays a central role in the definition, and causes the weights to have a completely different meaning.

Definition 4. [24] Let $\mathbf{w}$ be a weighting vector of dimension N; then, a mapping OWA: $\mathbb{R}^N \to \mathbb{R}$ is an *Ordered Weighting Averaging (OWA) operator* of dimension N if

$$\mathrm{OWA}_{\mathbf{w}}(a_1, \ldots, a_N) = \sum_{i=1}^{N} w_i a_{\sigma(i)},$$

where $\{\sigma(1), \ldots, \sigma(N)\}$ is a permutation of $\{1, \ldots, N\}$ such that $a_{\sigma(i-1)} \geq a_{\sigma(i)}$ for all $i = \{2, \ldots, N\}$ (i.e., $a_{\sigma(i)}$ is the ith largest element in the collection $a_1, \ldots, a_N$).

Note that in the case of the weighted mean, the weight p_1 is associated with the first argument (without taking into account whether this value is large, medium, or small with respect to the others). So, in the weighted mean, each weight is associated with an argument. That is, p_i corresponds to the weight of the ith argument. Then, if a_i is the data supplied by the ith information source or the ith criterion, we can say that p_i is the importance or reliability of this source or criterion. This is so because, naturally, the larger the weight p_i, the more the value a_i influences the outcome of the aggregation. In particular, if $p_i = 1$ (so, $p_j = 0$ for all $j \neq i$), the outcome of the aggregation is just a_i (i.e., a maximum importance to the ith information source, implies that all other values are completely disregarded). Therefore, as stated above, the weights in the weighted mean permits us to weight the criteria.

In contrast to that, in the case of the OWA the weight w_1 is always assigned to the largest value, while w_N is always assigned to the smallest value. So, w_i corresponds to the weight or importance of the value that occupies the ith position after the ordering process. Note that this is independent of the information source or criterion that has supplied this value. In this way, the weights in the OWA measure the importance of the values themselves. So, we can give importance to large values, or to small ones. Understanding the weights in this way, and from the point of view of multicriteria decision making, we can interpret the weights of the OWA in terms of compensation. That is, if we permit compensation of *bad* criteria with some good criteria, we would give larger importance to large values than to smaller ones. Instead, if no compensation is given, then the smallest values of $a_1, \ldots, a_N$ would have the largest weights.

Mathematically, compensation is measured with the *orness* measure. Formally, the orness of an operator establishes how similar is an operator to the maximum. Note that the maximum operator achieves maximum compensation because all *bad* values are compensated by the largest one. We give below the definition of the orness, and give a simpler expression for some operators. We include the definition for the case of the OWA.

Definition 5. Let $\mathbb{C}$ be an aggregation operator with parameters P; then, the *orness* of $\mathbb{C}_P$ is defined by

$$orness(\mathbb{C}_P) := \frac{AV(\mathbb{C}_P) - AV(\min)}{AV(\max) - AV(\min)}, \quad (1)$$

and the *andness* of $\mathbb{C}_P$ is defined by

$$andness(\mathbb{C}_P) := 1 - orness(\mathbb{C}_P) = \frac{AV(\max) - AV(\mathbb{C}_P)}{AV(\max) - AV(\min)},$$

where AV is the *average value* of $\mathbb{C}_P$ defined as

$$AV(\mathbb{C}_P) := \int_0^1 \dots \int_0^1 C_P(a_1, \dots, a_N)\, da_1 \dots da_N.$$

Proposition 1. *The following equalities can be proven for the average value:*

- $AV(\min) = N/(N+1)$
- $AV(\max) = 1/(N+1)$
- $AV(AM) = 1/2$

Proposition 2. *The following equalities can be established:*

- $orness\ (\max) = 1$
- $orness\ (\min) = 0$
- $orness\ (AM) = 1/2$
- $orness\ (OWA_{\mathbf{w}}) = \frac{1}{N-1} \sum_{i=1}^{N} (N-i) w_i$

3.2 The WOWA Operator

Due to the fact that in some decision-making problems the importance of the sources and the compensation degree are of relevance, the WOWA operator was defined. This operator combines a set of data with respect to two weighting vectors. We will denote them by p and w. p is interpreted as the weight of the weighted mean, and w is understood as the weight of the OWA operator. That is, p permits us to represent the importance of the criteria, and w permits us to represent the compensation degree.

The WOWA operator is defined as follows.

Definition 6. [15] Let $\mathbf{p}$ and $\mathbf{w}$ be two weighting vectors of dimension N; then, a mapping WOWA: $\mathbb{R}^N \to \mathbb{R}$ is a *Weighted Ordered Weighted Averaging (WOWA) operator* of dimension N if

$$WOWA_{\mathbf{p},\mathbf{w}}(a_1, \dots, a_N) = \sum_{i=1}^{N} \omega_i a_{\sigma(i)},$$

where σ is defined as in the case of OWA (i.e., $a_{\sigma(i)}$ is the ith largest element in the collection $a_1, \ldots, a_N$), and the weight ω_i is defined as

$$\omega_i = w^*\left(\sum_{j \leq i} p_{\sigma(j)}\right) - w^*\left(\sum_{j<i} p_{\sigma(j)}\right),$$

with w^* being a nondecreasing function that interpolates the points

$$\left\{\left(i/N, \sum_{j \leq i} w_j\right)\right\}_{i=1,\ldots,N} \cup \{(0,0)\}.$$

The function w^* is required to be a straight line when the points can be interpolated in this way.

Several properties can be proven for this operator. Some of the most important ones are the ones that relate it with the operators defined above. It can be proven that the WOWA operator generalizes both the weighted mean and the OWA operator. Formally, when $p = (1/N, \ldots, 1/N)$ we have $WOWA_{p,w} = OWA_w$. That is, the WOWA operator is equivalent to the OWA when all the criteria have the same importance. Similarly, when $w = (1/N, \ldots, 1/N)$ we have $WOWA_{p,w} = WM_p$. That is, the WOWA operator is equivalent to the weighted mean when we require an average compensation. Note that the vector corresponding to an average compensation, i.e., $w = (1/N, \ldots, 1/N)$, if used with the OWA operator, corresponds to an orness equal to 0.5.

So far, we have reviewed so far the operators moving from simpler ones to more complex ones. We have seen that the more complex operators generalize the simpler ones. That is, the WOWA generalizes both OWA and weighted mean, and both OWA and weighted mean generalize the arithmetic mean. We will now review the definition of the Choquet integral that generalizes all the operators seen so far.

3.3 The Choquet Integral

From a formal point of view, the Choquet integral integrates a function with respect to a fuzzy measure. To do so, we need to introduce the function to be integrated and the fuzzy measure. The function will play the role of the data being integrated, and the fuzzy measure will play the role of the parameters (e.g., weighting vectors) that have appeared in the previous operators.

Let us first consider the function. Let us recall that in our framework, we are considering one of the following two expressions as equivalent, and representing the aggregation of the utilities of the criteria.

- $u_{\mathbb{C}_P} = \mathbb{C}(a_1, \ldots, a_M)$
- $u_{\mathbb{C}_P} = \mathbb{C}(u(c_1), \ldots, u(c_M))$.

In the case of using the Choquet integral, the second expression is more appropriate as we have a function u over the criterion c_i. That is, we will consider the Choquet integral of the function u defined over the set of criteria C (i.e., $u : C \to [0, 1]$).

The second element in consideration is the fuzzy measure. The introduction of fuzzy measures in multicriteria decision making is based on the fact that we want to extend the *importance* of a particular criterion to the importance of sets of criteria. Note that in the weighted mean, we interpret p_i as the weight of the ith criterion. That is, we can understand p_i as follows: $p_i = p(c_i)$. From this point of view, we can consider p a set function, and then define $p(\{c_1, c_5\})$. Fuzzy measures permit us to model this situation. These functions, usually denoted by μ, need to satisfy some conditions. They are monotonic (the larger the set, the larger its importance), and bounded (the importance of the whole set is one – similar to the property that weights in a weighting vector add to one – and the importance of the empty set is zero).

The formal definition of fuzzy measures follows.

Definition 7. A fuzzy measure μ on a set X is a set function $\mu : \wp(X) \to [0, 1]$ satisfying the following axioms:

(i) $\mu(\emptyset) = 0$, $\mu(X) = 1$ (boundary conditions)
(ii) $A \subseteq B$ implies $\mu(A) \leq \mu(B)$ (monotonicity)

Several families of fuzzy measures have been defined in the literature. Such measures satisfy the previous requirements and some additional ones. We will use the term unconstrained fuzzy measures for the ones solely satisfying the conditions of Definition 7, when we need to make a difference between them. Probability measures are an example of constrained fuzzy measures. They correspond to additive fuzzy measures, and they are formally defined as fuzzy measures satisfying $\mu(A \cup B) = \mu(A) + \mu(B)$ for $A \cap B = \emptyset$. Decomposable fuzzy measures [23], Sugeno λ-measures [14], hierarchically decomposable fuzzy measures [16], distorted probabilities [4], m-dimensional distorted probabilities [11], and k-order additive fuzzy measures [7] are some other of the families of fuzzy measures. See [19] for details and a classification of these families.

There are several properties of interest about fuzzy measures (see e.g. [19] for details). Here, we only review the Möbius transformation because it is of interest when learning fuzzy measures from examples. Its definition is as follows.

Definition 8. Let μ be a fuzzy measure; then, its Möbius transform m is defined as

$$m_{\mu}(A) := \sum_{B \subseteq A} (-1)^{|A|-|B|} \mu(B) \quad (2)$$

for all $A \subset X$.

An important property of this transform is that given a function m that is a Möbius transform, we can reconstruct the original measure using the following expression:

$$\mu(A) = \sum_{B \subseteq A} m(B)$$

for all $A \subseteq X$.

Now, once we have defined fuzzy measures we can define the Choquet integral of a function with respect to a fuzzy measure. The definition is as follows.

Definition 9. Let μ be a fuzzy measure on X; then, the *Choquet integral* of a function $f : X \to \mathbb{R}^+$ with respect to the fuzzy measure μ is defined by

$$(C) \int f \mathrm{d}\mu = \sum_{i=1}^{N} [f(x_{s(i)}) - f(x_{s(i-1)})] \mu(A_{s(i)}), \qquad (3)$$

where $f(x_{s(i)})$ indicates that the indices have been permuted so that $0 \leq f(x_{s(1)}) \leq \cdots \leq f(x_{s(N)}) \leq 1$, and where $f(x_{s(0)}) = 0$ and $A_{s(i)} = \{x_{s(i)}, \ldots, x_{s(N)}\}$.

Several properties can be proved for this integral. See [19] for a detailed description of the properties and for an interpretation of the integral.

4 Learning Preferences

The process of learning preferences can be formalized in different alternative ways, according to the information available. In this chapter, we discuss some of the approaches. They are based on the assumptions we have described earlier. That is, we have a finite set of alternatives, and each alternative is described in terms of a set of criteria. In addition, in the learning process, we will assume that the utility functions are known. We also assume to know which kind of aggregation operator we want to use to aggregate the values of the utility functions. That is, for example, that we know whether we want to apply a weighted mean or a Choquet integral. See [19] for details and also for alternative types of problems.

Taking into account the available information, we reconsider our notation. As the alternative plays an important role in preference learning, we will include the alternative in our notation in an explicit way. Recall that we used a_i to denote the value of the utility function. We will now use a_i^j to denote the value of this utility function for the ith criterion when object x_j is considered.

Once this framework is fixed, two main approaches to learning exist.

One approach presumes that the only information we have about our preferences is the final ordering of the alternatives. That is, we have an order for the alternatives, their numerical evaluation for each of the criteria, and the aggregation operator to be used. Then, the learning process corresponds to learning the weights from this information. We refer to this type of problem as the case of parameter learning with preferences or partial orders.

Table 1 Data for preference learning

u_{c_1}	u_{c_2}	$\dots$	u_{c_N}	$R_{\mathbb{C}}$	$u_{\mathbb{C}}$
a_1^1	a_2^1	$\dots$	a_N^1	p^1	b^1
a_1^2	a_2^2	$\dots$	a_N^2	p^2	b^2
$\vdots$	$\vdots$		$\vdots$	$\vdots$	$\vdots$
a_1^M	a_2^M	$\dots$	a_N^M	p^M	b^M

The other approach presumes that we have not only the final ordering, but also an estimation of the overall criterion. That is, we know for each alternative an estimation of the aggregated value. We refer to this problem as the case of parameter learning with expected outcome.

Table 1 represents the information available for two problems. In both problems, the values a_i^j are assumed to be known. Then, in the first problem, we assume that we know the rank of the alternatives. The column $R_{\mathbb{C}}$ included in the table corresponds to this information. p^i would be natural numbers corresponding to the position of the ith element of our ranking. In the second approach, we have the values of $u_{\mathbb{C}}$. This is also represented in the table.

4.1 Parameter Learning with Expected Outcome

Parameter learning with expected outcome can be mathematically formalized as an optimization problem, once we have a way to measure the difference between the estimated value and the *correct* outcome. Here, the *correct* outcome corresponds to $u_{\mathbb{C}}$. In addition, as the aggregation operator requires some parameters, and these parameters should not be arbitrary but satisfy some constraints, the optimization problem is a constrained one.

We give below a formulation of the problem, assuming that we use the Euclidean distance to evaluate the estimated value with respect to the desired one.

$$\begin{aligned} &\text{Minimize } D_{\mathbb{C}}(P) = \sum_{j=1}^{M} (\mathbb{C}_P(a_1^j, \dots, a_N^j) - b^j)^2 \\ &\text{Subject to logical constraints on } \; P \end{aligned} \tag{4}$$

This problem can be rewritten for any of the operators described in Sect. 3. We will discuss how to solve it for some of them. We begin with the weighted mean as it is the simplest problem. That is, $\mathbb{C}$ is the weighted mean, and, therefore, its parameter P is a weighting vector. Formally, as weighting vectors are defined in terms of positive components that add to one, we have that the parameter is of the form $\mathbf{p} = (p_1, \dots, p_N)$, and such that $\sum_i p_i = 1$ and $p_i \geq 0$. Taking this into account, the optimization problem given above is rewritten as follows:

$$\begin{aligned}
&\text{Minimize } D_{WM}(\mathbf{p} = (p_1, \ldots, p_N)) = \sum_{j=1}^{M} (\sum_{i=1}^{N} p_i a_i^j - b^j)^2 \\
&\text{Subject to} \\
&\qquad \sum_{i=1}^{N} p_i = 1 \\
&\qquad p_i \geq 0
\end{aligned} \tag{5}$$

This optimization problem is of the class of quadratic problems with linear constraints. This is so because the optimization function, using appropriate transformations, can be rewritten as $(1/2)p'Q'p + r'p$, and the two constraints are linear with respect to the weights p_i. Quadratic problems with linear constraints can be solved using optimization techniques (as e.g. active set methods), and the optimal solution can be found easily. See e.g. [19, 25] for details.

An alternative problem is when, instead of the weighted mean, we consider the OWA operator. This problem is formulated as follows.

$$\begin{aligned}
&\text{Minimize } D_{OWA}(\mathbf{w} = (w_1, \ldots, w_N)) = \sum_{j=1}^{M} (\sum_{i=1}^{N} w_i a_{\sigma_j(i)}^j - b^j)^2 \\
&\text{Subject to} \\
&\qquad \sum_{i=1}^{N} w_i = 1 \\
&\qquad w_i \geq 0
\end{aligned} \tag{6}$$

This problem is similar to the one of the weighted mean. The only difference is that we have here the permutation σ. It should be noticed that this permutation is not the same for all alternatives, but it is alternative-dependent. That is, each alternative needs to define each own permutation as the values are ordered according to the permutation. That is why we are using σ_j in the problem instead of just σ. Due to all this, and, especially, because the values a_i^j are assumed to be known just before the learning process, we can order all the elements before the learning step. In this way, we can learn the weights of the OWA using the same approach used for the weights of the weighted mean. That is, we first order for each row the elements in decreasing order, and then we apply the approach for the weighted mean to the ordered set. The resulting weights are the ones for the OWA.

This approach permits us to obtain the OWA weights that correspond to a global optimum of the optimization function.

We will not discuss it here, but there are situations in which the columns of the matrix are not independent. In this case, for some nonindependent columns, the problem has no unique solution due to singularities in the matrices involved in the problem. This problem was studied in some detail in [17, 18]. Indeed, it was shown that not all nonindependent matrices lead to singularities.

Note that the problem of nonindependent columns is of relevance in the case of multicriteria decision making, as this corresponds to the case of non-independent criteria.

Although standard optimization approaches ensure a global optimum, other nonoptimal approaches have been defined. Yager [5] proposed a method based on gradient descent. His approach, which can be applied either to the weighted mean or to the OWA, consists of removing the constraints for the weights and applying then the gradient descent. The removal of the logical constraints for the weights is achieved by considering the unconstrained vector $\boldsymbol{\Lambda} = (\lambda_1 \dots \lambda_N)$. The weights p_i, if the weights being learned are the ones of the weighted mean, are extracted from $\boldsymbol{\Lambda}$ as follows:

$$\mathbf{p}_{\Lambda} = \left(\frac{e^{\lambda_1}}{\sum_{j=1}^{N} e^{\lambda_j}} \cdots \frac{e^{\lambda_N}}{\sum_{j=1}^{N} e^{\lambda_j}} \right). \tag{7}$$

This approach has the advantage that any vector $\boldsymbol{\Lambda} \in \mathbb{R}^N$ leads to a valid weighting vector. So, given an arbitrary $\boldsymbol{\Lambda}$, we have that $\mathbf{p}_{\Lambda}$, constructed using (7), satisfies $p_i \geq 0$ and $\sum_i p_i = 1$. That is why constraints on p can be removed. Taking this into account, we obtain the following unconstrained optimization problem.

$$\text{Minimize } D_{\text{WM}}(\Lambda) = \sum_{j=1}^{M} \left(\sum_{i=1}^{N} \frac{e^{\lambda_i}}{\sum_{j=1}^{N} e^{\lambda_j}} a_i^j - b^j \right)^2. \tag{8}$$

Note that this optimization problem is formulated for the weighted mean. A similar expression would be obtained for the OWA but in this case, we need the permutation σ. All applies in a similar way to the OWA operator.

The application of gradient descent to this problem is based on an iterative process. In each step, an example is selected, and then weights are updated according to the error. Algorithm 1 details this process. In this algorithm, β is the learning rate, a small value $0 \leq \beta \leq 1$. In this algorithm, the selection of example j corresponds to selecting a row in Table 1.

4.1.1 The WOWA Operator

The optimization problem in the case of the WOWA is not a quadratic one. Due to this, it is not so easy to find the corresponding weights. A valid approach for this problem is to apply a combination of the two techniques described above. That is, to use the approach for finding a global optimal for the weighted mean and the OWA, together with the gradient descent approach. As the WOWA generalizes both the OWA and the weighted mean, it is clear that from optimal solutions for such aggregation operators we can easily find feasible solutions of the WOWA. This is formalized in Algorithm 2.

Algorithm 1 Gradient descent for the weighted mean

Algorithm GradientDescentWM (A, b: Examples) **returns** weighting vector **is**
begin
  int t:=0;
  define $\Lambda(t) := (1 \dots 1)$;
  while no convergence **do**
    Select example j
    $\hat{b}^j := WM_{\lambda(t)}(a_1^j, \dots, a_N^j)$
    $e^j = \left(\hat{b}^j - b^j\right)^2$
    compute, for $i = 1, \dots, N$
      $\lambda_i(t+1) := \lambda_i(t) - \beta \frac{e^{\lambda_i(t)}}{\sum_{j=1}^N e^{\lambda_j(t)}}(a_i^j - \hat{b}^j)(\hat{b}^j - b^j)$,
    t:=t+1;
  end while
  Use Equation 7 to find weights **p** from Λ.
  return **p**;
end

Algorithm 2 Hybrid approach for the WOWA operator

Algorithm GradientDescentWOWA (A, b: Examples) **returns** weighting vectors **is**
begin
  p := GradientDescentWM(A, b); $D_{WM}(\mathbf{p}) := error(WM, \mathbf{p}, A, b)$;
  w := GradientDescentOWA(A, b); $D_{OWA}(\mathbf{w}) := error(OWA, \mathbf{w}, A, b)$;
  $D_{WOWA} := error(WOWA, \mathbf{p}, \mathbf{w}, A, b)$;
  $D_{AM} := error(AM, A, b)$;
  select the pair $(\mathbf{p}, \mathbf{w})$ with minimal D among $\{D_{WOWA}, D_{WM}, D_{OWA}, D_{AM}\}$
  define Λ^p from **p** and Λ^w from **w** so that Equation 7 holds
  apply gradient descent for the WOWA operator and obtain Λ
  compute **p** and **w** from Λ.
end

In this algorithm, the definition of p_i and w_i from Λ, unless there is one p_i or w_i equal to zero, is achieved defining $\lambda_i^p = \log p_i$ and $\lambda_i^w = \log w_i$. Zero weights can only be approximated with a large enough negative value for λ. The gradient descent for the WOWA operator requires the computation of e^j in terms of the WOWA operator. Then, the computation of the new $\lambda_i(t+1)$ from $\lambda_i(t)$ (for each of the two vectors) is based on a numerical approximation of the derivative in:

$$\lambda_i(t+1) := \lambda_i(t) - \beta \left. \frac{\partial e^j}{\partial \lambda_i} \right|_{\lambda_i = \lambda_i(t)}$$

4.1.2 The Choquet Integral

Parameter determination for the Choquet integral corresponds to the problem of determining the appropriate fuzzy measure. The formalization of this problem is similar to the one of learning the weights for the weighted mean (that is, Equation (5)). The formalization uses the Möbius transform of the fuzzy measure instead of using the fuzzy measure itself (i.e., learning the function m defined in Definition 8 instead of μ).

The minimization of the function is subject to the constraints that make μ a proper fuzzy measure. That is, that the measure of the empty set is zero, the measure of the whole reference set X is one, and the monotonicity conditions. These constraints can be expressed as linear constraints. Due to this, the whole transformation of the problem corresponds to a quadratic problem with linear constraints. Therefore, using optimization algorithms the optimal solution can be found.

The approach described here is about learning nonconstrained fuzzy measures. When the problem is about learning constrained ones, some specific algorithms are required. This is so because the constraints on the measures can lead to nonlinear expressions in the formalization of the problem. This is the case of distorted probabilities. A Choquet integral with respect to a distorted probability corresponds to the WOWA operator. Note that we have discussed this problem above and recall that, in such a case, it corresponds to a nonquadratic optimization problem.

4.1.3 Other Operators

Besides the operators reviewed in this work, there are others that could be suitable for constructing an overall preference. Some of them are reviewed in [19]. Beliakov et al. [2] discusses the case of some operators not satisfying unanimity as t-norms and t-conorms. In some of the cases, the operator causes the problem to be nonlinear, whatever be the constraints considered for the parameters. This is the case of the Sugeno integral. This fuzzy integral, which has some similarities with the Choquet integral, leads to problems that are nonlinear for any family of fuzzy measures involved. Some alternatives to standard optimization algorithms have been proposed for this problem, as e.g. genetic algorithms [22].

4.2 Parameter Learning with Preferences or Partial Orders

The formulation of this problem is based on the order relation defined on the alternatives. Let $\succ$ represent this relation, and let S represent the pairs (r, t) such that $x_r \succ x_t$ (i.e., the rth alternative is preferred to the tth alternative). Then, given an aggregation operator $\mathbb{C}$, the goal is to find its parameter P such that for all $(r, t) \in S$ it holds

$$\mathbb{C}_P(a_1^r, a_2^r, \ldots, a_N^r) > \mathbb{C}_P(a_1^t, a_2^t, \ldots, a_N^t)$$

or, equivalently,

$$\mathbb{C}_P(a_1^r, a_2^r, \ldots, a_N^r) - \mathbb{C}_P(a_1^t, a_2^t, \ldots, a_N^t) > 0.$$

As it is usually the case that there are no parameters P that make this equation true for all $(r,t) \in S$, it is preferable to account for some error and minimize the error. Formally, we rewrite the last equation considering terms $y_{(r,t)} \geq 0$ (the error). That is,

$$\mathbb{C}_P(a_1^r, a_2^r, \ldots, a_N^r) - \mathbb{C}_P(a_1^t, a_2^t, \ldots, a_N^t) + y_{(r,t)} > 0.$$

In this way, the problem is formalized as finding the parameter P that minimizes the number of violations when all $(r,t) \in S$ are considered. That is, the following term is minimized:

$$\sum_{(r,t)\in S} y_{(r,t)}.$$

All together, the problem to minimize is as follows:

$$\begin{aligned}
&\text{Minimize } \textstyle\sum_{(r,t)\in S} y_{(r,t)} \\
&\text{Subject to} \\
&\qquad \mathbb{C}_P(a_1^r, a_2^r, \ldots, a_N^r) - \mathbb{C}_P(a_1^t, a_2^t, \ldots, a_N^t) + y_{(r,t)} > 0 \\
&\qquad\qquad y_{(r,t)} \geq 0 \\
&\qquad \text{logical constraints on } P
\end{aligned} \tag{9}$$

When the operator is the weighted mean, the problem is rewritten as:

$$\begin{aligned}
&\text{Minimize } \sum_{(r,t)\in S} y_{(r,t)} \\
&\text{Subject to} \\
&\qquad \sum_{i=1}^{N} p_i(a_i^r - a_i^t) + y_{(r,t)} > 0 \\
&\qquad y_{(r,t)} \geq 0 \\
&\qquad \sum_{i=1}^{N} p_i = 1 \\
&\qquad p_i \geq 0
\end{aligned} \tag{10}$$

The optimal solution of this problem can be found using the same optimization techniques considered for the problem of expected outcomes. In this case we have as before linear constraints, and, in addition, the expression to be optimized is linear. Similar procedure applies when we use the Choquet integral as the selected aggregation operator, instead of the weighted mean.

There are a few variations of these models. For example, some require that the difference between two preferred examples is larger than a certain threshold (defined as constant), and then maximize the minimum difference between the two alternatives. See [19] for details.

Note that although the optimal solution can be found using optimization techniques, this does not imply that the cost of obtaining this solution is negligible. The Simplex algorithm, to be used for linear problems with linear constraints, has an exponential worst case time complexity.

The problem stated here is related to object ranking. A survey on object ranking can be found in this book [9]. In both problems, we have a set of alternatives or objects, and the ordering $\prec$ on them. The application of object ranking taking into account the utility functions, that is $\{(a_1^j, \ldots, a_N^j)\}_{j=1,\ldots,M}$ instead of just $x_1, \ldots x_M$, would lead to alternative methods for what has been described here.

5 Conclusions

In this chapter, we have reviewed some of the methods to be used for learning preferences when we can assign an utility value for each pair of alternative and criterion. We have also described algorithms to be applied when we can assign an overall utility value for each of the alternatives, and also when we can give an ordering of the alternatives.

Acknowledgements Partial support by the Generalitat de Catalunya (2005 SGR 00446 and 2005-SGR-00093) and by the Spanish MEC (projects ARES – CONSOLIDER INGENIO 2010 CSD2007-00004 – and eAEGIS – TSI2007-65406-C03-02) is acknowledged.

References

1. E. Armengol, F. Esteva, L. Godo, V. Torra, On learning similarity relations in fuzzy case-based reasoning, in *Transactions on Rough Sets II: Rough Sets and Fuzzy Sets, Lecture Notes in Computer Science*, vol. 3135, ed. by J.F. Peters, A. Skowron, D. Dubois, J. Grzymala-Busse, M. Inuiguchi, L. Polkowski (2004), pp. 14–32
2. G. Beliakov, R. Mesiar, L. Valaskova, Fitting generated aggregation operators to empirical data, Int. J. Unc. Fuzz. Knowl. Based Syst. **12**, 219–236, (2004)
3. T. Calvo, A. Kolesárová, M. Komorníková, R. Mesiar, Aggregation operators: properties, classes and construction methods, in *Aggregation Operators*, ed. by T. Calvo, G. Mayor, R. Mesiar (Physica-Verlag, 2002), pp. 3–104
4. W. Edwards, Probability-preferences in gambling. Am. J. Psychol. **66**, 349–364 (1953)
5. D.P. Filev, R.R. Yager, On the issue of obtaining OWA operator weights. Fuzzy Sets Syst. **94**, 157–169 (1998)
6. J. Fodor, M. Roubens, *Fuzzy Preference Modelling and Multicriteria Decision Support* (Kluwer, 1994)
7. M. Grabisch, k-order additive discrete fuzzy measures and their representation. Fuzzy Sets Syst. **92**(2), 167–189 (1997)
8. E. Hüllermeier, J. Fürnkranz, W. Cheng, K. Brinker, Label ranking by learning pairwise preferences. Artif. Intell. **172**, 1897–1916 (2008)
9. T. Kamishima, H. Kazawa, S. Akaho, A survey and empirical comparison of object ranking methods, in *Preference Learning*, ed. by J. Fürnkranz, E. Hüllermeier (Springer, 2010)
10. E.P. Klement, R. Mesiar, E. Pap, *Triangular Norms* (Kluwer, 2000)
11. Y. Narukawa, V. Torra, Fuzzy measure and probability distributions: distorted probabilities. IEEE Trans Fuzzy Syst. **13**(5), 617–629 (2005)
12. A. Rapoport, *Decision Theory and Decision Behavior* (Kluwer, 1989)
13. B. Roy, *Multicriteria Methodology for Decision Aiding* (Kluwer, 1996)
14. M. Sugeno, Theory of Fuzzy Integrals and its Applications, Ph.D. Dissertation, Tokyo Institute of Technology, Tokyo, Japan, 1974

15. V. Torra, The weighted OWA operator. Int. J. Intel. Syst. **12**, 153–166 (1997)
16. V. Torra, On hierarchically S-decomposable fuzzy measures. Int. J. Intel. Syst. **14**(9), 923–934 (1999)
17. V. Torra, On the learning of weights in some aggregation operators. Mathware Soft Comput. **6**, 249–265 (1999)
18. V. Torra, Learning weights for the Quasi-Weighted Mean. IEEE Trans. Fuzzy Syst. **10**(5), 653–666 (2002)
19. V. Torra, Y. Narukawa, *Modeling decisions: information fusion and aggregation operators* (Springer, 2007)
20. V. Torra, Y. Narukawa, *Modelització de decisions: fusió d'informació i operadors d'agregació* (UAB, 2007)
21. A. Valls, *ClusDM: A Multiple Criteria Decision Making Method for Heterogeneous Data Sets* (Monografies de l'IIIA, 2003)
22. A. Verkeyn, D. Botteldooren, B. De Baets, G. De Tré, Sugeno integrals for the modelling of noise annoyance aggregation, in *Proceedings of the 10th IFSA Conference, Lecture Notes in Artificial Intelligence*, vol. 2715 (2003), pp. 277–284
23. S. Weber, $\perp$-decomposable measures and integrals for archimedean t-conorms $\perp$. J. Math. Anal. Appl. **101**, 114–138 (1984)
24. R.R. Yager, On ordered weighted averaging aggregation operators in multi-criteria decision making. IEEE Trans. Syst. Man Cybern. **18**, 183–190 (1988)
25. http://www.mdai.cat/ifao/ifao.models.r

Part V
Preferences in Information Retrieval

Evaluating Search Engine Relevance with Click-Based Metrics*

Filip Radlinski, Madhu Kurup, and Thorsten Joachims

Abstract Automatically judging the quality of retrieval functions based on observable user behavior holds promise for making retrieval evaluation faster, cheaper, and more user centered. However, the relationship between observable user behavior and retrieval quality is not yet fully understood. In this chapter, we expand upon, Radlinski et al. (How does clickthrough data reflect retrieval quality, In Proceedings of the ACM Conference on Information and Knowledge Management (CIKM), 43–52, 2008), presenting a sequence of studies investigating this relationship for an operational search engine on the arXiv.org e-print archive. We find that none of the eight absolute usage metrics we explore (including the number of clicks observed, the frequency with which users reformulate their queries, and how often result sets are abandoned) reliably reflect retrieval quality for the sample sizes we consider. However, we find that paired experiment designs adapted from sensory analysis produce accurate and reliable statements about the relative quality of two retrieval functions. In particular, we investigate two paired comparison tests that analyze clickthrough data from an interleaved presentation of ranking pairs, and find that both give accurate and consistent results. We conclude that both paired comparison tests give substantially more accurate and sensitive evaluation results than the absolute usage metrics in our domain.

---

*Based on Filip Radlinski, Madhu Kurup, and Thorsten Joachims. How does clickthrough data reflect retrieval quality. In Proceedings of the ACM Conference on Information and Knowledge Management (CIKM), pages 43–52, 2008 ©.

F. Radlinski (✉)
Microsoft Research, Cambridge, UK
e-mail: filiprad@microsoft.com

M. Kurup
Amazon, Seattle, WA, USA
e-mail: madhuk@amazon.com

T. Joachims
Cornell University, Ithaca, NY, USA
e-mail: tj@cs.cornell.edu

J. Fürnkranz and E. Hüllermeier (eds.), *Preference Learning*,
DOI 10.1007/978-3-642-14125-6_16, © Springer-Verlag Berlin Heidelberg 2010

1 Introduction

Most search engine evaluation uses the traditional Cranfield methodology, where relevance judgments are provided manually by trained experts. For each query in a test set, the experts provide a label that specifies the relevance of each document in a corpus on a graded relevance scale. Given a ranking produced in response to these queries, the judgments for the top ranked documents can then be aggregated to assess the quality of the ranking. Averaging over many queries yields average performance scores such as Normalized Discounted Cumulative Gain, Mean Average Precision and Precision at K [21].

However, this Cranfield approach presents a number of challenges. First, the process of obtaining expert relevance judgments is time consuming [8] and thus expensive. The associated cost and turnaround times of the Cranfield approach make it economical only in large domains such as non-personalized Web search. Other retrieval applications from Desktop Search, to searching Wikipedia, to Intranet Search usually demand more flexible and efficient evaluation methods.

Second, queries are often ambiguous. The largest Cranfield style test collections generally used to evaluate the performance of ranking algorithms are produced as part of the annual TREC competition [29]. In the TREC setting, each query includes a long description, which is obtained before the judgments are created. When using arbitrary real queries, it can be difficult for expert relevance judges to reliably infer such intents from queries, as they are usually too short to unambiguously identify the users' information needs (e.g., [27]). Consequently, there is a danger that the labels provided may not match the extent to which the documents address the users' actual information needs.

Third, the experts must be knowledgable on the information needs relevant to the collection. When designing Web search systems for specialized subgroups of the general population (for instance, academic audiences) or on specialized document collections (for instance, digital libraries), the cost of obtaining relevance judgments for evaluation can thus become even more prohibitive.

Finally, even when reliable expert judgments are available for computing standard performance metrics, some of the standard metrics have been shown to not always correlate with user-centric performance measures [28].

One promising approach to address these challenges is evaluation based on implicit judgments from observable user behavior, such as clicks, query reformulations, and response times. The potential advantages are clear: Unlike expert judgments, usage data can be collected at essentially zero cost, is available in real time, and it reflects the values of the users, not those of judges far removed from the users' context at the time of the information need. The key problem with retrieval evaluation based on usage data lies in its proper interpretation. In particular, understanding how certain observable statistics relate to retrieval quality. We shed light onto this relationship through a user study with an operational search engine we deployed on the arXiv.org e-print archive. The study follows a controlled experiment design that is unlike previous evaluations of implicit feedback, which mostly investigated document-level relationships between (expert or user annotated) relevance and user behavior (e.g., [1,9,12]). Instead, we construct multiple retrieval functions for which we know their relative retrieval quality by construction (e.g., comparing a standard retrieval function versus the same function with some results randomly swapped within the top five positions). Fielding these retrieval functions to real users of our search engine, we test how implicit feedback statistics reflect the difference in retrieval quality. We ask whether there are universal properties of user behavior that can be used to evaluate ranking quality.

Specifically, we compare two evaluation methodologies, which we term "absolute metrics" and "paired comparison tests". Using absolute metrics for evaluation follows the hypothesis that retrieval quality impacts observable user behavior in an absolute sense (such as better retrieval leads to higher-ranked clicks, or better retrieval leads to faster clicks). We formulate eight such absolute metrics and hypothesize how they will change with improved retrieval quality. We then test whether these hypotheses hold in our search engine. The second evaluation methodology, paired comparison tests, was first proposed for retrieval evaluation by Joachims [14,15]. He follows experiment designs from the field of sensory analysis. When, for example, studying the taste of a new product, subjects are rarely asked to independently rate the product on an absolute scale, but are instead usually given a second product and asked to express a preference between the two (see [19] for a discussion of sensory analysis).

Joachims [14,15] proposed a method for interleaving the rankings from a pair of retrieval functions to mirror such paired comparisons. In this setting, clicks provide a blind preference judgment. We call the algorithm he proposed Balanced Interleaving. In this chapter, we evaluate the accuracy of Balanced Interleaving on arXiv.org, and also propose a new Team-Draft Interleaving method that overcomes potential problems of Balanced Interleaving for rankings that are close to identical.

The findings of our user study can be summarized as follows. None of the eight absolute metrics reflect retrieval performance in a significant, easily interpretable, and reliable way with the sample sizes we consider. In contrast, both interleaving tests accurately reflect the known differences in retrieval quality, inferring consistent and in most cases significant preferences in the correct direction given the same amount of user behavior data.

2 Related Work

From the perspective of facilitating evaluation of ranking strategies, a number of researchers have considered how to reduce the amount of labeling effort necessary for Cranfield-style evaluation (including [4, 6, 7, 26]), or how to obtain evaluation datasets more representative of realistic usage scenarios (e.g., [25]). However, our focus is on alternative evaluation methodologies, in particular using relevance feedback provided by users.

Two general strategies have been used for obtaining relevance judgements from users: explicitly asking for relevance judgments, or implicitly inferring judgments from user behavior. Asking users for explicit relevance judgments is onerous, as it essentially requires users to be expert relevance judges. Without an appropriate incentive, users have little motivation to provide high-quality relevance judgments. Hence, evaluating with explicit judgments is generally limited to settings such as movie ranking, where users are more willing to provide reliable judgments in return for personalized movie recommendations (e.g., [10, 24]). Evaluations using implicit feedback, based on observing natural user interactions with the search engine, are more practical in general search settings. This is motivated by the simplicity of recording user behavior such as querying and clicking.

Numerous proposals for evaluating ranking quality based on user behavior have previously been explored. Kelly and Teevan give an overview of many previously studied behavioral metrics [17]. Most of these fall into the category of absolute metrics, which we will evaluate in our user study. For instance, Fox et al. [12] learned to predict whether users were satisfied with specific search results using implicitly collected feedback. They found a number of particularly indicative features, such as time spent on result pages and how the search session was terminated (e.g., by closing the browser window or by typing a new Internet address). However, many of the most informative features they identified cannot be collected unless users are using a modified Web browser. In our work, we focus on measures that can be collected from all Web users without requiring additional downloads or browser instrumentation. A number of researchers have studied how to transform clicks into relevance judgments over documents, which could then be aggregated to evaluate ranking performance, including [1, 2, 11, 15, 22]. With a focus on evaluation with just clicks, Carterette and Jones [9] looked at whether they can identify the better of two ranking functions. They found that by training a probabilistic click model, they can predict the probability of relevance for each result. Aggregating over entire rankings, they were able to reliably predict the better of two rankings in terms of NDCG. Using clicks and other implicit feedback directly for evaluation of rankings was also studied by [3, 5, 13, 20]. Most directly related to this work, clicks were first used for evaluation using the Balanced Interleaving paired comparison test described here by Joachims [14, 15], although without a controlled study to compare the effectiveness of different evaluation metrics.

In contrast to all the previous studies, we present a controlled real-world experiment evaluating how user behavior given real user-generated queries changes in response to known changes in ranking quality. We measure how user-based evaluation would pick up the differences in quality between five different ranking functions, as measured by eight different absolute click metrics and two pairwise comparison algorithms. A somewhat shorter description of this work was presented in [23].

3 Design of the User Study

We implemented a search engine over the arXiv.org e-print archive.[1] This archive consists of a collection of several hundred thousand academic articles. It is used daily by many thousands of users, predominantly scientists from the fields of physics, mathematics, and computer science. Hundreds of these users use our search engine on any particular day.

The basic design of our study can be summarized as follows. Starting with an initial (hand-tuned) ranking function f_1, we derive several other ranking functions by artificially degrading their retrieval quality compared to f_1. In particular, we constructed triplets of ranking functions $f_1 \succ f_2 \succ f_3$, using the notation $f_i \succ f_j$ to indicate that the retrieval quality of ranking function f_i is better than that of f_j. For each such triplet of ranking functions, we know by construction that f_1 outperforms f_2, and that both outperform f_3. We then expose the users of arxiv.org to these three ranking functions as detailed below, and analyze whether, and under which types of exposure, their observable behavior reflects the known differences in retrieval quality.

Over four one-month periods we fielded triplets of ranking functions in the arXiv.org search engine. Our users were unaware of the experiments being conducted. As the users interacted with the search engine, we recorded the queries issued, and the results clicked on. We then performed aggregate analyses of the observed behavior, leading to the results reported below.

3.1 Constructing Comparison Triplets

We start by describing how we created two sets of ranking functions with known relative retrieval performance. Given that our document collection consisted of academic articles with rich metadata, we started with an original ranking function, called ORIG, that scores each document by computing a sum of the match between the query and each of the following document fields: authors, title, abstract, full text, arXiv identifier, article category, article area, article submitter, any journal reference and any author-provided comments. The first four fields are usually most important in matching results to queries. Note that this retrieval function weights, for example, words in the title more heavily, since these title words occur in multiple fields (specifically, in the title field and in the full text field). Our search engine was implemented on top of Lucene,[2] which implements a standard cosine similarity matching function.

[1] Made available at http://search.arxiv.org/

[2] Available at http://lucene.apache.org/

3.1.1 "ORIG≻FLAT≻RAND"-Comparison

To create the first triplet of ranking functions, we first eliminated much of the metadata available, then randomized the top search results. Specifically, the first degraded ranking function, FLAT, only computes the sum of the matches in the article full text, author list and article identifier. Note that while the abstract and title are included in the full text, by not scoring contributions on each field independently, we reduced the weight placed on these (usually particularly important) fields. Second, ranking function RAND reordered the top 11 results returned by FLAT completely at random. Since the nonrandomized ranking was of reasonable quality, randomization reduces the ranking quality. The documents below rank 11 were presented unchanged. By construction, we now have a triplet of ranking functions where it is safe to assume that ORIG $\succ$ FLAT $\succ$ RAND. In fact, our subjective impression is that these three ranking functions deliver substantially different retrieval quality – especially ORIG $\succ$ RAND – and any suitable evaluation method should be able to detect this difference.

3.1.2 "ORIG≻SWAP2≻SWAP4"-Comparison

To create a second triplet of ranking functions that shows a more subtle difference in retrieval quality, we degraded performance in a different way. Starting again with our ranking function ORIG, SWAP2 randomly selects two documents in the top 5 positions and swaps them with two random documents from ranks 7–11. This swapping pattern is then replicated on all later result pages (i.e., swapping two documents between ranks 11 and 15 with two originally ranked between 17 and 21, etc.). For instance, if ORIG returned the ten documents $(d_1, d_2, d_3, d_4, d_5, d_6, d_7, d_8, d_9, d_{10})$, SWAP2 might present the user with $(d_1, d_7, d_3, d_9, d_5, d_6, d_2, d_8, d_4, d_{10})$. Increasing the degradation, SWAP4 is constructed identically to SWAP2, except randomly selecting four documents to swap. This gives us a second triplet of ranking functions, where by construction we know that ORIG $\succ$ SWAP2 $\succ$ SWAP4. We believe the quality differences in this triplet are smaller than in the previous triplet. In particular, this is because the top 11 results always contain the same set of documents for all three ranking functions, just in a different order. In contrast, RAND and FLAT return different top 11 documents than ORIG.

3.2 Users and System Design

Figure 1 illustrates the user interface of the search engine. It takes a set of keywords as a query, and returns a ranking of 10 results per page. For each result, we show authors, title, year, a query-sensitive snippet, and the arXiv identifier of the paper. We register a "click" whenever a user follows a hyperlink associated with a result. These clicks lead to a metadata page from where a PDF is available for download.

Fig. 1 Screenshot of how results are presented

3.2.1 User Characterization

Given the nature of the arXiv document collection, consisting mostly of scientific articles from the fields of Physics, Mathematics, Computer Science, and to a lesser extent Nonlinear Sciences, Quantitative Biology and Statistics, we suspect that many of our users are researchers and students from these disciplines. On average, our search engine received about 700 queries per day from about 300 distinct IP addresses, registering about 600 clicks on results.

We identify users by their IP address. Since this definition of user is primarily used for identifying spammers and bots, as will be described below, we find it sufficient even though in some cases it may conflate multiple people working on the same computer or accessing arXiv.org through a proxy. The IP address is also used to (pseudo) randomly assign users to various experimental conditions in our study (e.g., the condition "users who receive the results from FLAT"). In particular, we segment the user population based on an MD5-hash of IP address and user agent reported by the browser. Moreover, for the rankings that involve randomizing the results returned, the random number generator is seeded with the same information. This method of assignment and seeding ensures that if a user repeats a query within a short time frame, he or she will be shown exactly the same results, making for a consistent user experience.

3.2.2 Data Collection

We recorded queries submitted, as well as clicks on search results for all queries. Each record included the experimental condition, the time, IP address, browser, a unique session identifier, and a unique query identifier.

We define a session as a sequence of interactions (clicks or queries) between a user and the search engine where less than 30 minutes pass between subsequent interactions. When attributing clicks to query results, we only counted clicks occurring within the same session as the query. This was necessary to eliminate clicks that appeared to come from saved or cached search results. Note that it is still possible for clicks to occur hours after the query if the user was continuously interacting with the search engine.

3.2.3 Quality Control and Testing

To test the system and our experiment setup, we conducted a test run between November 3rd and December 5th, 2007. Based on this data, we refined our methods for data cleaning and spam detection (described below), refined the system and experiment design, and validated the correctness of the software. For all crucial parts of data analysis, the first two authors of this chapter each independently implemented analysis code then compared their results to detect potential errors.

4 Experiment 1: Absolute Metrics

We can now ask: Do absolute metrics reflect retrieval quality? We define absolute metrics as search engine usage statistics that can be hypothesized to monotonically change with retrieval quality. We explore eight such metrics that quantify the clicking and session behavior of users.

4.1 Absolute Metrics and Hypotheses

We measured the following metrics. Many of them were previously suggested as performance metrics in the literature, as they reflect the key actions that users can choose to perform after issuing a query: clicking, reformulating, or abandoning the search.

Abandonment Rate	The fraction of queries for which no results were clicked on.
Reformulation Rate	The fraction of queries that were followed by another query during the same session.
Queries per Session	The mean number of queries issued by a user during a session.
Clicks per Query	The mean number of results that are clicked for each query.
Max Reciprocal Rank†	The mean value of $1/r$, where r is the rank of the highest ranked result clicked on.
Mean Reciprocal Rank†	The mean value of $\sum 1/r_i$, summing over the ranks r_i of all clicks for each query.
Time to First Click†	The mean time from query being issued until first click on any result.
Time to Last Click†	The mean time from query being issued until last click on any result.

When computing the metrics marked with †, we exclude queries with no clicks to avoid conflating the metrics with abandonment rate. For each metric, we hypothesize how we expect the metric to change as retrieval quality decreases. The explanation for the hypothesized direction of change is noted on the right.

Metric	Hypothesized change as ranking gets worse
Abandonment rate	Increase (more bad result sets)
Reformulation rate	Increase (more need to reformulate)
Queries per session	Increase (more need to reformulate)
Clicks per query	Decrease (fewer relevant results)
Max reciprocal rank	Decrease (top results are worse)
Mean reciprocal rank	Decrease (more need for many clicks)
Time to first click	Increase (good results are lower)
Time to last click	Decrease (fewer relevant results)

Even if the hypothesized directions of change are incorrect, we at least expect these metrics to change monotonically with retrieval quality. We now test these hypotheses for ORIG $\succ$ FLAT $\succ$ RAND and ORIG $\succ$ SWAP2 $\succ$ SWAP4.

4.2 Experiment Setup

We evaluate the absolute metrics in two phases. Data for the triplet of ranking functions ORIG $\succ$ SWAP2 $\succ$ SWAP4 was collected from December 19th, 2007 to January 25th, 2008 (Phase I); for the ranking functions ORIG $\succ$ FLAT $\succ$ RAND, it was collected from January 27th to February 25th, 2008 (Phase II). During each phase, each of the three ranking functions were assigned one experimental condition, receiving 1/6th of search engine visitors. This means that in Phase I, 1/6th of the users saw the results from ORIG, another 1/6th saw the results from FLAT, and yet another 1/6th got the results from RAND. In Phase II, the assignment was done analogously for ORIG, SWAP2, and SWAP4. The remaining 50% of the visitors were assigned to paired comparison conditions described in Sect. 5.

During our test run prior to these evaluations, we noticed that bots and spammers throw off our results. To compute the absolute metrics robustly, we processed the raw logs as follows. First, we eliminated all users (IP addresses) who clicked on more than 100 results on any day of our study. This eliminated under 10 users in each condition. We then computed each metric for every user, averaging over all queries submitted by that user. Finally, we computed the median (for the time to click metrics) or mean (for the others) across all users.[3] This simple per-user aggregation is fairly robust to spammers and bots, much more so than naive per-query aggregation. For instance, suppose we have 99 users and one spammer (or bot). Suppose the spammer ran 100 queries and always clicked on all top 10 results, while each of the 99 normal users ran just one query and clicked on one result. The average number of clicks per query that we compute is $(1 \times 10 + 99 \times 1)/100 = 1.09$, rather than $(100 \times 10 + 99 \times 1)/199 = 5.5$ as it would be with query-based averaging.

4.3 Results and Discussion

The measured values (± two standard errors/95% confidence intervals) are reported in Table 1 for each absolute metric and each ranking function. The column labeled $\mathcal{H}_1$ indicates our hypothesized change in the metric if retrieval quality is decreased. Upon inspection, one observes that none of the metrics consistently follows the hypothesized behavior. The number of pairs $A \succ B$ where the observed value follows (✓) or opposes ($\lightning$) the hypothesized change is summarized in the "weak" columns of Table 2. It shows that, for example, the abandonment rate agrees with our hypothesis for four pairs of ranking functions (ORIG $\succ$ FLAT, FLAT $\succ$ RAND, ORIG $\succ$ RAND, and SWAP2 $\succ$ SWAP4). However, for the remaining two pairs, it changes in the opposite direction. Even more strongly, none of the absolute metrics even changes strictly monotonically with retrieval quality.

[3] In other words, we report macro-averages rather than micro-averages.

Table 1 Absolute metrics for the "ORIG≻ FLAT≻ RAND" and the "ORIG ≻ SWAP2 ≻ SWAP4" comparisons (± two standard errors/95% confidence intervals). The second column shows the hypothesized change when retrieval quality is degraded

		ORIG ≻ FLAT ≻ RAND		
	$\mathcal{H}_1$	ORIG	FLAT	RAND
Abandonment Rate (Mean)	<	0.680±0.021	0.725±0.020	0.726±0.020
Reformulation Rate (Mean)	<	0.247±0.021	0.257±0.021	0.260±0.021
Queries per Session (Mean)	<	1.925±0.098	1.963±0.100	2.000±0.115
Clicks per Query (Mean)	>	0.713±0.091	0.556±0.081	0.533±0.077
Max Reciprocal Rank (Mean)	>	0.554±0.029	0.520±0.029	0.518±0.030
Mean Reciprocal Rank (Mean)	>	0.458±0.027	0.442±0.027	0.439±0.028
Time (s) to First Click (Median)	<	31.0±3.3	30.0±3.3	32.0±4.0
Time (s) to Last Click (Median)	>	64.0±19.0	60.0±14.0	62.0±9.0
		ORIG ≻ SWAP2 ≻ SWAP4		
	$\mathcal{H}_1$	ORIG	SWAP2	SWAP4
Abandonment Rate (Mean)	<	0.704±0.021	0.680±0.021	0.698±0.021
Reformulation Rate (Mean)	<	0.248±0.021	0.250±0.021	0.248±0.021
Queries per Session (Mean)	<	1.971±0.110	1.957±0.099	1.884±0.091
Clicks per Query (Mean)	>	0.720±0.098	0.760±0.127	0.734±0.125
Max Reciprocal Rank (Mean)	>	0.538±0.029	0.559±0.028	0.488±0.029
Mean Reciprocal Rank (Mean)	>	0.444±0.027	0.467±0.027	0.394±0.026
Time (s) to First Click (Median)	<	28.0±2.2	28.0±3.0	32.0±3.5
Time (s) to Last Click (Median)	>	71.0±19.0	56.0±10.0	66.0±15.0

Table 2 Comparing the number of correct ("✓") and false ("↯") preferences implied by the absolute metrics, aggregated over the "ORIG ≻ FLAT ≻ RAND" and the "ORIG ≻ SWAP2 ≻ SWAP4" comparison. A preference is weakly correct/false, if observed value follows/contradicts our hypothesized direction of change. A preference is significantly correct/false, if the difference between the observed values is statistically significant (95%) in the respective direction

Absolute metric signals	Weak		Significant	
	✓	↯	✓	↯
Abandonment Rate (Mean)	4	2	2	0
Reformulation Rate (Mean)	4	2	0	0
Queries per Session (Mean)	3	3	0	0
Clicks per Query (Mean)	4	2	2	0
Max Reciprocal Rank (Mean)	5	1	3	0
Mean Reciprocal Rank (Mean)	5	1	2	0
Time (s) to First Click (Median)	4	1	0	0
Time (s) to Last Click (Median)	4	2	1	1

The lack of consistency with the hypothesized change could partly be due to measurement noise, since the elements of Table 2 are estimates of a population mean/median. The column "significant" of Table 2 shows for how many pairs $A \succ B$ we can significantly (95% one-tailed confidence t-test for mean, χ^2-test for median) reject our hypothesis $\mathcal{H}_1$ (↯) or reject its negation (✓). We do not see a significant difference in the hypothesized direction for more than three out of the six pairs $A \succ B$ for any of the absolute metrics. With the exception of Max Reciprocal Rank, not even the "large difference" pairs ORIG $\succ$ RAND and ORIG $\succ$ SWAP4 are consistently significant for any of the metrics. This suggests that, at best, we need substantially more data to use these absolute metrics reliably, making them unsuitable for low-volume search applications such as desktop search, personalized Web search, and intranet search.

Figures 2 and 3 present a more detailed view of these metrics, giving some insight into how the estimates developed as more data was collected. The plots show the respective estimate after the first n distinct users (i.e., distinct IP addresses) were observed. Each datapoint represents a different cutoff date on which we computed the metric over all prior data. The error bars indicate one standard error/66% confidence interval. For example, the first point corresponds to roughly the first day of data, after which we had seen 50 distinct users in each experimental condition. The second data point corresponds to taking roughly the first two days, after which we had seen 100 distinct users. As time progressed we saw fewer new distinct IP addresses per day, hence each data point should not be considered as one day. The total experiment duration for each plot was one month.

Consider, as an example absolute metric, the mean abandonment rate for the "ORIG $\succ$ FLAT $\succ$ RAND" and "ORIG $\succ$ SWAP2 $\succ$ SWAP4" experiments as a function of the number of users who have visited the search engine. Note that for "ORIG $\succ$ FLAT $\succ$ RAND", the original (best) ranking function has the lowest abandonment rate, while for "ORIG $\succ$ SWAP2 $\succ$ SWAP4" the original ranking function has the highest abandonment rate.[4] This not only breaks our intuition about abandonment rate, but also indicates that different differences between ranking functions can have different effects on the abandonment rate, making it an unreliable indicator as to the relative quality of ranking functions if our assumed relative ordering of the ranking functions holds.

In general, we see that many of the curves still cross toward the end, indicating that the estimates have indeed not yet converged with sufficient precision. Second, the plots show that the (Gaussian) error bars are reasonable as confidence intervals for the mean, and therefore the t-test is also reasonable. In particular, the curves do indeed terminate within the two standard error interval of most prior datapoints. This also suggests that there are no substantial temporal changes (e.g., bot or spam attacks that we do not catch in our pre-processing) within each of the experiments. However, note that in Table 1 the Abandonment Rate and the Time to First Click of ORIG are significantly different between the data collected in December/January and the data collected in February. Our conjecture is that this is due to differences in user population and context (e.g., academic break vs. semester). It appears that the impact of these population differences on some of the absolute metrics can be of similar magnitude as the differences observed due to retrieval quality, confirming that only data collected during the same time period can be meaningfully compared.

[4] While the two "original ranking function" curves represent the same ranking function, they were collected on two different months thus explaining the variation between them.

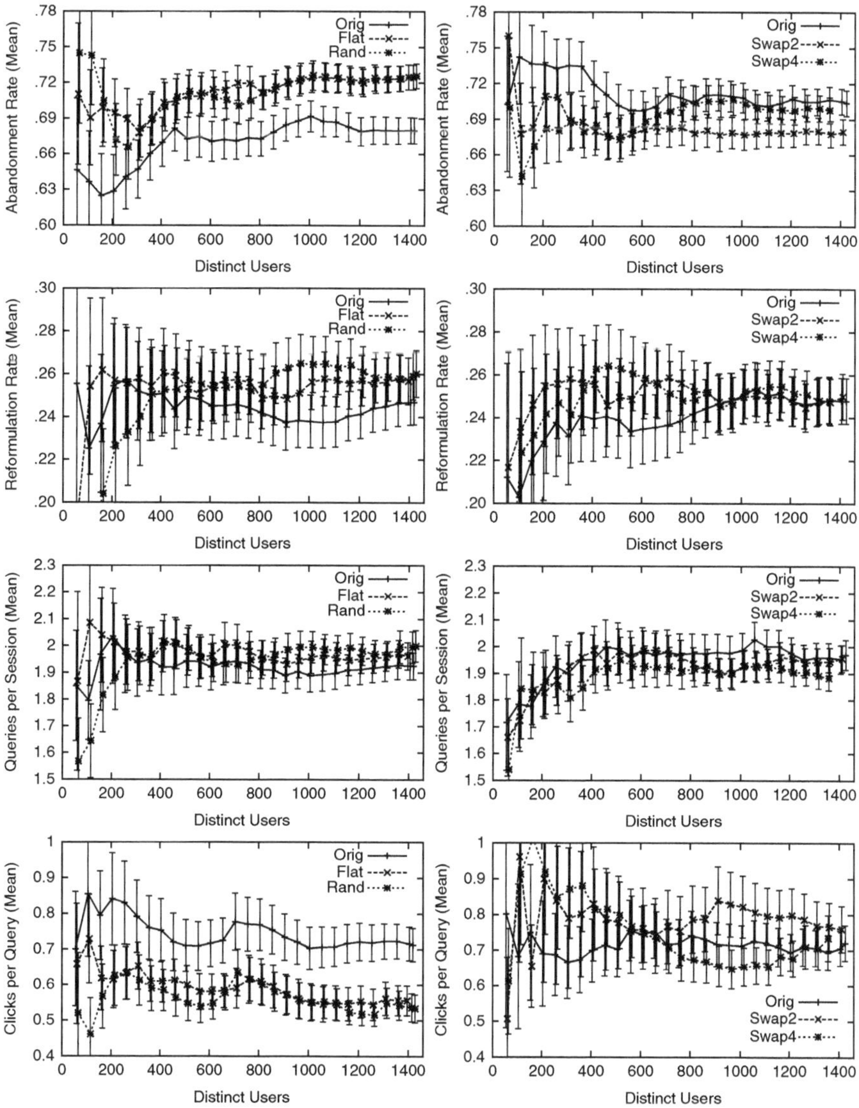

Fig. 2 Measurements of the first four absolute performance metrics, for ORIG ≻ FLAT ≻ RAND on the left, and ORIG ≻ SWAP2 ≻ SWAP4 on the right. The error bars indicate one standard error/66% confidence interval

5 Experiment 2: Paired Comparison Tests

Paired comparison tests are one of the central experiment designs used in sensory analysis [19]. When testing a perceptual quality of an item (e.g., taste, sound), it is recognized that absolute (Likert scale) evaluations are difficult to make. Instead,

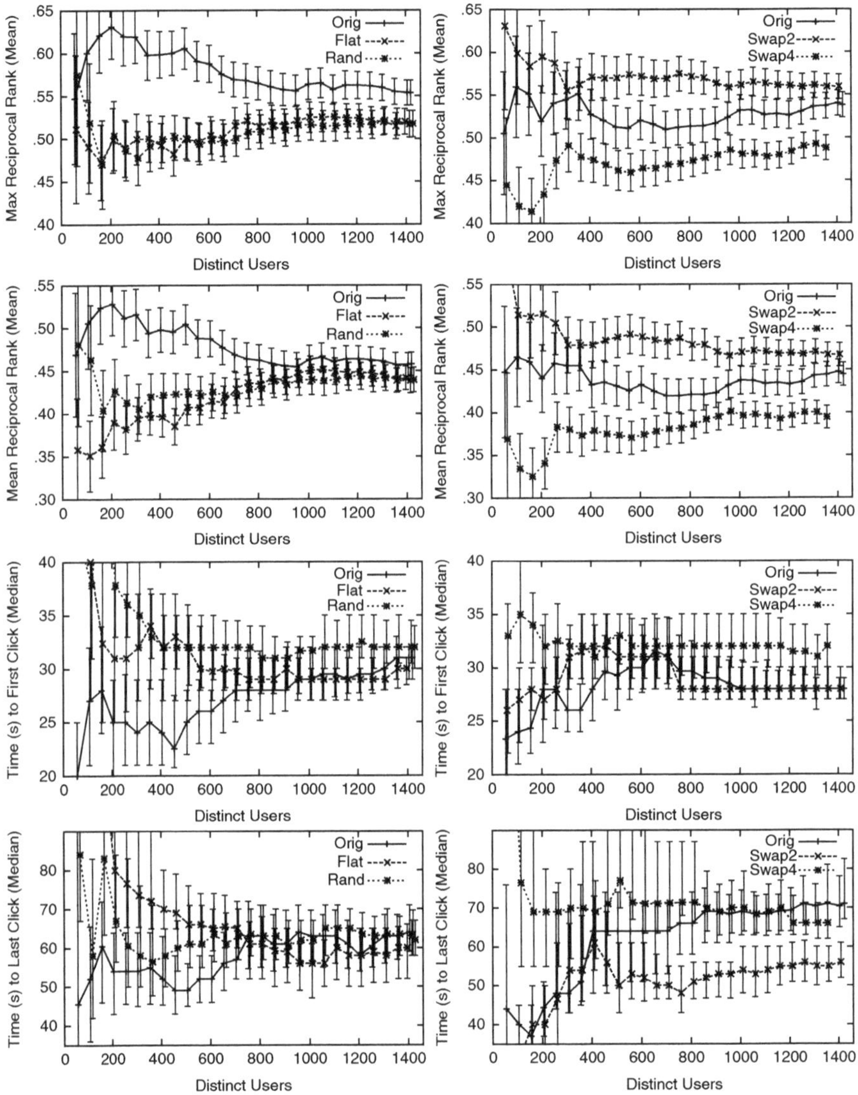

Fig. 3 Measurements of the last four absolute performance metrics, for ORIG ≻ FLAT ≻ RAND on the left, and ORIG ≻ SWAP2 ≻ SWAP4 on the right. The error bars indicate one standard error/66% confidence interval

subjects are presented with two or more alternatives and are asked to identify a difference or state a preference. In the simplest case, subjects are given two alternatives and are asked which of the two they prefer. For the evaluation of retrieval functions, this experiment design was first explored by Joachims [14, 15]. In particular, Joachims proposed a method for presenting the results from two retrieval

Algorithm 1 Balanced Interleaving

Input: Rankings $A = (a_1, a_2, \ldots)$ and $B = (b_1, b_2, \ldots)$
$I \leftarrow (); k_a \leftarrow 1; k_b \leftarrow 1;$
$AFirst \leftarrow RandBit()$ .... *decide which ranking gets priority*
while $(k_a \leq |A|) \wedge (k_b \leq |B|)$ **do** .... *if not at end of A or B*
  if $(k_a < k_b) \vee ((k_a = k_b) \wedge (AFirst = 1))$ **then**
    if $A[k_a] \notin I$ **then** $I \leftarrow I + A[k_a]$ .... *append next A result*
    $k_a \leftarrow k_a + 1$
  else
    if $B[k_b] \notin I$ **then** $I \leftarrow I + B[k_b]$ .... *append next B result*
    $k_b \leftarrow k_b + 1$
  end if
end while
Output: Interleaved ranking I

functions so that clicks indicate a user's preference between the two. In contrast to the absolute metrics discussed so far, paired comparison tests do not assume that observable user behavior changes with retrieval quality on some absolute scale, but merely that users can identify the preferred alternative in a direct comparison.

5.1 Balanced Interleaving Method

The key design issue for a paired comparison test between two retrieval functions is the method of presentation. As outlined in [14], the design should be (a) blind to the user with respect to the underlying conditions, (b) it should be robust to biases in the user's decision process that do not relate to retrieval quality, (c) it should not substantially alter the search experience, and (d) it should lead to clicks that reflect the user's preference. The naïve approach of simply presenting two rankings side by side would clearly violate (c), and it is not clear whether biases in user behavior would actually lead to meaningful clicks.

To overcome these problems, Joachims [14, 15] proposed a presentation where the results present in two rankings A and B are interleaved into a single ranking I in a balanced way. The interleaved ranking I is then presented to the user. This particular method of interleaving A and B ensures that any top k results in I always contain the top k_a results from A and the top k_b results from B, where k_a and k_b differ by at most 1. Intuitively, a user reading the results in I from top to bottom will have always seen an approximately equal number of results from each of A and B.

It can be shown that such an interleaved ranking always exists for any pair of rankings A and B, and that it is computed by Algorithm 1 [14]. The algorithm constructs this ranking by maintaining two pointers, namely k_a and k_b, and then interleaving greedily. The pointers are set to always point at the highest ranked result in the respective original ranking that is not yet in the combined ranking. To construct I, the lagging pointer among k_a and k_b is used to select the next result to add to I. Ties are broken randomly.

	Input Ranking		Interleaved Rankings					
			Balanced		*Team-Draft*			
Rank	A	B	A first	B first	AAA	BAA	ABA	...
1	a	b	a	b	a^A	b^B	a^A	
2	b	e	b	a	b^B	a^A	b^B	
3	c	a	e	e	c^A	c^A	e^B	
4	d	f	c	c	e^B	e^B	c^A	
5	g	g	d	f	d^A	d^A	d^A	
6	h	h	f	d	f^B	f^B	f^B	
⋮	⋮	⋮	⋮	⋮	⋮	⋮	⋮	

Fig. 4 Examples illustrating how the Balanced and the Team-Draft methods interleave input rankings A and B for different outcomes of the random coin flips. Superscript for the Team-Draft interleavings indicates team membership

Two examples of such combined rankings are presented in the column "Balanced" of Fig. 4. The left column assumes ranking A wins a tie-breaking coin toss, while the right column assumes that ranking B wins the toss.

Given an interleaving I of two rankings presented to the user, one can derive a preference statement from user clicks. In particular, let us assume that the user reads results from top to bottom (as supported by eye-tracking studies [16]), and that the number of links l viewed in I is known and fixed a priori. This means the user has l choices to click on, and an almost equal number came from A and from B. So, a randomly clicking user has approximately an equal chance of clicking on a result from A as from B. If we see more clicks on results from one of the two retrieval functions, we can infer a preference.

More formally, let $A = (a_1, a_2, \ldots)$ and $B = (b_1, b_2, \ldots)$ be two input rankings we wish to compare. Let $I = (i_1, i_2, \ldots)$ be the combined ranking computed by the Balanced Interleaving algorithm, and let $c_1, c_2, \ldots$ be the ranks of the clicks with respect to I. To estimate l, [14] proposes to use the lowest ranked click, namely $l \approx c_{\max} = \max\{c_1, c_2, \ldots\}$. Furthermore, to derive a preference between A and B, one compares the number of clicks in the top

$$k = \min\{j : (i_{c_{\max}} = a_j) \vee (i_{c_{\max}} = b_j)\} \tag{1}$$

results of A and B. In particular, the number h_a of clicks attributed to A and the number h_b of clicks attributed to B is computed as

$$h_a = |\{c_j : i_{c_j} \in (a_1, \ldots, a_k)\}| \tag{2}$$
$$h_b = |\{c_j : i_{c_j} \in (b_1, \ldots, b_k)\}|. \tag{3}$$

If $h_a > h_b$ we infer a preference for A, if $h_a < h_b$ we infer a preference for B, and if $h_a = h_b$ we infer no preference.

To further illustrate how preferences are derived from clicks in the interleaved ranking, suppose the user clicked on documents b and e in either of the two balanced interleavings shown in Fig. 4. Here, $k = 2$, since the top 3 documents in I were constructed by combining the top 2 results from A and B. Both clicked documents are in the top 2 of ranking B, but only one (b) is in the top 2 or ranking A. Hence, the user has expressed a preference for ranking B.

Over a sample of queries and users, denote with *wins(A)* the number of times A was preferred, and with *wins(B)* the number of times B was preferred. Using a binomial sign test, we can test whether one ranking function was preferred significantly more often.

5.2 Team-Draft Interleaving Method

Unfortunately, using (1) to estimate the number of results seen from each ranking can potentially lead to biased results for Balanced Interleaving in some cases, especially when rankings A and B are almost identical up to a small shift or insertion. For example, suppose we have two rankings, $A = (a, b, c, d)$ and $B = (b, c, d, a)$. Depending on which ranking wins the tie breaking coin toss in Algorithm 1, interleaving will produce either $I = (a, b, c, d)$ or $I = (b, a, c, d)$. Note that in both cases, a user who clicks uniformly at random on one of the results in I would produce a preference for B more often than for A, which is clearly undesirable. This is because all the documents except a are ranked higher by ranking B, and k is defined as the minimum cutoff that includes all documents. We now describe a new interleaving approach that does not suffer from this problem.

The new interleaving algorithm, called Team-Draft Interleaving, follows the analogy of selecting teams for a friendly team-sports match. One common approach is to first select two team captains, who then take turns selecting players for their team. We can use an adapted version of this algorithm for creating interleaved rankings. Suppose each document is a player, and rankings A and B are the preference orders of the two team captains. In each round, captains pick the next player by selecting their most preferred player that is still available, add the player to their team and append the player to the interleaved ranking I. We randomize which captain gets to pick first in each round. The algorithm is summarized in Algorithm 2, and the column "Team-Draft" of Fig. 4 gives three illustrative examples (e.g., the column "BAA" indicates that captain B picked first in the first round, and that captain A picked first in the second and third rounds).

To derive a preference between A and B from the observed clicking behavior in I, again denote the ranks of the clicks in the interleaved ranking $I = (i_1, i_2, \ldots)$ with $c_1, c_2, \ldots$. We then attribute the clicks to ranking A or B based on which ranking selected the clicked results (or, in the team sport analogy, which team that player was playing for). In particular,

$$h_a = |\{c_j : i_{c_j} \in Team\ A\}| \tag{4}$$

$$h_b = |\{c_j : i_{c_j} \in Team\ B\}|. \tag{5}$$

Algorithm 2 Team-Draft Interleaving

Input: Rankings $A = (a_1, a_2, \dots)$ and $B = (b_1, b_2, \dots)$
Init: $I \leftarrow ()$; $TeamA \leftarrow \emptyset$; $TeamB \leftarrow \emptyset$;
while $(\exists i : A[i] \notin I) \wedge (\exists j : B[j] \notin I)$ **do** .... *if not at end of A or B*
  if $(|TeamA| < |TeamB|) \vee$
    $((|TeamA| = |TeamB|) \wedge (RandBit() = 1))$ **then**
    $k \leftarrow \min_i \{i : A[i] \notin I\}$ .... *top result in A not yet in I*
    $I \leftarrow I + A[k]$; .... *append it to I*
    $TeamA \leftarrow TeamA \cup \{A[k]\}$ .... *clicks credited to A*
  else
    $k \leftarrow \min_i \{i : B[i] \notin I\}$ .... *top result in B not yet in I*
    $I \leftarrow I + B[k]$ .... *append it to I*
    $TeamB \leftarrow TeamB \cup \{B[k]\}$ .... *clicks credited to B*
  end if
end while
Output: Interleaved ranking I, $TeamA$, $TeamB$

If $h_a > h_b$ we infer a preference for A, if $h_a < h_b$ we infer a preference for B, and if $h_a = h_b$ we infer no preference. For the example in Fig. 4, a user clicking on b and e in the AAA ranking will click two members of *TeamB* ($h_b = 2$) and none in *TeamA* ($h_a = 0$). This generates a preference for B. Note that the randomized alternating assignment of documents to teams and ranks in I ensures that, unlike for Balanced Interleaving, a randomly clicking user will always produce equally many preferences for A as for B in expectation. This avoids the problem of Balanced Interleaving.

5.3 Experimental Evaluation

To compare the effectiveness of absolute metrics with paired comparison evaluation, we assigned one experimental condition to each pair of retrieval functions for each triplet of ranking functions studied. To avoid differences due to temporal effects, we conducted the evaluation of the Balanced Interleaving test at the same time as the evaluation of the absolute metrics. This means that data for Balanced Interleaving of ORIG $\succ$ SWAP2 $\succ$ SWAP4 was collected between December 19th, 2007 and January 25th, 2008 (Phase I); data for Balanced Interleaving of ORIG $\succ$ FLAT $\succ$ RAND was collected between January 27th and February 25th, 2008 (Phase II). Data for Team-Draft Interleaving was collected between March 15th, 2008, and April 20th, 2008 (Phase III), for both triplets at the same time. In all cases, each experimental condition was assigned 1/6th of the users.

We performed the same data cleaning as for the absolute metrics. However, in addition to user-based aggregation that was essential for estimating the absolute metrics robustly, we also evaluate the paired comparison tests in a query-based fashion.[5] Unlike for absolute performance metrics, it does not matter if some users run more queries than others: it simply gives heavier users more input into the experiment outcome. Despite this, random spammers or bots will not bias the outcome. For example a user who always clicks on the top result for thousands of queries would reduce the signal in our results, but clicking on an equal number from each "team" would not bias the final outcome. We call this query-based evaluation, and it simply follows the methods described above, where each query contributes a preference (or tie). We will compare the results of this query-based evaluation with user-based evaluation, where each user has exactly one "vote" per condition and that vote is determined by the majority of the individual click preferences of that user.

5.4 Paired Comparison Results

Table 3 shows how frequently each ranking functions receives a favorable preference (i.e., "win") in each pairwise comparison for both Balanced Interleaving and Team-Draft Interleaving. We do not count cases when the user did not click at all. For both interleaving methods and also for both query-based and user-based aggregation, the sign of $\Delta_{AB} = (wins(A) - wins(B))/(wins(A) + wins(B))$ perfectly reflects the true ordering in both ORIG ≻ FLAT ≻ RAND and

Table 3 Results of the paired comparison tests for the "ORIG ≻ FLAT ≻ RAND" and the "ORIG ≻ SWAP2 ≻ SWAP4" comparison. Wins and losses are counted on a per-query basis (left) or on a per-user basis (right). We only consider users and queries with at least one click, and their number is given in the table. The remaining percentage of queries/users are ties. Pairs where A (the higher-quality retrieval function) wins significantly (95%) more often than B (the lower-quality retrieval function) are printed in bold

Interleaving	Comparison Pair	Query Based			User Based		
Algorithm	A ≻ B	A wins	B wins	# queries	A wins	B wins	# users
Balanced	ORIG ≻ FLAT	**30.6%**	**21.9%**	857	**33.3%**	**23.8%**	538
	FLAT ≻ RAND	**28.0%**	**22.9%**	907	**31.8%**	**23.3%**	529
	ORIG ≻ RAND	**40.9%**	**30.1%**	930	**41.0%**	**27.1%**	553
	ORIG ≻ SWAP2	**18.1%**	**14.6%**	1035	**23.1%**	**17.1%**	589
	SWAP2 ≻ SWAP4	**33.6%**	**27.5%**	1061	35.1%	30.0%	606
	ORIG ≻ SWAP4	**32.1%**	**24.5%**	1173	**37.7%**	**26.7%**	591
Team-Draft	ORIG ≻ FLAT	**47.7%**	**37.3%**	1272	**49.6%**	**36.0%**	667
	FLAT ≻ RAND	**46.7%**	**39.7%**	1376	**46.3%**	**36.8%**	646
	ORIG ≻ RAND	**55.6%**	**29.8%**	1095	**58.7%**	**28.6%**	622
	ORIG ≻ SWAP2	44.4%	40.3%	1170	**44.7%**	**37.4%**	693
	SWAP2 ≻ SWAP4	44.2%	40.3%	1202	45.1%	39.8%	703
	ORIG ≻ SWAP4	**47.7%**	**37.8%**	1332	**47.2%**	**35.0%**	697

[5] in other words, using micro-averaging.

Table 4 Comparing the number of correct ("✓") and false ("↯") preferences implied by the interleaving methods aggregated over the "ORIG ≻ FLAT ≻ RAND" and the "ORIG ≻ SWAP2 ≻ SWAP4" comparison. A preference is weakly correct/false, if interleaving attributes more wins/losses to the better retrieval functions. A preference is significantly correct/false, if the number of wins is significantly (95%) greater than the number of losses

Paired Comparison Signals	Weak		Significant	
	✓	↯	✓	↯
Balanced Interleaving (per query)	6	0	6	0
Balanced Interleaving (per user)	6	0	5	0
Team-Draft Interleaving (per query)	6	0	4	0
Team-Draft Interleaving (per user)	6	0	5	0

ORIG ≻ SWAP2 ≻ SWAP4. As summarized in Table 4, in no case do any of the paired tests suggest a preference in the wrong direction. More formally, we statistically test whether the number of wins for the better retrieval function is indeed significantly larger by using a binomial test against $P(A \text{ wins over } B) \leq 0.5$. The significant differences are bolded in Table 3, and 20 out of the 24 pairs are significant. While the remaining four pairs fail the 95% significance level, they are significant at the 90% level. This supports our hypothesis that the paired comparison tests are able to identify a higher-quality retrieval function reliably.

Table 3 does not give substantial evidence that one interleaving or data aggregation method is preferable over the other. They each seem to be equally accurate and of comparable statistical power. However, note that Team-Draft Interleaving forces a strict preference more often than Balanced Interleaving. For example, any query with a single click always produces a strict preference in Team-Draft Interleaving, even if the input rankings are identical. While this does not change the mean, it might lead to larger variance in the individual preference votes than when using Balanced Interleaving, especially for retrieval functions that produce very similar rankings. It appears that the potential problem of Balanced Interleaving identified in Sect. 5.2 was not an issue in this evaluation.

Interestingly, not only does the sign of Δ_{AB} correspond to the correct ordering by retrieval quality, but the magnitude of this difference appears reasonable as well. In particular, for all tests of a triplet $A \succ B \succ C$, Table 3 shows that $\Delta_{AC} > \max\{\Delta_{AB}, \Delta_{BC}\}$, indicating Strong Stochastic Transitivity [18].

6 Discussion and Limitations

As in any controlled experiment, we were able to explore only a few aspects of the problem while keeping many variables in the environment fixed. In this section, we describe in more detail some of the assumptions that are inherent in the evaluation methods we compare, some of which apply to both absolute and paired comparison evaluation.

6.1 Search Setting

Most obviously, online retrieval of scientific documents is only one domain for information retrieval and other domains may have substantially different properties. In particular, we believe that most of our users were educated researchers and students using the system in a research context. It is possible that our users, for example, consider each result returned more carefully than most Web search users, and delve deeper into the result sets returned. Web search, intranet search, desktop search, online purchasing, and mobile search have a much broader and more diverse user base, as well as a different distribution of queries.

However, as our experiment design is not limited to arXiv.org, it will be interesting to conduct similar studies in those domains as well. The resulting set of studies would give a more complete view of the relationship between user behavior and retrieval quality than the single data point we provide here. From a practical perspective, such an evaluation can be performed in any of these settings without necessitating the creation of a custom search engine as we have implemented for the experiments reported here. In particular, through the simple use of a proxy implemented between users and any general purpose search engine, all of the experiments described here could be performed.

6.2 Click Filtering

For the sake of simplicity, we focused largely on "raw" clicks as feedback signal, with simple heuristics for removing potentially noisy clicks in the case of absolute metrics. This ignores that some clicks may be made in error (e.g., due to a misleading snippet). A more differentiated interpretation of clicks (e.g., based on dwell-time, use of the back button, etc.) may provide a cleaner signal. Additionally, for some queries the desired information is already presented in the snippet, which obviates the need for a click. Analyzing additional actions such as copy/paste and scan-paths collected via eyetracking may provide additional information.

However, it is important to observe that such additional information could be incorporated into *both* absolute metrics as well as paired comparison tests. If clicks followed by a long dwell time on the results are indeed more informative than raw clicks, filtering for such clicks would be expected to improve the strength of the signal observed in all the absolute metrics as well as the strength of the signal observed from interleaved evaluation.

Additionally, if some clicks are malicious, this again may obscure any signal observed. Apart from a few bots (and possibly some vanity searches), arXiv.org is a domain relatively free of click-spam. While many domains are similarly free of click-spam (e.g., personal information search, intranet search), it will be interesting to see how the paired comparison tests perform under more substantial click-spam attacks.

6.3 Snippets versus Documents

One particular assumption in using raw clicks is that clicks on the short snippets presented to users on results pages tell us about the relevance of the actual documents. The success of the paired comparison tests suggests that users of arXiv.org were able to make somewhat reliable relevance judgments of the articles retrieved based on the snippets generated. To assess the effect of snippets on the experiment results, we also repeated the paired comparison experiment for the ORIG $\succ$ FLAT and ORIG $\succ$ RAND pairs of ranking functions using alternative snippet generation algorithms during a fourth month-long experimental phase. We found that showing normal snippets (about 300 characters), longer snippets (about 450 characters) and simply showing the beginning of the article abstracts resulted in ratios of preference judgments that did not differ in a statistically significant manner from those reported in Table 3. This suggests that the relevance of the articles retrieved can be reliably conveyed in abbreviated form, probably because titles and author names are already very informative in the arXiv.org domain.

However, generating meaningful snippets might be more challenging in other domains (e.g., due to maliciously designed web pages). Furthermore, one has to be careful that snippet generation is not biased toward any particular retrieval function (e.g., in terms of abstract length or quality). In particular, this is one reason why completely independent search engines are difficult to compare with either absolute or interleaving tests. For instance, if results obtained from Web search engine A were simply interleaved with results obtained from Web search engine B, more clicks on the results from A may simply indicate that A produces more misleadingly good snippets, rather than that A is better. Similarly, we could see that A may have a higher abandonment rate in an absolute metric test because the snippets are a little shorter.

6.4 Absolute Metric Choices

While we strove for a set of absolute metrics that covers the majority of easily observable user behavior, there may be other absolute metrics that are more indicative of ranking quality. For example, there may be sophisticated combinations of various absolute metrics that are more reliable than any single metric [9, 12]. Furthermore, for many of the absolute metrics, the observed differences were not statistically significant given the amount of data we could practically collect. In domains like general Web search, where orders of magnitude more data is available, some of these absolute metrics might indeed make accurate predictions without the necessity of performing paired comparison tests.

6.5 *Scale of Differences*

In constructing artificially degraded retrieval functions, we aimed to design both large and small differences in ranking quality. However, further studies are needed to see how fine a difference paired comparison tests can detect. In particular, it would be interesting to explore whether Strong Stochastic Transitivity holds in other settings, and with even smaller quality differences. If some form of (approximate) stochastic transitivity holds, it is plausible that large numbers of retrieval functions could be reliably evaluated with far fewer than $O(n^2)$ comparisons using methods from tournament design, which also has implications for automatically learning improved retrieval functions based on paired comparison tests [30, 31].

Additionally, absolute metrics by their nature provide an absolute performance score. This score can then be optimized over time, providing information about long-term improvements to search systems. In contrast, paired comparison tests simply provide information about which ranking is preferred by users without necessarily indicating how much better the preferred ranking function is. It would be interesting to perform studies that measure how the strength of an interleaving signal compares with an absolute measure of ranking performance such as mean average precision or normalized discounted cumulative gain [21].

6.6 *Interactive Evaluation Limitations*

Finally, interleaved evaluation inherently requires interactive evaluation of ranking functions. This means that a dataset collected for one interleaving evaluation cannot later be reused to evaluate other ranking functions. In particular, as interleaving dynamically creates the ranking presented to users based on two input ranking functions, it would be difficult to infer which results would have been presented and clicked had one of the input functions been different. This differs from some absolute metric evaluations. For instance, if we were to measure mean reciprocal rank, assuming that relevant results tend to be clicked on, a new ranking function that tends to position the previously clicked results closer to the top of the ranking could be assumed to be better than the original one.

7 Summary and Conclusions

We explored and contrasted two possible approaches to retrieval evaluation based on implicit feedback, namely absolute metrics and paired comparison tests. In a real-world user study where we know the relative retrieval quality of several ranking functions by construction, we investigated how accurately these two approaches predict retrieval quality. None of the absolute metrics gave reliable results for the sample size collected in our study. In contrast, both paired comparison algorithms, namely Balanced Interleaving and the new Team-Draft Interleaving method we proposed, gave consistent and mostly significant results. Further studies are needed to extend these results to other search domains beyond the arXiv.org e-print archive.

Acknowledgements Many thanks to Paul Ginsparg and Simeon Warner for their insightful discussions and their support of the arXiv.org search. The first author was supported by a Microsoft Ph.D. Student Fellowship. This work was also supported by NSF Career Award No. 0237381, NSF Award IIS-0812091 and a gift from Google.

References

1. E. Agichtein, E. Brill, S. Dumais, R. Ragno, Learning user interaction models for prediction web search results preferences, in *Proceedings of the ACM Conference on Research and Development in Information Retrieval (SIGIR)* (2006), pp. 3–10
2. R. Agrawal, A. Halverson, K. Kenthapadi, N. Mishra, P. Tsaparas, Generating labels from clicks, in *Proceedings of ACM International Conference on Web Search and Data Mining (WSDM)* (2009), pp. 172–181
3. K. Ali, C. Chang, On the relationship between click-rate and relevance for search engines, in *Proceedings of Data Mining and Information Engineering* (2006)
4. J.A. Aslam, V. Pavlu, E. Yilmaz, A sampling technique for efficiently estimating measures of query retrieval performance using incomplete judgments, in *ICML Workshop on Learning with Partially Classified Training Data* (2005)
5. J. Boyan, D. Freitag, T. Joachims, A machine learning architecture for optimizing web search engines, in *AAAI Workshop on Internet Based Information Systems* (1996)
6. C. Buckley, E.M. Voorhees, Retrieval evaluation with incomplete information, in *Proceedings of the ACM Conference on Research and Development in Information Retrieval (SIGIR)* (2004), pp. 25–32
7. B. Carterette, J. Allan, R. Sitaraman, Minimal test collections for retrieval evaluation, in *Proceedings of the ACM Conference on Research and Development in Information Retrieval (SIGIR)* (2006), pp. 268–275
8. B. Carterette, P.N. Bennett, D.M. Chickering, S.T. Dumais, Here or there: Preference judgements for relevance, in *Proceedings of the European Conference on Information Retrieval (ECIR)* (2008), pp. 16–27
9. B. Carterette, R. Jones, Evaluating search engines by modeling the relationship between relevance and clicks, in *Proceedings of the International Conference on Advances in Neural Information Processing Systems (NIPS)* (2007), pp. 217–224
10. K. Crammer, Y. Singer, Pranking with ranking, in *Proceedings of the International Conference on Advances in Neural Information Processing Systems (NIPS)* (2001), pp. 641–647
11. G. Dupret, V. Murdock, B. Piwowarski, Web search engine evaluation using clickthrough data and a user model, in *WWW Workshop on Query Log Analysis* (2007)
12. S. Fox, K. Karnawat, M. Mydland, S. Dumais, T. White, Evaluating implicit measures to improve web search, ACM Trans. Inf. Sci. (TOIS) **23**(2), 147–168 (2005)
13. S.B. Huffman, M. Hochster, How well does result relevance predict session satisfaction? in *Proceedings of the ACM Conference on Research and Development in Information Retrieval (SIGIR)* (2007), pp. 567–573
14. T. Joachims, Evaluating retrieval performance using clickthrough data, in *Text Mining*, ed. by J. Franke, G. Nakhaeizadeh, I. Renz (Physica Verlag, 2003)
15. T. Joachims, Optimizing search engines using clickthrough data, in *Proceedings of the ACM International Conference on Knowledge Discovery and Data Mining (KDD)* (2002), pp. 132–142

16. T. Joachims, L. Granka, B. Pan, H. Hembrooke, F. Radlinski, G. Gay, Evaluating the accuracy of implicit feedback from clicks and query reformulations in web search. ACM Trans. Inf. Sci. (TOIS) **25**(2) (2007), Article 7
17. D. Kelly, J. Teevan, Implicit feedback for inferring user preference: A bibliography. ACM SIGIR Forum **37**(2), 18–28 (2003)
18. J. Kozielecki, *Psychological Decision Theory* (Kluwer, 1981)
19. D. Laming, *Sensory Analysis* (Academic, 1986)
20. Y. Liu, Y. Fu, M. Zhang, S. Ma, L. Ru, Automatic search engine performance evaluation with click-through data analysis, in *Proceedings of the International World Wide Web Conference (WWW)* (2007)
21. C.D. Manning, P. Raghavan, H. Schuetze, *Introduction to Information Retrieval* (Cambridge University Press, 2008)
22. F. Radlinski, T. Joachims, Query chains: Learning to rank from implicit feedback, in *Proceedings of the ACM International Conference on Knowledge Discovery and Data Mining (KDD)* (2005)
23. F. Radlinski, M. Kurup, T. Joachims, How does clickthrough data reflect retrieval quality, in *Proceedings of the ACM Conference on Information and Knowledge Management (CIKM)* (2008), pp. 43–52
24. S. Rajaram, A. Garg, Z.S. Zhou, T.S. Huang, Classification approach towards ranking and sorting problems, in *Lecture Notes in Artificial Intelligence* (2003), pp. 301–312
25. J. Reid, A task-oriented non-interactive evaluation methodology for information retrieval systems. Inf. Retr. **2**, 115–129 (2000)
26. I. Soboroff, C. Nicholas, P. Cahan, Ranking retrieval systems without relevance judgments, in *Proceedings of the ACM Conference on Research and Development in Information Retrieval (SIGIR)* (2001), pp. 66–73
27. A. Spink, D. Wolfram, M. Bernard, J. Jansen, T. Saracevic, Searching to web: The public and their queries. J. Am. Soc. Inf. Sci. Technol. **52**(3), 226–234 (2001)
28. A. Turpin, F. Scholer, User performance versus precision measures for simple search tasks, in *Proceedings of the ACM Conference on Research and Development in Information Retrieval (SIGIR)* (2006), pp. 11–18
29. E.M. Voorhees, D.K. Harman (eds.), *TREC: Experiment and Evaluation in Information Retrieval* (MIT, 2005)
30. Y. Yue, T. Joachims, Interatively optimizing information systems as a dueling bandits problem, in *NIPS 2008 Workshop on Beyond Search: Compuitations Intelligence for the Web* (2008)
31. Y. Yue, T. Joachims, Interatively optimizing information retrieval systems as a dueling bandits problem, in *Proceedings of the International Conference on Machine Learning (ICML)* (2009)

Learning SVM Ranking Functions from User Feedback Using Document Metadata and Active Learning in the Biomedical Domain

Robert Arens

Abstract Information overload is a well-known problem facing biomedical professionals. MEDLINE, the biomedical bibliographic database, adds hundreds of articles daily to the millions already in its collection. This overload is exacerbated by the lack of relevance-based ranking for search results, as well as disparate levels of search skill and domain experience of professionals using systems designed to search MEDLINE. We propose to address these problems through learning ranking functions from user relevance feedback. Simple active learning techniques can be used to learn ranking functions using a fraction of the available data, with performance approaching that of functions learned using all available data. Furthermore, ranking functions learned using metadata features from the Medical Subject Heading (MeSH) terms associated with MEDLINE citations greatly outperform functions learned using textual features. An in-depth investigation is made into the effect of a number of variables in the ranking round, while further investigation is made into peripheral issues such as users providing inconsistent data.

1 Introduction

MEDLINE, the National Library of Medicine's (NLM) bibliographic database, is a resource used globally by biomedical researchers and professionals. It contains over 16 million references, with thousands more added every week [28]. The use of this resource is ubiquitous throughout the biomedical community and beyond, by researchers, clinicians, and amateurs interested in the field. Users often encounter information overload when searching MEDLINE. In part, this is due to the enormity of the database. However, a greater issue is the lack of relevance-based ranking of these results. This means that the results most relevant to the user's query may be buried under thousands of irrelevant results, forcing the user to sort through them manually.

R. Arens
University of Iowa, Iowa City, IA, 52242, USA
e-mail: robert-arens@uiowa.edu

J. Fürnkranz and E. Hüllermeier (eds.), *Preference Learning*,
DOI 10.1007/978-3-642-14125-6_17,

Deeper issues exacerbate the problem. To avoid manually searching through potentially thousands of search results to find the citations they need, medical professionals must spend years developing the expertise necessary to use PubMed, the search engine front-end to MEDLINE, efficiently [4]. Even experienced users may find themselves unable to construct these efficient queries if they lack deep knowledge regarding the subject of their search. Improvements to PubMed, whether focusing on improving the retrieval system itself (e.g., [16, 23, 26]) or on presentation of search results (e.g., [20, 24, 29, 32]) fail to strike a balance between retrieval power and user burden.

We propose to address these problems by learning ranking functions from user feedback. A ranking function learned in this way will put the most relevant results above the less relevant, making search more convenient for the user. Our ranking functions will be used using a ranking version of the support vector machine (SVMs) algorithm, a learning method that has been shown effective and efficient at such tasks. Learning from feedback does not require a great amount of training, so it can be done by a novice user. Furthermore, a user does not need deep domain knowledge; giving positive feedback on citations that "look right" ensures that similar citations will be ranked highly. Finally, each ranking function will be tailored to the user training it, while traditional query-based ranking methods such as tf*idf would produce the same ordering for any given query. Joachims [22] argued that, "experience shows that users are only rarely willing to give explicit feedback". We agree with this point in general, but experience has shown us that biomedical professionals are highly motivated to give feedback if the overhead of feedback is offset by utility they gain from the system.

Our hypothesis is that a ranking function learned from feedback given on a small percentage of intelligently chosen examples from a retrieval set will perform comparably to a ranking function learned from the entire retrieval set. We will explore how to choose these examples and how much feedback to request from the user.

We further propose to learn our ranking functions from document metadata, as opposed to textual features. Citations in MEDLINE are annotated using the NLM's controlled vocabulary of Medical Subject Headings, called MeSH. We hypothesize that learning from these features will be superior to learning from query-based textual features. We will evaluate the quality of the rankings produced by our method, along with the amount of feedback required to produce that ranking.

2 Biomedical Document Collections and Data Sets

2.1 MEDLINE

As stated in the introduction, MEDLINE is a database of bibliographic references in biomedicine and health sciences, collected and curated by the NLM. Citations are taken from approximately 5,200 sources (mostly journals) in 37 languages, with the majority of journals selected for inclusion by a committee set up by the National

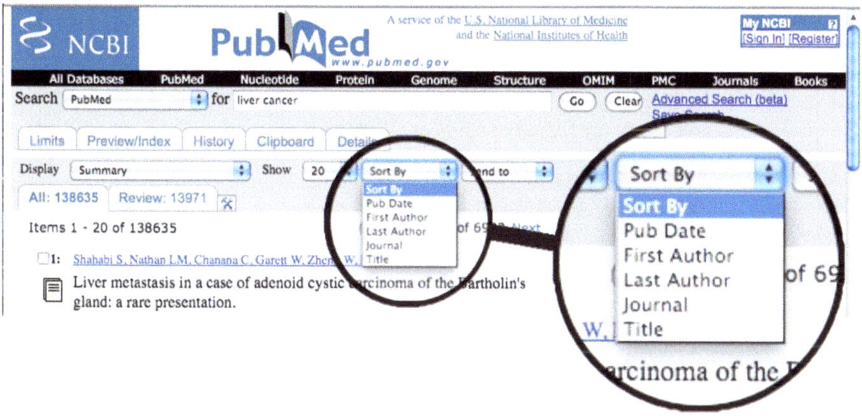

Fig. 1 Sample view of the results page from PubMed. Inset: document ranking options

Institutes of Health (NIH). Other sources are included based on NLM or NIH priorities (e.g., AIDS research). Dates covered range from 1949 to the present, with some older material. The database currently contains over 16 million references, with between two and four thousand added each week. In 2007, 670,000 references were added.

MEDLINE's subject coverage is extensive. It covers areas ranging from public heath policy to clinical care to bioengineering research, from humans to plants to bacteria. The target audience for MEDLINE is the biomedical community at large, including researchers, clinicians, educators, and amateurs. Most references include full citation data, as well as the abstract for the reference. Many references also contain links to the full article.

Searching MEDLINE is done most often via Entrez PubMed, the NLM's search engine operating over the database. Figure 1 shows an example of a results page from the PubMed search interface. While it is a robust retrieval system, PubMed lacks relevance-based ranking of search results. Users are limited in their choice of results sorting to date of publication, author names, or journal of publication. This leads to information overload for users of the system. As mentioned before, this is partly due to the size of the MEDLINE database; for example, a search for "heart disease" returns over eight hundred thousand results. This can lead to a "search boomerang", where users alter their search queries with more specific criteria (resulting in too few results), then relaxing their criteria (resulting in too many), repeating until a reasonably sized result set is obtained.

2.1.1 MeSH Terms

To assist users in navigating through MEDLINE's wealth of information, each reference in the database is tagged with a number of MeSH terms. The preface

- Body Regions [A01]
  - ...
  - Extremities [A01.378]
    - ...
    - Lower Extremity [A01.378.610]
      - Buttocks [A01.378.610.100]
      - Foot [A01.378.610.250]
        - Ankle [A01.378.610.250.149]
        - Forefoot, Human [A01.378.610.250.300]
          - Metatarsus [A01.378.610.250.300.480]
          - Toes [A01.378.610.250.300.792]
            - Hallux [A01.378.610.250.300.792.380]
        - Heel [A01.378.610.250.510]
      - Hip [A01.378.610.400]
      - Knee [A01.378.610.450]
      - Leg [A01.378.610.500]
      - Thigh [A01.378.610.750]

Fig. 2 Sample MeSH hierarchy

to [27] states that, "The Medical Subject Headings (MeSH) thesaurus is a controlled vocabulary produced by the National Library of Medicine and used for indexing, cataloging, and searching for biomedical and health-related information and documents." The terms themselves are heterogeneous, and are of three types. *Descriptors*, or *main headings*, cover the content of the reference, as well as metainformation such as the publication type and components (e.g., clinical trial, editorial, historical article, etc.) and the country of the reference's origin. *Qualifiers*, or *subheadings*, are used with descriptors to group together references dealing with a particular aspect of the descriptor; for example, pairing the qualifier "abnormalities" with the descriptor "heart" would indicate that the reference in question is concerned with congenital defects in the heart, as opposed to the heart itself. Finally, *Supplementary Concept Records*, or SCRs, catalogue specific drugs and chemicals referenced in the article. Currently, there are 25,186 MeSH descriptors, with 83 qualifiers and over 180,000 SCRs. References are manually tagged by human annotators.

MeSH is arranged hierarchically in 16 trees, grouped by the most general category of the descriptor. There is a tree for descriptors relating to parts of the anatomy, another for organisms, another for techniques and equipment, etc. As one descends a tree, the descriptors become increasingly specific. Figure 2 shows a sample from subtree A01 – Body Regions, the first subtree in tree A – Anatomy. The position of the entry "Toes" indicates that is more general than "Hallux", but more specific than "Forefoot, Human", which is itself more specific than "Foot".

Searching MEDLINE with MeSH terms can be done by entering the MeSH term into a PubMed search, just as one would any other search term. A mapping of 160,000 common terms to their synonymous MeSH headings is used along with the PubMed query system to expand and refine the user's query. Figure 3 shows the full PubMed translation of the query "diabetes".

```
"diabetes mellitus"[MeSH Terms] OR ("diabetes"[All
Fields] AND "mellitus"[All Fields]) OR "diabetes
mellitus"[All Fields] OR "diabetes"[All Fields] OR
"diabetes insipidus"[MeSH Terms] OR ("diabetes"[All
Fields] AND "insipidus"[All Fields]) OR "diabetes
insipidus"[All Fields]
```

Fig. 3 Full translation of the query "diabetes"

2.2 *OHSUMED*

OHSUMED is a collection of MEDLINE citations, created to carry out information retrieval experiments on MEDLINE [19]. It is composed of the results of 106 queries run against a five-year span of MEDLINE documents. Queries were generated from a questionnaire filled out by participants in the study, over a ten-month period, filtering out duplicate topics and author searches, as well as queries with inadequate information for replication by annotators. Eleven medical librarians and eleven physicians, all familiar with searching MEDLINE, replicated the searches and judged the retrieved references for relevance. Of 348,566 available references from 270 medical journals, 16,140 query-document pairs were retrieved, with the 12,565 unique pairs annotated for relevance. Pairs were classified as definitely, possibly, or not relevant, with 69.3% annotated as not relevant, 16.3% partially relevant, and 14.3% relevant. Five queries returned no citations judged relevant by the annotators. Inter-annotator agreement, a measure of how closely any two annotators agreed in their annotations, was calculated by having 11% of the documents annotated by two people. A kappa statistic of 0.41 was calculated for agreement, which the authors claim is comparable with other similar experiments.

2.3 *LETOR*

LETOR is a benchmark dataset created for information retrieval rank learning applications [25]. It consists of two parts; the first is based on the .gov dataset from the 2003 and 2004 TREC Web Track [11], and the second is based on OHSUMED (see Sect. 2.2). We will limit our discussion to the OHSUMED section of LETOR.

LETOR assigns a feature vector to each query-document pair in the OHSUMED collection. These feature vectors are composed of classical and more recently developed measures. Both high- and low-level features are calculated, where "[l]ow-level features include term frequency (tf), inverse document frequency (idf), document length, (dl) and their combinations" [25], and high-level features include measures such as BM25 [31]. Vectors contain 25 features, 10 calculated from the title, 10 from the abstract, and 5 from combined title-abstract text for each of the documents.

3 Learning Ranking Functions from User Feedback

Presenting retrieved documents according to the likelihood of their relevance to a user's query is a standard practice in information retrieval [1]. Here, we present a framework for learning a function to produce such a relevance ranking from feedback provided by the user. As this is an online task, two factors beyond raw system performance must be addressed. First, the system must run quickly enough to provide the user a reasonable search experience. Second, the amount of feedback required for learning must be a reasonable fraction of the number of search results. If either of these factors are not well addressed, the system will provide no benefit to users over traditional search engines. We choose to employ ranking SVMs for rank function learning because of their speed, and their performance in learning ranking functions [6, 22, 35]. To ensure users will have to provide as little feedback as possible, we employ active learning to choose examples that will be most useful for learning [10].

3.1 Initial Retrieval

We take initial retrieval as given. PubMed is an excellent retrieval system, employing many recall-boosting strategies such as term expansion and synonym matching, which we do not care to replicate. For our experiments, we use the existing OHSUMED dataset (described in Sect. 2.2) and perform no retrieval on our own. Our production system will use the Entrez eUtils,[1] which will allow users to submit

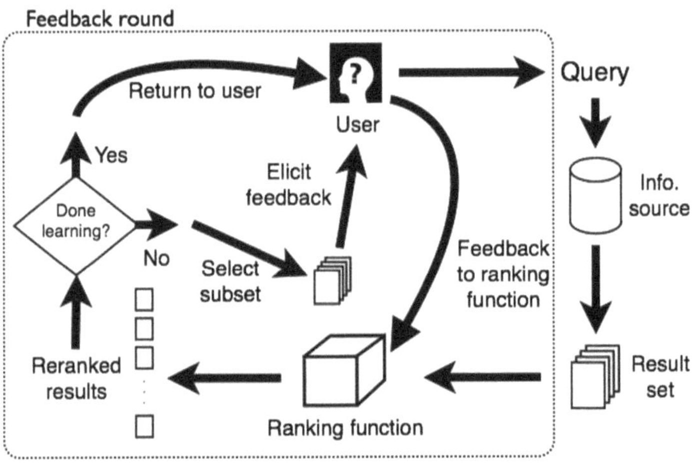

Fig. 4 Illustration of learning ranking functions from user feedback

[1] http://www.ncbi.nlm.nih.gov/entrez/query/static/eutils_help.html

queries to our system just as they would to PubMed itself. Document abstracts and their associated feature vectors will be stored locally.

3.2 The Feedback Round

A *feedback round* is one iteration of choosing examples for feedback, requesting feedback from the user, learning from the feedback, and checking the stopping criterion. Future references will be made to this sequence of steps as a performance measure, indicating one measure of how much overhead the user has incurred using the system. The number of rounds multiplied by the number of examples seen per round gives the total number of examples seen, which is the other overhead measure used.

3.3 Ranking

As previously stated, we learn and rank using the ranking SVM algorithm. Given a collection of data points ranked according to preference relation R^* over a document set $\mathcal{D}$ with two objects $\boldsymbol{d}_i, \boldsymbol{d}_j \in \mathcal{D}$, and a linear learning function f, we can say

$$\boldsymbol{d}_i \succ \boldsymbol{d}_j \Rightarrow f(\boldsymbol{d}_i) > f(\boldsymbol{d}_j), \tag{1}$$

where $\succ$ indicates that $\boldsymbol{d}_i$ is preferred over $\boldsymbol{d}_j$. We can define the function f as $f(\boldsymbol{d}) = \boldsymbol{w} \cdot \boldsymbol{d}$, where

$$f(\boldsymbol{d}_i) > f(\boldsymbol{d}_j) \Leftrightarrow \boldsymbol{w} \cdot \boldsymbol{d}_i > \boldsymbol{w} \cdot \boldsymbol{d}_j. \tag{2}$$

The vector $\boldsymbol{w}$ can be learned via the standard SVM learning method using slack variables [7],

$$\begin{aligned} &\text{minimize } \langle \boldsymbol{w} \cdot \boldsymbol{w} \rangle + C \sum_{i,j \in |R|} \xi_{ij} \\ &\text{subject to } \forall (\boldsymbol{d}_i, \boldsymbol{d}_j) \in \mathcal{D} : \boldsymbol{w} \cdot \boldsymbol{d}_i \geq \boldsymbol{w} \cdot \boldsymbol{d}_j + 1 - \xi_{ij} \\ &\qquad\qquad \forall (i, j) : \xi_{ij} \geq 0. \end{aligned} \tag{3}$$

Discovery of the support vectors and the generalization of the ranking SVM is done differently [22]. For data that are linearly separable, the ξ_{ij} are all equal to 0. In this case, we can view the ranking function as projecting the data points onto the separating hyperplane. In this case, the support vectors are the pairs of vectors $(\boldsymbol{d}_i, \boldsymbol{d}_j)$ nearest each other on the hyperplane. Generalization is achieved by calculating $\boldsymbol{w}$ to maximize the distance between these closest pairs. The distance between these points is calculated as

$$\frac{\boldsymbol{w}(\boldsymbol{d}_i - \boldsymbol{d}_j)}{\|\boldsymbol{w}\|} \tag{4}$$

Taking this as our margin γ, we can, as with the classification SVM algorithm [7], maximize the margin by minimizing $\|w\|$ for all such pairs.

3.4 Features for Learning

Two sets of features are used for learning. For the first set, we use the features from LETOR (see Sect. 2.3). We leave these features intact, with no modification. These textual features model the content of the query-document pair. The second set is built from the MeSH descriptors (see Sect. 2.1) for the OHSUMED documents. Each vector is composed of 25,186 binary features, with each feature indicating inclusion of the MeSH descriptor. These features model the metadata assigned to the documents by the MeSH annotators. It should be noted that MeSH terms offer an improvement over the textual features, in that annotators assigning MeSH terms to citations in MEDLINE have access to the entire article, while LETOR features are limited to the title and abstract.

We hypothesize that ranking functions learned from MeSH data will outperform those learned from LETOR data, both in ranking performance and the overhead required to reach similar performance levels.

3.5 Choosing Examples

In order to learn a ranking function, we require a training set. This set will be provided to us by the user in the form of explicit preference feedback on a subset of the retrieved documents. In order to ensure that we are asking for user feedback on as few examples as possible, we choose this subset via active learning.

Active learning describes a learning method wherein the learning algorithm itself has some control over which examples are added to its training set. Specifically, we need to ask a user to provide labels for some number of unlabeled examples. The learner chooses these examples based on some measure of learning utility; for example, choosing examples which will decrease the region of uncertainty in a learned function [10]. Repeated rounds of active learning improve both the learned function and the examples chosen for learning. Taking our previous measure as an example, the reduction in the region of uncertainty produces a function with better generalization power; however, reducing the region of uncertainty has an added benefit of leaving behind only examples which continue to contribute to uncertainty.

Two active learning strategies were employed for example selection. We used random sampling, simply choosing unseen examples at random, as a baseline against which these two methods will be compared. The first active learning method is top sampling. As discussed in [6], ranking functions should have their performance optimized toward top-ranked documents. Therefore, top sampling chooses unseen documents ranked highest by the current ranking function at each round for feedback. The other active learning method is mid sampling. Similar to Cohn et al. [10], we wish to reduce uncertainty in our ranking function. A learned ranking function will rank the best and worst documents with great confidence, but less so those in the middle. These middle-ranked documents are the ones ranked with the least confidence; therefore, learning from them should result in a stronger model.

We hypothesize that both top and mid sampling will outperform random sampling, both in ranking performance and overhead cost. We further hypothesize that top sampling will outperform mid sampling in ranking performance, as mid sampling is training to improve overall performance as opposed to focusing on the performance of highly ranked documents.

3.6 Eliciting Feedback

Feedback is elicited by asking the user to rate a document's relevance to his/her information need. We allow users to express preference on a neutral point scale ("yes", "maybe", "no"), rather than using a forced-choice ("yes or no") method, as the former shows higher measurement reliability [8]. This facilitates simulation using OHSUMED (see Sect. 2.2), allowing the user to rate a document as "definitely relevant", "possibly relevant", or "not relevant". Levels of relevance were expressed numerically, i.e., a document judged to be "definitely relevant" was given a score of 2, "possibly relevant" a score of 1, and "not relevant" a score of 0. These scores were used both for rank learning and calculating NDCG scores (see Sect. 4.1). The numeric value of the preference is associated with the query-document pair's feature vector, allowing the documents to be ordered for the rank learner (Sect. 3.3)

The number of examples presented for feedback in each round may influence both how quickly the ranking function is learned, and the quality of the ranking function. We investigated varying between one and five examples per round.

We hypothesize that functions learned from more feedback per round will have better ranking performance than those learned from fewer examples per round. This is an obvious hypothesis to make; more examples per round means more total training examples. However, we further hypothesize that learning from more examples per round will require users to look at fewer total examples. Our intuition is that since each round of training will produce a stronger ranking function, the active learning will be better at each round compared to ranking functions trained with fewer examples.

3.7 Stopping Threshold

At some point, feedback rounds must terminate. Rather than arbitrarily choosing a number of rounds or amount of feedback required before termination, feedback ends when the lists produced by the ranking function appear to be converging toward a stable list. Our convergence threshold is based on the Kendall's tau rank correlation coefficient, calculated between the current and immediately previous orderings produced by the ranking function. Once the correlation between these rankings exceeds a certain threshold, feedback rounds terminate. Our intuition is that highly correlated orderings indicate that rankings produced by the ranking function are converging, and we are therefore not learning any new information from feedback. Thresholds between 0.9 and 0.5 were investigated.

We hypothesize that higher thresholds will produce better ranking performance, but the overhead required to meet the threshold will increase.

4 Simulation Using OHSUMED

We test our framework experimentally via simulation, using the OHSUMED data set (described in Sect. 2.2). The five queries which returned no relevant citations have been excluded from the simulations. All experiments were run ten times per query, and the results averaged.

4.1 Metrics

We evaluate ranking performance using normalized discounted cumulative gain (NDCG) [21], a commonly used measure when multiple levels of relevance are considered. Discounted cumulative gain (DCG) at position i in a ranked list of documents is calculated as

$$\text{DCG@}i = \begin{cases} r_i & \text{if } i = 1 \\ \text{DCG@}(i\text{-}1) + \frac{r_i}{\log_2 i} & \text{otherwise} \end{cases} \tag{5}$$

where r_i is the relevance score of the document at position i. For our evaluation, relevant documents receive a score of 2, possibly relevant documents receive a score of 1, and irrelevant documents receive a score of 0. NDCG is calculated by dividing the DCG vector by an ideal DCG vector, DCG_I, calculated from an ideally ranked list (all documents scoring 2, followed by documents scoring 1, followed by documents scoring 0). Perfect ranking scores an NDCG of 1.0 at all positions. We compute NDCG@10 for our evaluation. We evaluate user overhead by counting the number of feedback rounds to produce a given ranking. Both metrics are averaged over the 101 queries used for simulation.

4.2 Results

Learning from MeSH features clearly outperformed learning from LETOR features in ranking performance. An upper bound for performance comparison was calculated by ranking documents for each OHSUMED query using a ranking SVM learned from all documents in the query, for both MeSH and LETOR feature vectors. Table 1 shows ranking SVMs learned from MeSH terms yielded nearly perfect ranking performance, vastly outperforming ranking SVMs learned from LETOR features. A performance gain is to be expected, as the MeSH terms are tailored to this data; however, we did not expect the gain to be this great. This trend continues in the active learning experiments. As shown in Tables 2–4, across all sampling methods, thresholds, and examples per round, ranking performance of ranking SVMs learned from MeSH features outperform their LETOR counterparts. For thresholds above 0.5, though LETOR SVMs consistently reached convergence before MeSH SVMs (indicating a much lower overhead), the performance was consistently poor.

Top sampling produced better ranking functions than the other two methods. Curiously, mid sampling performed worse than random sampling. This may be due

Table 1 NDCG calculated across all queries at positions 1–10 for ranking SVMs trained on all data available for a query

	@1	@2	@3	@4	@5	@6	@7	@8	@8	@10
MeSH	0.995	0.993	0.993	0.992	0.995	0.995	0.995	0.995	0.995	0.995
LETOR	0.624	0.634	0.622	0.617	0.606	0.604	0.596	0.593	0.596	0.596

Table 2 Performance for random sampling method, for all examples per round and thresholds. Top value in each row is NDCG@10, middle value is number of rounds until the convergence threshold is met, bottom value is total number of examples seen until convergence

Threshold	Random sampling									
	MeSH					LETOR				
	1	2	3	4	5	1	2	3	4	5
0.5	0.446	0.490	0.529	0.559	0.574	0.359	0.369	0.384	0.393	0.405
	7.197	4.551	3.542	3.287	2.975	7.329	4.823	3.804	3.386	3.088
	7.200	9.100	10.63	13.15	14.88	7.329	9.646	11.41	13.54	15.44
0.6	0.464	0.500	0.532	0.566	0.604	0.358	0.373	0.388	0.401	0.411
	6.866	4.722	3.886	3.490	3.393	7.461	4.533	3.858	3.540	3.190
	6.870	9.440	11.66	13.96	16.96	7.461	9.065	11.58	14.16	15.95
0.7	0.474	0.510	0.563	0.592	0.626	0.361	0.372	0.392	0.407	0.425
	7.591	5.250	4.573	4.201	4.108	7.672	4.788	4.223	3.736	3.601
	7.591	10.50	13.72	16.80	20.54	7.672	9.576	12.67	14.94	18.00
0.8	0.484	0.543	0.606	0.648	0.687	0.358	0.376	0.405	0.412	0.422
	8.411	6.197	5.881	5.861	5.845	7.760	5.284	4.612	4.326	4.144
	8.411	12.39	17.64	23.45	29.22	7.760	10.57	13.84	17.3	20.72
0.9	0.521	0.604	0.668	0.726	0.770	0.370	0.384	0.419	0.440	0.456
	10.94	9.593	10.06	10.87	10.90	9.205	7.073	6.681	6.640	6.618
	10.94	19.19	30.18	43.48	54.52	9.205	14.15	20.04	26.56	33.09

Table 3 Performance for mid sampling method, for all examples per round and thresholds. Top value in each row is NDCG@10, middle value is number of rounds until the convergence threshold is met, bottom value is total number of examples seen until convergence

Threshold	Mid sampling									
	MeSH					LETOR				
	1	2	3	4	5	1	2	3	4	5
	0.455	0.488	0.504	0.533	0.549	0.346	0.367	0.381	0.389	0.399
0.5	7.205	4.142	3.489	3.182	2.943	7.250	4.395	3.564	3.167	3.039
	7.205	8.280	10.47	12.73	14.71	7.250	8.790	10.69	12.67	15.19
	0.451	0.495	0.519	0.535	0.561	0.358	0.370	0.385	0.400	0.396
0.6	7.065	4.701	3.853	3.324	3.269	7.172	4.564	3.776	3.312	3.122
	7.065	9.402	11.56	13.30	16.35	7.172	9.129	11.33	13.25	15.61
	0.465	0.505	0.531	0.559	0.576	0.358	0.370	0.380	0.392	0.401
0.7	7.430	5.064	4.200	3.961	3.794	6.957	4.662	3.979	3.703	3.460
	7.430	10.13	12.60	15.85	18.97	6.957	9.324	11.94	14.81	17.30
	0.475	0.529	0.551	0.590	0.607	0.361	0.380	0.395	0.395	0.408
0.8	8.450	5.961	5.340	5.017	4.967	7.987	5.195	4.395	4.168	4.068
	8.450	11.92	16.02	20.07	24.84	7.987	10.39	13.19	16.67	20.34
	0.510	0.567	0.612	0.635	0.655	0.362	0.381	0.407	0.427	0.441
0.9	10.54	9.471	9.303	9.079	8.866	9.195	6.574	6.413	6.014	5.975
	10.54	18.94	27.91	36.32	44.33	9.195	13.15	19.24	24.06	29.88

Table 4 Performance for top sampling method, for all examples per round and thresholds. Top value in each row is NDCG@10, middle value is number of rounds until the convergence threshold is met, bottom value is total number of examples seen until convergence

Threshold	Top sampling									
	MeSH					LETOR				
	1	2	3	4	5	1	2	3	4	5
	0.463	0.522	0.562	0.616	0.669	0.374	0.389	0.412	0.443	0.457
0.5	6.956	4.425	3.779	3.446	3.464	7.420	4.571	3.774	3.391	3.192
	6.956	8.850	11.34	13.78	17.32	7.420	9.143	11.32	13.56	15.96
	0.481	0.543	0.600	0.666	0.724	0.373	0.395	0.427	0.447	0.468
0.6	7.332	4.992	4.443	4.184	4.122	7.433	4.862	3.944	3.677	3.330
	7.332	9.984	13.33	16.74	20.61	7.433	9.725	11.83	11.71	16.65
	0.490	0.580	0.667	0.742	0.802	0.382	0.406	0.438	0.464	0.472
0.7	7.666	5.958	5.642	5.548	5.451	7.677	4.922	4.163	3.785	3.560
	7.666	11.92	16.93	22.19	27.26	7.677	9.844	12.49	15.14	17.80
	0.536	0.660	0.767	0.827	0.866	0.387	0.413	0.456	0.483	0.502
0.8	8.862	8.228	8.195	7.975	7.761	8.208	5.469	4.822	4.501	4.414
	8.862	16.46	24.58	31.90	38.81	8.208	10.94	14.47	18.00	22.07
	0.648	0.799	0.866	0.897	0.918	0.407	0.461	0.498	0.538	0.541
0.9	14.97	14.94	14.75	14.05	12.83	9.432	7.502	6.890	6.843	6.285
	14.97	29.88	44.25	56.18	64.14	9.432	15.00	20.67	27.37	31.43

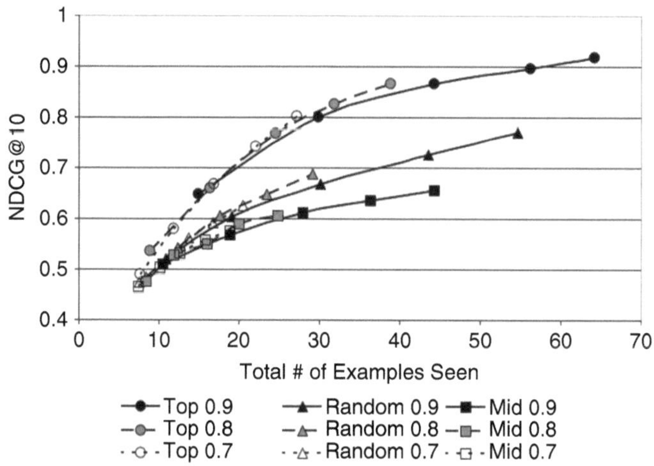

Fig. 5 Comparison of the total number of examples seen to NDCG@10 for all sampling methods, at thresholds 0.7, 0.8, and 0.9. Markers indicate number of examples per round, from one to five

to the fact that mid sampling is more likely to encounter documents ranked as possibly relevant as opposed to definitely relevant or irrelevant than random sampling. Mid sampling did incur less overhead than the other active learning methods, but it appears as though it converged to a poor final ranking.

In all cases, a greater number of examples per round produced better ranking performance. This is to be expected, as more examples per round yields a larger set of training data (see Fig. 5 and Sect. 6 for more discussion on this topic). More examples per round also decreased rounds to convergence; however, the decrease in the number of rounds was never great enough to lead to a decrease in the total number of examples seen.

As expected, higher thresholds for convergence resulted in higher ranking performance, at the cost of more feedback rounds. While performance climbed steadily, there was a marked jump in overhead between thresholds of 0.8 and 0.9.

5 Simulation with a Gradient Oracle

OHSUMED judgments are used as an oracle, or perfect knowledge, source for the simulations, as the judgments were made by a panel of professionals in the biomedical field. While actual users of this system are likely to be biomedical professionals as well, it is possible that due to simple human error they will not provide feedback perfectly in line with their preferences. Furthermore, we would like nonprofessionals to be able to use the system as well; their feedback may include guesses as to the relevance of documents regarding subjects with which they are uninformed. In order to explore the effects of incorrect feedback on system performance, we conducted a

Table 5 Results for the gradient oracle experiments

Gradient Oracle	1.0	0.95	0.90	0.85	0.80	0.75	0.70	0.65	0.60
NDCG@10	0.918	0.828	0.743	0.689	0.614	0.574	0.524	0.447	0.425

experiments using a *gradient oracle*. The gradient oracle provides correct answers with a certain probability, and incorrect answers for the remaining probability, i.e., a 0.9 gradient oracle giving feedback with an oracle judgment of 2 would produce a 2 with a 90% probability, a 1 with a 5% probability, and a 0 with a 5% probability. We repeated the experiment using top sampling, at a threshold of 0.9 with 5 examples per round, using gradient oracle values from 1.0 to 0.6 at intervals of 0.05.

5.1 Results

Table 5 shows NDCG@10 for the gradient oracle experiments, showing a clear degradation in performance as the oracle's quality decreases. A linear interpolation of the points results in a slope of 1.2236, indicating that a drop in feedback quality creates an even greater drop in performance. A likely reason for this is that poor feedback will be used to generate poor examples in future rounds, which cripples the sampling process. The primary implication of this finding is that the system will have to implement some form of consistency check if it is to be used by nonexperts and would be a benefit as well to even the most knowledgeable and careful professionals.

6 Discussion

Overall, our results are encouraging. We have achieved ranking performance within 8.2% of perfect ranking after an average of 12.8 feedback rounds and 64.14 examples seen, using top sampling with five examples per round and a convergence threshold of 0.9.

Experimental results have shown that our hypotheses regarding ranking performance are correct, with the exception of the performance of mid sampling. However, our assumptions regarding the overhead incurred to reach convergence as it relates to features for learning and sampling methods seem to have been incorrect. Poor performance was linked with less overhead, with better performance always demanding more overhead. While rounds to convergence fell slightly as the number of examples per rounds increased, this small decrease is insignificant as the overall number of examples seen by the user increased.

This led us to investigate whether the number of examples seen was the dominant predictor of performance. As shown in Fig. 5, however, sampling method played a greater role in performance. Top sampling provided better ranking performance for

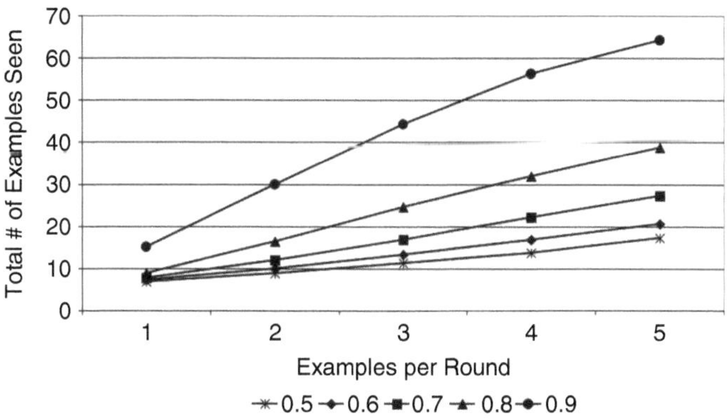

Fig. 6 The effect of increasing the number of examples per round on the total number of examples seen, across all thresholds, for top sampling

any number of examples seen, in many cases requiring fewer than half the number of examples to reach performance similar to the other active learning methods.

A note must be made regarding the stopping criterion. Since termination is determined as a function of learning, it effectively falls to the active learning technique to ensure that termination is not premature, as choosing uninformative examples will cause little to no shift in the ranking function. If this were the case, the learning process would not fulfill its potential, denied the chance to exhaust its stock of "good" examples to learn from. The effect of this would be that an active learning method which could potentially perform as well as another method would have worse performance and fewer total examples seen than a method which did not end prematurely. Examples of this happening may be present in this work, especially at high thresholds looking at 4 or 5 examples per round.

We argue that this effect is likely to be minimal. It is clear that at lower thresholds and lower examples per round, the active learning method itself is the dominant factor for performance. In Fig. 6, we see that each active learning method tends to improve as the number of examples increases; however, at no point does it appear that a method would "catch up" to a higher performing method if allowed to continue learning.

The remainder of our analysis focuses on factors affecting top sampling. Something to note in Fig. 5 is that performance gains began leveling off after reaching an NDCG@10 of around 0.8, requiring increasingly more examples for smaller gains in performance. Considering the OHSUMED queries returned 152.27 documents on average, it may appear that decent performance requires feedback on an unreasonable percentage of the returned data. Recall, however, that queries to MEDLINE often result in thousands of results. Further investigation is required to see whether queries which return such large results sets require feedback on a similar percentage of documents, a similar number of documents, or something in between.

We see in Fig. 6 that increasing examples per round increased the total number of examples seen before the convergence threshold was reached. This ran counter to one of our hypotheses; we expected that seeing more examples in each feedback

Table 6 Standard deviations in performance for top sampling. *Italics* indicates significantly larger standard deviations, while **boldface** indicates significantly smaller standard deviations

	Top sampling				
	1	2	3	4	5
0.5	0.208	0.216	0.216	0.222	0.213
0.6	0.207	0.213	0.224	0.221	0.215
0.7	0.206	0.223	0.215	0.220	0.215
0.8	0.217	0.220	*0.224*	0.207	**0.191**
0.9	*0.237*	*0.225*	0.218	**0.176**	**0.160**

round would reduce the total number of examples required to meet the convergence threshold. As this was not the case, and since examples per round had only a small effect on rounds until convergence, rounds until convergence must be dependent almost entirely on the convergence threshold.

We must initially conclude, therefore, that examples per round and the convergence threshold may be largely immaterial to the learning process. If ranking performance is tied only to the active learning method and number of examples seen, there may simply be a lower bound on the number of rounds required for effective active learning to take place, requiring a number of examples per round equal to the number of examples needed to reach the desired ranking performance divided by this number of rounds. Further investigation is required to determine this lower bound, if indeed it exists.

However, this initial conclusion requires further investigation. Recall that performance has been calculated as an average over all OHSUMED queries. Table 6 shows the standard deviations in the means of the NDCG scores calculated across all queries, for all thresholds and examples per round for the top sampling method. We find a mean standard deviation of around 0.212, with this mean having a standard deviation of around 0.016. At thresholds 0.8 and 0.9, we find standard deviations that deviate strongly from the mean. At four and five examples per round, the standard deviations are significantly lower than the mean, indicating that the performance is more stable across all queries.

7 Related Work

SVMs have proven useful for information retrieval tasks similar to the one proposed here. Drucker et al. [12] compared the use of SVMs for relevance feedback to the Rocchio and Ide algorithms, and found that SVMs outperformed both. Cao et al. [6] looked specifically at the problem of using SVMs for retrieval and ranking of documents. One important finding of theirs relevant to this work is that while an information retrieval system ought to be optimized for ranking the most preferable documents, SVMs optimize generally for both high and low document rankings.

Ranking SVMs have also been used to learn ranking functions from implicitly generated feedback [22,30]. In [22], Joachims learned ranking functions from search engine log files, using the clickthrough data from users as a way to implicitly gather preference data. Results showed that models learned in this fashion were effective at improving retrieval.

Much of the literature on active learning for SVMs has focused on classification, as opposed to ranking [13, 33]. Brinker [3] applied active learning to learning ranking SVMs; however, this research focused on learning label ranking functions, which is a fundamentally different task from document ranking. Tong and Chang [33] used pool-based active learning for image retrieval. Their method of selecting examples for labeling was based on the finding that by choosing examples which shrink the size of the version space in which the optimal weight vector $\boldsymbol{w}$ can lie, the SVM learned from those examples will approximate $\boldsymbol{w}$. Therefore, examples are chosen which will most nearly bisect the version space they occupy. This was achieved in practice by choosing examples based on their proximity to the SVM boundary; examples close to the boundary are likely to be more centrally located in the version space, and are thus more likely to bisect it.

Yu [35] constructed a system similar to the one presented here, operating over real estate data, and noted that methods such those in [33] could not be extended to learning ranking SVMs. As the ranking problem is more complex than the classification problem, active learning for learning ranking SVMs was conducted by selecting the most ambiguously ranked examples for labeling. This was done by noting that ambiguity in ranking could be measured based on how similarly the examples were ranked, with the closest pairs of examples being the most ambiguous. This method for selection is directly analogous to that in [33], even though it does not address reduction in version space; just as the support vectors in a classifying SVM are those examples closest to the SVM boundary, the support vectors in a ranking SVM are those examples that are most closely ranked. You and Hwang [34] used a similar framework and data set to learn ranking in a context-sensitive manner. Both of these works focused on general data retrieval, however, as opposed to focusing on document retrieval.

7.1 Other Methods for Learning to Rank from Preferences

While our work builds on methods employing ranking SVMs for learning preference ranking functions, other strategies to address the problem exist. Cohen, Schapire, and Singer [9] modeled the ranking problem as a directed graph $G = (V, E)$ with instances as the graph's vertices and the weight on an edge between vertices $E_{u,v}$ representing the strength of the preference of u over v. The problem was split into two steps, i.e., learning the weights and then constructing a ranking from the graph, and two methods for addressing this latter step were introduced. Independent of ranking was the learning method, based on the Hedge algorithm, which learned by updating weights based on expert input [14]. The first ranking algorithm was a greedy algorithm which computed the difference between the total weight leaving a vertex and the total weight entering it, with larger values indicating greater overall preference. The second ranking algorithm improved upon the first by separating strongly connected vertices into separate ranking problems, and combining the resulting rankings.

Burges et al. [5] used neural networks to rank using preferences by modeling a probabilistic cost function for pairs of ranked instances. *Cost* in this case was modeled as the cross-entropy cost between the known ranking of the pair and the probability that the model will produce that ranking. The authors then showed how these probabilities can be combined, allowing for computation of a cost function for preference pairs. A neural network training algorithm was then implemented, using the gradient of this function as an updating rule for the network's weights.

Har-Peled et al. [17] formulated the problem in terms of constraint classification using pairwise preferences. Specifically, given a set of instances and a set of labels, the task is to find a function which will produce an ordering of the labels for each instance. The authors achieve this by learning a binary classification model $\mathcal{M}_{i,j}$ for each pair of labels $\mathcal{Y}_i$, $\mathcal{Y}_j$, with $\mathcal{M}_{i,j}$ predicting $\mathcal{Y}_i \succ \mathcal{Y}_j$. The output from each classifier is tallied as a vote for the preferred label, and the labels are ranked according to these votes.

Though certainly not an exhaustive survey of the available work, the approaches mentioned here provide context and motivation for using ranking SVMs in our solution. An approach similar to [9] would not work well in our active learning system, as it is not clear how well the approach performs on unseen data. The authors of [5] provide data on how long their system took to train, and with a training time that had to be measured in hours as opposed to seconds, this approach would be cumbersome in an online system. [17] appears to be addressing a fundamentally different problem; however, the results from related work by Fürnkranz and Hüllermeier indicating that performance can be preserved in the face of inconsistent data may prove useful in the future (see Sect. 5).

7.2 *Improvements to PubMed Search*

Ours is far from the first work attempting to improve the PubMed search experience. Improvements to the search functionality itself [16, 23] spare users the task of learning the PubMed search system, at the cost of losing access to its powerful feature set. Other improvements retain the power of PubMed search and still hide its complexity, but doing so removes access to features some users may require [26]. Improvements to results presentation often focus on ranking results, either by "importance-based" measures [20, 29] or by similarity to a user-defined set of documents [32]. While some users' preference may be importance-based, many others may find such a ranking inadequate to reflect their preferences. And while allowing users to define a set of documents allows for user-based ranking, requiring users to create such a set every time they query PubMed is an unreasonable burden.

MeSH terms have, however, been used as a tool to aid biomedical professionals both in performing searches in PubMed, and in analyzing search results [2, 18, 24]. In particular, Lin et al. [24] used MeSH terms along with keywords to generate labels for documents clustered together using textual features, while Blott et al. [2] clustered based on the MeSH terms themselves. To the best of our knowledge, ours is the only existing work, which attempts document ranking using solely MeSH terms.

8 Conclusion and Future Work

We have presented a framework for learning ranking functions from user feedback, designed for biomedical professionals searching MEDLINE. We have shown that learning these functions using MeSH metadata is superior to learning from textual features, and that by employing active learning we can achieve near perfect ranking performance using less than half of the available data. Questions remain regarding whether the amount of data required to achieve this performance is proportional to the size of the number of documents retrieved, and whether there is a lower bound on the number of feedback rounds required to gain the benefit of active learning.

Future work will investigate these questions. It will also include implementation of the system as a web-based search utility. A user study will be conducted with parameters similar to experiments presented here to assess its performance on "live" queries. System capabilities will be expanded to allow users to save results of previous searches along with their associated ranking functions, as well as applying previously learned ranking functions to new search results.

The introduction of human users to the system introduces the possibility of incorrect or inconsistent feedback. We have shown that this poses a serious threat to system performance. Future work will address this issue as well. Potential areas for investigation include using a window of preference data, e.g., training a ranking model at each round from feedback obtained from the previous n feedback rounds. Another area of investigation is to ignore some user feedback. Results from [15] showed that ignoring potentially erroneous feedback results in better performance over models which included the potentially erroneous feedback.

The system presented here will also be adapted to other domains, particularly that of legal discovery. As discovery is often carried out over collections containing millions of documents, we can expect to see feedback on hundreds or thousands of documents, instead of dozens. However, we clearly cannot expect users to give feedback on a percentage of retrieved documents similar to the percentage used in these experiments. Answering the question of how much feedback is required to learn a reasonable ranking function will be central to the feasibility of application to this domain.

References

1. R. Baeza-Yates, B. Ribeiro-Neto, *Modern Information Retrieval* (Addison-Wesley, 1999)
2. S. Blott, F. Camous, C. Gurrin, G.J.F. Jones, A.F. Smeaton, On the use of clustering and the MeSH controlled vocabulary to improve MEDLINE abstract search, in *Proceedings Conference on Information Research and Applications (CORIA '05)* (2005)
3. K. Brinker, Active learning of label ranking functions, in *Proceedings of the International Conference on Machine Learning (ICML 2004)* (2004)
4. K.A. Bronander, P.H. Goodman, T.F. Inman, T.L. Veach, Boolean search experience and abilities of medical students and practicing physicians. Teach. Learn. Med. **16**(3), 284–289 (2004)
5. C. Burges, T. Shaked, E. Renshaw, A. Lazier, M. Deeds, N. Hamilton, G. Hullender, Learning to rank using gradient descent, in *Proceedings of the 22nd International Conference on Machine Learning* (2005)
6. Y. Cao, J. Xu, T.-Y. Liu, H. Li, Y. Huang, , H.-W. Hon, Adapting ranking SVM to document retrieval, in *Proceedings of the ACM SIGIR International Conference on Information Retrieval (SIGIR '06)* (2006)
7. N. Christianini, J. Shawe-Taylor, *An Introduction to Support Vector Machines and Other Kernel-Based Learning Methods* (Cambridge University Press, 2000)
8. G.A. Churchill, J. Peter, Research design effects on the reliability of rating scales: A meta-analysis. J. Mark. Res. **21**(4), 360–375 (1984)
9. W.W. Cohen, R.E. Schapire, Y. Singer, Learning to order things. J. Artif. Intell. Res. **10**, 243–270 (1999)
10. D. Cohn, L. Atlas, R. Ladner, Improving generalization with active learning. Mach. Learn. **15**(2), 201–221 (1994)
11. N. Craswell, D. Hawking, Overview of the TREC 2004 web track, in *Proceedings of the Text Retrieval Conference (TREC '04)* (2004)
12. H. Drucker, B. Shahrary, D.C. Gibbon, Support vector machines: Relevance feedback and information retrieval. Inf. Process. Manag. **38**, 305–323 (2002)
13. S. Ertekin, J. Huang, L. Bottou, C.L. Giles, Learning on the border: Active learning in imbalanced data classification, in *Proceedings of the ACM Conference on Information and Knowledge Management (CIKM '07)* (2007)
14. Y. Freund, R.E. Schapire, A decision-theoretic generalization of on-line learning and an application to boosting. J. Comput. Syst. Sci. (1997)
15. J. Fürnkranz, E. Hüllermeier, Pairwise preference learning and ranking, in *Proceedings of the 14th European Conference on Machine Learning (ECML-03)* (Springer, 2003), pp. 145–156
16. T. Goetz, C.-W. von der Lieth, PubFinder: A tool for improving retrieval rate of relevant pubmed abstracts. Nucleic Acids Res. **33**, W774–W778 (2005). Web Server issue
17. S. Har-Peled, D. Roth, D. Zimak, Constraint classification: A new apporach to multiclass classification, in *Algorithmic Learning Theory* (Springer Berlin/Heidelberg, 2002)
18. R.B. Haynes, K.A. McKibbon, N.L. Wilczynski, S.D. Walter, S.R. Werre, Optimal search strategies for retrieving scientifically strong studies of treatment from MEDLINE: analytical survey. Br. Med. J. **330**(7501), 1179 (2005)
19. W. Hersh, C. Buckley, T. Leone, D. Hickam, OHSUMED: An interactive retrieval evaluation and new large test collection for research, in *Proceedings of the ACM SIGIR International Conference on Information Retrieval (SIGIR '94)* (1994)
20. J.R. Herskovic, E.V. Bernstam, Using incomplete citation data for MEDLINE results ranking, in *Proceedings of the Annual Symposium of the American Medical Informatics Association (AMIA '05)* (2005), pp. 316–320
21. K. Järvelin, J. Kekäläinen, Cumulated gain-based evaluation of IR techniques. ACM Trans. Inf. Syst. **20**(4), 422–446 (2002)
22. T. Joachims, Optimizing search engines using clickthrough data, in *Proceedings of the ACM SIGKDD International Conference on Knowledge Discovery and Data Mining (SIGKDD '02)* (2002), pp. 133–142

23. J. Lewis, S. Ossowski, J. Hicks, M. Errami, H.R. Garner, Text similarity: an alternative way to search MEDLINE. Bioinformatics **22**(18), 2298–2304 (2006)
24. Y. Lin, W. Li, K. Chen, Y. Liu, A document clustering and ranking system for exploring MEDLINE citations. J. Am. Med. Inf. Assn. **14**(5), 651–661 (2007)
25. T.-Y. Liu, J. Xu, T. Qin, W. Xiong, H. Li, Letor: Benchmark dataset for research on learning to rank for information retrieval, in *Proceedings of the ACM SIGIR International Conference on Information Retrieval (SIGIR '07)* (2007)
26. M. Muin, P. Fontelo, Technical development of PubMed interact: an improved interface for MEDLINE/PubMed searches. BMC Bioinformatics **6**(36), (2006)
27. National Library of Medicine. *Introduction to MeSH.* http://www.nlm.nih.gov/mesh/introduction.html (2009)
28. National Library of Medicine. *MEDLINE fact sheet.* http://www.nlm.nih.gov/pubs/factsheets/medline.html (2009)
29. M.V. Plikus, Z. Zhang, C.-M. Chuong, Pubfocus: Semantic MEDLINE/PubMed citations analytics through integration of controlled biomedical dictionaries and ranking algorithm. BMC Bioinformatics **7**, 424 (2006)
30. F. Radlinski, T. Joachims, Query chains: Learning to rank from implicit feedback, in *Proceedings of the ACM International Conference on Knowledge Discovery and Data Mining (SIGKDD '05)* (2005)
31. S. Robertson, S. Walker, S. Jones, M. Hancock-Beaulieu, M. Gatford, Okapi at TREC-3, in *Proceedings of the 3rd Text Retrieval Conference (TREC-3)* (1995)
32. B.P. Suomela, M.A. Andrade, Ranking the whole MEDLINE database according to a large training set using text indexing. BMC Bioinformatics **6**, 75 (2005)
33. S. Tong, E. Chang, Support vector machine active learning for image retrieval, in *Proceedings of the ACM International Conference on Multimedia (MM '01)* (2001)
34. G. You, S. Hwang, Personalized ranking: A contextual ranking approach, in *Proceedings of the ACM Symposium on Applied Computing (SAC '07)* (2007)
35. H. Yu, SVM selective sampling for ranking with application to data retrieval, in *Proceedings of the International ACM SIGKDD Conference on Knowledge Discovery and Data Mining (SIGKDD '05)* (2005)

Part VI
Preferences in Recommender Systems

Learning Preference Models in Recommender Systems

Marco de Gemmis, Leo Iaquinta, Pasquale Lops, Cataldo Musto, Fedelucio Narducci, and Giovanni Semeraro

Abstract As proved by the continuous growth of the number of web sites which embody recommender systems as a way of personalizing the experience of users with their content, recommender systems represent one of the most popular applications of principles and techniques coming from Information Filtering (IF). As IF techniques usually perform a progressive removal of nonrelevant content according to the information stored in a user profile, recommendation algorithms process information about user interests – acquired in an explicit (e.g., letting users express their opinion about items) or implicit (e.g., studying some behavioral features) way – and exploit these data to generate a list of recommended items. Although each type of filtering method has its own weaknesses and strengths, preference handling is one of the core issues in the design of every recommender system: since these systems aim to guide users in a personalized way to interesting or useful objects in a large space of possible options, it is important for them to accurately capture and model user preferences.

The goal of this chapter is to provide a general overview of the approaches to learning preference models in the context of recommender systems. In the first part, we introduce general concepts and terminology of recommender systems, giving a brief analysis of advantages and drawbacks for each filtering approach. Then we will deal with the issue of learning preference models, show the most popular techniques for profile learning and preference elicitation, and analyze methods for feedback gathering in recommender systems.

1 Introduction

How many times did you find a lot of unwanted mails opening your mailbox? How many times did you search something on the Web and you were not able to find what you were looking for?

M. de Gemmis (✉), L. Iaquinta, P. Lops, C. Musto, F. Narducci, and G. Semeraro
Department of Computer Science, University of Bari "Aldo Moro", Italy
e-mail: {degemmis,iaquinta,lops,cataldomusto,narducci,semeraro}@di.uniba.it,
http://www.di.uniba.it/

J. Fürnkranz and E. Hüllermeier (eds.), *Preference Learning*,
DOI 10.1007/978-3-642-14125-6_18,

The existence of a large quantity of information, in combination with the dynamic and heterogeneous nature of the Web, makes retrieval a hard task for the average user, who is usually overwhelmed by the abundant amount of information.

In this context (we usually refer to this as *Information Overload* problem), the role of user modeling and personalized information access is becoming crucial: although it is too soon to deeply understand the long-term effects of this surplus of information in our habits and in daily life, it is clear that users need a personalized support in sifting through large amounts of available information according to their interests and preferences.

Information Filtering systems, such as Recommender Systems, relying on this idea, adapt their behavior to individual users by learning their tastes during the interaction to construct a profile that can be later exploited to select relevant items. Nowadays these systems represent the main solution to the information overload problem, because they are able to gather and exploit heterogeneous information about users, emerging as one of the most useful tools to achieve a more intelligent information access.

In the workflow of a typical recommendation process, *learning user preferences* is a primary step: catching and modeling user interests in an effective way can be a key issue for personalization goals. Gathering user characteristics, acquired through an explicit (e.g., directly asking to the user) or implicit process (e.g., observing the user behavior), can produce a user model to be exploited to enable adaptivity mechanisms during the interaction with an information system.

The problem of recommending items has been studied extensively, and two main paradigms have emerged. *Content-based* recommendation systems try to recommend items similar to those a given user has liked in the past, whereas systems designed according to the *collaborative* recommendation paradigm identify users whose preferences are similar to those of the given user and recommend items they have liked [4]. Further, in the literature we found also other noteworthy paradigms: *demographic recommenders*, whose aim is to categorize the user starting from personal attributes making recommendation based on demographic classes; *knowledge-based* systems, which exploit knowledge about how a particular item meets a particular user need (such as case-based reasoning that solve a problem retrieving a past similar solved one [30];) *hybrid* systems, at last, combine different recommendation techniques trying to exploit their advantages and reducing at the same time their drawbacks. Each of above paradigms has particular methods to elicit user interests and preferences: most of them are related to machine learning area (probabilistic models, bayesian or neural networks, decision trees, association rules), but there are also some other techniques (so-called *heuristics*) which learn user profiles by exploiting preferences expressed by similar users (usually referred to as "neighbours") or processing textual contents describing the items liked. The goal of this chapter is to provide a general overview of the approaches to learning preference models in the context of recommender systems, by showing advantages and drawbacks of each technique and finally giving a brief analysis of the state of the art.

This chapter is organized as follows. Section 2 introduces general concepts and terminology about recommender systems. Preference learning issues in the area of recommender systems are presented in Section 3, where we also introduce the feedback gathering problem and some machine learning techniques used to acquire and infer user preferences. Conclusions are drawn in the last section.

2 Basics of Recommender Systems

Nowadays it is very important for people to be supported in their decisions, due to the exponential increase of available information.

Everyday we get advice from other people: "Hey, check out this Web site", "I saw this book, you will like it", "That restaurant is very good!". When making a choice in the absence of decisive first-hand knowledge, choosing as other like-minded people have chosen in the past may be a good strategy. Recommender systems have the same role as human recommenders: they present information that they perceive to be useful and worth trying out.

These systems are used in several application domains to support users in taking decisions, to help them in managing the exponential increase of information and, in general, to provide a more *intelligent form of information access*.

The creation and management of personalized recommendations require mainly three distinct and important components: a user profile, an algorithm to update the profile given usage/input information, and an adaptive tool that exploits the profile to provide personalization.

First, the system needs to be able to store relevant information about users that will be used to infer their preferences and needs. Such information is stored in an individual user profile. Second, if the system has to adapt with the user over time, some mechanism is needed to keep the profile up-to-date. This could happen through explicit data input or implicit recording of user behavior as she interacts with the system, or a combination of them. Third, the system needs some way to exploit the current profile data in making recommendations to the user. The types of information stored in the profile will depend on the goals of the system and the algorithms it employs to provide recommendations. Different approaches to recommendation will require different pieces of information about the user, thus the profile structure will differ from system to system.

In this section, we provide a complete overview of the latter step, showing the main recommending approaches and explaining the benefits and weaknesses of each one, while in Section 3.1 we analyze thoroughly the techniques used to acquire and infer user preferences through explicit or implicit feedbacks.

2.1 Content-Based Recommender Systems

The core of the content-based approach is the processing of the contents describing the items to be recommended. The items can be very different depending on the number and type of attributes used to describe them. Each item can be described by the same small number of attributes with known sets of values, but this is not appropriate for items, such as Web pages, news or documents, described by means of unstructured text. In this case, there are no attributes with well-defined values, and the use of document modeling techniques with roots in information retrieval [3,45] and information filtering [5] research is desirable.

A method to represent unstructured data is the Vector Space Model (VSM). The VSM [50] is a spatial representation of text documents. In this model, each document is represented by a vector in a n-dimensional space, where each dimension corresponds to a term from the overall vocabulary of a given document collection. Formally, every document is represented as a vector of term weights, where each weight indicates the degree of association between the document and the term.

The content-based approach can be applied only in the domains where we can provide some textual source describing the items: for example, text recommendation systems like the newsgroup filtering system *NewsWeeder* [26] uses the words of their texts as features. Otherwise, in the domain of movies, the attributes can be movie genre (comedy, horror, drama, etc), main actor and actress, producer, director, etc.

A content-based recommender learns a profile of the user interests based on the features present in the objects the user rated. For example, if a feature (e.g., Inter, or football) occurs in some news the user previously liked, we can expect that he will like other news where this feature often occurs. In this case, a text document may be recommended based on a comparison between the content of the document and the user profile. The system exploits the user profile to suggest relevant items by matching the profile representation against that of items to be recommended. The result of this matching is a binary or continuous relevance judgment, the latter case resulting in a ranked list of potentially interesting items. If data are represented by the VSM, the matching might be realized by computing the cosine similarity between the prototype vector and the item vectors. In some cases, the user is asked for feedback after the document has been shown to her. If the user likes the recommendation, the weights of the words extracted from the document are increased. This process is called *relevance feedback*.

The adoption of the content-based recommendation paradigm has several advantages when compared to the collaborative one:

- USER INDEPENDENCE: Content-based recommenders exploit solely ratings provided by the active user to build her own profile. Instead, collaborative filtering methods need ratings from other users to find the "nearest neighbors" of the active user, i.e., users that have similar tastes (rated the same items similarly). Then, only the items that are most liked by the neighbors of the active user would be recommended;
- TRANSPARENCY: Explanations of recommended items can be provided by explicitly listing content features or descriptions that caused an item to be recommended. These features are indicators to consult to decide when to trust a recommendation and when to doubt one. On the other hand, collaborative systems are

black boxes since the only motivation for an item recommendation is that users with similar tastes liked that item;
- NEW ITEM: Content-based recommenders are capable of recommending items not yet rated by any user. As a consequence, they do not suffer from the new user problem, which affects collaborative recommenders relying solely on users' preferences to make recommendations. Therefore, until the new item is rated by a substantial number of users, the system would not be able to recommend it.

On the other hand, content-based systems have the following shortcomings.

2.1.1 Limited Content Analysis

Content-based techniques are limited by the features that are associated either automatically or manually with the objects that these systems recommend. No content-based recommendation system can provide good suggestions if the content does not contain enough information to distinguish items the user likes from items the user does not like. Some representations capture only certain aspects of the content, but there are many others that would influence a user's experience. For instance, there often is not enough information in the word frequency to model the user interests in jokes or poems, while techniques for affective computing would be most appropriate. Again, for Web pages, feature extraction by using techniques for text representation completely ignores aesthetic qualities and all multimedia information.

To sum up, both automatic extraction and manually assignment of features to items could not be sufficient to define the distinguishing aspects of items able to elicit the user interests.

2.1.2 Overspecialization

Content-based recommenders have no inherent method for finding something unexpected. The system recommends only items scoring highly against the user profile, hence the user is limited to being recommended items similar to those already rated. This drawback is also called *serendipity* problem. To give an example, when a user has only rated movies based on novels by Stephen King, she will be recommended just this kind of movies. A "perfect" content-based technique would never find anything *novel*, limiting the range of applications for which it would be useful.

2.1.3 New User

Enough ratings have to be collected before a content-based recommender system can really understand user preferences and provide accurate recommendations. Therefore, when few ratings are available, such as for a new user, the system would not be able to provide reliable recommendations.

2.2 Collaborative Recommender Systems

We can think of the Collaborative Filtering (CF) paradigm as a computerized process of *word of mouth*. For instance, when looking for a restaurant we usually rely on friends advice, or when looking for a book to read we ask friends who have the same taste. Similarly in collaborative recommender systems, *user opinions* are used to choose what items the user likes. In CF systems, recommendations are based on evaluations of users who share similar interests among them. The idea behind these systems is that a set of users which liked the same items in the past probably share the same preferences. Thus, picking a user from this set, we can suggest her all the unseen items which other users with similar tastes showed to like in the past.

Opinions on items can be expressed as explicit user ratings on some scale ranging from bad to good, or as implicit ratings given by logging user actions. As an example of the latter, viewing or skipping items could be interpreted as positive and negative ratings, respectively.

CF systems analyze opinions of other users on items, thus they provide a liking degree not based on the nature of the item, but on human judgment. Because of this characteristic, CF systems are generally perceived to be more useful than IF based systems [19].

The main advantage of collaborative methods is that items in different product categories can be recommended. Movies, images, art, and text items are all represented by opinions of users and thus they can be recommended by the same system.

In collaborative filtering, a user profile simply consists of the data the user has specified. These data are compared to those of other users to find overlaps in interests among users. For example, the nearest neighbor approach, used in some collaborative recommender system [29], represents the preferences by the items rated (or purchased) by the user. The profile is represented by the user-item matrix [34], where for each cell (u,i) we have the rating of the user u on the item i. Thus, the recommender algorithm exploits the matrix to identify for each user the set of nearest neighbors. In this case, the recommender algorithm performs three tasks: it finds similar users, creates the nearest neighbors set for each user, and infers the like degree for an unseen item based on the nearest neighbors behavior. For example, in an e-commerce scenario, when the user u puts the "Shining" movie into her basket, the system recommends her the "Silence of the lambs" book because more users (who share similar tastes with her, i.e., the nearest neighbors of u) purchased them together.

Terveen and Hill [58] claim three essentials are needed to support collaborative filtering: many people must participate (increasing the likelihood that any one person will find other users with similar preferences), there must be an easy way to represent the user interests in the system, and the algorithms must be able to match people with similar interests. These three elements are not that easy to develop, and produce the main shortcoming of collaborative filtering systems. Following are the main limitations of collaborative systems [4, 27].

2.2.1 New User Problem

In order to make accurate recommendations, the system must first learn the preferences of the user from her ratings. Several techniques have been proposed to address this problem. Most of them use a hybrid recommendation approach, which combines content-based and collaborative techniques.

2.2.2 New Item Problem (Early Rater)

New items are added regularly to recommender systems. Collaborative systems rely only on users preferences to make recommendations. Therefore, until the new item is rated by a substantial number of users, the recommender system would not be able to recommend it. As extreme case of the early rater problem, when a collaborative filtering system first begins, every user suffers from the early rater problem for every item. This problem can also be addressed using hybrid recommendation approaches.

2.2.3 Sparsity Problem

In any recommender system, one of the biggest problem to find recommendations is the extreme sparsity of data in the database. The number of ratings obtained is usually very small compared to the number of ratings to be predicted. Effective prediction of ratings from a small number of examples is important. Also, the success of the collaborative recommender system depends on the availability of a critical mass of users. For example, in a movie recommendation system there might be many movies that have been rated only by few people and these movies would be recommended very rarely, even if those few users gave high ratings to them. One way to overcome the problem of rating sparsity is to use user profile information when calculating user similarity. That is, two users could be considered similar not only if they similarly rated the same items, but also if they belong to the same demographic segment. For example, Pazzani uses gender, age, area code, education, and employment information of users in a restaurant recommendation application [39]. This extension of traditional collaborative filtering techniques is sometimes called demographic filtering.

2.2.4 Grey Sheep Problem (Unusual User)

In a small or even medium community of users, there are individuals who would not benefit from pure collaborative filtering systems because their opinions do not consistently agree or disagree with any group of people. These individuals will rarely, if ever, receive accurate predictions, even after the initial start-up phase for the user and the system [13].

The majority of users falls into the class of the so-called white sheep, those who have high correlation with many other users and who will therefore, in theory, be easy to find recommendations for. The opposite type of people are the black sheep, those for whom there are no or few people who they correlate with. This makes it very difficult to make recommendations for them. On the positive side, for statistical reasons, as the number of users of a system increases the chance of finding other people with similar tastes increases and so better recommendations can be provided.

2.2.5 Scalability Problem

Collaborative filtering systems require a lot of computational resources with the increasing number of users and items. Collaborative filtering systems require data from a large number of users before being effective as well as requiring a large amount of data from each user. The critical dependency on the size and composition of the user population also influences a users group of nearest neighbors. In a situation in which feedback fails to cause this group of nearest neighbors to change, expressing dislike for an item will not necessarily prevent the user from receiving similar items in the future. Furthermore, the lack of access to the content of items prevents similar users from being matched unless they have rated the exact same items.

2.3 Demographic Recommender Systems

These systems aim to categorize the user starting from personal attributes making recommendation based on demographic classes. *Grundy* [47], for example, recommends books by gathering personal information through an interactive dialog matching users responses against a library of manually assembled user stereotypes. *LifeStyle Finder* [25] tries to identify to which cluster a user belongs tailoring recommendations exploiting preferences of the other users in the cluster. Pazzani [39] uses machine learning techniques to obtain a classifier based on demographic data. The representation of demographic information in a user model can vary greatly. *Grundy* system uses hand-crafted attributes with numeric confidence values, while Pazzani extracts features from users' home pages.

The benefit of a demographic approach is that it may not require a history of user ratings of the type needed by collaborative and content-based techniques. However, up to our knowledge, there are not many recommender systems using demographic data because this form of information is difficult to collect: till some years ago, indeed, users were reluctant to share a big amount of personal information with a system. Nowadays with the exponential growth of social network and the continuous expansion of Web 2.0 platforms like Flickr and YouTube, the situation is changed: people's point of view is evolving toward a more open perspective, with users more trustful to sharing of information. Despite this, still today demographic approaches notice less success than others.

2.4 Knowledge-Based Recommender Systems

These systems uses a knowledge-based approach to generate recommendations.

All recommendation techniques make some kind of inference. Knowledge-based approaches are distinguished in that they have functional knowledge: they have knowledge about how a particular item meets a particular user need, and can therefore reason about the relationship between a need and a possible recommendation [11]. The user profile can be any knowledge structure that supports this inference. In the simplest case, as in Google, it may simply be the query that the user has formulated. In others, it may be a more detailed representation of the user needs [59].

A particular kind of knowledge-based recommender systems implement *case-based reasoning (CBR)*. This recommender solves a new problem by retrieving a known solution to a similar problem. In [30], four main steps of a CBR recommender are identified: *retrieve, reuse, adaptation*, and *retain*. The first step looks in the knowledge-base for a case similar to the new problem, then reuses the retrieved solution (making some adaptation, if necessary). Finally, the new adapted case is stored in the caselibrary. In this system, there is not a user preference elicitation because the main task of the recommendation algorithm is to retrieve the case most similar to the problem to solve. From the point of view of the system, the search for a product to recommend is similar to diagnose a disease. In the first case, the system retrieves a product with particular requirements, in the latter case, it retrieves a disease with particular symptoms.

A knowledge-based recommender system avoids some of the drawbacks of other recommendation techniques. It does not have a *ramp-up* problem ("early rater" problem and the "sparse ratings" problem) since its recommendations do not depend on a base of user ratings. As stated above, it does not have to gather information about a particular user because system judgments are independent of individual tastes. These characteristics make knowledge-based recommenders not only valuable systems on their own, but also highly complementary to other types of recommender systems [10].

2.5 Hybrid Recommender Systems

They combine two or more recommender algorithms (the more frequent approach is to combine collaborative filtering with content-based filtering) to emphasize their strengths and to level out their corresponding weaknesses.

Robin Burke proposed a very analytical classification of hybrid systems [11], listing a number of hybridization methods to combine pairs of recommender algorithms.

- WEIGHTED: In weighted hybrid recommenders, the score (or votes) of a recommended item is computed from the results of all of the available recommendation techniques present in the system. This means that the scores of several recommendation techniques are combined together to produce a single recommendation. The simplest combined hybrid would be a linear combination of recommendation scores.
- SWITCHING: A switching hybrid uses some criterion to switch between recommendation techniques. Switching hybrids introduce additional complexity into the recommendation process since the switching criteria must be determined, and this introduces another level of parameterization.
- MIXED: Recommendations from several different recommenders are presented at the same time. This may be possible where it is practical to make large number of recommendations simultaneously.
- FEATURE COMBINATION: Features from different recommendation sources are thrown together into a single recommendation algorithm. For example, content and collaborative techniques might be merged treating collaborative information as simply additional feature data associated with each example and using content-based techniques over this augmented data set.
- CASCADE: The cascade hybrid involves a staged process because one recommender refines the recommendations given by another one. This means that one recommendation technique is employed first to produce a coarse ranking of candidates and a second technique refines the recommendation from the candidate set.
- FEATURE AUGMENTATION: Output from one technique is used as an input feature to another. This means that one technique is employed to produce a rating or classification of an item and that information is then incorporated into the processing of the next recommendation technique.
- META-LEVEL: The model learned by one recommender is used as input to another. This differs from feature augmentation: in an augmentation hybrid, we use a learned model to generate features for input to a second algorithm; in a meta-level hybrid, the entire model becomes the input.

In order to complete the survey, we should also mention some hybrid recommender systems combining collaborative and content-based methods, such as *Fab* [4], *WebWatcher* [21], *P-Tango* [13], *ProfBuilder*, [60], *PTV* [56], *Content-boosted Collaborative Filtering* [31], and *CinemaScreen* [49].

3 Learning User Preferences in Recommender Systems

Before entering in technical details concerning methods for preference acquisition and especially techniques for learning user profiles, we need to take a step backward to thoroughly analyze other important issues: *what are preferences?*

As stated by [9], a *preference* is an ordering relation between two or more items that lets us to characterize which, among a set of possible choices, is the one that best fits our tastes. *Preferences* are something able to guide our choices, discriminating items we like from those we do not like (or we like the least). In other terms, learning user preferences is a way to find the solution of a research (or optimization, in some cases) problem whose space of possible solutions is represented by the set of the items the user can enjoy (namely, in recommender systems, the set of items that can be recommended). Although the semantics of the concept of preference is pretty clear, acquiring user preferences and working with them is a more difficult task. Indeed, the complexity of the problem of preference learning is strictly related to the number of dimensions we used to represent the set of possible choices. We can think at a simple example: choosing a mobile phone. If the only feature to consider is its price, ordering the set of phones and suggesting user the preferred one (namely, the cheaper one) is a simple task. However, if we also add only one more feature (e.g., camera zoom) ordering process becomes more complex and hardly to manage by users in an effective way. When we choose a restaurant, in the same way, we need to find a trade-off between a lot of aspects such as price, service, distance, available time, quality of food, and so on.

So, to generate a user profile, we need to *gather user feedback* to catch information about user preferences and *model* them using a specific representation. Next, this information can be processed (e.g., through machine learning-related approaches) in order to *learn user profiles* to be exploited in the recommendation process.

3.1 Feedback Gathering

The information filtering and information retrieval systems rely on relevance feedback (RF) to capture an appropriate snapshot of user information needs in order to allow the user to directly express her notion of relevance with respect to individual documents [5]. RF has been employed in several classes of personalization systems. Driven by the need for better representation of information needs, RF was initially introduced to support basic query expansion [48]. However, its success in inferring the user's notion of relevance on a per-document basis has lead to a subsequent adoption by information filtering and recommendation systems. RF approaches are based on a feedback gathering scheme, either explicit or implicit. In the former, object ratings of predefined scale are provided explicitly by users, while implicit feedback gathering techniques infer object relevance in a transparent fashion, by monitoring user interaction with the system.

3.1.1 Explicit Ratings

The use of explicit ratings is common in everyday life; ranging from grading students' work to assessing competing consumer goods (see Alton-Scheidl et al. [2] for a review). Although some forms of rating are made in free text form (e.g., book reviews), it is frequently the case that ratings are made on an agreed discrete scale (e.g., star ratings for restaurants, marks out of ten for films, etc). Ratings made on these scales allow these judgments to be processed statistically to provide averages, ranges, distributions, etc.

Several online systems have adopted the explicit ratings approach. For instance, Grouplens [45] provides the collaborative filtering of Internet news. Grouplens users rate articles after having read them and the system aggregates ratings and analyses for future use. MovieLens [33] follows a similar technique to provide movie recommendation services to their members by creating a user profile based on subjective rating of films. Since both these systems originate directly from user explicit judgments, they lead to an accurate estimation of information requirements.

A central feature of explicit ratings is that the evaluator has to examine an item and assign it a value on the rating scale. This imposes a cognitive cost on the evaluator to assess the performance of an object [37]. Indeed, the act of rating alters the user behavior from her normal interaction pattern and, consequently, even less noticeable explicit feedback approaches are considered expensive. Since the results may not become immediately apparent, users tend to skip the rating task [18].

Also, explicit RF techniques disregard user knowledge on the current topic. Users are often unclear about their search interests. They browse for more information to clarify their need and re-formulate their query accordingly. The uncertainty in their search episodes increases the cognitive load during explicit RF, as users must decide on the relevance of a document possibly with a lack of confidence.

Finally, the use of explicit ratings imposes privacy issues that have to be resolved [22]. Irrespective of the underlying reason, users are not always comfortable in providing direct indications of their interests. Due to the obtrusive nature of explicit ratings, not many users are willing to provide them. Hence, the performance of profile capturing and recommendation algorithms of such systems degrades, due to the dearth of ratings. In social filtering systems based on explicit feedback gathering policies, the sparsity of RF judgments can often render such systems unusable, since there are few previous assessments to learn from.

Explicit RF can relying also on critiquing examples. For instance, SmartClient [42] is a tool for planning travel arrangements. Users are required to criticize examples of possible solutions. For instance, "the arrival time of this flight leg is too late." The interaction is cyclical: (1) the system provides example solutions, (2) the user examines any of them and may state a critique on any aspect of it, (3) the critique becomes an additional preference in the model, and (4) the system refines the solution set. Ricci and Nguyen [46] propose a similar critiquing interaction to provide recommendations of restaurants in a mobile context.

As discussed in Pu and Chen [41], the motivation for this methodology is that people usually cannot state preferences in advance but construct their preferences as they see the available options. However, because the critiques come from the user in response to the shown examples, the current solutions can hinder the user from refocusing the search in another direction (the anchoring effect). A complete preference model can be acquired only if the system is able to stimulate the user by showing diverse examples.

3.1.2 Implicit Ratings

Implicit RF gathering techniques are proposed as unobtrusive alternative or supplement to explicit ratings to state (indirect) assessment about usefulness of any individual item. Such techniques passively monitor user interactions with the system to estimate user interests [36]. Click-throughs, time spent viewing a document and mouse gestures are among the possible sources of implicit feedback [23]. The main benefits of implicit feedback, over explicit ratings, are that they remove the cognitive cost of providing relevance judgments explicitly and they can be gathered in large quantities and aggregated to infer item relevance. Since implicit judgments are derived transparently, they contain less indicative value than explicit ratings. Although the accuracy of implicit approaches has been questioned [37], recent studies have shown that they can be effectively adopted to state relevance feedback [61].

There are several types of feedback that can be implicitly captured. For instance, whether a message was read or ignored, whether it was saved or deleted, and whether or not a follow-up message was posted are utilized as an implicit feedback source in conjunction with explicit rating by InfoScope [57] to filter Internet discussion groups. Monitoring the reading time of a document could also be used. Morita and Shinoda [36] concluded that the time spent reading documents on the web is closely related to the degree it suits the needs of each user. An alternative measure of implicit feedback is to assume that all printed documents are relevant and therefore try to detect the user profile from this kind of behavior [24].

Nichols [37] presented a list of potential types of user behaviors that could be exploited as sources for implicit feedback. Kelly and Teevan [23] extended a classification of observable feedback behaviors according to two axes, *Behavior Category*[1] and *Minimum Scope*[2] to categorize actions that can be observed during user information seeking episodes. Their work has also focused on classifying existing scientific literature on implicit feedback according to Behavior Category and Minimum Scope. Unsurprising, a lot of analyzed works concerns examination with object scope, i.e., click-through or scrolling measures are largely investigated and exhibit a strong positive correlation with the explicit ratings. Such data can be easily captured in realtime at no considerable computational cost, while user behaviors that fall in the "Reference", "Annotate", and "Create" require a more precise control over individual services and applications and, thus, are hard to capture and benefits for estimating user interests are not fully clear.

[1] The Behavior Category (Examine, Retain, Reference, Annotate, and Create) refers to the underlying purpose of the observed behavior.

[2] Minimum Scope (Segment, Object, and Class) refers to the smallest possible scope of the item being acted upon.

3.2 Modeling User Preferences

Feedback gathering techniques let us collect as much information as possible about user tastes and interests. However, before this information can be exploited, it needs to be represented in a way that facilitates the learning of preference models. Techniques for modeling information (we usually refer to this as *items*, in recommender systems) can be split depending on the kind of data which will be stored in the user profile. If we have to handle unstructured data (the ones kind usually exploited by content-based recommenders), it is necessary to process them through some information retrieval-related techniques (such as stemming, lemmatization, indexing, and so on), which allow us to shift from a textual source to a structured one. For structured data, such as generic ratings or some well-defined attribute-value pairs (e.g., demographic data), instead, it is possible to represent them through a matrix, as it usually happens in collaborative recommender systems. In both cases, all the information provided by the user, apart from their nature, can be also represented in a more complex way (semantic or neural networks, probabilistic models, etc.) so that we can use them as input for learning user profiles.

In the next section, we will focus our attention on machine learning techniques showing how we can learn a user profile and adapt it gathering user feedback on recommended items. We will complete the analysis trying to give also a complete overview of the state of the art in this area, showing how each approach was implemented in a real recommender system.

3.3 Techniques for Learning User Profiles

Most systems learn user profiles using an online learning approach, building and updating the model to make recommendations in real time. Offline learning methods, instead, fit better in domains where, as stated in [34], user preferences change slowly with respect to the time needed to build the model.

The application of machine learning techniques is a typical way to fulfill the task of learning user profiles in model-based recommender systems. A common approach is to learn the user profile by building a *classifier*. In [34], a classifier is defined as a model able to assign a category to a specific input.

In the machine learning approach to categorization, an inductive process automatically builds a classifier by learning from a *training set* (items labeled with the categories they belong to) the features of the categories. In this approach, the problem of learning user profiles is considered as a binary categorization task: each item has to be classified as interesting or not with respect to the user preferences. Therefore, the set of categories is $C = \{c_+, c_-\}$, where c_+ is the positive class (user-likes) and c_- the negative one (user-dislikes).

Classifiers may be implemented using many different machine learning strategies including probabilistic approaches, neural networks, decision trees, association rules, and Bayesian networks. In this section, we provide a general overview of these techniques.

3.3.1 Naïve Bayes

It is the most used probabilistic algorithm and belongs to the general class of Bayesian classifiers.

These approaches generate a probabilistic model based on previously observed data. It is usually used in content-based recommender systems where the items to recommend are represented by textual documents. Thus, the model estimates the *a posteriori* probability, $P(c|d)$, of document d belonging to class c. This estimation is based on the a priori probability, $P(c)$, the probability of observing a document in class c, $P(d|c)$, the probability of observing the document d given c and, $P(d)$, the probability of observing the instance d. Using these probabilities, the Bayes theorem is applied to calculate $P(c|d)$:

$$P(c|d) = \frac{P(c)P(d|c)}{P(d)} \tag{1}$$

To classify the document d, the class with the highest probability is chosen:

$$c = \text{argmax}_{c_j} \frac{P(c_j)P(d|c_j)}{P(d)}$$

$P(d)$ is generally removed as it is equal for all c_j. As we do not know the value for $P(d|c)$ and $P(c)$, we estimate them by observing the training data. However, estimating $P(d|c)$ in this way is problematic, as it is very unlikely to see the same document more than once: the observed data is generally not enough to be able to generate good probabilities. The naïve Bayes classifier overcomes this problem by simplifying the model by making the independence assumption: all the words or tokens in the observed document d are conditionally independent of each other given the class. Individual probabilities for the words in a document are estimated one by one rather than the complete document as a whole. The conditional independence assumption is clearly violated in real-world data, however, despite these violations, empirically the naïve Bayes classifier does a good job of classifying text documents [6, 28].

Although naïve Bayes classifiers are not as good as probability estimators, it has been shown that they can perform surprisingly well in the classification tasks where the computed probability is not important [17]. Another advantage of the naïve Bayes approach is that it is very efficient and easy to implement compared to other learning methods.

The naïve Bayes classifier has been used in several content-based recommendation systems, such as *Syskill and Webert* [38, 40], *NewsDude* [7], *Daily Learner* [8], *LIBRA* [35], and *ITR* [16, 54].

3.3.2 Rocchio's Method

Some linear classifiers consist of an explicit profile (or prototypical document) of the category [53]. The Rocchio's method is used for inducing linear, profile-style classifiers. It relies on an adaptation to text categorization of the well-known Rocchio's formula for relevance feedback in the VSM [48].

This algorithm represents documents as vectors so that documents with similar content have similar vectors. Each component of such a vector corresponds to a term in the document, typically a word. The weight of each component is computed using the TF-IDF [51] term weighting scheme. Learning is achieved by combining document vectors (of positive and negative examples) into a prototype vector for each class in the set of classes C. To classify a new document d, the similarity between the prototype vectors and the corresponding document vector representing d are calculated for each class (e.g., by using the cosine similarity measure), then d is assigned to the class with which its document vector has the highest similarity value. More formally, Rocchio's method computes a classifier $\overrightarrow{c_i} = (\omega_{1i}, \ldots, \omega_{|T|i})$ for category c_i (T is the *vocabulary*, that is the set of distinct terms in the training set) by means of the formula:

$$\omega_{ki} = \beta \cdot \sum_{\{d_j \in \mathrm{POS}_i\}} \frac{\omega_{kj}}{|\mathrm{POS}_i|} - \gamma \cdot \sum_{\{d_j \in \mathrm{NEG}_i\}} \frac{\omega_{kj}}{|\mathrm{NEG}_i|}, \tag{2}$$

where ω_{kj} is the TF-IDF weight of the term t_k in document d_j, POS_i and NEG_i are the set of positive and negative examples in the training set for the specific class c_j, β and γ are control parameters that allow setting the relative importance of *all* positive and negative examples.

To assign a class $\tilde{c}$ to a document d_j, the similarity between each prototype vector $\overrightarrow{c_i}$ and the document vector $\overrightarrow{d_j}$ is computed and $\tilde{c}$ will be the c_i with the highest value of similarity. Relevance feedback has been used in several content-based recommendation systems, such as *YourNews* [1], *Fab* [4], and *NewT* [55].

3.3.3 Decision Trees Learners

Decision trees are trees in which internal nodes are labeled by terms, branches departing from them are labeled by tests on the weight that the term has in the test document, and leaves are labeled by categories. Decision trees are built by recursively partitioning training data, that is text documents, into subgroups, until those subgroups contain only instances of a single class. The test for partitioning data is run on the weights that the terms labeling the internal nodes have in the document. The choice of the term on which to operate the partition is generally made according to an information gain or entropy criterion [62]. Decision trees are used in the *Syskill and Webert* [38,40] recommender system. The most widely used decision tree learner applied to profiling is ID3 [43].

3.3.4 Decision Rule Classifiers

Decision rules are similar to decision trees, because they are learned in a similar way to the recursive data partitioning approach described above. An advantage of rule learners is that they tend to generate more compact classifiers than decision trees learners. Rule learning methods usually attempt to select from all the possible covering rules (i.e., rules that correctly classify all the training examples) the "best" one according to some minimality criterion. Some examples of inductive learning techniques are Ripper [14], Slipper [15], CN2 [12], and C4.5rules [44].

3.3.5 Neural Networks

Like the approaches seen above, neural networks have a training phase to learn the user profile. These networks model complex relationships between input and output cells. The user interests are represented by the output cells and each of them are achievable by a specific pattern in the network. When an error occurs, there is a backward propagation until the responsible cell is achieved, so the cell parameters are adjusted. Jennings and Higuchi employed a neural network for constructing a users profile [20].

3.3.6 Bayesian Network

It represents a probability distribution by a direct acyclic graph. There are random variables (nodes) and relations among them (arcs). The nodes represent attributes and the arcs represent probability correlations. In [52], a method integrating Case Based Reasoning and Bayesian Network for the user profiling task is shown. Bayesian Network is employed to model quantitative and qualitative relationships between items that users have liked. Bayesian Network is generally used in those situations where user interests change slowly.

3.3.7 Nearest Neighbor Algorithms

These algorithms, also called lazy learners, simply store training data in memory, and classify a new unseen item by comparing it to all stored items by using a similarity function. The "nearest neighbor" or the "k-nearest neighbors" items are determined, and the class label for the unclassified item is derived from the class labels of the nearest neighbors. A similarity function is needed, for example the cosine similarity measure is adopted when items are represented using the VSM. Nearest neighbor algorithms are quite effective, albeit the most important drawback is their inefficiency at classification time, since they do not have a true training phase and thus defer all the computation to classification time. *Daily Learner* [8] and *Quickstep* [32] use the nearest neighbor algorithm to create a model of the user short-term interest and for associating semantic annotation of papers with class names within the ontology, respectively.

4 Conclusions

In this chapter, we surveyed the issue of learning user preferences in recommender systems area. First, we introduced the topic of recommender systems by showing the main approaches presented in literature: we described the content-based and collaborative approaches, showing also the features of some other models such as demographic, knowledge-based, and hybrid ones. Although each kind of recommender system has its own weaknesses and strengths, all of them are joined by a common goal: filtering a set of items, identifying which ones the user will like more by exploiting the information stored in her profile.

In the second part of the chapter, we shift our attention on the core of the recommendation process investigating the issues of learning and modeling user preferences. Several manners to gather user information are exposed, in particular we focus on techniques to get user feedback in implicit and explicit way. Thus, techniques for learning user profiles are analyzed. In particular, we presented machine learning methods, such as Naïve Bayes, Rocchio's method, etc.

We hope that the survey presented in this chapter will contribute to stimulate the research community about the next generation of recommendation technologies and can provide the basis for researches toward new methods for user preferences gathering and modeling in recommender system area.

References

1. J. Ahn, P. Brusilovsky, J. Grady, D. He, S. Yeon Syn, Open user profiles for adaptive news systems: help or harm? in *Proceedings of the 16th International Conference on World Wide Web*, ed. by Carey L. Williamson, Mary Ellen Zurko, Peter F. Patel-Schneider, Prashant J. Shenoy (ACM, 2007), pp. 11–20
2. R. Alton-Scheidl, R. Schumutzer, P.P. Sint, G. Tscherteu. *Voting and rating in Web4Groups* (R. Oldenbourg, 1997), pp. 13–103
3. R. Baeza-Yates, B. Ribeiro-Neto, *Modern Information Retrieval* (Addison-Wesley, 1999)
4. M. Balabanovic, Y. Shoham, Fab: content-based, collaborative recommendation. Commun. ACM **40**(3), 66–72 (1997)
5. N. Belkin, B. Croft, Information filtering and information retrieval:two sides of the same coin? Commun. ACM **35**(12), 29–37 (1992)
6. D. Billsus, M. Pazzani, Learning probabilistic user models, in *Proceedings of the Workshop on Machine Learning for User Modeling* (Chia Laguna, IT, 1997)
7. D. Billsus, M.J. Pazzani, A hybrid user model for news story classification, in *Proceedings of the Seventh International Conference on User Modeling* (Banff, Canada, 1999)
8. D. Billsus, M.J. Pazzani, User modeling for adaptive news access. User Model. User-Adapt. Interact. **10**(2-3), 147–180 (2000)
9. R. Brafman, C. Domshlak, Preference handling - an introductory tutorial. AI Mag. **30**(1), 58–86 (2009)
10. R. Burke, Knowledge-based recommender systems, in *Encyclopedia of Library and Information Systems*, vol. 69, no. 32, ed. by A. Kent (2000)
11. R. Burke, Hybrid recommender systems: survey and experiments. User Model. User-Adapt. Interact. **12**(4), 331–370 (2002)
12. P. Clark, T. Niblett, The cn2 induction algorithm. Mach. Learn. **3**, 261–283 (1989)

13. M. Claypool, A. Gokhale, T. Miranda, P. Murnikov, D. Netes, M. Sartin, Combining content-based and collaborative filters in an online newspaper, in *Proceedings of ACM SIGIR Workshop on Recommender Systems* (1999)
14. W.W. Cohen, Fast effective rule induction, in *Proceedings of the 12th International Conference on Machine Learning (ICML'95)* (1995), pp. 115–123
15. W.W. Cohen, Y. Singer, A simple, fast, and efficitive rule learner, iIn *AAAI/IAAI* (1999), pp. 335–342
16. M. Degemmis, P. Lops, G. Semeraro, A content-collaborative recommender that exploits WordNet-based user profiles for neighborhood formation. User Model. User-Adapt. Interact. J. Per. Res. (UMUAI), **17**(3), 217–255 (Springer Science + Business Media B.V., 2007)
17. P. Domingos, M.J. Pazzani, On the Optimality of the Simple Bayesian Classifier under Zero-One Loss. Mach. Learn. **29**(2-3), 103–130 (1997)
18. J. Grudin, Groupware and social dynamics: eight challenges for developers. Commun. ACM **37**(1), 92–105 (1994)
19. J.L. Herlocker, J.A. Konstan, A. Borchers, J. Riedl, An algorithmic framework for performing collaborative filtering, in *Proceedings of the 22nd Annual International ACM SIGIR Conference on Research and Development in Information Retrieval*, Theoretical Models (1999), pp. 230–237
20. A. Jennings, H. Higuchi, A user model neural network for a personal news service. User Model. User-Adapt. Interact. **3**(1), 1–25 (1993)
21. T. Joachims, D. Freitag, T.M. Mitchell, Web Watcher: A Tour Guide for the World Wide Web, in *15th International Joint Conference on Artificial Intelligence* (1997), pp. 770–777
22. K. Keenoy, M. Levene, Personalisation of web search. in *Intelligent Techniques for Web Personalization*, vol. 3169 *Lecture Notes in Computer Science*, ed. by B. Mobasher, S.S. Anand (Springer, 2003), pp. 201–228
23. D. Kelly, J. Teevan, Implicit feedback for inferring user preference: a bibliography. SIGIR Forum **37**(2), 18–28 (2003)
24. J. Kim, D.W. Oard, K. Romanik, Using implicit feedback for user modeling in internet and intranet searching. Technical report, University of Maryland at College Park, 2000
25. B. Krulwich, Lifestyle finder: Intelligent user profiling using large-scale demographic data. Artif. Intell. Mag. **18**(2), 37–45 (1997)
26. K. Lang, Newsweeder: Learning to filter news, in *Proceedings of the 12th International Conference on Machine Learning* (Lake Tahoe, CA, 1995), pp. 331–339
27. W.S. Lee, Collaborative learning for recommender systems, in *Procedings of the International Conference on Machine Learning* (2001)
28. D.D. Lewis, M. Ringuette, A comparison of two learning algorithms for text categorization, in *Proceedings of SDAIR-94, 3rd Annual Symposium on Document Analysis and Information Retrieval* (Las Vegas, US, 1994), pp. 81–93
29. G. Linden, B. Smith, J. York, Amazon.com recommendations: item-to-item collaborative filtering. IEEE Internet Comput. **7**(1), 76–80 (2003)
30. F. Lorenzi, F. Ricci, Case-based recommender systems: a unifying view, in *ITWP*, vol. 3169, Lecture Notes in Computer Science, ed. by B. Mobasher, S.S. Anand (Springer, 2003), pp. 89–113
31. P. Melville, R.J. Mooney, R. Nagarajan, Content-boosted collaborative filtering for improved recommendations, in *Proceedings of the Eighteenth National Conference on Artificial Intelligence and Fourteenth Conference on Innovative Applications of Artificial Intelligence (AAAI/IAAI-02)* (AAAI, Menlo Parc, CA, USA, 2002), pp. 187–192
32. S.E. Middleton, N.R. Shadbolt, D.C. De Roure, Ontological User Profiling in Recommender Systems. ACM Trans. Inf. Syst. **22**(1), 54–88 (2004)
33. B.N. Miller, I. Albert, S.K. Lam, J.A. Konstan, J. Riedl, Movielens unplugged: experiences with an occasionally connected recommender system, In *IUI '03: Proceedings of the 8th international conference on Intelligent user interfaces* (ACM, New York, NY, USA, 2003), (pp. 263–266)
34. M. Montaner, B. Lopez, J.L. De La Rosa, A Taxonomy of Recommender Agents on the Internet. Artif. Intell. Rev. **19**(4), 285–330 (2003)

35. R.J. Mooney, L. Roy, Content-Based Book Recommending Using Learning for Text Categorization, in *Proceedings of the 5th ACM Conference on Digital Libraries* (ACM, San Antonio, US, New York, US, 2000), pp. 195–204
36. M. Morita, Y. Shinoda, Information filtering based on user behavior analysis and best match text retrieval, in *SIGIR '94: Proceedings of the 17th annual international ACM SIGIR conference on Research and development in information retrieval* (New York, NY, USA, 1994), pp. 272–281
37. D.M. Nichols, Implicit rating and filtering, in *Proceedings of Fifth DELOS Workshop on Filtering and Collaborative Filtering* (ERCIM, 1997), pp. 31–36
38. M. Pazzani, D. Billsus, Learning and Revising User Profiles: The Identification of Interesting Web Sites. Mach. Learn. **27**(3), 313–331 (1997)
39. M.J. Pazzani, A framework for collaborative, content-based and demographic filtering. Artif. Intell. Rev. **13**(5-6), 393–408 (1999)
40. M.J. Pazzani, J. Muramatsu, D. Billsus, Syskill and Webert: Identifying interesting web sites, in *Proceedings of the Thirteenth National Conference on Artificial Intelligence and the Eighth Innovative Applications of Artificial Intelligence Conference* (AAAI/MIT, Menlo Park, 1996), pp. 54–61
41. P. Pu, L. Chen, User-involved preference elicitation for product search and recommender systems. AI Mag. **29**(4), 93–103 (2008)
42. P. Pu, B. Faltings, Enriching buyers' experiences: the smartclient approach, in *CHI '00: Proceedings of the SIGCHI conference on Human factors in computing systems* (ACM, New York, NY, USA, 2000), pp. 289–296
43. J.R. Quinlan, Learning efficient classification procedures and their application to chess end games, in *Machine Learning. An Artificial Intelligence Approach* (1983), pp. 463–482
44. J.R. Quinlan, The minimum description length principle and categorical theories, in *ICML* (1994), pp. 233–241
45. P. Resnick, N. Iacovou, M. Suchak, P. Bergstrom, J. Riedl, GroupLens: An Open Architecture for Collaborative Filtering of Netnews, in *Proceedings of ACM 1994 Conference on Computer Supported Cooperative Work* (ACM, Chapel Hill, North Carolina, 1994), pp. 175–186
46. F. Ricci, Q.N. Nguyen, Acquiring and revising preferences in a critique-based mobile recommender system. IEEE Intell. Syst. **22**(3), 22–29 (2007)
47. E. Rich, User Modeling via Stereotypes. Cogn. Sci. **3**, 329–354 (1979)
48. J.J. Rocchio, *Relevance Feedback in Information Retrieval* (Prentice Hall, Englewood, Cliffs, New Jersey, 1971)
49. J. Salter, N. Antonoupoulos, CinemaScreen Recommender Agent: Combining collaborative and content-based filtering, IEEE Intell. Syst. **21**(1), 35–41 (2006)
50. G. Salton, A. Wong, C.S. Yang, A vector space model for automatic indexing. Commun. ACM **18**(11), 613–620 (1975)
51. G. Salton, C. Buckley, Term-weighting approaches in automatic text retrieval. Technical report, 1988
52. S.N. Schiaffino, A. Amandi, User profiling with case-based reasoning and bayesian networks, in *IBERAMIA-SBIA 2000 Open Discussion Track* (2000), pp. 12–21
53. F. Sebastiani, Machine Learning in Automated Text Categorization. ACM Comput. Surv. **34**(1), 1–47 (2002)
54. G. Semeraro, P. Basile, M. de Gemmis, P. Lops, User Profiles for Personalizing Digital Libraries, in *Handbook of Research on Digital Libraries: Design, Development and Impact*, ed. by Yin-Leng Theng, Schubert Foo, Dion Goh Hoe Lian, Jin-Cheon Na (IGI Global, 2009), pp. 149–158. ISBN 978-159904879-6
55. B. Sheth, P. Maes, Evolving agents for personalized information filtering, in *Proceedings of the Ninth Conference on Artificial Intelligence for Applications* (IEEE Computer Society, 1993), pp. 345–352
56. B. Smith, P. Cotter, A Personalized TV Listings Service for the Digital TV Age. Knowl. Based Syst. **13**, 53–59 (2000)
57. F.C. Stevens, Knowledge-based assistance for accessing large, poorly structured information spaces. PhD thesis, Boulder, CO, USA, 1993

58. L. Terveen, W. Hill, Human-Computer Collaboration in Recommender Systems, in *HCI on the new Millennium*, ed. by J. Carroll (Addison Wesley, 2001)
59. B. Towle, C. Quinn, Knowledge based recommender systems using explicit user models, in *Papers from the AAAI Workshop, AAAI Technical Report WS-00-04* (AAAI, Menlo Park, CA, 2000), pp. 74–77
60. A.M. Wasfi, Collecting user access patterns for building user profiles and collaborative filtering, in *Proceedings of the International Conference on Intelligent User Interfaces* (1999), pp. 57–64
61. R. White, I. Ruthven, J.M. Jose, The use of implicit evidence for relevance feedback in web retrieval, in *Proceedings of the 24th BCS-IRSG European Colloquium on IR Research* (Springer, London, UK, 2002), pp. 93–109
62. Y. Yang, J.O. Pedersen, A Comparative Study on Feature Selection in Text Categorization, in *Proceedings of ICML-97, 14th International Conference on Machine Learning*, ed. by Douglas H. Fisher (Morgan Kaufmann, Nashville, San Francisco, US, 1997), pp. 412–420

Collaborative Preference Learning

Alexandros Karatzoglou and Markus Weimer

Abstract Every recommender system needs the notion of preferences of a user to suggest one item and not another. However, current recommender algorithms deduct these preferences by first predicting an actual rating of the items and then sorting those. Departing from this, we present an algorithm that is capable of directly learning the preference function from given ratings.

The presented approach combines recent results on preference learning, state-of-the-art optimization algorithms, and the large margin approach to capacity control. The algorithm follows the matrix factorization paradigm to collaborative filtering. Maximum Margin Matrix Factorization (MMMF) has been introduced to control the capacity of the prediction to avoid overfitting.

We present an extension to this approach that is capable of using the methodology developed by the Learning to Rank community to learn a ranking of unrated items for each user. In addition, we integrate several recently proposed extensions to MMMF into one coherent framework where they can be combined in a mix-and-match fashion.

1 Introduction

Recommender systems are used by many websites to suggest content, products, or services to their visitors. However, suggesting the right items is a highly nontrivial task: (1) There are many items to choose from. (2) Customers are willing to consider only a small number of recommendations (typically in the order of ten). Thus, the

A. Karatzoglou (✉)
INSA de Rouen, LITIS, France
e-mail: alexis@ci.tuwien.ac.at

M. Weimer
Yahoo! Labs, Santa Clara, California
e-mail: weimer@acm.org

J. Fürnkranz and E. Hüllermeier (eds.), *Preference Learning*,
DOI 10.1007/978-3-642-14125-6_19,

recommender system needs to have a good grasp of the user's preferences to suggest one item and not another.

Recommender algorithms often deduct these preferences by first predicting an actual *rating* of the items and then sorting those. This does not do justice to the way the resulting rankings are usually used: Users are presented only a limited subset of the items. Or a ranked list is shown where the top k (usually 10) items are presented on the first page and all others are hidden on subsequent pages. The procedure outlined above does not guarantee to do well in these scenarios, for example because it puts equal emphasis on all predictions as opposed to just the top k.

The question of how to evaluate a predicted ranking is nontrivial in itself which results in a plethora of different ranking measures. Recently, the machine learning community developed approaches to optimize the prediction directly according to one of those measures. This is generally known as "Learning To Rank". While the initial approaches have been measure-specific, there is a growing interest in unifying the optimization for all measures in a single learning framework where the measure is merely a parameter [4, 10].

In this chapter, we show how to transfer these results from the Learning To Rank community to recommender systems.

Related Work

A popular approach to recommender systems is collaborative filtering. Collaborative filtering addresses the recommendation problem by learning the suggestion function for a user from ratings provided by this and other users on items.

A common approach to collaborative filtering is to fit a factor model to the data. For example by extracting a feature vector for each user and item in the data set such that the inner product of these features minimizes an explicit or implicit loss functional (e.g., [7] following a probabilistic approach). The underlying idea behind these methods is that both user preferences and item properties can be modeled by a number of factors.

Matrix factorization approaches build upon this idea: The known data can be thought of as a sparse $n \times m$ matrix Y of rating/purchase information, where n denotes the number of users and m is the number of items. In this context, Y_{ij} indicates the rating of item j given by user i. The rating is often given on a five-star scale and thus $Y \in \{0, \ldots, 5\}^{n \times m}$, where the value 0 indicates that a user did not rate an item. In this sense, 0 is special since it does *not* indicate that a user dislikes an item but rather that data is missing.

The basic idea of matrix factorization approaches is to fit the original matrix Y with a low rank approximation F. More specifically, the goal is to find such an approximation that minimizes the sum of the squared distances between the known entries in Y and their predictions in F. One possibility of doing so is by using a singular value decomposition of Y and by using only a small number of the vectors obtained by this procedure. In the information retrieval community, this numerical operation is commonly referred to as Latent Semantic Indexing.

Note, however, that this method does not do justice to the way Y was formed. An entry $Y_{ij} = 0$ indicates that we did not observe a $(user, object)$ pair. It does, however, not indicate that user i disliked object j. In [15], an alternative approach is suggested which is the basis of the method described in this chapter. We aim to find two matrices U and M where $U \in R^{n \times d}$ and $M \in R^{d \times m}$ such that $F = UM$ with the goal to approximate the *observed* entries in Y rather than approximating all entries at the same time.

In general, finding a globally optimal solution of the low rank approximation problem is unrealistic: in particular, the approach proposed by [15] for computing a weighted factorization, which is relevant in collaborative filtering settings, requires semidefinite programming, which is feasible only for hundreds, at most, thousands of terms. Departing from the goal of minimizing the rank, Maximum Margin Matrix Factorization (MMMF) aims at minimizing the Frobenius norms of U and M, resulting in a set of convex problems when taken in isolation and thus tractable by current optimization techniques. It was shown in [16, 17] that optimizing the Frobenius norm is a good proxy for optimizing the rank in its application to model complexity control. Similar ideas based on matrix factorization have been also proposed in [12, 18].

In the remainder of this chapter, we show how to extend the MMMF approach to ranking. Additionally, we discuss several extensions to the MMMF model that work equally well for ranking and rating tasks.

The chapter is organized as follows: Sect. 2 describes the general MMMF model, its generalization to ranking, and the use of state-of-the-art optimization methods to train the model. Section 3 describes extensions to that model. In Sect. 4, we discuss experimental evaluations, and Sect. 5 concludes the chapter with remarks on future work.

2 Maximum Margin Matrix Factorization

2.1 *Optimization Problem*

MMMF computes a dense approximation F of the sparse matrix Y which forms the training data. The approximation is based on the modeling assumption that any particular rating of item j by user i is a linear combination of item and user features. Thus, the approximation can be written as $F = UM$. Here, U_{i*} represents the feature vector for user i and M_{*j} is the feature vector for item j. The predicted rating of item j by user i is then the inner product between these feature vectors:

$$F_{ij} = \langle U_{i*}, M_{*j} \rangle$$

Finding the appropriate matrices U and M is achieved by minimizing the regularized loss functional where the Frobenius norm ($\|U\|_F^2 = \operatorname{tr} U U^\top$) of U and M

is used for capacity control and thus overfitting prevention. The Frobenius norm has been introduced to the MMMF framework and shown to be a proper norm on F [17]. This leads us to the following optimization problem:

$$\underset{U,M}{\text{minimize}}\, L(F,Y) + \frac{\lambda_m}{2}\|M\|_F^2 + \frac{\lambda_u}{2}\|U\|_F^2 \tag{1}$$

Here, λ_m, λ_u are the regularization parameters for the M and U matrix, respectively, and $F = UM$. Moreover, $L(F,Y)$ is a loss measuring the discrepancy between Y and F.

The optimization problem can be solved exactly by using a semidefinite reformulation [16]. However, this dramatically limits the size of the problem to several thousand users/movies. Instead, we exploit the fact that the problem is convex in U and M, respectively, when the other variables are fixed to perform subspace descent [11].

2.2 Loss Functions

Squared Loss

In the original MMMF formulation, $L(F,Y)$ was chosen to be the sum of the squared errors [16]:

$$L(F,Y) = \frac{1}{2}\sum_{i=0}^{n}\sum_{j=0}^{m} S_{ij}(F_{ij}-Y_{ij})^2 \text{ where } S_{ij} = \begin{cases} 1 & \text{if user } i \text{ rated item } j \\ 0 & \text{otherwise} \end{cases} \tag{2}$$

This loss decomposes for the nonzero elements of Y and, consequently, it is amenable to efficient minimization by repeatedly solving a linear system of equations for each row/column of U and M separately (i.e., in parallel) – the objective function in (1) is convex quadratic in U and M, respectively, whenever the other term is fixed.

The gradient of $L(F,Y)$ with respect to F can be computed efficiently, since $\partial_{F_{ij}} L(F,Y) = S_{ij} F_{ij} - Y_{ij}$. This means that we have

$$\partial_F L(F,Y) = S.{*}(F-Y), \tag{3}$$

where $.*$ implies element-wise multiplication of S with $F-Y$. In other words, the gradient of the loss is a sparse matrix.

Non Separable Loss

This decomposition into losses, depending on Y_{ij} and F_{ij} alone, fails when dealing with structured losses that take an entire row of predictions, i.e., all predictions for a given user into account. Such losses are closer to what is needed in recommender systems, since users typically want to get good recommendations about which movies they are interested in. A fairly accurate description of which movies they hate is probably less desirable. The recent paper [21] describes an optimization procedure that is capable of dealing with such problems. In general, a non-separable loss takes on the following form:

$$L(F,Y) := \sum_{i=1}^{n} l(F_{i*}, Y_{i*}) \tag{4}$$

Gradients of $L(F,Y)$ decompose immediately into $\partial_{F_{i*}} l(F_{i*}, Y_{i*})$. This allows for efficient gradient computation.

We will now present two ranking scores with their respective loss functions and gradients: The ordinal regression and the NDCG score. For simplicity of notation we only study a row-wise loss $l(f, y)$, where we assume that $f := F_{i*}$ and $y := Y_{i*}$ have already been compressed to contain only nonzero entries in Y_{i*} with the corresponding entries of F_{i*} having been selected accordingly.

Ordinal Loss

The ordinal regression score [6] is based on the notion that predicted rankings can be scored based on the number of pairwise mis-orderings. For each pair of items rated by the user, we incur a loss if the predicted ordering of the two items is not the ordering the user gave.

In a more formal way, this loss can be described as follows: Assume that y is of length m containing m_j movies of score j, that is $\sum_j m_j = m$. For a given pair of movies (u, v), we consider them to be ranked correctly whenever $y_u > y_v$ implies that $f_u > f_v$. A loss of 1 is incurred whenever this implication does not hold. That is, we count

$$\sum_{y_u > y_v} C(y_u, y_v)\,\{f_u \leq f_v\}\,. \tag{5}$$

Here, $C(y_u, y_v)$ denotes the cost of confusing a movie with score y_u with one of score y_v. As there are $n = \frac{1}{2}\left[m^2 - \sum_j m_j^2\right]$ terms in the sum, we need to renormalize the error by n to render the losses among different users comparable. Moreover, we need to impose a soft-margin loss on the comparator $\{f_u \leq f_v\}$ to obtain a convex differentiable loss. This yields the loss

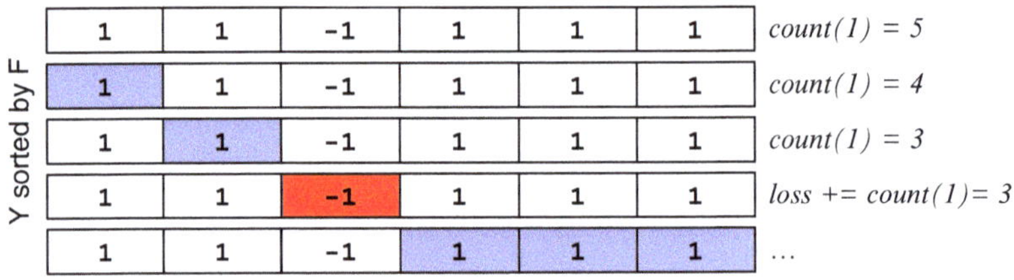

Fig. 1 The fast procedure to compute the ordinal regression loss

$$l(f, y) = 2\Big[m^2 - \sum_j m_j^2\Big]^{-1} \sum_{y_u > y_v} C(y_u, y_v) \max(0, 1 - f_u + f_v). \tag{6}$$

The gradient $\partial_f l(f, y)$ can be computed in a straightforward fashion via

$$\partial_f l(f, y) = -2 \frac{\sum_{y_u > y_v} C(y_u, y_v)}{m^2 - \sum_j m_j^2}. \tag{7}$$

In general, computing losses using preferences such as (6) is an $O(m^2)$ operation. However, the reasoning presented in [8] has been extended to more than binary scores to obtain an $O(m \log m)$ algorithm instead in [23]. The idea is to sort the values of y by the ordering induced by f. Additionally, one counts how often each rating is present in y as $count[i]$. Now, one can do one pass over the sorted y. For each element, the counter for that rating is decremented by one. If we pass an element where the counters for higher ratings are not zero, a loss is induced based on the counters of those elements. Figure 1 illustrates this procedure for only two possible ratings $\{+1, -1\}$.

NDCG

The perceived usefulness of a ranking is not only based on the ordering itself, but also on the position of possible errors. A user may very well tolerate an occasional error for items she does not like as opposed to items she likes.

This observation leads to the development of various position-dependent ranking measures. In the paper [21], the way to optimize the predictions for one of these measures, namely the Discounted Cumulative Gain (DCG), was shown, which is defined as:

$$\mathrm{DCG}(Y_{i*}, \pi)@k = \sum_{j=0}^{k} \frac{2^{Y_{i\pi[j]}} - 1}{\log_2(j+1)} \tag{8}$$

The permutation π is computed as the argsort of the predicted values: $\pi = \mathrm{argsort}(F_{i*})$.[1] The parameter k is a cut-off beyond which the actual ranking does no longer matter. This follows the intuition that typical recommender systems can only present a limited number of items to the user. Since the DCG depends on the possible values of Y, a normalized version is commonly used:

$$\mathrm{NDCG}(Y_{i*}, \pi)@k = \frac{\mathrm{DCG}(Y_{i*}, \pi)@k}{\mathrm{DCG}(Y_{i*}, \pi_s)@k}, \tag{9}$$

where π_s the perfect permutation is the `argsort` of the true ratings given by the user: $\pi_s = \mathrm{argsort}(Y_{i*})$. A NDCG of 1.0 indicates that the model sorts the items in the same order as the user.

NDCG is position-dependent measure: Higher positions have more influence on the score than lower positions. Optimizing DCG has gained much interest in the machine learning and information retrieval communities (e.g., [3]). We present an effort to optimize this measure for collaborative filtering.

Unlike classification and regression measures, DCG is defined on permutations, not absolute values of the ratings. Optimizing over the NDCG measure directly is not possible since the measure is not convex, it is in fact piecewise constant. One thus resorts to structured estimation techniques [19, 20] to compute a convex upper bound on the NDCG measure.

The conversion of the NDCG measure into a loss is a three-step process as shown in [21]:

1. The gain needs to be converted into a loss. This is done by using $\Delta(f, y) = 1 - \mathrm{NDCG}(y, \pi)$.
2. We obtain a linear mapping from f to π.
3. A convex upper bound for the loss is derived.

We now describe the second and third steps in more detail.

A linear mapping is created by employing the following inequality: For any two vectors $a, b \in \mathbb{R}^n$, the inner product $\langle a, b\rangle$ is maximized by sorting a and b in the same order. That is $\langle a, b\rangle \leq \langle \mathrm{sort}(a), \mathrm{sort}(b)\rangle$. This allows us to encode the permutation $\pi = \mathrm{argsort}(f)$ in the following fashion: denote by $c \in \mathbb{R}^n$ a decreasing nonnegative sequence, then the function

$$\psi(\pi, f) := \langle c, f_\pi\rangle \tag{10}$$

is linear in f and maximized with respect to π for $\mathrm{argsort}(f)$.

Finally, by adapting a result of [20] to finding convex upper bounds on nonconvex problems, it was shown in [21] that

$$l(f, y) := \max_\pi \Big[\Delta(\pi, y) + \langle c, f_\pi - f\rangle\Big] \tag{11}$$

[1] We assume $\mathrm{argsort}(f)$ to sort f decreasingly.

is a convex upper bound on the NDCG measure and constitutes a proper loss function. Here, ψ is defined as in (10) and $\pi^* := \text{argsort}(f)$ is the ranking induced by f.

In the same paper, the computation of the $\max_\pi[...]$ is shown to be an instance of the well-studied linear assignment problem. The gradient of this convex upper bound can be computed as:

$$\partial_f l(f, y) = c_{\bar{\pi}^{-1}} - c,$$

where $\bar{\pi}^{-1}$ is the inverted maximizer of (11).

2.3 *Optimization*

Although (1) is not *jointly convex* in U and M, it is still convex in U if M is kept fixed and convex in M if U is kept fixed. We thus resort to alternating subspace descent as proposed by [11] by keeping U fixed and minimizing over M and repeating the process for M with U fixed. We have the following procedure:

repeat
  For fixed M minimize (1) with respect to U.
  For fixed U minimize (1) with respect to M.
until no more progress is made or a maximum iteration count is reached.

As the overall problem is still nonconvex, the procedure cannot be guaranteed to converge to a global minimum. However, it proved to be rather efficient and scalable for problems of up to 10^8 nonzero entries in Y (Netflix) [21].

Each of the two minimization steps is convex and thus amenable to a wide range of optimization procedures, such as stochastic gradient descent (SGD) [2] and LBFGS [9, 25]. However, these methods are guaranteed to converge within $\frac{1}{\epsilon^2}$ steps at best to a solution that is within ϵ of the minimum.

Recently, bundle methods have been introduced with promising results for optimizing regularized risk functions in supervised machine learning and have been shown to converge within $\frac{1}{\epsilon}$ steps [13]. This makes them especially suited to the optimization problem at hand. Each step of the optimization requires a loss and gradient computation. As we have shown above, this can be a costly operation, especially for losses, such as NDCG stemming from the structured estimation framework [20].

The key idea behind bundle methods is to compute successively improving linear lower bounds of an objective function through first order Taylor approximations as shown in Fig. 2. Several lower bounds from previous iterations are bundled to gain more information on the global behavior of the function. The minimum of these lower bounds is then used as a new location where to compute the next approximation, which leads to increasingly tighter bounds and convergence.

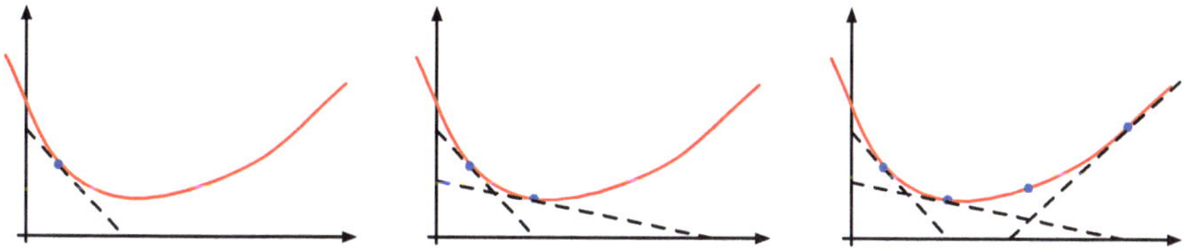

Fig. 2 A convex function (*solid*) is bounded from below by Taylor approximations of first order (*dashed*). Adding more terms improves the bound

Algorithm 1 Optimization over U

input Matrix U and M, data Y
output Matrix U
for $i = 1$ **to** n **do**
  Select $\mathtt{idx}$ as the nonzero set of Y_{i*}
  Initialize $w = U_{i*}$
  $U_{i,\mathtt{idx}} = \mathrm{argmin}_w \, l(wM_{\mathtt{idx},*}, Y_{i,\mathtt{idx}}) + \frac{\lambda_u}{2} \|w\|^2$
end for

Algorithm 2 Computation of $\partial_M L$

input Matrix U and M, data Y
output $\partial_M L = D^\top M$
for $i = 1$ **to** n **do**
  Update $w \leftarrow U_{i*}$
  Find index *ind* where $Y_{i*} \neq 0$
  $X \leftarrow M[ind, :]$
  Update $D_{i*} \leftarrow \partial_F L(wX, Y_{i*}[ind])$
end for
return $\partial_M L = D^\top M$

The main computational cost in using gradient-based solvers is the computation of the gradients with respect to M and U. Using the chain rule yields

$$\partial_M L(F, Y) = U^\top \partial_F L(F, Y) \text{ and } \partial_U L(F, Y) = [\partial_F L(F, Y)]^\top M. \tag{12}$$

This computation is to parallelize, since terms in $\partial_F L(F, Y)$ with respect to each user can be computed separately.

We assume that the loss decomposes per user. Thus, we can optimize U by optimizing each row of U on its own as shown in Algorithm 1. Note that we effectively construct and solve a regularized risk model for each user for a dense data matrix X and parameters w.

When minimizing with respect to M, we need to deal with the entire loss jointly. The main issue to solve is to compute the loss and its gradient with respect to M. Algorithm 2 shows an efficient way to compute the gradient which decomposes again for all users besides the final multiplication.

Analysis

The only interface of the algorithm to the actual rating data is through the loss function. All it needs for building the model is a loss value and the gradient for each iteration. Both these quantities can be computed independently for each user. This is important for several different reasons:

First of all, users need not to share their rating data with their service provider. Instead, they can just exchange losses and gradients. This may reassure privacy cautious users. Additionally, it moves one of the costliest parts of the computation off the service provider's systems and onto those of the users. Communication hardly seems like an issue, as the amount of exchanged data is dominated by the number of rated items per user.

From a similar point of view, the idea of collaborative filtering as a business to business service becomes feasible. In such a setting, an online service provider may be reluctant to share one of his key assets, the rating data. This reluctance currently leads to the problem that the quality of the predictions is mostly determined by the number of customers a service provider has. Using the algorithm described by us, service providers could keep their rating data private while still enjoying the benefit of a way better estimation of M.

Finally, these properties make it trivial to parallelize the algorithm onto a cluster of compute nodes. There, each node would be responsible for a certain number of users and computes the loss and the gradient thereof for these users. This allows the algorithm to be run with loss functions that would otherwise be prohibitively expensive.

3 Extensions

After describing the generalized MMMF model and procedures to optimize it, we now discuss extensions of the model to take prior knowledge about the function class into account.

Offset

Individual users may have different standards when it comes to rating movies. For instance, some users may rarely award a 5 while others are quite generous with it. Likewise, movies have an inherent quality bias. For instance, "Plan 9 from Outer Space" will probably not garner high ratings with any movie buff while other movies may prove universally popular. This can be taken into account by means of an offset per movie. This can be incorporated via

$$F_{ij} = \langle U_{i*}, M_{j*} \rangle + u_i + m_j. \tag{13}$$

Here, u and m are bias vectors for movies and users alike. In practice, we simply extend the dimensionalities of U and M by one for each bias while pinning the corresponding coordinate of the other matrix to assume the value of 1. In this form, no algorithmic modification for the U and M optimization is needed. The relative computational cost of this extension is near zero, since the dimensions of the feature vectors of the convex optimization problem are extended by only one or two.

Please note that this offset is different from a simple normalization of the input data and is not meant to replace preprocessing procedures altogether. The offset is learned for the loss function in use and for each user and movie, while it would be tricky to find a normalization that does cater for both appropriately at the same time.

Graph Kernels

So far we ignored a crucial piece of information, namely the fact that the ratings themselves are not random. For instance, knowing that a user *rated* "Die Hard", "Die Hard 2", and "Top Gun" makes it likely that this user is interested in action movies. This information can be gained without even looking at the score matrix Y. One would expect that we should be able to take advantage of this structural information in addition to the actual scores.

One possibility, proposed by [1] is to use the inner product between the movies two users rent as a kernel for comparing two different users. Denote by $S_{ij} = \{C_{ij} > 0\}$. In this case, they define the kernel between users i and i' to be $\langle S_{i*}, S_{i'*} \rangle$. It is well known that such a model is equivalent to using a linear model with user-features given by S. We can use this to improve the user matrix U to $U + SA$ for a suitably chosen feature matrix A.

Independently, [12] recently developed a related line of thought by assuming that the user matrix is given by $U + \bar{S}A$, where U and A are normally distributed and $\bar{S}$ is a row-normalized version of S, that is $\bar{S}_{i*} = \|S_{i*}\|_1^{-1} S_{i*}$. While their optimization strategy is very different (they use Markov Chain Monte Carlo sampling), it should already be clear at this point that the outcome is very similar to that of [1].

We now show that both approaches, which are approximately equivalent (barring the normalization of S to $\bar{S}$), are also equivalent to the use of graph kernels on the bipartite ranking graph defined between users and movies. For this purpose, we require the following lemma:

Lemma 1. *Denote by $f : \mathbb{R}^n \to \mathbb{R}$ some function and let $A \in \mathbb{R}^{n \times d}$. Moreover, let $U \in \mathbb{R}^{n \times d}$ and $S \in \mathbb{R}^{n \times m}$. Then the following problems are equivalent:*

$$\underset{U,A}{\text{minimize}} \; f(V) + \|U\|^2 + \|A\|^2 \tag{14}$$

$$\underset{V}{\text{minimize}} \; f(V) + U^\top (\mathbf{1} + SS^\top)^{-1} U \tag{15}$$

Proof. Denote by (U^*, A^*) the optimal solution of (14). Clearly in this case for $V := U^* + SA^*$ the optimization problem

$$\underset{A}{\text{minimize}}\ f(V) + \|V - SA\|^2 + \|A\|^2 \tag{16}$$

needs to have A^* as its solution. What remains is to express A as a function of V and to show that in this case (14) and (15) are equivalent. Taking derivatives of (15) with respect to A yields

$$\partial_A \left[\|V - SA\|^2 + \|A\|^2\right] = 2S^\top(SA - V) + 2A.$$

Hence, the gradient of the objective function vanishes for $A^* = (\mathbf{1} + S^\top S)^{-1} S^\top V$. Plugging this back into (15) yields the objective function

$$\begin{aligned} & f(V) + \left\|\left[\mathbf{1} - S(\mathbf{1} + S^\top S)^{-1} S^\top\right] V\right\|^2 + \left\|(\mathbf{1} + S^\top S)^{-1} S^\top V\right\|^2 \\ = \ & f(V) + \left\|(\mathbf{1} + SS^\top)^{-1} V\right\|^2 + \left\|S^\top(\mathbf{1} + SS^\top)^{-1} V\right\|^2 \\ = \ & f(V) + V^\top(\mathbf{1} + SS^\top)^{-1} V. \end{aligned}$$

Here, we have used the Sherman–Morrison–Woodbury identity to transform the second term in the second line. The third term follows from the fact that left and right singular vectors associated with S constitute the eigenvectors of $(\mathbf{1} + SS^\top)$ and $(\mathbf{1} + S^\top S)$, respectively. Hence, we may "push" S to the left in the third term. The last equality follows by direct calculation.

This lemma shows that the parametrization $U + SA$ (or $U + \bar{S}A$, respectively) is equivalent to using a kernel $(1 + SS^\top)^{-1}$ as regularization. The latter is well known as the inverse Laplacian kernel, since $SS^\top$ encodes the undirected graph obtained by connecting all users which watched the same movie. The connection strength in $SS^\top$ denotes the number of movies both users shared.

The net result of this reparametrization is that (14) is a computationally more efficient way of dealing with such symmetries rather than computing the inverse of $(\mathbf{1} + SS^\top)$. Since we may have millions of users the latter would be computationally infeasible.

Note also the connection to the spectral theory of graphs [14]: the eigenvalues and eigenvectors of $\mathbf{1} + \bar{S}\bar{S}^\top$ are close to those of the bipartite graph Laplacian, and can be used for clustering between movies and users, respectively. This means that for similar users and movies, we end up using similar parameters.

4 Experiments

We present evaluation results from the papers [21, 24] and [23]. Evaluations have been done on two aspects of the model described above. First, the impact of the model extensions has been studied empirically. Second, the performance of the different loss functions (NDCG, Ordinal, and Regression) has been evaluated. All

Table 1 Data set statistics

Data set	Users	Movies	Ratings
EachMovie	61,265	1,623	2,811,717
MovieLens	983	1,682	100,000
Netflix	480,189	17,770	100,480,507

evaluations have been conducted on data sets that are well known in the recommender systems literature. The statistics of these can be found in Table 1. All these data sets are obtained from movie recommender services where the movies are rated by the users on a five-star scale.

Evaluation Measures

To evaluate the predictive performance of the different variants of the MMMF model, we resort to two *evaluation measures*. The rating performance is evaluated with the root mean squared error (RMSE) measure:

$$L(F, Y) = \frac{1}{2} \sum_{i=0}^{n} \sum_{j=0}^{m} S_{ij}(F_{ij} - Y_{ij})^2 \text{ where } S_{ij} = \begin{cases} 1 & \text{if user } i \text{ rated item } j \\ 0 & \text{otherwise} \end{cases},$$

where F_{ij} is the model prediction and Y_{ij}, the value in the test data.

As the focus of this chapter is on ranking, we also evaluated using a ranking measure. We choose the *Normalized Discounted Cumulative Gain (NDCG)* score as defined in 8. In all our experiments, we evaluated using NDCG@10.

Evaluation Procedure

Following [26] and [21], we distinguish two different evaluation scenarios: *strong* and *weak generalization*.

Weak generalization: For each user, a number n of ratings is sampled from the known data. The system is then trained on these ratings and evaluated on the remaining items. We present results for $n = 10, 20, 50$. This evaluation resembles a system with an established user base that recommendations are generated for.

Strong generalization: This evaluation procedure has been suggested in [26]: Movies with less than 50 ratings are discarded. The 100 users with the most rated movies are selected as the test set, and the methods are trained on the remaining users. In evaluation, 10, 20, or 50 ratings are sampled from the test users. We then use these ratings to learn a new user feature matrix U_{strong} by performing a single iteration of the user phase. The remaining ratings are used as the test set.

Note that sampling a small number of ratings for the training set mimics a real-world recommender setting where ratings are very scarce compared to the total number of movies.

In all experiments, the regularization parameters λ_u and λ_m were fixed to 10 and not formally tuned. The dimension of U and M was set to $d = 10$ for the evaluation of the extensions and to $d = 100$ for the evaluation of the different losses. We did not observe significant performance fluctuations when comparing the performance of the system using $d = 10$, $d = 100$ in [21] or $d = 30$ in [12]. This is an interesting observation in its own right: The preferences of a user for movies can be described by a relatively small number (in the order of 10) of real numbers.

All experiments were performed ten times with different random draws of the train and test set from the data sets. In total, we report results from more than 1,000 experiments.

4.1 Results: Model Extensions

In this section, we present results only for the least squares regression loss to isolate the influence of the extensions.

Weak Generalization

Table 2 contains the results of the weak generalization scenario experiments. We observe that adding the offset terms yields significant improvements in the performance of the model. On the other hand, enabling the graph kernel does not seem to significantly improve performance. This might be attributed to the fact that the regularization factors were not tuned. Enabling the graph kernel adds a large set of additional parameters to the model, which might lead to overfitting without the adjustment of the regularization parameters. Nonetheless, the graph kernel did add further improvements together with the offsets.

Table 2 The NGDC@10 accuracy over ten runs and the standard deviation for the weak generalization evaluation

	Method	$N = 10$	$N = 20$	$N = 50$
EachMovie	Plain	0.625 ± 0.000	0.639 ± 0.000	0.641 ± 0.000
	Offset	$\mathbf{0.646 \pm 0.000}$	$\mathbf{0.653 \pm 0.000}$	$\mathbf{0.647 \pm 0.000}$
	GraphKernel	0.583 ± 0.000	0.585 ± 0.000	0.590 ± 0.001
	OffsetGK	0.576 ± 0.000	0.597 ± 0.000	0.580 ± 0.001
MovieLens	Plain	0.657 ± 0.000	0.658 ± 0.000	0.686 ± 0.000
	Offset	$\mathbf{0.678 \pm 0.000}$	0.680 ± 0.000	$\mathbf{0.701 \pm 0.000}$
	GraphKernel	0.624 ± 0.001	0.644 ± 0.000	0.682 ± 0.000
	OffsetGK	0.670 ± 0.001	$\mathbf{0.681 \pm 0.000}$	0.682 ± 0.000

Strong Generalization

For the strong generalization setting, the generalized MMMF models were compared to Gaussian Process Ordinal Regression (GPOR) [5] Gaussian Process Regression (GPR), their collaborative extensions (CPR, CGPOR) as well as the original MMMF implementation [21, 26]. Table 3 shows the results for the generalized MMMF models compared to the ones from [26].

For the Movielens data, the model with the offset and graph kernel extensions outperforms the other systems. Additionally, the system with both extensions performs consistently better than the ones with only one extension. On the Eachmovie data, the generalized MMMF model performs the best with the offset extension enabled. It appears that the graph kernel in the Eachmovie data set does not improve the performance but again this could be attributed to a poor choice of the regularization parameters for this data set.

The generalized MMMF model performance is particularly convincing in the strong evaluation setting. This is especially notable as the systems we compare to do use external features of the movies. These features can be obtained by crawling the Internet Movie Database (IMDB) for information on these movies. The paper [22] shows how to extend the MMMF models discussed here to use features as well, but does not provide evaluations on the data sets used here.

The good performance of the generalized MMMF model can be attributed to the fact that the system performs alternate convex optimization steps over item and user features. Once a "good" set of item features is obtained, there is reason to believe

Table 3 The NGDC@10 accuracy over ten runs and the standard deviation for the strong generalization evaluation

	Method	$N = 10$	$N = 20$	$N = 50$
EachMovie	Plain	0.615 ± 0.000	0.633 ± 0.000	0.636 ± 0.000
	Offset	$\mathbf{0.641 \pm 0.000}$	$\mathbf{0.647 \pm 0.000}$	$\mathbf{0.644 \pm 0.000}$
	GraphKernel	0.574 ± 0.000	0.581 ± 0.000	0.596 ± 0.000
	OffsetGK	0.568 ± 0.000	0.594 ± 0.000	0.579 ± 0.000
	GPR	0.456 ± 0.015	0.485 ± 0.007	0.538 ± 0.009
	CGPR	0.573 ± 0.014	0.599 ± 0.012	0.634 ± 0.011
	GPOR	0.369 ± 0.002	0.368 ± 0.003	0.366 ± 0.002
	CGPOR	0.379 ± 0.011	0.378 ± 0.006	0.377 ± 0.004
	MMMF	0.475 ± 0.034	0.479 ± 0.014	0.549 ± 0.021
MovieLens	Plain	0.587 ± 0.001	0.644 ± 0.001	0.630 ± 0.001
	Offset	0.583 ± 0.000	0.444 ± 0.000	0.690 ± 0.000
	GraphKernel	0.613 ± 0.000	0.634 ± 0.000	0.637 ± 0.001
	OffsetGK	$\mathbf{0.684 \pm 0.000}$	$\mathbf{0.691 \pm 0.000}$	$\mathbf{0.692 \pm 0.000}$
	GPR	0.494 ± 0.011	0.502 ± 0.009	0.509 ± 0.014
	CGPR	0.510 ± 0.008	0.525 ± 0.007	0.544 ± 0.006
	GPOR	0.499 ± 0.004	0.500 ± 0.005	0.501 ± 0.005
	CGPOR	0.505 ± 0.005	0.509 ± 0.004	0.505 ± 0.004
	MMMF	0.552 ± 0.018	0.613 ± 0.018	0.665 ± 0.019

Table 4 Results for the weak generalization experiments. We report the NDCG@10 accuracy for various numbers of training ratings used per user. For most results, we report the mean over ten runs and the standard deviation. We also report the p-values for the best vs. second best score

	Method	$N = 10$	$N = 20$	$N = 50$
EachMovie	NDCG	0.656 ± 0.001	0.664 ± 0.002	0.641 ± 0.004
	Ordinal	$\mathbf{0.673 \pm 0.031}$	$\mathbf{0.724 \pm 0.002}$	$\mathbf{0.721 \pm 0.008}$
	Regression	0.611 ± 0.022	0.640 ± 0.035	0.569 ± 0.043
MovieLens	NDCG	0.640 ± 0.006	0.631 ± 0.006	0.608 ± 0.008
	Ordinal	0.623 ± 0.004	0.669 ± 0.006	$\mathbf{0.717 \pm 0.006}$
	Regression	$\mathbf{0.642 \pm 0.025}$	0.651 ± 0.019	0.658 ± 0.019
	MMMF	0.606 ± 0.004	$\mathbf{0.694 \pm 0.004}$	0.699 ± 0.005
Netflix	NDCG	0.608	0.620	
	Regression	0.608	0.629	

that it is a good representation of the items, even for new users. We believe that this is an important benefit of the generalized MMMF model in many applications, as it allows for fast accurate predictions for new users without the need to retrain the whole system.

4.2 Results: Ranking Losses

To compare the performance of the different loss functions, the system was trained without the extensions to allow for the analysis of the loss function's impact in isolation. Experiments were performed with the NDCG, Ordinal Regression, and Least Squares Regression loss functions. In addition, we also report results obtained with the original MATLAB implementation of the MMMF model where possible.[2]

Weak Generalization

Table 4 contains the results of the experiments. The Ordinal Regression loss performs best overall with some exceptions where the Least Squares Regression Loss performs slightly better.

Strong Generalization

For the strong generalization setting, the NDCG scores are compared to those reported in [26] (Table 5).

[2] The implementation did not scale to the bigger data sets Eachmovie and Netflix.

Table 5 The NGDC@10 accuracy over ten runs and the standard deviation for the strong generalization evaluation

	Method	$N = 10$	$N = 20$	$N = 50$
EachMovie	NDCG	**0.638 ± 0.001**	**0.662 ± 0.002**	**0.677 ± 0.002**
	GPR	0.456 ± 0.0105	0.485 ± 0.007	0.538 ± 0.009
	CGPR	0.573 ± 0.014	0.599 ± 0.012	0.634 ± 0.011
	GPOR	0.369 ± 0.002	0.368 ± 0.003	0.366 ± 0.002
	CGPOR	0.379 ± 0.011	0.378 ± 0.006	0.377 ± 0.004
	MMMF	0.475 ± 0.034	0.479 ± 0.014	0.548 ± 0.021
MovieLens	NDCG	**0.624 ± 0.024**	**0.671 ± 0.007**	0.646 ± 0.010
	GPR	0.494 ± 0.011	0.502 ± 0.009	0.509 ± 0.014
	CGPR	0.510 ± 0.008	0.525 ± 0.007	0.544 ± 0.006
	GPOR	0.499 ± 0.004	0.500 ± 0.005	0.501 ± 0.005
	CGPOR	0.505 ± 0.005	0.509 ± 0.004	0.505 ± 0.004
	MMMF	0.552 ± 0.018	0.613 ± 0.018	**0.665 ± 0.019**

The generalized MMMF model with the NDCG loss performs strongly compared to most of the other tested methods. Particularly in the strong generalization setting, it outperforms the existing methods in almost all of the settings. Again, note that all methods except MMMF use additional extracted features, which are either provided with the data set or extracted from the `IMDB`.

4.3 *Observations*

The variance over the ten runs on different data samples in all experiments is surprisingly low, especially given the fact that a nonconvex function is optimized. The same is true for the variance on the objective function. The low variance may mean that the same local minimum is always reached or that this minimum is indeed a global one.

The model extensions described above are capable of improving the predictive performance of the MMMF models. Note that the regularization parameters were not tuned, which might have improved the performance even further. This is especially true when using the graph kernel which adds an additional set of parameters to be learned. We also observe that optimizing the NDCG-based loss function does not necessarily provide for better test NDCG values. This can be attributed to the fact that the NDCG loss function for the sake of convexity provides only an upper bound to the actual NDCG measure. Note that we did not provide results for the Netflix dataset for $N = 50$ due to restrictions in the computational resources.

5 Conclusion

In this chapter, we have presented a model for collaborative filtering based on the well-established Maximum Margin Matrix Factorization (MMMF). We have shown how to extend the model and the optimization procedure thereof to accommodate recent results from the Learning to Rank community. To this end, we have discussed the direct optimization of the Ordinal Regression and NDCG loss functions.

In addition, we have described several extensions to the model recently introduced in the literature. We have introduced offset terms and a graph kernel on the recommender graph. We have also shown that recent extensions to MMMF [12] as well as well-known approaches [1] are both instances of a common graph kernel formulation.

The software developed to evaluate the methods described in this paper is available on http://www.cofirank.org.

Acknowledgements Markus Weimer has been funded under Grant 1223 by the German Science Foundation (DFG). Alexandros Karatzoglou was funded by a grant of the ANR - CADI project. We gratefully acknowledge support by the Frankfurt Center for Scientific Computing in running our experiments. We would also like to thank Alex Smola for important discussions and support.

References

1. J. Basilico, T. Hofmann, Unifying collaborative and content-based filtering, in *Proceedings of the 21st International Conference on Machine Learning (ICML)* (ACM, New York, NY, 2004), pp. 65–72
2. L. Bottou, Stochastic learning, in *Advanced Lectures on Machine Learning, Lecture Notes in Artificial Intelligence, LNAI 3176*, ed. by O. Bousquet, U. von Luxburg (Springer, Berlin, 2004), pp. 146–168
3. C.J. Burges, Q.V. Le, R. Ragno, Learning to rank with nonsmooth cost functions, in *Advances in Neural Information Processing Systems (NIPS)*, vol. 19, ed. by B. Schölkopf, J. Platt, T. Hofmann (2007), pp. 193–200
4. O. Chapelle, Q.V. Le, A. Smola, Large margin optimization of ranking measures, in *NIPS Workshop: Machine Learning for Web Search* (2007)
5. W. Chu, Z. Ghahramani, Gaussian processes for ordinal regression. J. Mach. Learn. Res. **6**, 1019–1041 (2005)
6. R. Herbrich, T. Graepel, K. Obermayer, Large margin rank boundaries for ordinal regression, in *Advances in Large Margin Classifiers*, ed. by A.J. Smola, P.L. Bartlett, B. Schölkopf, D. Schuurmans (MIT, Cambridge, MA, 2000), pp. 115–132
7. T. Hofmann, Latent semantic models for collaborative filtering. ACM Trans. Inf. Syst. (TOIS) **22**(1), 89–115 (2004)
8. T. Joachims, Training linear SVMs in linear time, in *Proceedings of the ACM Conference on Knowledge Discovery and Data Mining (KDD)* (ACM, 2006), pp. 217–226
9. J. Nocedal, S.J. Wright, *Numerical Optimization, Springer Series in Operations Research* (Springer, 1999)
10. T. Qin, T.-Y. Liu, H. Li, A general approximation framework for direct optimization of information retrieval measures. Technical Report MSR-TR-2008-164, Microsoft Research, November 2008

11. J. Rennie, N. Srebro, Fast maximum margin matrix factoriazation for collaborative prediction, in *Proceedings of the 22nd International Conference on Machine Learning (ICML)* (2005), pp. 713–719
12. R. Salakhutdinov, A. Mnih, Probabilistic matrix factorization, in *Advances in Neural Information Processing Systems (NIPS)*, vol. 20 (MIT, Cambridge, MA, 2008)
13. A. Smola, S.V.N. Vishwanathan, Q. Le, Bundle methods for machine learning, in *Advances in Neural Information Processing Systems (NIPS)*, vol. 20 (MIT, Cambridge, MA, 2008)
14. A.J. Smola, I.R. Kondor, Kernels and regularization on graphs, in *Proceedings of the Annual Conference on Computational Learning Theory (COLT)*, Lecture Notes in Computer Science, ed. by B. Schölkopf, M.K. Warmuth (Springer, Heidelberg, Germany, 2003), pp. 144–158
15. N. Srebro, T. Jaakkola, Weighted low-rank approximations, in *Proceedings of the 20th International Conference on Machine Learning (ICML 2003)* (AAAI, 2003), pp. 720–727
16. N. Srebro, J. Rennie, T. Jaakkola, Maximum-margin matrix factorization, in *Advances in Neural Information Processing Systems (NIPS)*, vol. 17, ed. by L.K. Saul, Y. Weiss, L. Bottou (MIT, Cambridge, MA, 2005), pp. 1329–1336
17. N. Srebro, A. Shraibman, Rank, trace-norm and max-norm, in *Proceedings of the Annual Conference on Computational Learning Theory (COLT)*, vol. 3559, Lecture Notes in Artificial Intelligence, ed. by P. Auer, R. Meir (Springer, 2005), pp. 545–560
18. G. Takács, I. Pilászy, B. Németh, D. Tikk, Major components of the gravity recommendation system. SIGKDD Explor. Newslett. **9**(2), 80–83 (2007)
19. B. Taskar, C. Guestrin, D. Koller, Max-margin Markov networks, in *Advances in Neural Information Processing Systems (NIPS)*, vol. 16, ed. by S. Thrun, L. Saul, B. Schölkopf (MIT, Cambridge, MA, 2004), pp. 25–32
20. I. Tsochantaridis, T. Joachims, T. Hofmann, Y. Altun, Large margin methods for structured and interdependent output variables. J. Mach. Learn. Res. **6**, 1453–1484 (2005)
21. M. Weimer, A. Karatzoglou, Q. Le, A. Smola, Cofirank - maximum margin matrix factorization for collaborative ranking, in *Advances in Neural Information Processing Systems (NIPS)* vol. 20 (MIT, Cambridge, MA, 2008)
22. M. Weimer, A. Karatzoglou, A. Smola, Adaptive collaborative filtering, in *Proceedings of ACM Recommender Systems 2008* (2008), pp. 275–282
23. M. Weimer, A. Karatzoglou, A. Smola, Improving maximum margin matrix factorization. Mach. Learn. **72**(3), 263–276 (2008)
24. M. Weimer, A. Karatzoglou, A. Smola, Improving maximum margin matrix factorization, in *Machine Learning and Knowledge Discovery in Databases*, vol. 5211, *LNAI*, ed. by W. Daelemans, B. Goethals, K. Morik (Springer, 2008), pp. 14–14
25. J. Yu, S.V.N. Vishwanathan, S. Günter, N.N. Schraudolph, A quasi-Newton approach to nonsmooth convex optimization, In *Proceedings of the 25th International Conference on Machine Learning (ICML)*, ed. by A. McCallum, S. Roweis (Omnipress, 2008.), pp. 1216–1223
26. S. Yu, K. Yu, V. Tresp, H.P. Kriegel, Collaborative ordinal regression, in *Proceedings of the 23rd International Conference on Machine Learning (ICML)*, ed. by W.W. Cohen, A. Moore (ACM, 2006), pp. 1089–1096

Discerning Relevant Model Features in a Content-based Collaborative Recommender System

Alejandro Bellogín, Iván Cantador, Pablo Castells, and Álvaro Ortigosa

Abstract Recommender systems suggest users information items they may be interested in. User profiles or usage data are compared with some reference characteristics, which may belong to the items (content-based approach), or to other users in the same context (collaborative filtering approach). These items are usually presented as a ranking, where the more relevant an item is predicted to be for a user, the higher it appears in the ranking. In this scenario, a preferential order has to be inferred, and therefore, preference learning methods can be naturally helpful. The relevant recommendation model features for the learning-based enhancements explored in this work comprise parameters of the recommendation algorithms, and user-related attributes. In the researched approach, machine learning techniques are used to discover which model features are relevant in providing accurate recommendations. The assessment of relevant model features, which is the focus of this paper, is envisioned as the first step in a learning cycle in which improved recommendation models are produced and executed after the discovery step, based on the findings that result from it.

1 Introduction

A recommender system suggests to a user products or services he might be interested in. Tastes, interests, and goals are explicitly declared by the user, or implicitly inferred by the system, based on the user's behavior. User profiles and usage data are then compared to some reference characteristics, which might belong to the recommended items (in content-based approaches) [22], to the user's social environment (in collaborative filtering approaches, CF) [18, 19], or to both information sources (in hybrid approaches) [6]. These comparisons usually result in numeric preference values that are used to rank (order) the suggested items for the user. The

A. Bellogín (✉), I. Cantador, P. Castells, and Á. Ortigosa
Universidad Autónoma de Madrid 28049 Madrid, Spain
e-mail: alejandro.bellogin, ivan.cantador, pablo.castells, alvaro.ortigosa@uam.es

J. Fürnkranz and E. Hüllermeier (eds.), *Preference Learning*,
DOI 10.1007/978-3-642-14125-6_20,

recommendation process can thus be considered as an information-ranking problem, where a suitable preference model, consisting of user interests, item content features, and system settings, has to be built.

In this context, research efforts to date can be said to have mainly focused on the study of the improvement of the recommendation algorithms by using all the available knowledge and profiling information. However, few studies have addressed the issue of finding out which of the preference model characteristics are actually most significant when accurate and nonaccurate recommendations are generated. If these characteristics were identified, recommendation strategies could be enhanced by reinforcing or turning down their dependencies with specific stereotypes of users and items.

We thus envision the construction of a recommender system as a virtuous cycle with three main steps. First, an initial recommendation model is created with all the available information. This model is used to compute suggestions for the given user and item repositories. Next, the obtained outputs are analyzed to identify links between specific (input) model characteristics and the quality of recommendations. Finally, considering and adapting the identified characteristics, a new recommendation model, which is expected to generate more accurate results, is produced. The main challenge in this cycle, which is the focus of the research presented here, is in the second step, namely, how to discern (learn) those relevant preference model characteristics based on sets of system inputs, outputs, and user feedback.

In our proposed approach, Machine Learning (ML) techniques are used as a tool to determine which user and system characteristics are shared by most of the top items in a recommendation ranking. Specifically, for each recommendation evaluated (rated) by the user, a training sample is created. The attributes of the sample are the characteristics we aim to analyze, and their values are obtained from log information databases. The class of the training example can be assigned two possible values, correct and incorrect, depending on whether the user evaluated the corresponding recommendation as relevant or irrelevant. By classifying these examples, an ML algorithm facilitates the analysis of the above preference characteristics.

We have tested this proposal with News@hand [8], a news recommender system that suggests news articles according to several recommendation models, namely: (1) a personalized content-based model, the item suggestions from which are based on long-term user profiles [9], (2) a context-aware model that exploits user preferences which are not expressed in the user profile, but can be implicitly detected in the current user recommendation context [23], and (3) a collaborative model that finds and exploits implicit interest relations among users to provide enriched recommendations [7].

As described in the following, the identification of the user profile features and system settings from which each recommendation model should be executed is achieved by means of decision trees. The easy interpretability, the possibility of adding prior knowledge, and the selection of most informative attributes are the main advantages offered by the ML techniques to our recommendation mechanisms.

The rest of the chapter is organized as follows. Section 2 gives an overview of related works in which ML techniques have been applied to automatically learn preferences in personalized content retrieval, recommender and adaptive systems.

Section 3 introduces News@hand, the news recommender system in which our preference analysis proposal is evaluated. Along with this system, the base recommendation algorithms, and the attributes that have been chosen for the analyzed samples are also described. Section 4 briefly explains decision trees, the ML techniques used in our proposal. Section 5 reports on the conducted experiments to evaluate the proposed approach. Finally, Sect. 6 concludes with some discussion and future research lines.

2 Related Work

ML techniques are useful when huge amounts of data have to be classified and analyzed, which nowadays is a very common situation in many scenarios, such as web information exploitation [20]. They have also proved to be of use in adaptive e-learning environments, where student data is used to adapt a system to user preferences and capabilities to facilitate the learning process. Hence, for example, in Becker and Marquardt's work [3], students' logs are analyzed with the goal of finding patterns that reveal the system browsing paths followed by students. Talavera and Gaudioso [21] use classification techniques to analyze student behavior in a cooperative learning environment. Their main goal is to discover patterns that reflect the students' behavior, supporting tutoring activities on virtual learning communities.

Other authors have also investigated the application of these techniques to Recommender Systems [28], evaluating the performance of personalization mechanisms, particularly Adaptive Hypermedia Systems (AHS) and Adaptive Educational Systems (AES) [5]. For example, Zaïane proposed using association rules in AEH domains [27]. His work focuses on two basic points: the first point is to give automated support to students who take an online course proposing the use of advising systems; the second is to support the instructor in identifying student behavior patterns, based on the information that students provide when taking online courses. In the same context, Vialardi et al. [25] use data mining techniques to discover and present relevant pedagogic knowledge to the teachers. They propose to use classification trees and association rules to detect opportunities for improvement on the adaptation decisions of an AES. In [2], several examples where ML techniques are used to learn a user model (based on previous ratings) and classify unseen items are explained. A review of these techniques is also given by Adomavicius and Tuzhilin in [2], where decision trees, clustering, artificial neural networks and Bayesian classifiers are mentioned. Our system also takes into consideration the current user's interest context [23], which is similar to the idea of using short and long term profiles explained in [17].

Despite the above works, to our knowledge, there have been few attempts to use ML techniques as we propose here. In our approach, ML techniques are used to evaluate the system to make explicit improvement on its performance. In this way, we are more interested in the model generated (which variables are more informative, which can be discarded by the model, etc.) by the ML techniques than in the classification itself. This is different from the above approaches, where ML techniques are used as an integrated part of the (recommender, learning) system. Nevertheless, a similar idea can be seen in [24], where ML techniques find patterns for assisting adaptive hypermedia authors during design and evaluation phases. The authors build a model representing the student behavior on a particular course, and use it to obtain and exploit a vision of the behavior and performance of student groups.

3 News@hand: A News Recommender System

News@hand is a news recommender system that combines textual features and collaborative information to make news suggestions, and uses a controlled and structured vocabulary to describe user preferences and news contents. For this purpose, it makes use of Semantic Web technologies. News items and user profiles are represented in terms of concepts appearing in domain ontologies. For example, a news item about a particular football match could be annotated with general concepts as "football" and "match", or specific instances of football teams and players (e.g., *Real Madrid F.C.*, *Zinedine Zidane*).

More specifically, user preferences are described as vectors $\mathbf{u}_m = (u_{m,1}, u_{m,2}, \ldots, u_{m,K})$ where $u_{m,k} \in [-1, 1]$ measures the intensity of the interest of user $u_m \in \mathcal{U}$ for concept $c_k \in \mathcal{O}$ (a class or an instance) in a domain ontology $\mathcal{O}$, K being the total number of concepts in the ontology. Similarly, items $d_n \in \mathcal{D}$ are assumed to be annotated by vectors $\mathbf{d}_n = (d_{n,1}, d_{n,2}, \ldots, d_{n,K})$ of concept weights, in the same vector-space as user preferences.

Ontology concept-based preferences are more precise, and reduce the effect of the ambiguity caused by simple keyword terms. For instance, if a user states an interest for the keyword "java", a system might not have information to distinguish *Java, the programming language*, from *Java, the Pacific island*. However, a preference stated as "ProgrammingLanguage:Java" (this is read as the instance Java from the Programming Language class) lets the system understand unambiguously the preference of the user, and also allows the exploitation of more appropriate related semantics (e.g., synonym, hypernym, subsumption, etc.). This, together with disambiguation techniques, might lead to the effective recommendation of text-annotated items.

In News@hand (Fig. 1), news items are classified in 8 different sections: headlines, world, business, technology, science, health, sports and entertainment. When the user is not logged in the system, he can browse any of the previous sections, but the items are listed without any personalization criterion. He can only sort them by their publication date, source, or level of popularity (i.e., according to a classic rating-based CF mechanism). On the other hand, when the user is logged in the system, recommendation and profile edition functionalities are enabled, and the user can browse the news according to his and others' semantic preferences in different ways. Short- and long-term preferences are considered. Click history is used to define the short-term user preferences, and the resultant rankings can be adapted to the current context of interest.

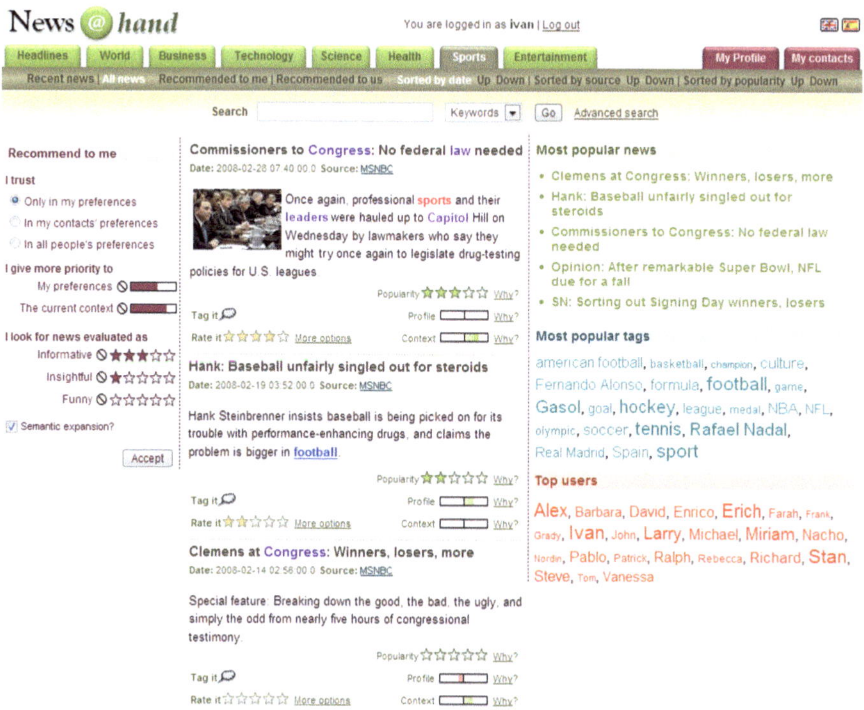

Fig. 1 A typical news recommendation page in News@hand system

Characteristics such as the topic section, the type of recommendation (personalized, context-aware, collaborative), and the number of the page in which accurate recommendations appear are analyzed by our preference learning proposal.

3.1 Semantic Expansion of Preference

Semantic relations among concepts are exploited to enrich the proposed ontology-based knowledge representations, and are incorporated within the recommendation processes. For instance, a user interested in animals (superclass of dog) is also recommended items about dogs. Inversely, a user interested in skiing, snowboarding, and ice hockey can be inferred with a certain confidence to be globally interested in winter sports. Also, a user keen on Spain can be assumed to like Madrid, through locatedIn transitive relation, assuming that this relation had been seen as relevant for inferring previous underlying user's interests.

We have developed [23] a semantic preference spreading mechanism that expands the initial set of preferences stored in user profiles through explicit semantic relations with other concepts in the ontology (Fig. 2). The approach is based on the so-called Constrained Spreading Activation (CSA) strategy [13–15]. The expansion is self-controlled by applying a decay factor to the intensity of preference each time

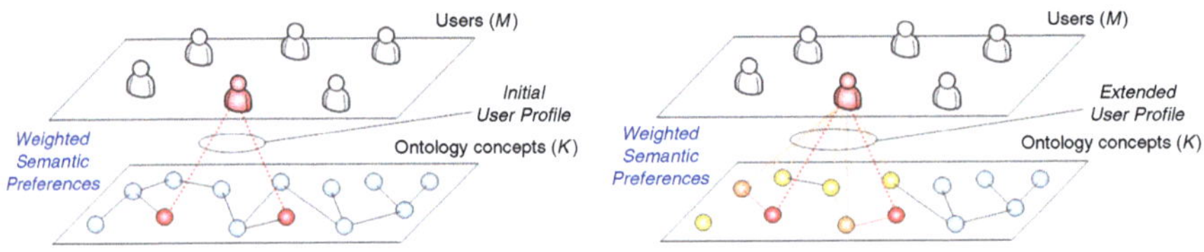

Fig. 2 Semantic preference extension

a relation is traversed, and taking into account constraints (threshold weights) during the spreading process.

News@hand recommendation models output ranked lists of content items taking into account not only the initial user profiles, but also the semantic extension of user preferences and item annotations. The question of whether our semantic expansion technique really benefits the obtaining of more accurate item suggestions is in fact one preference characteristic we analyze in this work.

3.2 Architecture

Figure 3 depicts how ontology-based item descriptions and user profiles are created in News@hand. News items are automatic and periodically retrieved from several online news services via RSS feeds. The title, summary, and category of the retrieved news are then annotated with concepts of the system domain ontologies. Thus, for example, all the news about actors, actresses, and similar terms might be annotated with the concept "actor". A TF-IDF technique is applied to assign weights to the annotated concepts.

With a client/server architecture, users utilize a web interface to receive online news recommendations, and update their profiles. A dynamic graphical interface allows the system to automatically store all the users' inputs, analyze their behavior, and adjust the news recommendations in real time. Explicit and implicit user interests are taken into account, via manual preferences, tags and ratings, and via automatic learning from the users' actions.

Deriving benefit from the semantically annotated news items, the defined ontology-based user profiles, and the knowledge represented by the domain ontologies, a set of recommendation algorithms is executed. Among other approaches, News@hand offers personalized [23], context-aware [9], and collaborative multi-facet recommendations [7]. Configurations and combinations of the above recommendation models are model feature characteristics included in the study presented herein.

3.3 Content-Based Recommendations

Our notion of personalized content retrieval is based on a matching algorithm that provides a relevance measure $\text{pref}(u_m, d_n)$ of anitem d_n. for a user u_m. This measure

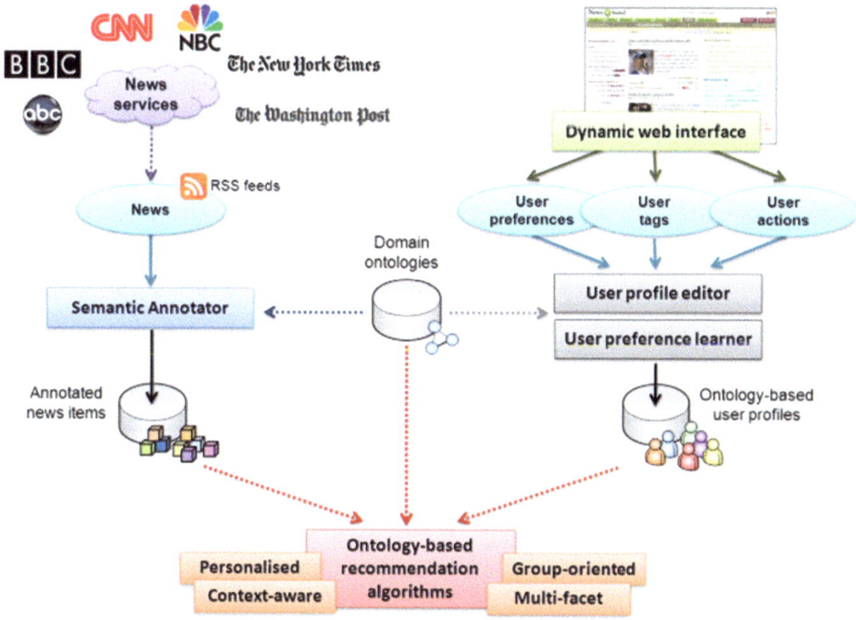

Fig. 3 Architecture of News@hand system

is set according to the semantic preferences of the user and the semantic annotations of the item and based on cosine-based vector similarities

$$\mathrm{pref}(d_n, u_m) = \cos(\mathbf{d}_n, \mathbf{u}_m) = \mathbf{d}_n \cdot \mathbf{u}_m / \|\mathbf{d}_n\| \times \|\mathbf{u}_m\|.$$

The formula matches two weighted-concept vectors, and produces a value in $[-1, +1]$. Values close to -1 are obtained when the vectors are dissimilar, and indicate that user preferences negatively match the content metadata. On the other hand, values close to $+1$ indicate that user preferences significantly match the content metadata, which means a potential interest of the user for the item.

The content-based recommendation results can be combined with query-based scores without personalization [12], and semantic context information, to produce combined rankings. This last approach is described in the next section.

In this model, the size of the user profile will be a reference characteristic to be studied when accurate recommendations are obtained.

3.4 *Context-Aware Recommendations*

We propose a particular notion of context, useful in semantic content retrieval: that of semantic runtime context, which we define as the background topics under which user activities occur within a given unit of time. A runtime context is represented in our approach [9,23] as a set of weighted concepts from the domain ontologies. This set is obtained by collecting the concepts that have been involved in the interaction of the user (e.g., accessed items) during a session.

The context is built in such a way that the importance (weight) of concepts fades away with time (number of accesses back when the concept was referenced) by a decay factor ξ in [0, 1]:

$$C_m^t[c_k] = \xi \cdot C_m^{t-1}[c_k] + (1 - \xi) \cdot \mathrm{Req}^t[c_k],$$

where $\mathrm{Req}^t[c_k]$ in $[0, 1]^K$ is a vector whose components measure the degree in which the concepts c_k are involved in the user's request at time t. This vector can be defined in multiple ways, depending on the application: a query concept-vector (if a request is expressed in term of a concept-based search query), a concept vector containing the most relevant concepts in a document (if a request is a "view document" request), the average concept-vector corresponding to a set of items marked as relevant by the user (if a request is a *relevance feedback* step), etc.

Once the context is built, a contextual activation of preferences is achieved by finding semantic paths linking preferences to context, as follows:

$$\begin{aligned} \mathrm{pref}_{C^t}(d_n, u_m) &= \lambda \cdot \mathrm{pref}(d_n, u_m) + (1 - \lambda) \cdot \mathrm{sim}(d_n, C_t) \\ &= \lambda \cdot \cos(d_n, EU_m) + (1 - \lambda) \cdot \cos(d_n, EC_t), \end{aligned}$$

where λ in [0, 1] measures the strength of the personalization component with respect to the current context. This parameter could be manually established by the user, or dynamically adapted by the system according to multiple factors, such as the current size of the context, the automatic detection of a change in the user's search focus, etc.

The perceived effect of contextualization is that user interests that are out of focus, under a given context, are disregarded, reinforcing those that are in the semantic scope of the ongoing user activity are considered for recommendation (see Fig. 4).

Analogously to the personalization model, where the size of the user profile is a critical aspect, the context-aware recommendation approach will be affected by the size and precision of the current semantic context. These characteristics will be also included in the analytical experiments.

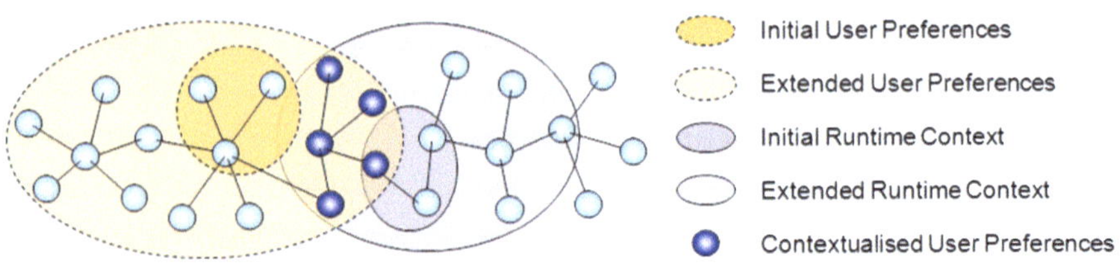

Fig. 4 Contextualization of user preferences

3.5 Collaborative Recommendations

Collaborative filtering techniques match people with similar preferences to make recommendations. Unlike content-based methods, collaborative recommender systems aim to predict the utility of items for a particular user according to the items previously evaluated by others [18, 19]. One of the main benefits of these approaches is the possibility to recommend items that do not share features with respect to the items rated in the past by the user. However, these approaches introduce certain problems [1]; for example, a new item cannot be recommended to a user until other users rate it.

The utility gain function $g(u_m, i_n)$ of item $i_n \subset \mathcal{I}$ for user $u_m \in \mathcal{U}$ is estimated based on the utilities $g(u_j, i_n)$ assigned to item i_n by those users u_j that are "similar" to user u_m. In this work, we use two different well-known collaborative filtering approaches: user-based and item-based [18, 19]. In the first situation, the following approach has been taken:

$$g(u_m, i_n) = \frac{\sum\limits_{u_j \in \hat{\mathcal{U}}_m} \mathrm{sim}(u_m, u_j) \times r_{j,n}}{\sum\limits_{u_j \in \hat{\mathcal{U}}_m} \left|\mathrm{sim}(u_m, u_j)\right|},$$

$$\mathrm{sim}(u_m, u_j) = \frac{\sum\limits_{i_n \in \mathcal{I}_{m,j}} (r_{m,n} - \bar{r}_m) \cdot (r_{j,n} - \bar{r}_j)}{\sqrt{\sum\limits_{i_n \in \mathcal{I}_{m,j}} (r_{m,n} - \bar{r}_m)^2} \sqrt{\sum\limits_{i_n \in \mathcal{I}_{m,j}} (r_{j,n} - \bar{r}_j)^2}},$$

where the similarity function is called Pearson correlation.

In the item-based situation, we use a similar formulation:

$$g(u_m, i_n) = \frac{\sum\limits_{i_j \in \hat{\mathcal{I}}_n} \mathrm{sim}(i_n, i_j) \times r_{m,j}}{\sum\limits_{i_j \in \hat{\mathcal{I}}_n} \left|\mathrm{sim}(i_n, i_j)\right|},$$

$$\mathrm{sim}(i_n, i_j) = \frac{\sum\limits_{u_m \in \mathcal{U}_{n,j}} (r_{m,n} - \bar{r}_n) \cdot (r_{m,j} - \bar{r}_j)}{\sqrt{\sum\limits_{u_m \in \mathcal{U}_{n,j}} (r_{m,n} - \bar{r}_n)^2} \sqrt{\sum\limits_{u_m \in \mathcal{U}_{n,j}} (r_{m,j} - \bar{r}_j)^2}}.$$

The predicted value $g(u_m, i_n)$ is a very solid information source to know whether the above algorithms would work in a real scenario, so we will study it in our experiments, along with the type of collaborative filtering technique used.

3.6 Log Database

The system monitors all the actions the user performs, and gathers them in a log database. Table 1 shows the attributes of the database tables.

In this work, we focus on the user evaluation and browsing tables, which store information about ratings and rated items, and system configurations for specific actions, respectively. The database tables share a session identifier that allows us to recognize relationships among actions. More specifically, given a row from the user evaluation table, we extract the session identifier, the rated item, and the action timestamp to infer which system configuration was at that moment, as follows:

1. Get all the browsing actions matching a given session identifier.
2. Select the actions with the same item identifier, previously extracted from the browsing table.
3. Use the timestamp to obtain the system configuration, such as user profile weight (0 if personalization is off), and context weight.

Table 1 Summary of the log database tables and attributes. Session id is an intertable identifier, whilst action id is an intratable attribute. Action type is a string distinguishing between different actions a table can contain (for instance, LOGIN and LOGOUT are stored in user accesses table)

Table	Attributes
Browsing	actionID, actionType, timestamp, sessionID, itemID, itemRankingPosition, itemRankingProfile, itemRankingContext, itemRankingCollaborative, itemRankingHybridUP, itemRankingHybridNUP, itemRankingHybridUPq, itemRankingHybridNUPq, topicSection, interestSituation, userProfileWeight, contextWeight, collaborative, scoreSearch
Context updates	actionID, actionType, timestamp, sessionID, context, origin, changeOfFocus
Queries	actionID, actionType, timestamp, sessionID, keywords, topicSection, interestSituation
Recommendations	actionID, actionType, timestamp, sessionID, recommendationType, userProfileWeight, contextWeight, collaborative, topicSection, interestSituation
User accesses	actionID, actionType, timestamp, sessionID
User evaluations	actionID, actionType, timestamp, sessionID, itemID, rating, userFeedback, tags, comments, topicSection, interestSituation, duration
User preferences	actionID, actionType, timestamp, sessionID, concept, weight, interestSituation
User profiles	actionID, actionType, timestamp, sessionID, userProfile
User sessions	sessionID, userID, timestamp

4 Decision Trees for Model Feature Learning

The main goal of our research is twofold: the creation of training samples that correspond to positive (relevant) and negative (nonrelevant) recommendation cases, and the analysis of these samples with ML techniques to determine which model features seem to be most significant to provide either positive or negative recommendations.

In this section, we describe the ML algorithms applied for our model feature learning purposes. We focus on one of these techniques: Decision Trees. However, a previous work also explored Attribute Selection technique [4]. Information for creating the samples is obtained from the log database introduced in Sect. 3.6.

Decision Trees apply a divide-and-conquer strategy for producing classifiers with the following benefits [16]:

- They are interpretable.
- They enable an easy attachment of prior knowledge from human expert.
- They tend to select the most informative attributes measuring their entropy, boosting them to their top levels.
- They are useful for nonmetric data (the represented queries do not require any notion of metric, as they can be asked in a "yes/no", "true/false" or other discrete value set representations).

However, despite these advantages, Decision Trees are usually overfitted and might not generalize well to independent test sets. Two possible solutions are applicable: stopped splitting and pruning. C4.5 is one of the most common algorithms to build Decision Trees, and uses heuristics for pruning based on statistical significance of splits. In the experiments, we make use of its well-known revision J48.

It is worth noting that in this paper we are interested in the model generated by this classifier, instead of its predictive power. Proceeding in this way, Decision Trees will show which attributes are more informative (those appearing at the top of the tree), and which of their values tend to classify an instance as positive or negative.

5 Experiments

The experiments have been conducted using News@hand system, presented in Sect. 3. In the following, a description of the system item database and knowledge repository is provided. We also explain the two different experiments performed (*stages* from now on), including the tasks and phases fulfilled by users during the evaluation, and conclude with the obtained results.

5.1 News Item Database and Knowledge Repository

For two months, RSS feeds were collected on a daily basis. A total of 9,698 news items were stored. With this dataset, we run our semantic annotation mechanism mentioned in Sect. 3.2, and a total of 66,378 annotations were obtained. For more details, see [11].

A set of 17 ontologies is used by the current version of the system. They are adaptations of the IPTC ontology[1], which contains concepts of multiple domains such as education, culture, politics, religion, science, technology, business, health, entertainment, sports, weather, etc. They have been populated with concepts appearing in the gathered news items using semantic information from Wikipedia, and applying a population mechanism explained in [11]. A total of 137,254 Wikipedia entries were used to populate 744 ontology classes with 121,135 instances.

5.2 Experimental Setup

Two different stages have been designed to discover which model features are relevant in providing accurate recommendations. The first one is focused on personalization functionalities, in particular: ontology-based content retrieval, and semantic context-aware personalization. Ontology-based content retrieval is tested against a keyword-based approach, whilst context-aware personalization is turned on and off to investigate its contribution to the user's experience. Another important part of these methods has also been evaluated: semantic expansion of preferences.

In the second stage, we analyze which features of our model are more influential when using a collaborative filtering algorithm. With this objective in mind, we have integrated two well-known, state-of-the-art collaborative filtering algorithms into the system, and studied their discriminative power for classifying a news item as relevant or irrelevant.

First Stage: Evaluation of Content-Based and Context-Aware Recommendation

In this section, we present a first experiment conducted to evaluate the precision of the personalization and the context-aware recommendation functionalities available in News@hand (Sects. 3.3 and 3.4). We also aimed to investigate the influence of each mechanism in the integrated system, measuring the precision of the recommendations when a combination of both models is used. Sixteen members of our department were requested to participate. There were 12 undergraduate/graduate students and 4 lecturers.

---

[1] IPTC ontology, http://nets.ii.uam.es/mesh/news-at-hand/news-at-hand_iptc-kb_v01.zip.

Table 2 Summary of the search tasks performed in the experiment

Profile	Section	Query		Task goal
1	World	$Q_{1,1}$	Pakistan	News about media: TV, Radio, Internet
Telecom	Entertainment	$Q_{1,2}$	music	News about software piracy, illegal downloads, file sharing
2	Business	$Q_{2,1}$	dollar	News about oil prices
Banking	Headlines	$Q_{2,2}$	fraud	News about money losses
3	Science	$Q_{3,1}$	food	News about cloning
Social care	Headlines	$Q_{3,2}$	internet	News about children, young people, child safety, child abuse

The experiment comprised two phases, each composed of two different tasks. In the first phase, only the personalization module was active, and the tasks were different in having the semantic expansion (see Sect. 3.1) enabled or disabled. In the second phase, the contextualization and semantic expansion functionalities were active. In its second task, the personalized recommendations were also enabled. More details are given in the next section.

A task was defined as finding and evaluating those news items that were relevant to a given goal. Each goal was framed in a specific domain, and we considered three domains: telecommunications, banking, and social care issues. For each domain, a user profile and two search goals were set as explained below. Table 2 shows a summary of the involved tasks.

To simplify the searching tasks, they were defined for a pre-established section and query. Hence, for example, the task goal of finding news items about software piracy, illegal downloads and file sharing, $Q_{1,2}$, was reduced to evaluate those articles existing in *Entertainment* section that were retrieved with the query "music".

To cover as many system configurations as possible with the available users, the assignment of the tasks was set according to the following principles:

- A user should not repeat a query during the experiment.
- The domains should be equally covered by each experiment phase.
- A user has to manually define a user profile once in the experiment.

For each phase, the combination of personalized and context-aware recommendations was established as a linear combination of their results using two weights $w_p, w_c \in [0, 1]$:

$$\text{score}(d_n, u_m) = w_p \cdot \text{pref}(d_n, u_m) + w_c \cdot \text{pref}(d_n, u_m, \text{context})$$

In the personalization phase, the contextualization was disabled (i.e., $w_c = 0$). Its first tasks were performed without semantic expansion, and its second tasks had the semantic expansion activated. In the contextualization phase, w_c was set to 1, and the expansion was enabled. Its first tasks were done without personalization ($w_p = 0$), and its second tasks were influenced by the corresponding profiles ($w_p = 0.5$).

As mentioned before, a fixed user profile was used for each domain. Some of them were predefined profiles, and others were created by the users during the experiment, using the profile editor of News@hand. In addition, some tasks were done with user profiles containing concepts belonging to all the three domains.

There is also an important issue about how the users rated. Every time the user read an item, he had to assess whether the item was relevant to the profile, to the current goal, or to both/neither of them. In each situation, a different rating criterion was defined:

- Rate with 1 star if the item was not relevant.
- Rate with 2 stars if the item was relevant to the current goal.
- Rate with 3 stars if the item was relevant to the profile.
- Rate with 4 stars if the item was relevant to the current goal and the profile.

These rating constraints gave us a bounded frame for evaluation. In the next sections, it will be shown that they also allowed us to have different criteria to set the class values of the training samples.

5.2.1 Content-Based Phase

The objective of the two tasks performed in the first experiment phase was to evaluate the importance of activating the semantic expansion of our recommendation models. The following are the steps the users had to do in these tasks:

- Launch the query with the personalization module deactivated.
- Rate the top 15 news items.
- Launch the query with the personalization module activated (and the semantic expansion enabled/disabled depending on the case).
- Rate again the top 15 news items.

At the end of this phase, each user had rated 30 items with expansion enabled and 30 with expansion disabled.

5.2.2 Contextualization Phase

The objective of the two tasks performed for the second experiment phase was to evaluate the quality of the results when the contextualization functionality is activated and combined with personalization. The steps done in this case are the following:

- Launch the query with the contextualization activated (semantic expansion enabled, and personalization enabled/disabled depending on the case).
- Rate the top 15 news items, and evaluate as relevant (clicking the title) the first item related to the task goal. Doing this the current semantic context is updated.
- Repeat the last two steps twice (the last time it is not necessary to update the context, since the evaluation will not continue).

At the end of this phase, each user had rated 45 items with personalization on and 45 items with personalization off. He had also evaluated as relevant 4 news items that were incorporated into the context.

5.2.3 Selection of Sample Attributes and Classes Based on Evaluation Parameters

Each user had to assign a rating depending on the four existing possibilities for each news item: relevant to the goal (2), the profile (3), both (4), and neither of them (1). Considering these four options, we defined three different criteria to classify an item (sample) as relevant:

- The item is relevant in general, if the user has rated it with 2, 3, or 4 stars.
- The item is relevant to the current goal, if the user has rated it with 2 or 4 stars.
- The item is relevant to the profile, if the user has rated it with 3 or 4 stars.

In this work, we focus on the second criterion, although a preliminary analysis with the first one is also tested because of its generality.

In addition to the sample classes, according to the evaluation made, we selected those attributes whose impact on the recommendations we wanted to analyze. For each item rating log entry, we chose several attributes that can be categorized as follows:

5.2.4 User-Based Features

- *Profile type*: A string attribute with two possible values: fixed or user-defined preferences (manual preferences).
- *Profile size*: An integer attribute indicating the total number of concepts included in the profile (number of nonzero components in the vector representation).
- *Context size*: An integer attribute indicating the number of concepts included in the current context.

5.2.5 Model-Based Features

- *Topic section*: Name of the news section in which the rated item appeared.
- *Ranking result page*: Number of the page in which the rated item appeared. Each page shows five news items.
- *Personalized recommendations*: a Boolean value indicating whether the personalized recommender was activated or deactivated.
- *Context-aware recommendations*: a Boolean value indicating whether the context-aware recommender was activated or deactivated.
- *Semantic preference expansion*: A Boolean value indicating whether the expansion of user preferences and item annotations was activated or not.

- *Context-aware phase*: A number indicating how many times the user has clicked as relevant an item when the context is activated. A value of -1 is given if the context-aware recommendations are off.

Second Stage: Evaluation of Collaborative Recommendation

A second experiment was conducted with News@hand to evaluate the collaborative recommendation models included in the system. One of the objectives of this experiment was to compare the relevance judgments given by the users with the recommendations obtained using the CF approach explained in Sect. 3.5. The comparison will be given by the model built applying the ML techniques explained in Sect. 4, in such a way that CF values (potential recommended items) should be correlated with the relevance judgments, at least, when certain model conditions are fulfilled.

The 16 members of our department who participated in the previous experiment were again requested to take part of the evaluation presented herein. Each user performed three different tasks, assessing news recommendations for three news sections: *Business*, *Sports*, and *World* (see below why we selected these sections). For each task, two subtasks were defined:

- In the first subtask, the users had to rate a number of news items from a random list.
- In the second subtask, the users had to rate several news items from a list generated with the personalization functionality activated.

Each subtask was defined as finding out and rating those news items that were "related to" a personal user profile. By "related to" we mean that a news item contains semantic annotations whose concepts appear in the user's profile.

Similarly to the experiment described in previous section, the evaluators were asked to define their preferences. However, in this case, they could only select preferences from a given list of semantic concepts. They were provided a form with a list of 128 semantic concepts, classified in 8 different domains. From this list, the users had to select a subset of concepts, and assign them negative/positive weights according to personal interests. Table 3 shows the concepts available for each domain, and the average number of preferences per user. On average, each profile was created with 7.8 preferences per domain, duplicating the preferences introduced by the users when they had to manually search the concepts in the ontology browser (first stage).

In the next sections, we explain in detail the different tasks performed by the users, and the data extracted from their interaction with the system to draw appropriate conclusions.

Table 3 Topics and concepts allowed for the user profiles in the evaluation of the hybrid recommenders

Domain	Concepts	#prefs	Avg. #pref./user
Computers Technology Telecommunications	Computer, digital, ebay, google, ibm, internet, mass, media, microsoft, networking, online, satellite, software, technology, video, website	135	8.4
Wars Armed conflicts	Al-qaeda, army, battle, combat, crime, kidnapping, kill, memorial, military, murder, peace, prison, strike, terrorism, war, weapons	104	6.5
Social issues	Aids, assassination, babies, children, death sentence, divorce, drugs, family, health, hospital, immigration, love, obesity, smoking, suburb, suicide	115	7.2
Television Cinema Music	Actor, bbc, cinema, cnn, film, grammy, hollywood, movie, music, musician, nbc, radio, rock, oscar, singer, television	129	8.1
Sports	Baseball, cricket, football, lakers, nascar, nba, new england patriots, new york giants, nfl, olympics, premier league, running, sports, soccer, super bowl, tennis	168	10.5
Politics	George bush, condolezza rice, congress, democracy, elections, government, hillary clinton, john maccain, barack obama, parliament, politics, president, senate, senator, voting, white house	104	6.5
Banking Economy Finance	Banking, business, cash, companies, earnings, economy, employment, finance, fraud, gas price, industry, marketing, markets, money, oil price, wall street	120	7.5
Climate Weather Natural disasters	Air, climate, earth, earthquake, electricity, energy, fire, flood, forecast, fuel, gas, pollution, sea, storm, weather, woods	128	8.0

5.2.6 Interaction with the System

The users had to perform three tasks, each of them in one of the following news sections: *Business*, *Sports*, and *World*. Successively, for each section, a user had to:

- Deactivate the personalization functionality, and display the news items of the section. The goal is to present to all the users the same set of news items, to obtain a "shared" group of rated items (this is very important for the collaborative filtering model, since this step reduces the sparsity).
- Rate 20 news items that are related (with negative or positive weights) to the user profile. Taking into account the similarities between item annotations with user preferences, assign a 1–5 start rating to the selected news items. No restriction is placed on which items have to be rated.

- Activate the personalization functionality, and display again the news items of the section. This time the order (ranking) of the news items is different to the one shown previously.
- Rate (as explained before) 50 news items not evaluated previously.

With this strategy, the 16 users provided a total of 3,360 ratings for 859 different news items.

5.2.7 Selection of Training Sample Attributes

The training sample creation was performed similarly as in the previous experiment, but since the experiment setup was different, the attributes to consider also changed. For example, this time, context was not evaluated, all the profiles were manually defined, and semantic preference expansion was always activated. After this simplification, the relevant sample attributes for this experiment are the following:

5.2.8 User-Based Features

- *Profile size*: An integer attribute indicating the total number of concepts included in the profile (the same as before).

5.2.9 Model-Based Features

- *Topic section*: Name of the news section in which the rated item appeared.
- *Ranking result page*: Number of the page in which the rated item appeared. Each page shows five news items.
- *Personalized recommendations*: a Boolean value indicating whether the personalized recommender was activated or deactivated.
- *Collaborative filtering algorithm*: Since we consider two different collaborative filtering algorithms, this is a nominal attribute: user-based and item-based. However, we have preferred to combine samples with values obtained from the same algorithm to facilitate the ML algorithm classification task. This means that we have two sets of samples: one for each collaborative filtering algorithm.
- *Collaborative filtering value*: A number indicating the value given by the recommender system for a particular pair of user and item, given the rest of user ratings (predicted value, see Sect. 3.5).

5.3 Results

This section presents the results obtained using ML techniques to analyze the evaluations previously described. These results are classified into three different categories, according to the consequences that can be drawn from them. First, we present the results related to the personalization phase (first stage), where the impact of the semantic expansion is considered. Second, the contextualization phase (first stage) results are presented, where the importance of context combined with personalization is studied. Finally, results related to the collaborative filtering phase are shown (second stage), where correlation between predicted and real ratings is analyzed. Furthermore, some conclusions about the evaluation itself are shown in the discussion section.

For developing these results, Weka ML toolkit [26] and Taste library[2] were used. The Decision Trees presented in the figures of this section were generated using different parameters according to the stage they refer. Specifically, in the first stage we generated the trees using the following parameters[3]: "*-C 0.3 -M 5*", whereas in the second stage we used "*-R -N 3 -M 25*". Although we tried with different configurations, we chose the ones presented here because they generate comprehensible but detailed trees.

Furthermore, we have to note that every decision tree presented in this work report more than a 69% of predictive accuracy in a tenfold cross-validation experiment (the maximum accuracy obtained is 80%). Based on these results, we can assume the induced models by the decision trees are trustworthy enough to obtain reliable results.

Learning Model Features from Personalized Recommendations

In the personalization phase, we wanted to investigate whether the semantic expansion helps the user to find relevant news. After using ML techniques, we found more useful user and system features:

- *Profile size.* In Figs. 5 and 6, it can be seen that this user feature is useful when retrieving relevant items, and is connected with expansion and activation of personalization. In Fig. 5 we can see that, in the *Business* section, if the profile size is between 5 and 12 concepts, it is very likely that user will find relevant news. On the other hand, in Fig. 6, it can be seen that a small profile produces more irrelevant news to be retrieved.
- *Ranking page.* Fortunately, the system retrieves relevant news in the first page (top 5 news items), as shown in Figs. 5 and 7. Because of that, our analysis focused on subtrees where the ranking page has a value of 1.
- *Expansion.* The importance of this model feature is shown in Fig. 7, where users find relevant news only when personalization and expansion are activated.

[2] http://taste.sourceforge.net/

[3] The meaning of these parameters is the following: '-C' sets confidence threshold for pruning, '-R' creates a decision tree using reduced error pruning, when this option is available, '-N' sets the number of folds used for reduced error pruning.

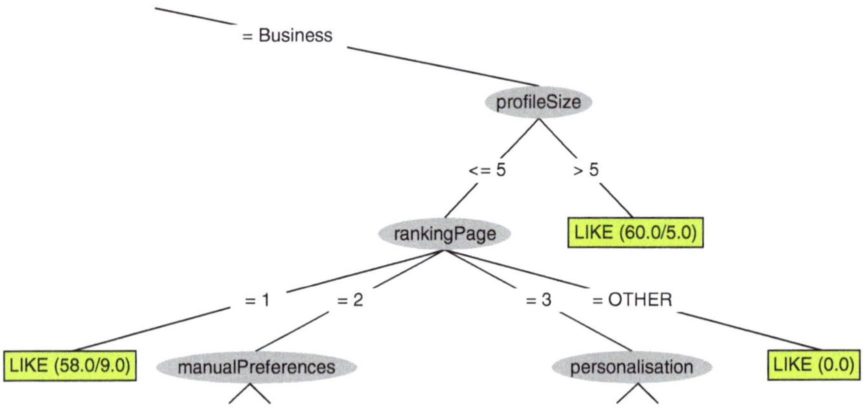

Fig. 5 Branch when profile size is less than 12, using all available logs (general evaluation)

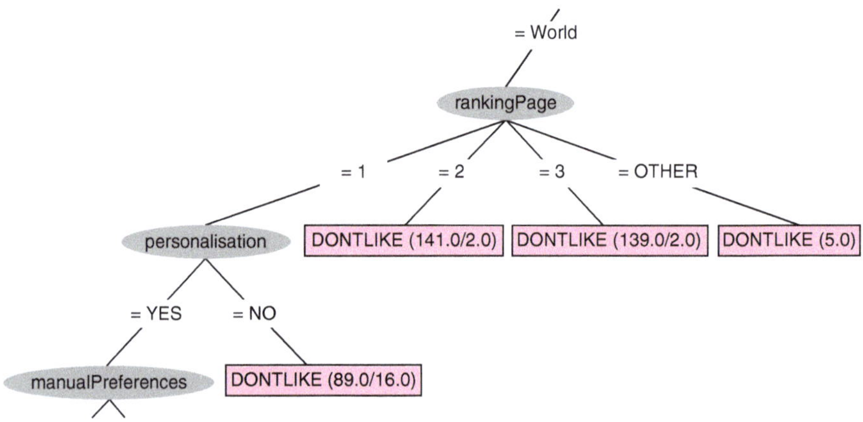

Fig. 6 Branch when profile size is less than 12, using all available logs (general evaluation)

In general, we have found that using personalization in combination with semantic expansion improves the performance in the first page. Although not all the news sections behave equally, this seems to be true in general sections such as Headlines, despite the fact that in the second and third pages, personalization improves little and needs the help of other strategies, such as contextualization (see next section).

Learning Model Features from Context-Aware Recommendations

In the experiments, we found some model features are more likely to help in context-aware recommendations. For instance, personalization was a well-performing system setting when it is combined with context. Although sometimes context alone

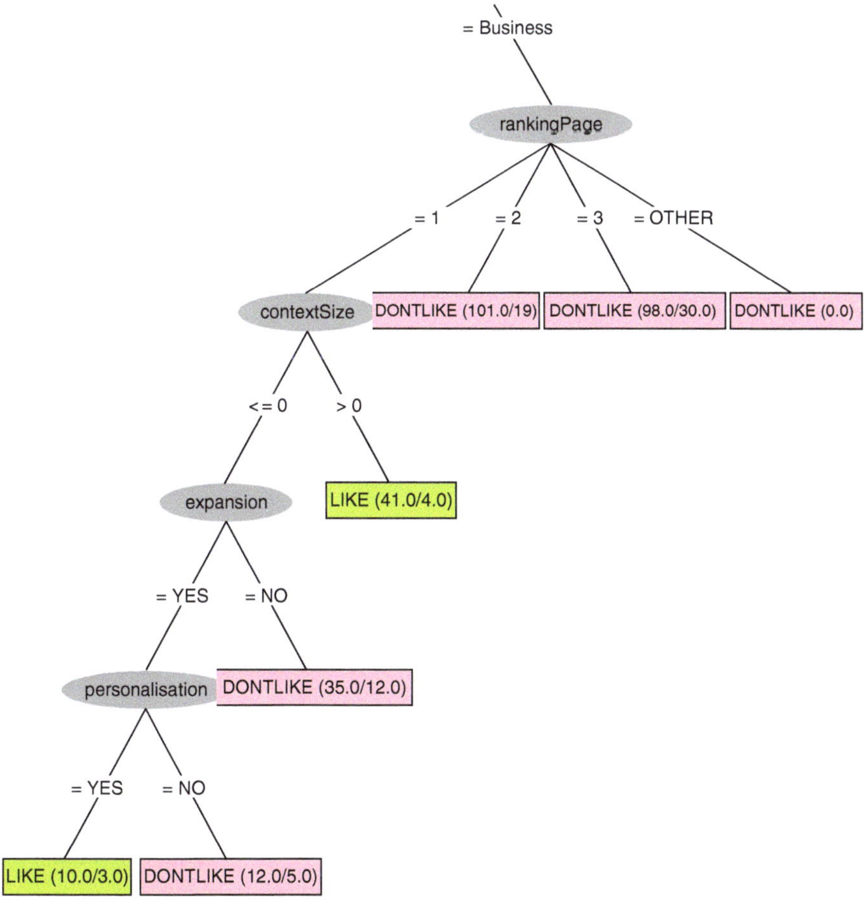

Fig. 7 Business branch and using all available logs (goal evaluation)

performs well (Fig. 7), in Fig. 8 we show an example where context needs personalization to obtain good results.

Another relevant indicator is the context size (Fig. 8). In previous experiments [4], it showed better discrimination power, but in the current ones, its main function is to distinguish between when the context was on or off. A model feature that does not have influence in context is the fact of having manual preferences or not, since the context has more to do with the short-term preferences, rather than long-term ones.

Learning Model Features from Collaborative Recommendations

The goal of the second stage was to investigate whether collaborative filtering predicted values correlates with human relevance judgments or not. A first conclu-

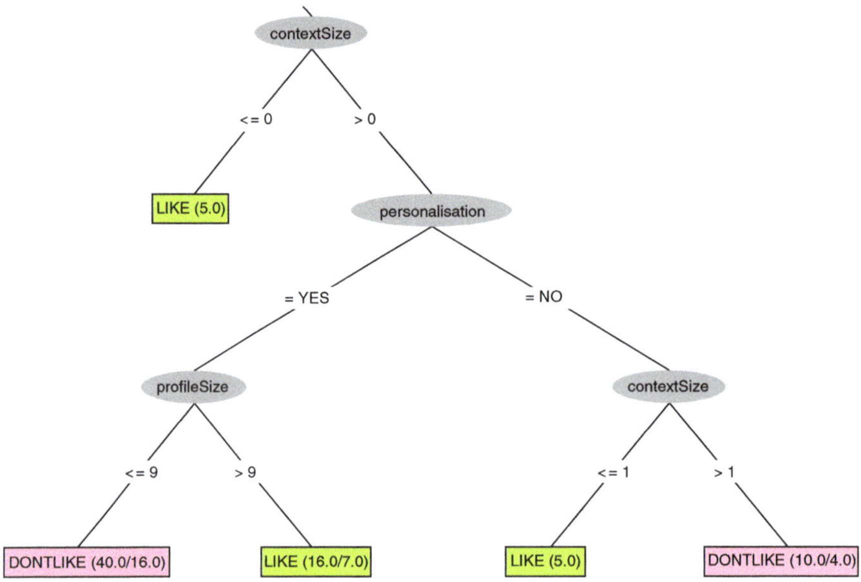

Fig. 8 Science branch in the third ranking page, using context-related logs (goal evaluation)

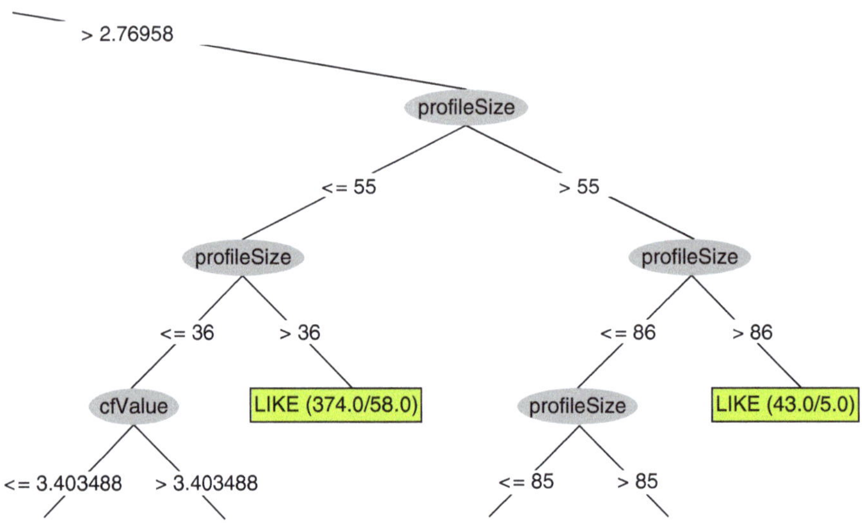

Fig. 9 Branch with CF value greater than 2.76, using a user-based algorithm

sion can be drawn after analyzing Figs. 9, 10, and 11: in most cases, collaborative filtering algorithms predict successfully the relevant class of a news item. However, a different behavior can be found between the two algorithms used here (item-based and user-based). If we focus on Figs. 10 and 11, we can see differences in the *Topic section* classification. *Business* is a very specific section, and most of its items are

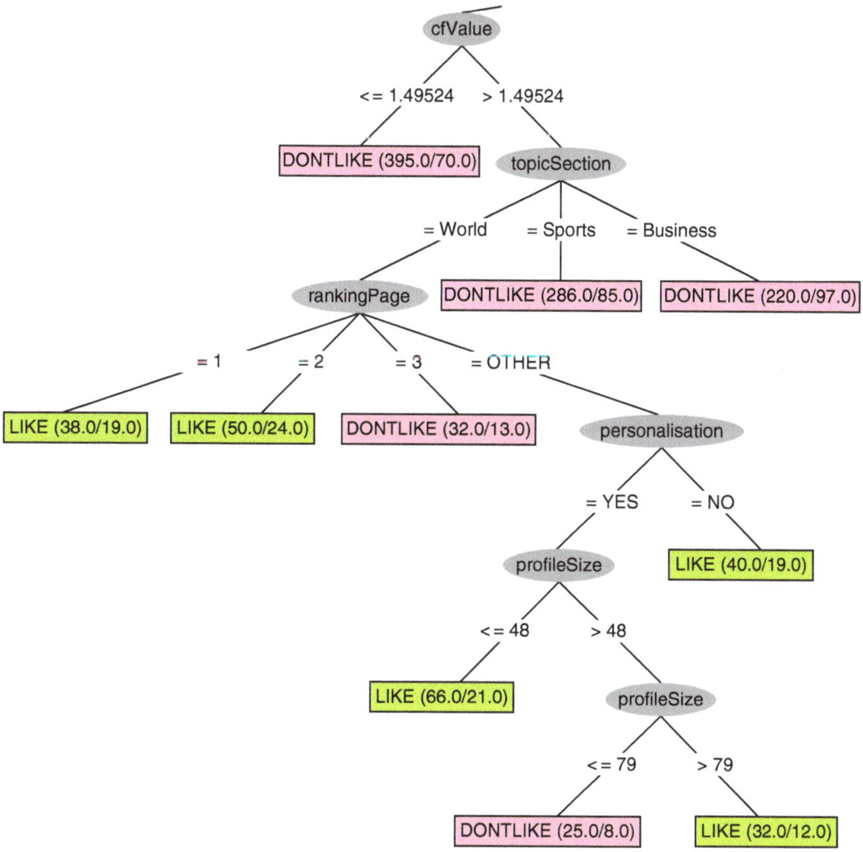

Fig. 10 Branch with CF value less than 2.76, using a user-based algorithm

very similar. Item-based algorithm lacks of information, and needs the assistance of other methods to be able to classify items as relevant or irrelevant. At the same time, *World* section is a very general section, containing objects of different types, which gives a lot of information to fulfill its goal. User-based algorithm behaves quite the opposite, which gives us the results shown in Figs. 9 and 10. *Sports* section has to be left aside, since most of the users choose no concepts related with this section, resulting in irrelevant news retrieved by the system very often.

The predictive power of these algorithms can be seen in Fig. 11, where, in this case, two threshold values can be inferred to guess if the news item will be relevant or not. In this figure, these two values are 2.752 (above this value a news item can be considered relevant) and 1.967 (below this value news are irrelevant with a great confidence). A similar situation happens in Fig. 10, where only the threshold to set irrelevant news is shown. This situation allows us to focus on a more limited set of news items: the ones located between the two thresholds. This set is not large, but ambiguous, since a value of 2.5 can be positive for one person but negative for

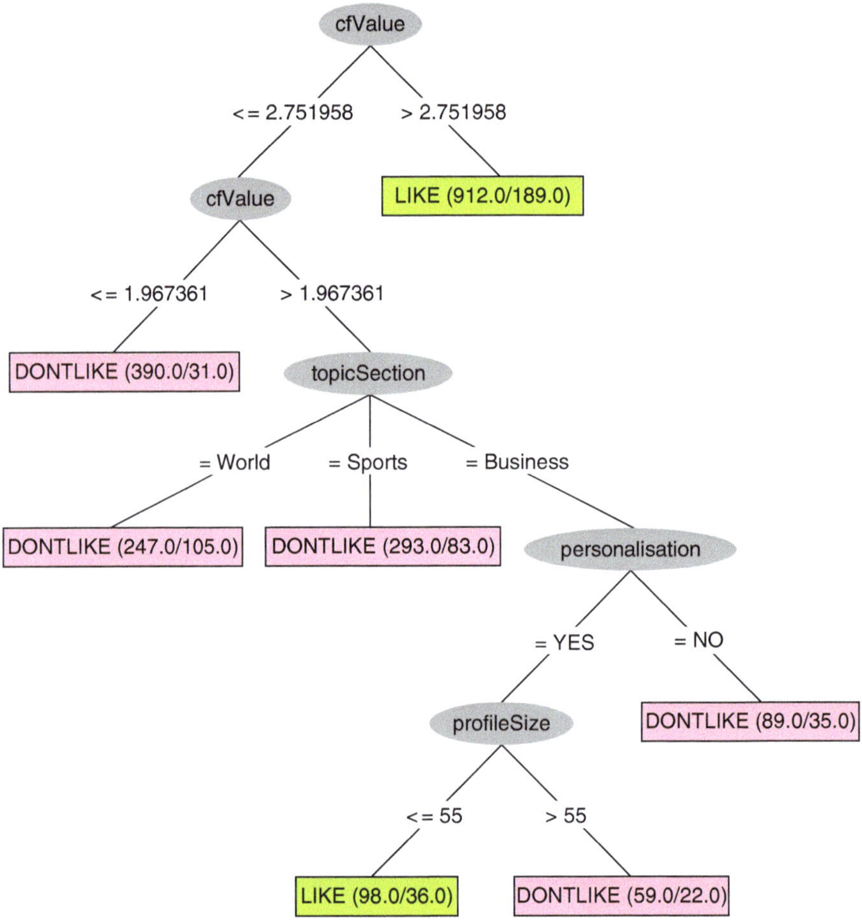

Fig. 11 Whole tree generated by using an item-based algorithm

another one. It is worthwhile noting that personalization methods are indeed helpful in discriminating relevant news items pertaining to this small set.

Profile size has been found to be a very informative feature (see Fig. 9). Another discriminative feature is the ranking page (Fig. 10), although in this experiment it has less classification power than in the previous ones already explained (compare with Fig. 7, for example).

Finally, personalization confirms its utility once again. For example, in Fig. 11, activated personalization helps the collaborative filtering algorithm to classify an item as relevant when the predicted value is ambiguous (and, in this case, along with a profile not too big).

In general, we have found that the values predicted by collaborative filtering algorithms are very close to real ones. Indeed, if a predicted value is extreme (above 3.5 or below 1.5), in most of the cases, we can be confident of that, and classify the item

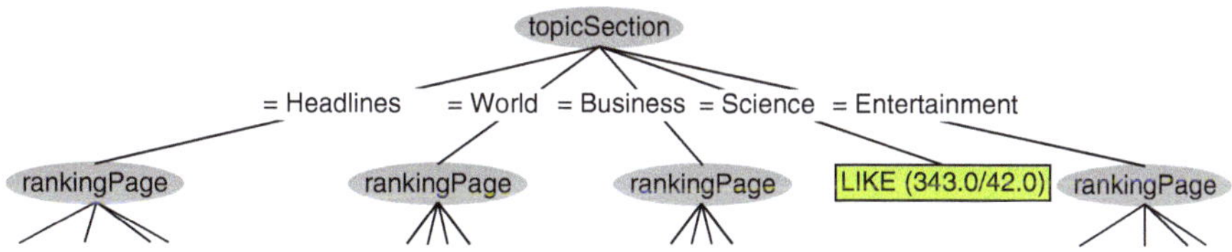

Fig. 12 Fragment of the decision tree with unbalance node load

accordingly. This situation is improved when it is combined with our personalization algorithms. In a future analysis, we have to verify whether such a combination with our context and expansion models also leads to similar improvements.

6 Discussion

We have presented a method for the automatic, iterative refinement of a recommender system by a virtuous cycle with three main steps. First, an initial recommendation model is run on a set of available input data to compute suggestions for a given user. Next, the obtained outputs are analyzed to identify latent dependencies between model characteristics recommendation quality, in order to single out the most relevant ones. Finally, adjusting the identified characteristics, a new recommendation model is produced, aiming to generate more accurate results. The work herein presented focuses on the identification of such relevant model characteristics.

The proposed approach applies ML techniques to learn the user and system features that favor correct recommendations by the system. Specifically, for every recommendation evaluated (rated) by the user a training sample is created. The attributes of the sample are the target characteristics for analysis, and their values are taken from log information databases. The training example is assigned one of the two possible classes, correct or incorrect, depending on whether the user evaluated the corresponding recommendation as relevant or irrelevant. The ML strategy consists of a classification algorithm on these examples.

The approach discussed above has been tested in the News@hand recommender system [8]. Further work is needed to measure the performance improvements obtained in that system after applying the proposed strategy, as well as to investigate other machine learning techniques, apart from decision trees, to select the most relevant model features.

Besides assessing the potential direct benefits on recommendation performance, further findings were drawn from the empiric experience with regards to the experimental methodology itself, identifying shortcomings and weaknesses. We found out that the first stage of our evaluation was unbalanced in terms of the difficulty to obtain news items relevant for each task. The decision tree in Fig. 12 is such an example. The classifier infers that most of the users liked (almost) every news item in a particular section, while in other sections this is conditional on other parameters such as the ranking page or other model characteristics. The task related to the Science section was identified as 'very easy' by the users, probably because the query used in this task biased the results to be relevant to the goal. We also observed that some tasks performed better when contextualization was activated. This could be caused by the fact that a particular goal was very specific, and there was no profile focused on that domain (in our case, *Business* section). A similar situation was the one in which the profiles had to be very specific to get some results. Since the users were not finding relevant news items, the context was useless (this happened in *Entertainment* section). Finally, another important conclusion concerns the manual profiles. When users create theirs profiles, they do not know anything about which will be their goals or queries, which makes very difficult for personalization algorithms to rank relevant news in the first pages.

Acknowledgements This research has been supported by the Spanish Ministry of Science and Education (TIN2007-64718 and TIN2008-06566-C04-02).

References

1. G. Adomavicius, R. Sankaranarayanan, S. Sen, A. Tuzhilin, Incorporating Contextual Information in Recommender Systems Using a Multidimensional Approach. ACM Trans. Inf. Syst. **23**(1), 103–145 (2005)
2. G. Adomavicius, A. Tuzhilin, Toward the Next Generation of Recommender Systems: A Survey of the State-of-the-Art and Possible Extensions. IEEE Trans. Knowl. Data Eng. **17**(6), 734–749 (2005)
3. K. Becker, C.G. Marquardt, D.D. Ruiz, A Pre-Processing Tool for Web Usage Mining in the Distance Education Domain, in *Proceedings of the 8th International Database Engineering and Applications Symposium (IDEAS 2004)* (2004), pp. 78–87
4. A. Bellogín, I. Cantador, P. Castells, A. Ortigosa, Discovering Relevant Preferences in a Personalised Recommender System using Machine Learning Techniques, in *Proceedings of the ECML/PKDD-08 Workshop on Preference Learning (PL 2008)*, pp. 82–96
5. P. Brusilovsky, Developing adaptive educational hypermedia systems: From design models to authoring tools, in *Authoring Tools for Advanced Technology Learning Environment* (2003), pp. 377–409
6. R. Burke, Hybrid Recommender Systems: Survey and Experiments. User Model. User-Adapt. Interact. **12**(4), 331–370 (2002)
7. I. Cantador, A. Bellogín, P. Castells, A Multilayer Ontology-based Hybrid Recommendation Model. AI Commun. **21**(2-3), 203–210 (2008)
8. I. Cantador, A. Bellogín, P. Castells, News@hand: A Semantic Web Approach to Recommending News, in *Proceedings of the 5th International Conference on Adaptive Hypermedia and Adaptive Web-Based Systems (AH 2008)* (2008), pp. 279–283
9. I. Cantador, A. Bellogín, P. Castells, Ontology-based Personalised and Context-aware Recommendations of News Items, in *Proceedings of the 2008 Web Intelligence Conference* (2008), pp. 562–565
10. I. Cantador, P. Castells, Extracting Multilayered Semantic Communities of Interest from Ontology-based User Profiles: Application to Group Modelling and Hybrid Recommendations. *Computers in Human Behavior* (Elsevier, 2008)
11. I. Cantador, M. Szomszor, H. Alani, M. Fernández, P. Castells, Enriching Ontological User Profiles with Tagging History for Multi-Domain Recommendations, in *Proceedings of the*

1st Intl. Workshop on Collective Intelligence and the Semantic Web (CISWeb 2008) (2008), pp. 5–19
12. P. Castells, M. Fernández, D. Vallet, An Adaptation of the Vector-Space Model for Ontology-Based Information Retrieval. IEEE Trans. Knowl. Data Eng. **19**(2), 261–272 (2007)
13. P.R. Cohen, R. Kjeldsen, Information Retrieval by Constrained Spreading Activation in Semantic Networks. Inf. Process. Manag. **23**(4), 255–268 (1987)
14. F. Crestani, Application of Spreading Activation Techniques in Information Retrieval. Artif. Intell. Rev. **11**(6), 453–482 (1997)
15. F. Crestani, P.L. Lee, Searching the Web by Constrained Spreading Activation. Inf. Process. Manag. **36**(4), 585–605 (2000)
16. R.O. Duda, P.E. Hart, D.G. Stork, *Pattern Classification* (Wiley-InterScience, 2000)
17. R. Rafter, B. Smyth, Conversational Collaborative Recommendation: An Experimental Analysis. Artif. Intell. Rev. **24**(3-4), 301–318 (2005)
18. P. Resnick, N. Iacovou, M. Suchak, P. Bergstrom, J. Riedl, Grouplens: An open architecture for collaborative filtering on netnews, in *Proceedings of the 1994 Conference on. Computer Supported Collaborative Work* (1994), pp. 175–186
19. B. Sarwar, G. Karypis, J. Konstan, J. Riedl, Item-based collaborative filtering recommendation algorithms, in *Proceedings of the 2001 WWW Conference* (2001), pp. 285–295
20. J. Srivastava, R. Cooley, M. Deshpande, P. Tan, Web Usage Mining: Discovery and Applications of Usage Patterns from Web Data. SIGKDD Explor. **1**(2), 12–23 (2000)
21. L. Talavera, E. Gaudioso, Mining Student Data to Characterize Similar Behavior Groups in Unstructured Collaboration Spaces, in *Proceedings Workshop on AI in CSCL* (2004), pp. 17–23
22. L. Terveen, W. Hill, Beyond Recommender Systems: Helping People Help Each Other, in *Human-Computer Interaction in the New Millennium* (2001), pp. 487–509
23. D. Vallet, P. Castells, M. Fernández, P. Mylonas, Y. Avrithis, Personalised Content Retrieval in Context Using Ontological Knowledge. IEEE TCSVT **17**(3), 336–346 (2007)
24. C. Vialardi, J. Bravo, A. Ortigosa, Empowering AEH Authors Using Data Mining Techniques, in *Proceedings of the 5th Int. Workshop on Authoring of Adaptive and Adaptable Hypermedia* (2007)
25. C. Vialardi, J. Bravo, A. Ortigosa, Improving AEH Courses through Log Analysis. J. Univers. Comput. Sci. **14**(17), 2777–2798 (2008)
26. I.H. Witten, E. Frank, *Data Mining: Practical Machine Learning Tools and Techniques* (Morgan Kaufmann Series in Data Management Systems, 2005)
27. O.R. Zaïane, Recommender System for E-learning: Towards Non-Instructive Web Mining, in *Data Mining in E-Learning*, ed. by C. Romero, S. Ventura (2006), pp. 79–96
28. T. Zhang, V.S. Iyengar, Recommender Systems Using Linear Classifiers J. Mach. Learn. Res. **2**, 313–334 (2002)

Subject Index

Author Index

MIX
Papier aus verantwortungsvollen Quellen
Paper from responsible sources
FSC® C105338

If you have any concerns about our products,
you can contact us on
ProductSafety@springernature.com

In case Publisher is established outside the EU,
the EU authorized representative is:
Springer Nature Customer Service Center GmbH
Europaplatz 3, 69115 Heidelberg, Germany

Printed by Libri Plureos GmbH
in Hamburg, Germany